NATURAL DISASTER RESEARCH, PREDICTION AND MITIGATION

THE BIG SHAKE: IMPLICATIONS OF A MAJOR EARTHQUAKE IN CALIFORNIA

NATURAL DISASTER RESEARCH, PREDICTION AND MITIGATION

Additional books in this series can be found on Nova's website under the Series tab.

Additional E-books in this series can be found on Nova's website under the E-book tab.

NATURAL DISASTER RESEARCH, PREDICTION AND MITIGATION

THE BIG SHAKE: IMPLICATIONS OF A MAJOR EARTHQUAKE IN CALIFORNIA

CHRISTOPHER D. MONAHAN
EDITOR

Nova Science Publishers, Inc.
New York

For permission to use material from this book please contact us:
Telephone 631-231-7269; Fax 631-231-8175
Web Site: http://www.novapublishers.com

LIBRARY OF CONGRESS CATALOGING-IN-PUBLICATION DATA

The big shake implications of a major earthquake in California / editor,
Christopher D. Monahan.
p. cm.
Includes index.
ISBN 978-1-61728-512-7 (hardcover)
1. Earthquake prediction--California. 2. Earthquakes--California. I.
Monahan, Christopher D.
QE538.8.B54 2009
551.2209794'9--dc22

2010025405

Published by Nova Science Publishers, Inc. † New York

CONTENTS

Preface		**vii**
Chapter 1	Earthquakes: Risk, Detection, Warning, and Research *Peter Folger*	**1**
Chapter 2	The ShakeOut Scenario *Lucile M. Jones, Richard Bernknopf, Dale Cox, James Goltz, Kenneth Hudnut, Dennis Mileti, Suzanne Perry, Daniel Ponti, Keith Porter, Michael Reichle, Hope Seligson, Kimberley Shoaf, Jerry Treiman and Anne Wein*	**31**
Chapter 3	Forecasting California's Earthquakes: What Can we Expect in the Next 30 Years? *U.S. Geological Survey*	**307**
Chapter 4	The ShakeOut Earthquake Scenario: A Story that Southern Californians are Writing *U.S. Geological Survey*	**315**
Chapter 5	Historical Seismicity in the South San Francisco Bay Region *William H. Bakun*	**337**
Chapter 6	The Hayward Fault: Is it Due for a Repeat of the Powerful 1868 Earthquake? *U.S. Geological Survey*	**373**
Chapter Sources		**383**
Index		**385**

PREFACE

California straddles the boundary between two of the Earth's tectonic plates. As a result, it is broken by numerous earthquake faults. Taking into account the earthquake histories and relative rates of motion on these many faults, the Uniform California Earthquake Rupture Forecast (UCERF) study concludes that there is a probability of more than 99% that in the next 30 years Californians will experience one or more magnitude 6.7 or greater quakes, potentially capable of causing extensive damage and loss of life. This book provides an overview of the implications, risk, detection, warning and research into a major earthquake in California within the next 30 years.

Chapter 1- The 1994 Northridge (CA) earthquake caused as much as $26 billion (in 2005 dollars) in damage and was one of the costliest natural disasters to strike the United States. The Federal Emergency Management Agency has estimated that earthquakes cost the United States over $5 billion per year. A hypothetical scenario for a magnitude 7.8 earthquake in southern California estimated a possibility of 1,800 fatalities and over $200 billion in economic losses. The January 12, 2010, magnitude 7.0 earthquake that struck Haiti only 15 miles from Port-au-Prince, the capital city, has caused an estimated 230,000 fatalities and 300,000 injuries. On February 27, 2010, a magnitude 8.8 earthquake occurred offshore of Chile and has caused massive damage, although far fewer fatalities than the Haiti earthquake. The Cascadia Subduction Zone megathrust fault, which stretches from southern British Columbia to northern California, has the potential to generate a massive earthquake and tsunami, similar to the Chilean earthquake, that could threaten Oregon, Washington, and northern California.

The United States faces the possibility of large economic losses from earthquake-damaged buildings and infrastructure. California alone accounts for most of the estimated annualized earthquake losses for the nation, and with Oregon and Washington the three states account for nearly $4.1 billion (77%) of the U.S. total estimated annualized loss. A single large earthquake, however, can cause far more damage than the average annual estimate.

An ongoing issue for Congress is whether the federally supported programs aimed at reducing U.S. vulnerability to earthquakes are an adequate response to the earthquake hazard. Under the National Earthquake Hazards Reduction Program (NEHRP), four federal agencies have responsibility for long-term earthquake risk reduction: the U.S. Geological Survey (USGS), the National Science Foundation (NSF), the Federal Emergency Management Agency (FEMA), and the National Institute of Standards and Technology (NIST). They variously assess U.S. earthquake hazards, send notifications of seismic events, develop measures to reduce earthquake hazards, and conduct research to help reduce overall U.S. vulnerability to earthquakes.

Congress reauthorized NEHRP in 2004 (P.L. 108-360) through FY2009. Appropriations for NEHRP from FY2005 to FY2009 did not meet authorized levels; the total funding enacted was $615.5 million, approximately 68% of the total amount of $902.4 million authorized by P.L. 108- 360. The American Recovery and Reinvestment Act (ARRA; P.L. 111-5) provided some additional funding for earthquake activities under NEHRP through FY20 10. In the 111th Congress, H.R. 3820 would reauthorize NEHRP through FY2014, authorizing total appropriations of $806 million over five years for the program, with 88% of the funding designated for the USGS and NSF, and the remainder for FEMA and NIST. The total authorized amounts are about $100 million less than what was authorized by P.L. 108-360 over five years. If future appropriations match the funding levels authorized under H.R. 3820, however, these funds would exceed the total cumulative amounts actually appropriated between FY2005 and FY2009.

What effect funding at the levels enacted through FY2009 under NEHRP has had on the U.S. capability to detect earthquakes and minimize losses after an earthquake occurs is not clear. It is also difficult to predict precisely how NEHRP reauthorized under H.R. 3820 would achieve a major goal of the bill: to reduce the loss of life and damage to communities and infrastructure through increasing the adoption of hazard mitigation measures. A perennial issue for Congress is whether activities under NEHRP can reduce the potential for catastrophic loss in the next giant earthquake to strike the United States.

Chapter 2 - This is the initial publication of the results of a cooperative project to examine the implications of a major earthquake in southern California. The study comprised eight counties: Imperial, Kern, Los Angeles, Orange, Riverside, San Bernardino, San Diego, and Ventura. Its results will be used as the basis of an emergency response and preparedness exercise, the Great Southern California ShakeOut, and for this purpose we defined our earthquake as occurring at 10:00 a.m. on November 13, 2008. As members of the southern California community use the ShakeOut Scenario to plan and execute the exercise, we anticipate discussion and feedback. This community input will be used to refine our assessment and will lead to a formal publication in early 2009.

Our goal in the ShakeOut Scenario is to identify the physical, social and economic consequences of a major earthquake in southern California and in so doing, enable the users of our results to identify what they can change now—*before* the earthquake—to avoid catastrophic impact after the inevitable earthquake occurs. To do so, we had to determine the physical damages (casualties and losses) caused by the earthquake and the impact of those damages on the region's social and economic systems. To do this, we needed to know about the earthquake ground shaking and fault rupture. So we first constructed an earthquake, taking all available earthquake research information, from trenching and exposed evidence of prehistoric earthquakes, to analysis of instrumental recordings of large earthquakes and the latest theory in earthquake source physics. We modeled a magnitude (*M*) 7.8 earthquake on the southern San Andreas Fault, a plausible event on the fault most likely to produce a major earthquake. This information was then fed forward into the rest of the ShakeOut Scenario.

Chapter 3 - In a new comprehensive study, scientists have determined that the chance of having one or more magnitude 6.7 or larger earthquakes in the California area over the next 30 years is greater than 99%. Such quakes can be deadly, as shown by the 1989 magnitude 6.9 Loma Prieta and the 1994 magnitude 6.7 Northridge earthquakes. The likelihood of at least one even more powerful quake of magnitude 7.5 or greater in the next 30 years is 46%— such a quake is most likely to occur in the southern half of the State. Building codes, earthquake insurance, and emergency planning will be affected by these new results, which highlight the urgency to prepare now for the powerful quakes that are inevitable in California's future.

Chapter 4 - The question is not *if* but *when* southern California will be hit by a major earthquake— one so damaging that it will permanently change lives and livelihoods in the region. How severe the changes will be depends on the actions that individuals, schools, businesses, organizations, communities, and governments take to get ready. To help prepare for this event, scientists of the U.S. Geological Survey (USGS) have changed the way that earthquake scenarios are done, uniting a multidisciplinary team that spans an unprecedented number of specialties. The team includes the California Geological Survey, Southern California Earthquake Center, and nearly 200 other partners in government, academia, emergency response, and industry, working to understand the long-term impacts of an enormous earthquake on the complicated social and economic interactions that sustain southern California society. This project, the ShakeOut Scenario, has applied the best current scientific understanding to identify what can be done now to avoid an earthquake catastrophe. More information on the science behind this project will be available in *The ShakeOut Scenario* (USGS Open-File Report 2008-1150; http://pubs.usgs.gov/of/2008/1150/).

The "what if?" earthquake modeled in the ShakeOut Scenario is a magnitude 7.8 on the southern San Andreas Fault. Geologists selected the details of this hypothetical earthquake by considering the amount of stored strain on that part of the fault with the greatest risk of imminent rupture. From this, seismologists and computer scientists modeled the ground shaking that would occur in this earthquake. Engineers and other professionals used the shaking to produce a realistic picture of this earthquake's damage to buildings, roads, pipelines, and other infrastructure. From these damages, social scientists projected casualties, emergency response, and the impact of the scenario earthquake on southern California's economy and society. The earthquake, its damages, and resulting losses are one realistic outcome, deliberately not a worst-case scenario, rather one worth preparing for and mitigating against.

Decades of improving the life-safety requirements in building codes have greatly reduced the risk of death in earthquakes, yet southern California's economic and social systems are still vulnerable to large-scale disruptions. Because of this, the ShakeOut Scenario earthquake would dramatically alter the nature of the southern California community. Fortunately, steps can be taken now that can change that outcome and repay any costs many times over. The ShakeOut Scenario is the first public product of the USGS Multi- Hazards Demonstration Project, created to show how hazards science can increase a community's resiliency to natural disasters through improved planning, mitigation, and response.

Chapter 5 - Locations (intensity centers) and moment magnitude **M** for 24 earthquakes (1858-1911) in the southern San Francisco Bay area are estimated from Modified Mercalli intensity (MMI) assignments. The uncertainties in location and **M** are generally large because there are few MMI assignments available. Preferred locations are selected to be consistent with a simple model (Oppenheimer *et al.*, 2008; Bakun *et al.*, 2008) for seismic activity on the central and southern Calaveras fault. Significant seismic activity can be explained by repetitive failure of the same fault areas in nearly identical earthquakes.

Significant earthquake activity occurred on both east- and west-Bay faults in the ten or so years before the 1868 east- Bay earthquake and before the1906 and 1989 west-Bay earthquakes.

Chapter 6 - In the early morning of October 21, 1868, seismic waves from a powerful earthquake raced through the fog-shrouded San Francisco Bay region. Frightened people ran out of their homes, and cattle and even fire-engine horses panicked and bolted. Strong shaking lasted more than 40 seconds, devastating several East Bay towns. Brick buildings, walls, and chimneys were also shaken down in Oakland, San Francisco, Santa Rosa, and San Jose, and there was serious damage in Napa and Hollister. Numerous witnesses reported seeing the ground move in waves. Shaking was felt as far away as Nevada, and aftershocks rattled the Bay Area for weeks. Even though the region was only sparsely populated at the time, the 1868 quake killed about 30 people and caused great property damage. It still ranks as one of the most destructive earthquakes in California's history—but this is not the end of the story. The Hayward Fault will rupture violently again, and perhaps very soon.

In: The Big Shake Implications of a Major Earthquake in … ISBN: 978-1-61728-512-7
Editor: Christopher D. Monahan

Chapter 1

EARTHQUAKES: RISK, DETECTION, WARNING, AND RESEARCH

Peter Folger

SUMMARY

The 1994 Northridge (CA) earthquake caused as much as $26 billion (in 2005 dollars) in damage and was one of the costliest natural disasters to strike the United States. The Federal Emergency Management Agency has estimated that earthquakes cost the United States over $5 billion per year. A hypothetical scenario for a magnitude 7.8 earthquake in southern California estimated a possibility of 1,800 fatalities and over $200 billion in economic losses. The January 12, 2010, magnitude 7.0 earthquake that struck Haiti only 15 miles from Port-au-Prince, the capital city, has caused an estimated 230,000 fatalities and 300,000 injuries. On February 27, 2010, a magnitude 8.8 earthquake occurred offshore of Chile and has caused massive damage, although far fewer fatalities than the Haiti earthquake. The Cascadia Subduction Zone megathrust fault, which stretches from southern British Columbia to northern California, has the potential to generate a massive earthquake and tsunami, similar to the Chilean earthquake, that could threaten Oregon, Washington, and northern California.

The United States faces the possibility of large economic losses from earthquake-damaged buildings and infrastructure. California alone accounts for most of the estimated annualized earthquake losses for the nation, and with Oregon and Washington the three states account for nearly $4.1 billion (77%) of the U.S. total estimated annualized loss. A single large earthquake, however, can cause far more damage than the average annual estimate.

An ongoing issue for Congress is whether the federally supported programs aimed at reducing U.S. vulnerability to earthquakes are an adequate response to the earthquake hazard. Under the National Earthquake Hazards Reduction Program (NEHRP), four federal agencies have responsibility for long-term earthquake risk reduction: the U.S. Geological Survey (USGS), the National Science Foundation (NSF), the Federal Emergency Management Agency (FEMA), and the National Institute of Standards and Technology (NIST). They variously assess U.S. earthquake hazards, send notifications of seismic events, develop measures to reduce earthquake hazards, and conduct research to help reduce overall U.S. vulnerability to earthquakes.

Congress reauthorized NEHRP in 2004 (P.L. 108-360) through FY2009. Appropriations for NEHRP from FY2005 to FY2009 did not meet authorized levels; the total funding enacted was $615.5 million, approximately 68% of the total amount of $902.4 million authorized by P.L. 108- 360. The American Recovery and Reinvestment Act (ARRA; P.L. 111-5) provided some additional funding for earthquake activities under NEHRP through FY20 10. In the 111th Congress, H.R. 3820 would reauthorize NEHRP through FY2014, authorizing total appropriations of $806 million over five years for the program, with 88% of the funding designated for the USGS and NSF, and the remainder for FEMA and NIST. The total authorized amounts are about $100 million less than what was authorized by P.L. 108-360 over five years. If future appropriations match the funding levels authorized under H.R. 3820, however, these funds would exceed the total cumulative amounts actually appropriated between FY2005 and FY2009.

What effect funding at the levels enacted through FY2009 under NEHRP has had on the U.S. capability to detect earthquakes and minimize losses after an earthquake occurs is not clear. It is also difficult to predict precisely how NEHRP reauthorized under H.R. 3820 would achieve a major goal of the bill: to reduce the loss of life and damage to communities and infrastructure through increasing the adoption of hazard mitigation measures. A perennial issue for Congress is whether activities under NEHRP can reduce the potential for catastrophic loss in the next giant earthquake to strike the United States.

INTRODUCTION

Close to 75 million people in 39 states face some risk from earthquakes. Earthquake hazards are greatest in the western United States, particularly California, but also Alaska, Washington, Oregon, and Hawaii. Earthquake hazards are also prominent in the Rocky Mountain region and the New Madrid Seismic Zone (a portion of the central United States), as well as portions of the eastern seaboard, particularly South Carolina. Given the potentially huge costs associated with a large, damaging earthquake in the United States, an ongoing issue for Congress is whether the federally supported earthquake programs are appropriate for the earthquake risk.

Under the National Earthquake Hazards Reduction Program (NEHRP), the federal government supports efforts to assess and monitor earthquake hazards and risk in the United States. Four federal agencies, responsible for long-term earthquake risk reduction, coordinate their activities under NEHRP: the U.S. Geological Survey (USGS), the National Science Foundation (NSF), the Federal Emergency Management Agency (FEMA), and the National Institute of Standards and Technology (NIST). Congress reauthorized NEHRP in 2004 (P.L. 108-360), and authorized appropriations through FY2009 for a total of $902.4 million over five years. In the 111th Congress, H.R. 3820 (Title I) would reauthorize NEHRP through FY2014.

This report discusses:

- NEHRP—the multi-agency federal program to reduce the nation's risk from earthquakes;
- earthquake hazards and risk in the United States;
- federal programs that support earthquake monitoring;
- the U.S. capability to detect earthquakes and issue notifications and warnings; and
- federally supported research to improve the fundamental scientific understanding of earthquakes with a goal of reducing U.S. vulnerability.

NATIONAL EARTHQUAKE HAZARDS REDUCTION PROGRAM (NEHRP)

In 1977 Congress passed the Earthquake Hazards Reduction Act (P.L. 95-124) establishing NEHRP as a long-term earthquake risk reduction program for the United States. The program initially focused on research, led by USGS and NSF, toward understanding and ultimately predicting earthquakes. Earthquake prediction has proved intractable thus far, and the NEHRP program shifted its focus to minimizing losses from earthquakes after they occur. FEMA was created in 1979 and President Carter designated it as the lead agency for NEHRP. In 1980, Congress reauthorized the Earthquake Hazards Reduction Act (P.L. 9 6-472), defining FEMA as the lead agency and authorizing additional funding for earthquake hazard preparedness and mitigation to FEMA and the National Bureau of Standards (now NIST).

A Shift in Program Emphasis to Hazard Reduction

In 1990, Congress reauthorized NEHRP (P.L. 101-6 14) and made substantive changes, to decrease the emphasis on earthquake prediction, clarify the role of FEMA, clarify and expand the program objectives, and require federal agencies to adopt seismic safety standards for new and existing federal buildings. In 2004, Congress reauthorized NEHRP through FY2009 (P.L. 108- 360) and shifted primary responsibility for planning and coordinating NEHRP from FEMA to NIST. It also established a new interagency coordinating committee and a new advisory committee, both focused on earthquake hazards reduction.

The current program activities are focused on four broad areas:

- developing effective measures to reduce earthquake hazards;
- promoting the adoption of earthquake hazards reduction measures by federal, state, and local governments, national building standards and model building code organizations, engineers, architects, building owners, and others who play a role in planning and constructing buildings, bridges, structures, and critical infrastructure or lifelines;
- improving the basic understanding of earthquakes and their effects on people and infrastructure, through interdisciplinary research involving engineering, natural sciences, and social, economic, and decision sciences; and

- developing and maintaining the Advanced National Seismic System (ANSS), the George E. Brown Jr. Network for Earthquake Engineering Simulation (NEES), and the Global Seismic Network (GSN).[1]

The House Science Committee report in the 108th Congress on H.R. 2608 (P.L. 108-360) noted that NEHRP has produced a wealth of useful information since 1977, but it also stated that the program's potential has been limited by the inability of the NEHRP agencies to coordinate their efforts.[2] The committee asserted that restructuring the program with NIST as the lead agency, directing funding towards appropriate priorities, and implementing it as a true interagency program would lead to improvement.

The 2004 reauthorization directed the Director of NIST to chair the Interagency Coordinating Committee. Other members of the committee include the directors of FEMA, USGS, NSF, the Office of Science and Technology Policy, and the Office of Management and Budget. The Interagency Coordinating Committee is charged with overseeing the planning, management, and coordination of the program. Primary responsibilities for the NEHRP agencies break down as follows (see also **Figure 1**):

- NIST supports the development of performance-based seismic engineering tools and works with other groups to promote the commercial application of the tools through building codes, standards, and construction practices.
- FEMA assists other agencies and private-sector groups to prepare and disseminate building codes and practices for structures and "lifelines",[3] and aids development of performance-based codes for buildings and other structures.
- USGS conducts research and other activities to characterize and assess earthquake risks, and (1) operates a forum, using the National Earthquake Information Center (NEIC), for the international exchange of earthquake information, (2) works with other NEHRP agencies to coordinate activities with earthquake reduction efforts in other countries, and (3) maintains seismic hazard maps in support of building codes for structures and lifelines, and other maps needed for performance-based design approaches.
- NSF supports research to improve safety and performance of buildings, structures, and lifelines using the large-scale experimental and computational facilities of NEES and other institutions engaged in research and implementation of NEHRP.

Table 1 shows the authorized and enacted appropriations for NEHRP from FY2005 through FY2009. The total enacted amount for FY2005-FY2009 was $615.5 million, or 68% of the $902.4 million total amount authorized in P.L. 108-360 over the five-year span. President Obama requested a total of $131.7 million for NEHRP in FY20 10, even though authorization for the program under P.L. 108-360 expired at the end of FY2009.

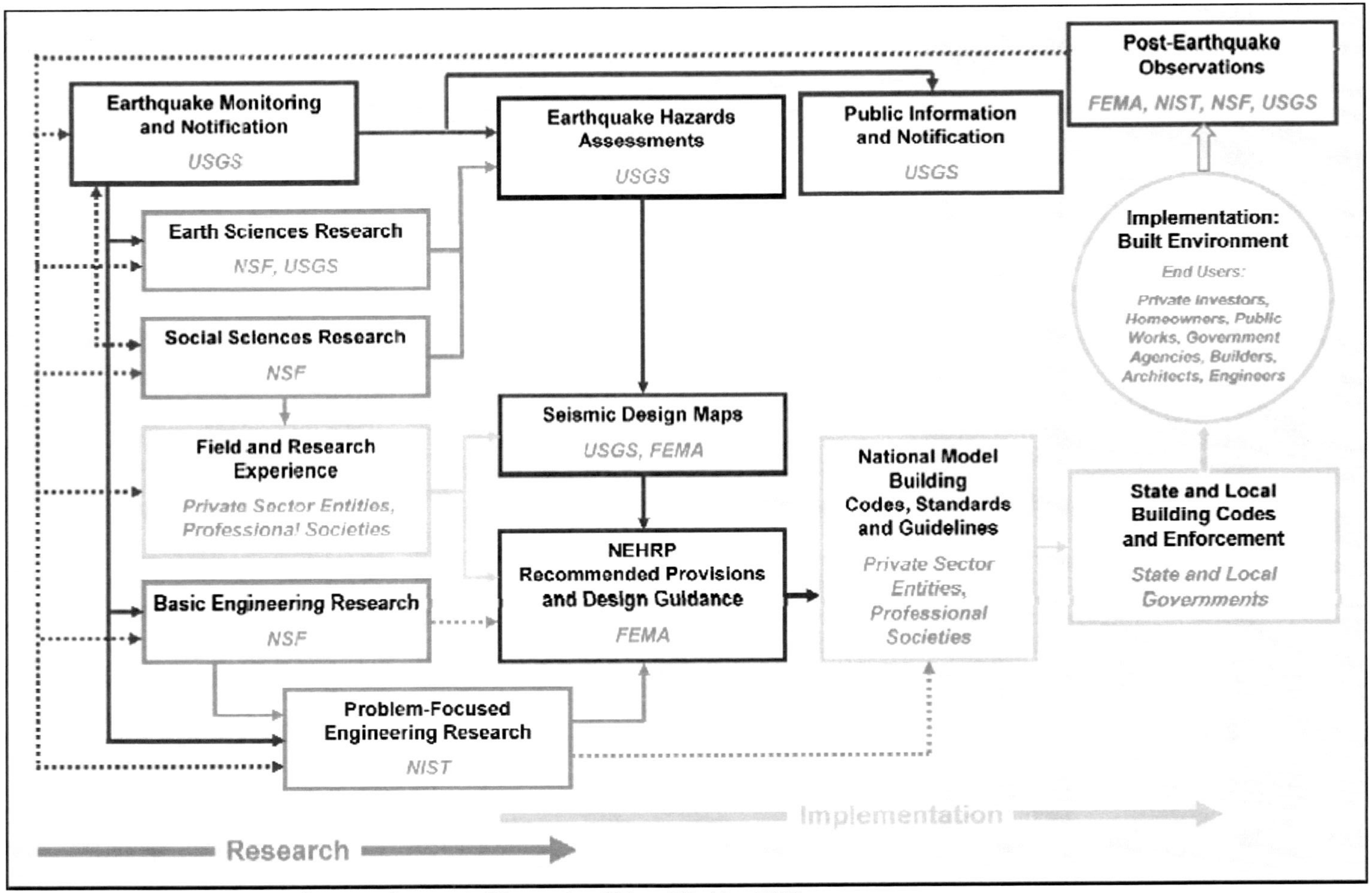

Source: NEHRP program office at http://www.nehrp.gov/pdf/ppt_sdr.pdf (modified by CRS).

Figure 1. NEHRP Agency Responsibilities and End Users of NEHRP Outcomes

Table 1. Authorized and Enacted Funding for NEHRP ($ millions)

		USGS	NSF	FEMA	NIST	Total
FY2005	Authorized	77.0	58.0	21.0	10.0	**166.0**
	Enacted	58.4	53.1	14.7	0.9	**127.1**
FY2006	Authorized	84.4	59.5	21.6	11.0	**176.5**
	Enacted	54.5	53.8	9.5	0.9	**118.7**
FY2007	Authorized	85.9	61.2	22.3	12.1	**181.5**
	Enacted	55.4	54.2	7.2	1.7	**118.5**
FY2008	Authorized	87.4	62.9	23.0	13.3	**186.6**
	Enacted	58.1	55.6	6.1	1.7	**121.5**
FY2009	Authorized	88.9	64.7	23.6	14.6	**191.8**
	Enacted	61.2	55.3	9.1	4.1	**129.7**
FY2010	Requested	61.5	57.0	9.1	4.1	**131.7**

Source: NEHRP program office, at http://www.nehrp.gov/pdf/ppt_sdr.pdf.

Notes: According to the NIST program office, the FEMA FY2010 requested budget is the estimated portion of the Administration's Department of Homeland Security budget request that would be allocated for FEMA NEHRP activities.

NEHRP Reauthorization in the 111th Congress

Title I of H.R. 3820, the Natural Hazards Risk Reduction Act of 2009, would reauthorize NEHRP through FY2014, retain NIST as the lead NEHRP agency, and authorize total appropriations of slightly more than $806 million over five years. Title II of H.R. 3820 would reauthorize the National Windstorm Impact Reduction Act (first enacted in 2004 as Title II of P.L. 108-360 and modeled after NEHRP), and Title III would create an interagency coordinating committee, chaired by the Director of NIST, that would oversee the planning and coordination of both the earthquake and wind hazards programs. The single interagency coordinating committee would replace the two separate interagency committees overseeing the current earthquake and wind hazards programs. The bill was ordered to be reported by the House Science and Technology Committee on October 21, 2009, but has not seen further action.

The interagency coordinating committee also would be given authority to "make proposals for planning and coordination of any other federal research for natural hazards mitigation that the Committee considers appropriate." The potentially broader mandate for the interagency coordinating committee—to embrace all natural hazards in its deliberations—could reflect an emphasis on natural hazard *mitigation* presented in the bill's "Findings" section. The bill finds that research is needed to better understand institutional, social, behavioral, and economic factors that influence how risk mitigation is implemented, and that a major goal of federally supported natural hazards-related research should be to increase the adoption of hazard mitigation measures. This theme is an aspect of an "all-hazards" approach to reducing risk, and could reflect observations presented during a June 11, 2009, hearing of the House Science and Technology Committee at which a witness noted that "there appear to be significant similarities in societal responses to different hazards."[4] In that testimony, the witness observed that there are limited opportunities to study earthquake emergency response and recovery because damaging earthquakes are relatively infrequent; however, there are important lessons to be learned from studying other, more frequent, hazards, such as tornadoes.[5]

Authorization of Appropriations in H.R. 3820

H.R. 3820 would authorize total appropriations for NEHRP of approximately $806 million for a five-year period ending in FY2014, with 88% of the funding authorized for the USGS and NSF, and the remainder for FEMA and NIST. (See **Table 2.**) The total authorized amounts are about $100 million less than what was authorized by P.L. 108-360 over five years. However, funding under H.R. 3820 would exceed the amounts actually appropriated between FY2005 and FY2009 for NEHRP if Congress enacted appropriations over five years that matched the authorized amounts. Compared to the previously enacted authorized funding, total authorized amounts would be less under H.R. 3820 for NIST, FEMA, and the USGS, but greater for NSF.

The USGS would receive the largest share of authorized appropriations under H.R. 3820, as under the previous reauthorization of NEHRP, although the total amount would be approximately $52 million less than the amount authorized for FY2005 through FY2009. Also, the previous reauthorization singled out the Advanced National Seismic System (ANSS) to receive a subset of authorized appropriations within the total USGS-authorized amount. Specifically, ANSS was authorized to receive $30 million in FY2005, and $36 million each year through FY2009, for a total of $174 million over five years. In contrast, H.R. 3820 does not specify a subset of authorized appropriations for ANSS, although it does state that appropriations for the USGS are to include ANSS. Funding for ANSS would therefore be left to the discretion of the agency under H.R. 3820. It is unclear whether (and probably unlikely that) the USGS would be able to attain the goal of installing 7,000 seismic stations originally envisioned for ANSS even if appropriations matched authorization levels over the next five years. (See "Advanced National Seismic System (ANSS)" below for further discussion.)

Table 2. NEHRP Authorization for Appropriations under H.R. 3820 ($ millions)

	FY 2010	FY 2011	FY 2012	FY 2013	FY 2014	Total FY2010-FY2014 auth.	Total FY2005-FY2009 auth.	Total FY2005-FY2009 enact.
USGS	70.0	72.1	74.3	76.5	78.8	***371.6***	***423.6***	***287.6***
NSF	64.1	66.1	68.0	70.1	72.2	***340.5***	**306.3**	**272.0**
FEMA	10.2	10.6	10.9	11.2	11.5	***54.4***	**111.5**	**46.6**
NIST	7.0	7.7	7.9	8.2	8.4	***39.2***	**61.0**	**9.3**
Total	***151.4***	***156.4***	***161.1***	**165.9**	***170.9***	***805.7***	***902.4***	***615.5***

Source: U.S. House of Representatives, H.R. 3820; and NEHRP program office, at http://www.nehrp.gov/pdf/ ppt_sdr.pdf.

Note: Totals may not sum due to rounding.

Other Changes Proposed in H.R. 3820

Section 105 of H.R. 3820 would shift a post-earthquake investigations program from the USGS to be led instead by NIST, in consultation with the other NEHRP agencies, and the program would be organized to study the implications of earthquakes in the areas of responsibility of each NEHRP agency.

The interagency coordinating committee that would be created under Title III of H.R. 3820 would be largely similar to the current interagency committee except that it would also include the National Oceanic and Atmospheric Administration (NOAA), and the committee would have the discretion to include the head of any other federal agency it considers appropriate. In addition to acting as a single coordinating committee for both the earthquake and wind hazards program, the committee is charged with developing a strategic plan for both programs,[6] providing progress reports to Congress, and developing coordinated budgets for each program. Also, the Director of NIST is required to establish advisory committees for both programs, similar to the current advisory committees established under P.L. 108-360.

Title III of H.R. 3820 also requires the Subcommittee on Disaster Reduction of the Committee on Environment and Natural Resources of the National Science and Technology Council to submit a report, within two years of enactment, identifying federal research, development, and technology transfer activities for natural disasters, common areas of research among the natural hazards, and opportunities to "create synergies between the research activities for the hazards."

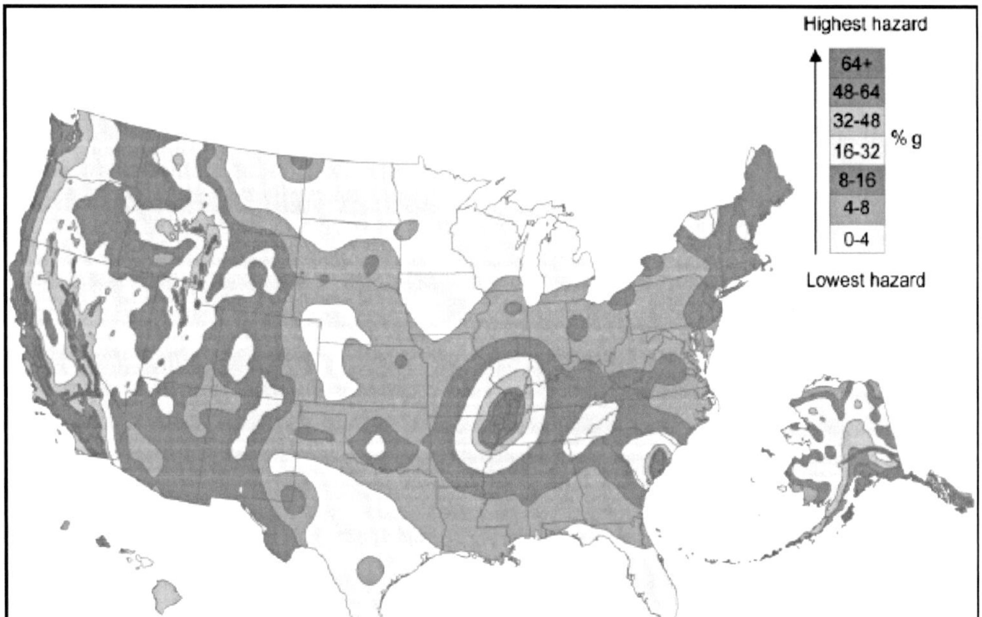

Source: USGS Fact Sheet 2008-3018 (April 2008), at http://earthquake.usgs.gov/research/hazmaps/products_data/images/nshm_us02.gif. Modified by CRS.

Note: The bar in the upper right shows the potential ground motion—expressed as a percentage of the acceleration due to gravity (g)—with up to a 1 in 10 chance of being exceeded over a 50-year period.

Figure 2. Earthquake Hazard in the United States

EARTHQUAKE HAZARDS AND RISK

Figure 2 indicates that detailed information exists on where earthquakes are likely to occur in the United States and how severe the earthquake magnitude and resulting ground shaking are likely to be. The map in **Figure 2** depicts the potential shaking hazard from future earthquakes. It is based on the frequency at which earthquakes occur in different areas and how far the strong shaking extends from the source of the earthquake. In **Figure 2**, the hazard levels indicate the potential ground motion—expressed as a percentage of the acceleration due to gravity (g). In a sense, the map shows the likelihood of where earthquakes could occur, and where the strongest shaking could take place.

All 50 states and the District of Columbia are vulnerable to earthquake hazards, although risks vary greatly across the country. Seismic hazards are greatest in the western United States, particularly California, Washington, Oregon, and Alaska and Hawaii. Alaska is the most earthquake-prone state, experiencing a magnitude 7 earthquake almost every year and a magnitude 8 earthquake every 14 years on average. (See box below for a description of earthquake magnitude.) Because of its low population and infrastructure density, Alaska has a relatively low risk for large economic losses from an earthquake. In contrast, California has more citizens and infrastructure at risk than any other state because of the state's frequent seismic activity combined with its high population.

EARTHQUAKE MAGNITUDE AND INTENSITY

Earthquake magnitude is a number that characterizes the relative size of an earthquake. It is often reported using the *Richter* scale (magnitudes in this report are generally consistent with the Richter scale). Richter magnitude is calculated from the strongest seismic wave recorded from the earthquake, and is based on a logarithmic (base 1 0) scale: for each whole number increase in the Richter scale, the ground motion increases by ten times. The amount of energy released per whole number increase, however, goes up by a factor of 32. The *moment magnitude* scale is another expression of earthquake size, or energy released during an earthquake, that roughly corresponds to the Richter magnitude and is used by most seismologists because it more accurately describes the size of very large earthquakes. *Intensity* is a measure of how much shaking occurred at a site based on observations and amount of damage. Intensity is usually reported on the Modified Mercalli Intensity Scale as a Roman numeral ranging from I (not felt) to XII (total destruction).

Figure 2 also shows relatively high earthquake hazard in the Rocky Mountain region, portions of the eastern seaboard—particularly South Carolina—and a part of the central United States known as the New Madrid Seismic Zone (see "The New Madrid Seismic Zone," below). Other portions of the eastern and northeastern United States are also vulnerable to moderate seismic hazard. According to the USGS, 75 million people in 39 states are subject to "significant risk."[7]

National Seismic Hazards Maps and Earthquake Forecast for California

On April 21, 2008, the USGS released National Seismic Hazards Maps that updated the version published in 2002. Compared to the 2002 version, the new maps indicate lower ground motions (by 10% to 25%) for the central and eastern United States, based on modifications to the ground-motion models used for earthquakes. The new maps indicate that estimates of ground motion for the western United States are as much as 30% lower for certain types of ground motion, called long-period seismic waves, which affect taller, multi-story buildings. Ground motion that affects shorter buildings of a few stories, called short-period seismic waves, is roughly similar to the 2002 maps. The new maps show higher estimates for ground motion for western Oregon and Washington compared to the 2002 maps, due to new ground motion models for the offshore Cascadia subduction zone. In formulating the 2008 maps, the USGS gave more weight to the probability of a catastrophic magnitude 9 earthquake occurring along the Cascadia subduction zone. The Cascadia subduction zone fault ruptures, on average, every 500 years, and has the potential to generate destructive earthquakes and tsunamis along the coasts of Washington, Oregon, and northern California.

According to a report released on April 14, 2008, California has a 99% chance of experiencing a magnitude 6.7 or larger earthquake in the next 30 years. The likelihood of an even larger earthquake, magnitude 7.5 or greater, is 46% and will likely occur in the southern part of the state. The fault with the highest probability of generating at least one earthquake of magnitude 6.7 or greater over the next 30 years is the San Andreas in southern California (59% probability); for northern California it is the Hayward-Rodgers Creek fault (31%). The earthquake forecasts are not predictions (i.e., they do not give a specific date or time), but represent probabilities over a given time period. In addition, the probabilities have variability associated with them. The earthquake forecasts are known as the "Uniform California Earthquake Rupture Forecast (UCERF)" and are produced by a working group composed of the USGS, the California Geological Survey, and the Southern California Earthquake Center.

Sources: USGS Fact Sheet 2008-3018, "2008 United States National Seismic Hazard Maps" (April 2008), at http://pubs.usgs.gov/fs/2008/3018/pdf/FS08-3018_508.pdf; USGS Fact Sheet 2008-3027, "Forecasting California's Earthquakes—What Can We Expect in the Next 30 Years?" (2008), at http://pubs.usgs.gov/fs/2008/3027/fs2008- 3027.pdf.

Shaking hazards maps, such as the one in **Figure 2**, are often combined with other data, such as the strength of existing buildings, to estimate possible damage in an area due to an earthquake. The combination of seismic risk, population, and vulnerable infrastructure can help improve the understanding of which urban areas across the United States face risks from earthquake hazards that may not be immediately obvious from the probability maps of shaking hazards alone. The USGS has identified 26 urban areas that face a significant seismic risk from the combination of population and severity of shaking. **Table 3** lists those areas at greatest risk.

Table 3. Urban Areas Facing Significant Seismic Risk (alphabetically by state for cities with at least 300,000 people)

State	City	State	City
Alaska	Anchorage	Nevada	Las Vegas
California	Fresno	Nevada	Reno
California	Los Angeles	New Mexico	Albuquerque
California	Sacramento	New York	New York
California	Salinas	Oregon	Eugene-Springfield
California	San Diego	Oregon	Portland
California	San Francisco-Oakland	Puerto Rico	San Juan
California	Santa Barbara	South Carolina	Charleston
California	Stockton-Lodi	Tennessee	Chattanooga-Knoxville
Idaho	Boise	Tennessee	Memphis
Indiana	Evansville	Utah	Provo-Orem
Massachusetts	Boston	Utah	Salt Lake City
Missouri	St. Louis	Washington	Seattle

Sources: USGS Fact Sheet 2006-30 16 (March 2006); USGS Circular 1188, Table 3.

Note: These areas are identified using a population-based risk factor based on 1999 population data. (William Leith, ANSS Coordinator, USGS, Reston, VA, telephone conversation, Nov. 15, 2006).

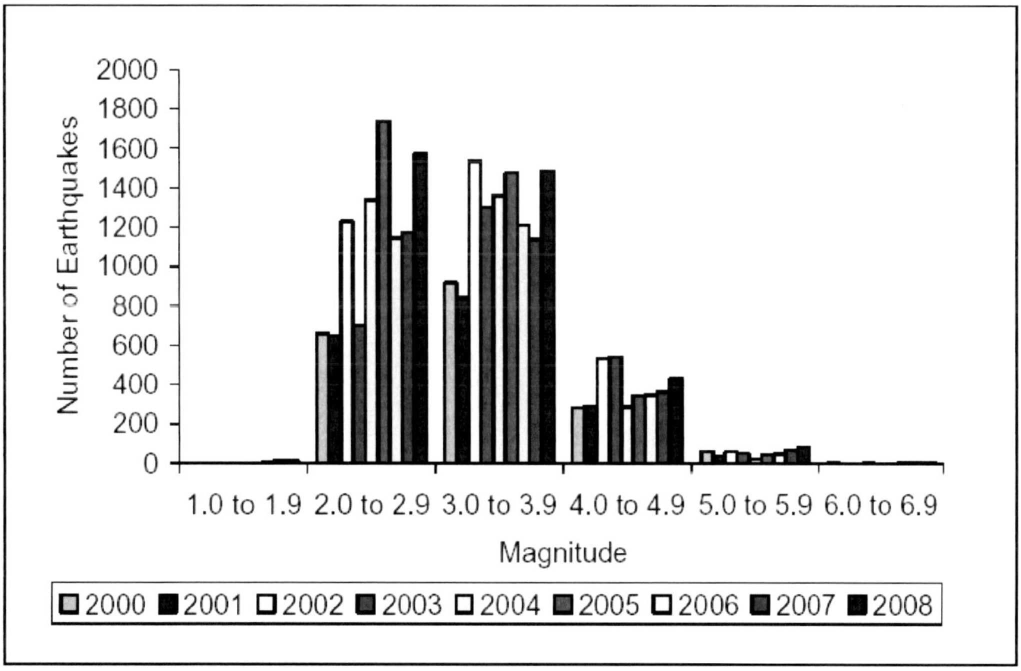

Source: USGS, "Earthquake Facts and Statistics," at http://neic.usgs.gov/neis/eqlists/eqstats.html; data as of December 7, 2009.

Note: Earthquakes greater than magnitude 7.0 and less than 1.0 are not shown. According to the USGS, 6 earthquakes of magnitude 7.0 or greater occurred in the United States between 2001 and 2007.

Figure 3. Histogram of the Number of U.S. Earthquakes from 2000 to 2008 by Magnitude (1.0 to 6.9)

Table 4. Earthquakes Responsible for Most U.S. Fatalities Since 1970

Date	Location	Magnitude	Deaths
February 9, 1971	San Fernando Valley, CA	6.6	65
October 18, 1989	Loma Prieta, CA	6.9	63
January 17, 1994	Northridge, CA	6.7	60

Source: USGS, http://earthquake.usgs.gov/regional/states/us_deaths

Note: Other sources report different numbers of fatalities associated with the Northridge earthquake.

The USGS estimates that several million earthquakes occur worldwide each year, but the majority are of small magnitude or occur in remote areas, and are not detectable. More earthquakes are detected each year as more seismometers[8] are installed in the world, but the number of large earthquakes (magnitude greater than 6.0)[9] has remained relatively constant. Between 2000 and 2008 there were between 2,261 and 3,876 earthquakes per year in the United States, according to the National Earthquake Information Center (NEIC). (See **Figure 3**.)

As **Figure 3** shows, about 98% of earthquakes detected each year by the NEIC are smaller than magnitude 5.0; only 59 earthquakes exceeded magnitude 6.0 for the nine-year period (less than 0.3% of the total earthquakes detected) for an average of less than seven earthquakes per year of at least 6.0 magnitude. Large earthquakes, although infrequent, cause the most damage and are responsible for most earthquake-related deaths. The great San Francisco earthquake of 1906 claimed an estimated 3,000 lives, as a result of both the earthquake and subsequent fires. Over the past 100 years, relatively few Americans have died as a result of earthquakes, compared to citizens in some other countries. [10] Since 1970, three major earthquakes in the United States were responsible for 188 of the 212 total earthquake-related fatalities (see **Table 4**).

Since 2000, only two deaths directly caused by earthquakes have occurred in the United States, both associated with falling debris in Paso Robles (CA) during the December 22, 2003, San Simeon earthquake of magnitude 6.5. In contrast, earthquakes have been directly or indirectly responsible for more than 685,000 fatalities in other countries since 2000.[11] Approximately 65% of those estimated deaths resulted from the December 2004 Indonesian earthquake (and resulting tsunami) of magnitude 9.1, and the January 2010 magnitude 7.0 earthquake in Haiti. Since the devastating Indonesian earthquake occurred, an October 2005 magnitude 7.6 earthquake in Pakistan killed approximately 86,000 people, and a May 2008 magnitude 7.9 earthquake struck Eastern Sichuan, China, causing as many as 87,000 fatalities (see box).

The 1994 Northridge earthquake was the nation's most damaging earthquake in the past 100 years, preceded five years earlier by the second most costly earthquake—Loma Prieta. **Table 5** shows the 10 costliest U.S. earthquakes in terms of insured and uninsured losses. Comparing losses between different earthquakes, and between earthquakes and other disasters such as hurricanes, can be difficult because of the different ways losses are calculated. Calculations may include a combination of insured losses, uninsured losses, and estimates of lost economic activity. For example, insured losses from Hurricane Katrina in 2005—mainly property—may be $41 billion, according to one estimate. [12] Total property damage would rise if uninsured property were included; and including interrupted economic activity in the calculation could bring the total loss for Hurricane Katrina to $100 billion, according to one estimate.[13]

THE MAY 12, 2008, SICHUAN EARTHQUAKE IN CHINA AND IMPLICATIONS FOR THE UNITED STATES

On May 12, 2008, a catastrophic earthquake of magnitude 7.9 struck Eastern Sichuan, China. The epicenter was located approximately 960 miles southwest of Beijing, and the earthquake was triggered approximately 12 miles below the earth's surface. Approximately 87,000 fatalities have been reported. The earthquake was felt in parts of eastern, southern, and central China, and as far away as Bangladesh, Taiwan, Thailand, and Vietnam. Several large aftershocks occurred after the main seismic event.

The May 12 earthquake resulted from movement along a northeast-trending *reverse* or *thrust* fault, reflecting stresses from the convergence of rocks of the Tibetan Plateau, to the west, against the crust underlying the Sichuan Basin and southeastern China. The region has experienced large earthquakes in the past; on August 25, 1933, a magnitude 7.5 earthquake struck the northwestern margin of the Sichuan Basin, resulting in approximately 9,300 fatalities.

Some concerns have been raised about the possibility of an earthquake of similar magnitude occurring in a seismically active region of the United States, such as southern California, where fault movement similar to the Eastern Sichuan earthquake may occur. On May 22, 2008, the USGS released a hypothetical scenario for a magnitude 7.8 southern California earthquake, called the ShakeOut Scenario. In the scenario, scientists hypothetically simulated the ground shaking and fault rupture associated with a magnitude 7.8 earthquake, and estimated the resulting damage to buildings and infrastructure. The scenario estimated approximately 1,800 fatalities and $213 billion in economic losses as a result of the earthquake. The report points to aggressive retrofitting programs that have increased the seismic resistance of buildings, highways, and other critical infrastructure in southern California as one reason why the number of possible fatalities is relatively low.

Some scientists have raised the possibility that earthquakes, such as the May 12, 2008, Sichuan event, may sometimes exhibit cascading behavior, where bursts of seismic energy are released along different places in a single fault, or jump between connected faults. Earthquakes that occur along the Sierra Madre fault in southern California, for example, could trigger a series of cascading seismic events along other faults, such as the San Andreas. Seismic hazard estimates may not fully account for the damage that could be caused by cascading earthquakes along a connected fault system. Scientists are hoping to examine the Sichuan earthquake in more detail to better understand the nature of cascading seismic events and how they affect the U.S. seismic hazard estimates.

Sources: Ken Hudnut, geophysicist, USGS, Pasadena, CA, phone conversation, June 11, 2008; USGS Earthquake Hazards Program, at http://earthquake.usgs.gov/eqcenter/eqinthenews/2008/us2008ryan/#summary; USGS Newsroom, *Earthquake Fatalities High in 2008*, at http://www.usgs.gov/newsroom/article.asp?ID=2101; and USGS, The ShakeOut Scenario, Open-File Report 2008-1150 (2008), at http://pubs.usgs.gov/of/2008/1150/.

Table 5. The 10 Most Damaging Earthquakes in the United States

Year	Location	Magnitude	$2005 (billions)
1994	Northridge, CA[a]	6.7	$26.0
1989	Loma Prieta, CA	6.9	$11.0
1964	Anchorage, AK	9.2	$3.1
1971	San Fernando, CA	6.5	$2.7
2001	Nisqually, WA	6.8	$2.5
1987	Whittier Narrows, CA	5.9	$0.62
1933	Long Beach, CA	6.3	$0.60
1953	Kern County, CA	7.5	$0.44
1992	Landers, CA	7.6	$0.13
1992	Cape Mendocino, CA	7.1	$0.092

Source: Insurance Information Institute, at http://www.iii.org/media.

Note: Includes insured and uninsured losses.

a. Estimates for total losses resulting from the Northridge earthquake vary; the Congressional Budget Office estimated $43 billion in total losses ($50 million in 2005 dollars). See *Federal Reinsurance for Disasters*, Congressional Budget Office (September 2002), p. 19.

JANUARY 12, 2010, MAGNITUDE 7.0 EARTHQUAKE STRIKES HAITI

On Tuesday, January 12, 2010, a magnitude 7.0 earthquake struck Haiti at 4:53 p.m. The epicenter was located approximately 15 miles southwest of Port-au-Prince, and the earthquake occurred at a depth of about 8 miles, according to initial USGS reports. The relatively shallow earthquake, and its close proximity to the capital city, exposed millions of Haitians to severe to violent ground shaking. The earthquake occurred along the EnriquilloPlantain Garden fault system, a major east-west trending strike-slip fault system that lies between the Caribbean tectonic plate and the North American tectonic plate; the Caribbean plate actively moves against the North American plate and shear stresses are created at the boundary. At a strike-slip fault, the rocks move past each other horizontally along the fault line (in contrast to a thrust fault, where rocks on one side of the fault move on top of the rocks on the other side). Earthquakes comparable to or stronger than the January 12 event have struck Haiti at least four times over the past three centuries. Earthquakes in 1751 and 1770—which probably occurred along the same fault system—destroyed Port-au-Prince. Other examples of strike-slip faults are the San Andreas fault in California and the Red River fault in China.

The January 12 earthquake caused widespread damage in the Port-au-Prince area, causing approximately 230,000 deaths and 300,000 injuries. Also, a series of aftershocks followed the main earthquake. There were 14 aftershocks greater than magnitude 5 and 36 greater than magnitude 4 within the first day following the magnitude 7.0 event. Aftershocks have the potential to cause further damage, especially to structures weakened by the initial large earthquake. In a February 23, 2010, statement, the USGS indicated that the frequency of aftershocks should diminish with time, but the threat of additional damaging earthquakes remains. The USGS noted that buildings in the Port-au- Prince area will continue to be at risk from strong earthquake shaking, and that the fault responsible for the January 12 earthquake still stores sufficient strain to be released as a large, damaging earthquake during the lifetime of structures built during the reconstruction effort.

The USGS based its probability estimates on techniques developed to assess earthquake hazards in the United States. Using these techniques, the USGS estimated that the probability of a magnitude 7 or greater earthquake occurring within the next 50 years along the Enriquillo fault near Port-au-Prince is between 5% and 15%. The range of probabilities reflects the current understanding of the seismicity and tectonics of the Haiti region. By comparison, the USGS has estimated that that the probability of a magnitude 7 or greater earthquake occurring within the next 50 years along the Hayward-Rodgers Creek fault east of San Francisco is about 15%.

Sources: USGS Earthquake Hazards Program, Significant Earthquakes: Magnitude 7.0—Haiti Region, at http://earthquake.usgs.gov/earthquakes/eqinthenews/2010/us2010rja6/; Michael Blanpied, Associate Coordinator for the USGS Earthquake Hazards Program, podcast, Jan. 13, 2010, at http://www.usgs.gov/corecast/details.asp?ep=117; email from David Applegate, Senior Science Advisor for Earthquake and Geologic Hazards, Jan. 13, 2010; USGS statement, "USGS Updates Assessment of Earthquake Hazard and Safety in Haiti and the Caribbean," February 23, 2010, at http://www.usgs.gov/newsroom/article.asp?ID=2413&from=rss_home. See also CRS Report R4 1023, *Haiti Earthquake: Crisis and Response*, by Rhoda Margesson and Maureen Taft-Morales.

Table 6. U.S. Metropolitan Areas with Estimated Annualized Earthquake Losses of More Than $10 Million (in millions)

Rank	Metro area	AEL	Rank	Metro area	AEL
1	Los Angeles-Long Beach-Santa Ana, CA	$1,312	23	Reno-Sparks, NV	$29
2	San Francisco-Oakland-Fremont, CA	$781	24	Charleston-North Charleston, SC	$22
3	Riverside-San Bernadino-Ontario, CA	$397	25	Columbia, SC	$22
4	San Jose-Sunnyvale-Santa Clara, CA	$277	26	Stockton, CA	$21
5	Seattle-Tacoma, WA	$244	27	Atlanta-Sandy Springs-Marietta, GA	$19
6	San Diego-Carlsbad-San Marcos, CA	$155	28	Bremerton-Silverdale, WA	$18
7	Portland-Vancouver-Carlsbad, OR	$137	29	Ogden-Clearfield, UT	$18
8	Oxnard-Thousand Oaks-Ventura, CA	$111	30	Salem, OR	$17
9	Santa Rosa-Petaluma, CA	$69	31	Eugene-Springfield, OR	$17
10	St. Louis, MO-IL	$59	32	Napa, CA	$16
11	Salt Lake City, UT	$52	33	San Luis Obispo-Paso Robles, CA	$16
12	Sacramento-Arden-Arcade-Roseville, CA	$52	34	Nashville-Davidson-Murfreesboro, TN	$15

Table 6. (Continued)

Rank	Metro area	AEL	Rank	Metro area	AEL
13	Vallejo-Fairfield, CA	$40	35	Albuquerque, NM	$15
14	Memphis, TN	$38	36	Olympia, WA	$14
15	Santa Cruz-Watsonville, CA	$36	37	Modesto, CA	$13
16	Anchorage, AK	$35	38	Fresno, CA	$13
17	Santa Barbara-Santa Maria-Goleta, CA	$34	39	Evansville, IN-KY	$12
18	Las Vegas-Paradise, NV	$33	40	Birmingham-Hoover, AL	$11
19	Honolulu, HI	$32	41	El Centro, CA	$11
20	Bakersfield, CA	$30	42	Little Rock-North Little Rock, AR	$11
21	New York-Northern New Jersey-Long Island, NY	$30	43	Provo-Orem, UT	$10
22	Salinas, CA	$29			

Source FEMA Publication 366, *HAZUS MH Estimated Annualized Earthquake Losses for the United States* (April 2008). Annualized earthquake losses (AEL) calculated in 2005 dollars.

Potential Losses from Earthquakes

The United States faces potentially large total losses due to earthquake-caused damage to buildings and infrastructure and lost economic activity. As urban development continues in earthquake-prone regions in the United States, concerns are increasing about the exposure of the built environment, including utilities and transportation systems, to potential earthquake damage.[14] One estimate of loss from a severe earthquake in the Los Angeles area is over $500 billion. An even higher estimate—approximately $900 billion—includes damage to the heavily populated central New Jersey-Philadelphia corridor if a 6.5 magnitude earthquake occurred along a fault lying between New York City and Philadelphia.[15]

Some studies and techniques combine seismic risk with the value of the building inventory[16] and income losses (e.g., business interruption, wage, and rental income losses) in cities, counties, or regions across the country to provide estimations of economic losses from earthquakes. An April 2008 report from FEMA calculated that the *annualized* loss from earthquakes nationwide is $5.3 billion, with California, Oregon, and Washington accounting for nearly $4.1 billion (77%) of the U.S. total estimated annualized loss. [17] **Table 6** shows metropolitan areas with estimated annualized U.S. earthquake losses over $10 million.

Annualized earthquake loss (AEL) addresses two components of seismic risk: the probability of ground motion and the consequences of ground motion. It enables comparison between different regions with different seismic hazards and different building construction types and quality. For example, earthquake hazard is higher in the Los Angeles area than in Memphis, but the general building stock in Los Angeles is more resistant to the effects of earthquakes. The AEL annualizes the expected losses by averaging them by year.

February 27, 2010, Magnitude 8.8 Earthquake in Chile; Is There a Similar Risk to the United States?

A massive magnitude 8.8 earthquake struck Chile on February 27, 2010, along a subduction zone plate boundary fault 70 miles north-northeast of the city of Concepcion and offshore of the Chilean coast. The earthquake occurred at a depth of approximately 22 miles below the seafloor, much deeper than the earthquake that struck Haiti near Port-au-Prince on January 12. Several cities, including Concepcion, experienced intensity VIII shaking on the Modified Mercalli Intensity Index, corresponding to severe perceived shaking. The capital city of Santiago, located 200 miles northeast of the epicenter, experienced intensity VII shaking corresponding to very strong perceived shaking. The earthquake has caused widespread damage that may amount to $3 to $8 billion in insured losses, and probably more in total economic losses. Over 700 deaths have been reported, many from the tsunami generated by the subsea earthquake, and possibly 2 million people have been affected. Far fewer fatalities are expected in Chile compared to the Haiti earthquake, even though the Chilean earthquake was much more powerful, in part because the epicenter was further from major population centers and the earthquake occurred 14 miles deeper than the Haitian event. However, details are beginning to emerge about the effectiveness of earthquake-resistant design and building construction in Chile compared to Haiti.

Because the earthquake occurred offshore, it generated a tsunami which struck parts of the Chilean coastline and offshore islands, causing damage and fatalities. Tsunami warnings were issued by the National Weather Service Pacific Tsunami Warning Center for Hawaii, Japan, and other regions bordering the Pacific Ocean that may have been vulnerable to a damaging tsunami wave, although most regions far from the epicenter did not experience any serious damage. A tsunami caused significant damage to the city of Hilo, Hawaii, following the May 1960 magnitude 9.5 earthquake that also occurred along the subduction zone fault about 143 miles south of the February 27, 2010, earthquake. Why the 1960 earthquake generated a tsunami that caused damage and fatalities in Hawaii, Japan, and the Philippines while the 2010 earthquake did not is not yet well understood and is being actively studied.

The magnitude 8.8 earthquake occurred along the boundary between the Nazca tectonic plate and the South American tectonic plate, which converge at a rate of about 3 inches per year. The Nazca plate is subducting under the South American plate, which rides over the top of the Nazca plate. In geologic terms, this is known as a thrust fault or megathrust, in contrast to a strike-slip fault, where the rocks on one side of the fault slide past each other. The San Andreas fault and the Enriquillo fault that caused the January 2010 Haiti earthquake are strike-slip faults. The Sumatran-Andaman megathrust fault, which triggered the December 2004 Indonesian earthquake and tsunami, is a subduction zone fault or megathrust geologically similar to the Nazca-South American tectonic plate subduction zone.

Subduction zone megathrust faults generate the largest earthquakes in the world. The Cascadia Subduction Zone megathrust that stretches from mid-Vancouver Island in southern British Columbia southward to Cape Mendocino in northern California has the potential to generate a very large earthquake, similar in magnitude to the February 27

Chilean earthquake. The fault's proximity to the northwestern United States coastline—approximately 50-100 miles offshore—also poses a significant tsunami hazard; destructive waves from a large earthquake along the fault could reach the coast of Oregon and Washington in less than an hour, possibly in tens of minutes. The Cascadia Subduction Zone fault forms the boundary between the subducting Juan de Fuca tectonic plate and the overriding North American plate, very similar to the relationship between the Nazca Plate and the South American Plate off the Chilean coast. If the Cascadia Subduction Zone megathrust were to "unzip" or rupture along a large section of its entire length, it would likely generate a megathrust earthquake near magnitude 9 or more, similar to the 1964 Alaskan earthquake, the 1960 and 2010 Chilean earthquakes, and the 2004 Indonesian earthquake. Scientists have documented that the last time this occurred along the Cascadia Subduction Zone fault was in 1700. The 1700 earthquake spawned a tsunami that traveled across the Pacific Ocean and struck Japan. Because of the tectonic similarities between the Cascadia Subduction Zone megathrust and the Nazca-South American plate megathrust, scientists hope to learn a great deal about the seismic hazard in the Pacific Northwest by studying the unique strong ground motion recordings from the 2010 Chilean magnitude 8.8 earthquake.

Sources: USGS Earthquake Hazards Program, *Magnitude 8.8–Offshore Maule, Chile*, at http://earthquake.usgs.gov/ earthquakes/eqinthenews/20 1 0/us20 1 0tfan/; Eqecat, "M8.8 Earthquake Offshore Maule, Chile," CatWatch, February 27, 2010, at http://www.eqecat.com/catWatchREV/secureSite/report.cfm?id=220 (Eqecat provides loss estimates for insurance and financial institutions); Business Week, *Chile Asks for International Aid as Looting Spreads*, March 2, 2010, at http://www.businessweek.com/news/2010-03-02/chile-asks-for-international-aid-as-looting-spreads-correct-.html; Brian F. Atwater et al., *The Orphan Tsunami of 1700—Japanese Clues to a Parent Earthquake in North America*, USGS, Professional Paper 1707, 2005, http://pubs.usgs.gov/pp/pp1707/; E-mail from Richard Aster, Professor of Geophysics, New Mexico Institute of Mining and Technology, March 2, 2010.

A Decrease in Estimated Loss?

In its most recent publication estimating earthquake losses, FEMA noted that the $5.3 billion in annualized earthquake loss nationwide was 21% higher than the $4.4 billion calculated in FEMA's previous report, published in February 2001.[18] However, the 2001 report calculated losses using 1994 dollars, and when adjusted to reflect 2005 dollars the earlier estimate increased to $5.6 billion, indicating a small decrease in nationwide annualized earthquake loss potential since the 2001 report was published. According to FEMA, this loss occurred even though the national building inventory increased by 50% over this same period.

What factors led to a decreased estimate in potential loss despite growth in building inventory? According to FEMA, two primary factors were responsible: (1) a slight decrease in estimated earthquake hazard in the western United States (namely California) except for some parts of Washington and Utah, and (2) a change in the distribution of building inventory in California, with an increase in wood frame buildings of 17% and a reduction in the amount of masonry (-6%), steel (-5.8%), and concrete (-3%) buildings in the state.[19] Wood frame buildings are less vulnerable to earthquake damage, generally, compared to other construction types. Because California accounts for 66% of the overall nationwide annualized earthquake loss, a 17% increase in wood frame buildings had a proportionally large effect. In fact, FEMA attributed 78% of the loss reduction between 2001 and 2008 to the change in building inventory distribution, and 22% to the decrease in earthquake hazard for California.[20]

A single large earthquake can cause far more damage than the average annual estimate. Annualized estimates, however, help provide comparisons of infrequent, high impact events like damaging earthquakes, with more frequently occurring hazards like floods, hurricanes, or other types of severe weather. The annualized earthquake loss values shown in **Table 6** represent future estimates, and are calculated by multiplying losses from potential future ground motions by their respective frequencies of occurrence, and then summing these values.[21]

Table 6 also shows that annualized earthquake losses in the New York-Northern New Jersey- Long Island metropolitan area are $30 million (ranked 21 out of 43 metropolitan areas with losses greater than $1 million per year), even though no destructive earthquakes have struck that area for generations.[22] This area has a relatively low seismic hazard, but also has an extensive infrastructure and is densely populated. That combination of seismic risk, extensive infrastructure, and dense population produces a significant risk to people and structures, according to some estimates.[23] In the absence of any significant or damaging earthquakes for that region in recent memory, however, the actual risk may be difficult to grasp intuitively.

The New Madrid Seismic Zone

The New Madrid Seismic Zone in the central United States is vulnerable to large but infrequent earthquakes. A series of large (magnitude greater than 7.0) earthquakes struck the Mississippi Valley over the winter of 1811-1812, centered close to the town of New Madrid, MO. Some of the tremors were felt as far away as Charleston, SC, and Washington, DC. The mechanism for the earthquakes in the New Madrid zone is poorly understood,[24] and no earthquakes of comparable magnitude have occurred in the area since these events.

Estimating earthquake damage is not an exact science and depends on many factors. As described above, these are primarily the probability of ground motion occurring in a particular area (see **Figure 2**), and the consequences of that ground motion, which are largely a function of building construction type and quality, and of the level of ground motion and shaking during the actual event. Such factors contribute to the difficulty of making a reasonable damage estimate for a low- frequency, high-impact event in the New Madrid region based on the probability of an earthquake of similar magnitude occurring. This uncertainty has implications for policy decisions to ameliorate risk, such as setting building codes, and for designing and building structures to withstand a level of shaking commensurate with the risk. Developers of building codes tend to err on the side of caution; presumably the same seismic hazard should lead to similar building codes in urban areas (e.g., compare the seismic hazard for the New Madrid area with parts of California shown in **Figure 2**).

Some researchers have questioned whether erring on the side of caution in the New Madrid Seismic Zone is justified.[25] These researchers question whether the benefits of building structures to conform with the earthquake probability estimates merit the costs, in light of the uncertainty in making those probability estimates.[26] These analyses may call into question whether the probability of ground motion estimates for the New Madrid Seismic Zone (the bulls-eye-shaped area shown in **Figure 2** that includes parts of Arkansas, Illinois, Tennessee, and Missouri), and other regions of the country that experience infrequent earthquakes, are too high.[27] An uncertainty analysis of the seismic hazard in the New Madrid Seismic Zone is beyond the scope of this report.

MONITORING

Congress authorized the USGS to monitor seismic activity in the United States in the 1990 reauthorization of the National Earthquake Hazards Reduction Act (P.L. 101-614). The USGS operates a nationwide network of seismographic stations called the Advanced National Seismic System (ANSS), which includes the National Strong-Motion Project (NSMP). Globally, the USGS and the Incorporated Research Institutions for Seismology (IRIS) operate 140 seismic stations of the Global Seismic Network (GSN) in more than 80 countries. The GSN provides worldwide coverage of earthquakes, including reporting and research.[28]

Advanced National Seismic System (ANSS)

According to the USGS, "the mission of ANSS is to provide accurate and timely data and information products for seismic events, including their effects on buildings and structures, employing modern monitoring methods and technologies."[29] If fully implemented, ANSS would encompass more than 7,000 earthquake sensor systems covering parts of the nation vulnerable to earthquake hazards. As envisioned, the system would consist of dense urban networks, regional networks, and backbone stations.

ANSS Funding

Congress first authorized the ANSS program in P.L. 106-503 at a level of $38 million for FY2002 and $44 million for FY2003. The 2004 reauthorization of NEHRP (P.L. 108-360) authorized $30 million for ANSS in FY2005 and then $36 million per year through FY2009. The USGS spent $1.6 million in FY2000 and $3.6 million in FY2001 on ANSS-directed funding, but expenditures have never reached authorized levels since Congress first authorized appropriations for ANSS. From FY2005 through FY2009, the USGS spent a total of approximately $42.5 million within its Earthquake Hazards Program on ANSS, or approximately 24% of the total authorized levels over the five-year period since NEHRP was last reauthorized.[30]

The FY2010 budget request stated that the USGS plans to install a cumulative total of 822 ANSS monitoring stations by the end of 2009. That would represent approximately 12% of the 7,000 seismic stations originally envisioned for the program. According to its budget justification, the USGS plans to devote its ANS S-directed resources to operating and maintaining the installed system.[31] Of the approximately $8.8 million for ANSS-directed funding in FY2009, about $1.5 million was devoted to the development, modernization, and expansion of the system; the remainder of FY2009 funding was used to operate the existing system.[32] However, the American Recovery and Reinvestment Act (ARRA, P.L. 111-5) provided approximately $19 million in ANS S-directed funding to be spent over FY2009-FY2011.[33] All of the ARRA funding would be provided for modernization and expansion of the current system, and when supplemented by base program funds over the three-year period, would double the number of ANSS stations.[34] A doubling of the number of current stations would total approximately 23% of the 7,000 stations originally planned for ANSS.

Dense Urban Networks

In the original conception for ANSS, approximately 6,000 of the planned stations would be installed in 26 high-risk urban areas to monitor strong ground shaking and how buildings and other structures respond. Currently, five high-risk urban areas have instruments deployed in sufficient density to generate the data to produce near real-time maps, called ShakeMaps, which can be used in emergency response during and after an earthquake.[35] (See "ShakeMap," below.)

Regional Networks

Approximately 1,000 new instruments would replace aging and obsolete stations in the networks that now monitor the nation's most seismically active regions. The current regional networks contain a mix of modern, digital, broadband, and high-resolution instruments that can provide real-time data; they are supplemented by older instruments that may require manual downloading of data. Universities in the region typically operate the regional networks and will likely continue to do so as ANSS is implemented.

Backbone Stations

Approximately 100 instruments comprise the existing "backbone" of ANSS, with a roughly uniform distribution across the United States, including Alaska and Hawaii. These instruments provide a broad and uniform minimum threshold of coverage across the country. The backbone network consists of USGS-deployed instruments and other instruments that serve both ANSS and the EarthScope project (described below, under "National Science Foundation").

National Strong-Motion Project (NSMP)

Under ANSS, the USGS operates the NSMP to record seismic data from damaging earthquakes in the United States on the ground and in buildings and other structures in densely urbanized areas. The program currently has 900 strong-motion[36] instruments in 701 permanent stations across the United States and in the Caribbean. The NSMP has three components: data acquisition, data management, and research. The near real-time measurements collected by the NSMP are used by other government agencies for emergency response and real-time warnings. If fully implemented, the ANSS program would deploy about 3,000 strong-motion instruments. Many of the current NSMP instruments are older designs and are being upgraded with modern seismometers.

Global Seismic Network (GSN)

The GSN is a system of broadband digital seismographs arrayed around the globe and designed to collect high-quality data that are readily accessible to users worldwide, typically via computer. Currently, 140 stations have been installed in 80 countries and the system is nearly complete, although in some regions the spacing and location of stations has not fully met the original goal of uniform spacing of approximately 2,000 kilometers. The system is currently providing data to the United States and other countries and institutions for earthquake reporting and research, as well as for monitoring nuclear explosions to assess compliance with the Comprehensive Test Ban Treaty. Funding for the GSN totaled approximately $9 million in FY2009.[37]

The Incorporated Research Institutions for Seismology (IRIS)[38] coordinates the GSN and manages and makes available the large amounts of data that are generated from the network. The actual network of seismographs is organized into two main components, each managed separately. The USGS operates two-thirds of the stations from its Albuquerque Seismological Laboratory, and the University of California-San Diego manages the other third via its Project IDA (International Deployment of Accelerometers). Other universities and affiliated agencies and institutions operate a small number of additional stations. IRIS, with funding from the NSF, supports all of the stations not funded through the USGS appropriations.

Detection, Notification, and Warning

Unlike other natural hazards, such as hurricanes, where predicting the location and timing of landfall is becoming increasingly accurate, the scientific understanding of earthquakes does not yet allow for precise earthquake prediction. Instead, notification and warning typically involves communicating the location and magnitude of an earthquake as soon as possible after the event to emergency response providers and others who need the information.

Some probabilistic earthquake forecasts are now available that give, for example, a 24-hour probability of earthquake aftershocks for a particular region, such as California. These forecasts are not predictions, and are currently intended to increase public awareness of the seismic hazard, improve emergency response, and increase scientific understanding of the short-term hazard.[39] In the California example, a time-dependent map is created and updated every hour by a system that considers all earthquakes, large and small, detected by the California Integrated Seismic Network,[40] and calculates a probability that each earthquake will be followed by an aftershock[41] that can cause strong shaking. The probabilities are calculated from known behavior of aftershocks and the possible shaking pattern based on historical data.

When a destructive earthquake occurs in the United States or in other countries, the first reports of its location, or epicenter,[42] and magnitude originate either from the National Earthquake Information Center (NEIC), or from one of the regional seismic networks that are part of ANSS. Other organizations, such as universities, consortia, and individual seismologists may also contribute information about the earthquake after the event. Products such as ShakeMap are assembled as rapidly as possible to assist in emergency response and damage estimation following a destructive earthquake.

National Earthquake Information Center (NEIC)

The NEIC, part of the USGS, is located in Golden, CO. Originally established as part of the National Ocean Survey (U.S. Department of Commerce) in 1966, the NEIC was made part of the USGS in 1973. With data gathered from the networks described above and from other sources, the NEIC determines the location and size of all destructive earthquakes that occur worldwide and disseminates the information to the appropriate national or international agencies, government public information channels, news media, scientists and scientific groups, and the general public.

With the advent of the USGS Earthquake Notification Service (ENS), notifications of earthquakes detected by the ANSS/NEIC are provided free to interested parties. Users of the service can specify the regions of interest, establish notification thresholds of earthquake magnitude, designate whether they wish to receive notification of aftershocks, and even set different magnitude thresholds for daytime or nighttime to trigger a notification.

The NEIC has long-standing agreements with key emergency response groups, federal, state, and local authorities, and other key organizations in earthquake-prone regions who receive automated alerts—typically location and magnitude of an earthquake—within a few minutes of an event in the United States. The NEIC sends these preliminary alerts by email and pager immediately after an earthquake's magnitude and epicenter are automatically determined by computer.[43] This initial determination is then checked by around-the-clock staff who confirm and update the magnitude and location data.[44] After the confirmation, a second set of notifications and confirmations are triggered to key recipients by email, pager, fax, and telephone.

For earthquakes outside the United States, the NEIC notifies the State Department Operations Center, and often sends alerts directly to staff at American embassies and consulates in the affected countries, to the International Red Cross, the U.N. Department of Humanitarian Affairs, and other recipients who have made arrangements to receive alerts.

ShakeMap

Traditionally, the information commonly available following a destructive earthquake has been epicenter and magnitude, as in the data provided by the NEIC described above. Those two parameters by themselves, however, do not always indicate the intensity of shaking and extent of damage following a major earthquake. Recently, the USGS developed a product called ShakeMap that provides a nearly real-time map of ground motion and shaking intensity following an earthquake in areas of the United States where the ShakeMap system is in place. **Figure 4** shows an example of a ShakeMap.

The maps produced portray the extent of damaging shaking and can be used by emergency response and for estimating loss following a major earthquake. Currently, ShakeMaps are available for northern California, southern California, the Pacific Northwest, Nevada, Utah, Hawaii, and Alaska.[45]

With improvements to the regional seismographic networks in the areas where ShakeMap is available, new real-time telemetry from the region, and advances in digital communication and computation, ShakeMaps are now triggered automatically and made available within minutes of the event via the Web. In addition, better maps are now available because of recent improvements in understanding the relationship between the ground motions recorded during the earthquake and the intensity of resulting damage. If databases containing inventories of buildings and lifelines are available, they can be combined with shaking intensity data to produce maps of estimated damage. The ShakeMaps have limitations, especially during the first few minutes following an earthquake before more data arrive from distributed sources. Because they are generated automatically, the initial maps are preliminary, and may not have been reviewed by experts when first made available. They are considered a work in progress, but are deemed to be very promising, especially as more modern seismic instruments are added to the regional networks under ANSS and computational and telecommunication abilities improve.

Prompt Assessment of Global Earthquakes for Response (PAGER)

Another USGS product that is designed to provide nearly real-time earthquake information to emergency responders, government agencies, and the media is the Prompt Assessment of Global Earthquakes for Response, or PAGER, system.[46] This automated system rapidly assesses the number of people, cities, and regions exposed to severe shaking by an earthquake, and generally makes results available within 30 minutes. Following the determination of earthquake location and magnitude, the PAGER system calculates the degree of ground shaking using the methodology developed for ShakeMap, estimates the number of people exposed to various levels of shaking, and produces a description of the vulnerability of the exposed population and infrastructure. The vulnerability includes potential for earthquake-triggered landslides, which could be devastating, as was the case for the huge May 12, 2008, earthquake in Sichuan, China. The automated and rapid reports produced by the PAGER system provide an advantage compared to the traditional accounts from eye-witnesses on the ground or media reports, because communications networks may have been disabled from the earthquake. Emergency responders, relief organizations, and government agencies could make plans based on PAGER system reports even before getting "ground-truth" information from eye-witnesses and the media.[47]

Figure 5 shows an example of PAGER output for the January 12, 2010, magnitude 7.0 earthquake in Haiti.

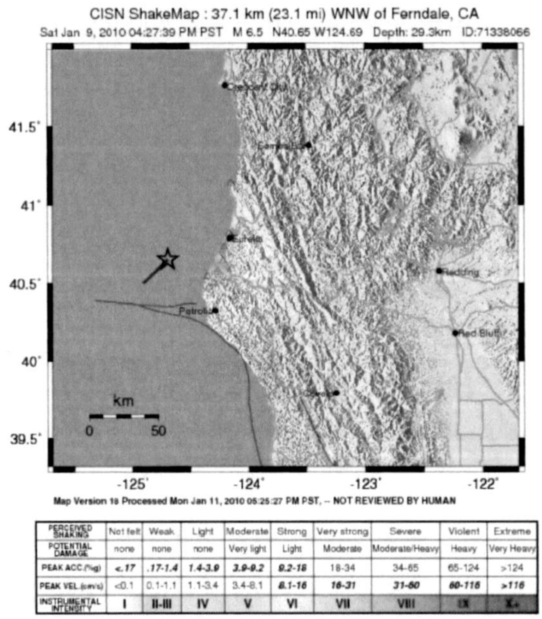

Source: USGS, http://earthquake.usgs.gov/eqcenter/shakemap/nc/shake/71338066/.

Note: Earthquake occurred 23.1 miles west-northwest of Ferndale, CA, at 4:27 p.m. on January 9, 2010, with a magnitude of 6.5. The star indicates the epicenter of the earthquake. Viewed on January 12, 2010. According to preliminary news reports, some damage estimates are nearly $30 million for the town of Eureka itself from the earthquake. (See http://www.times-standard.com/news/ci_14171082.)

Figure 4. Example of a ShakeMap

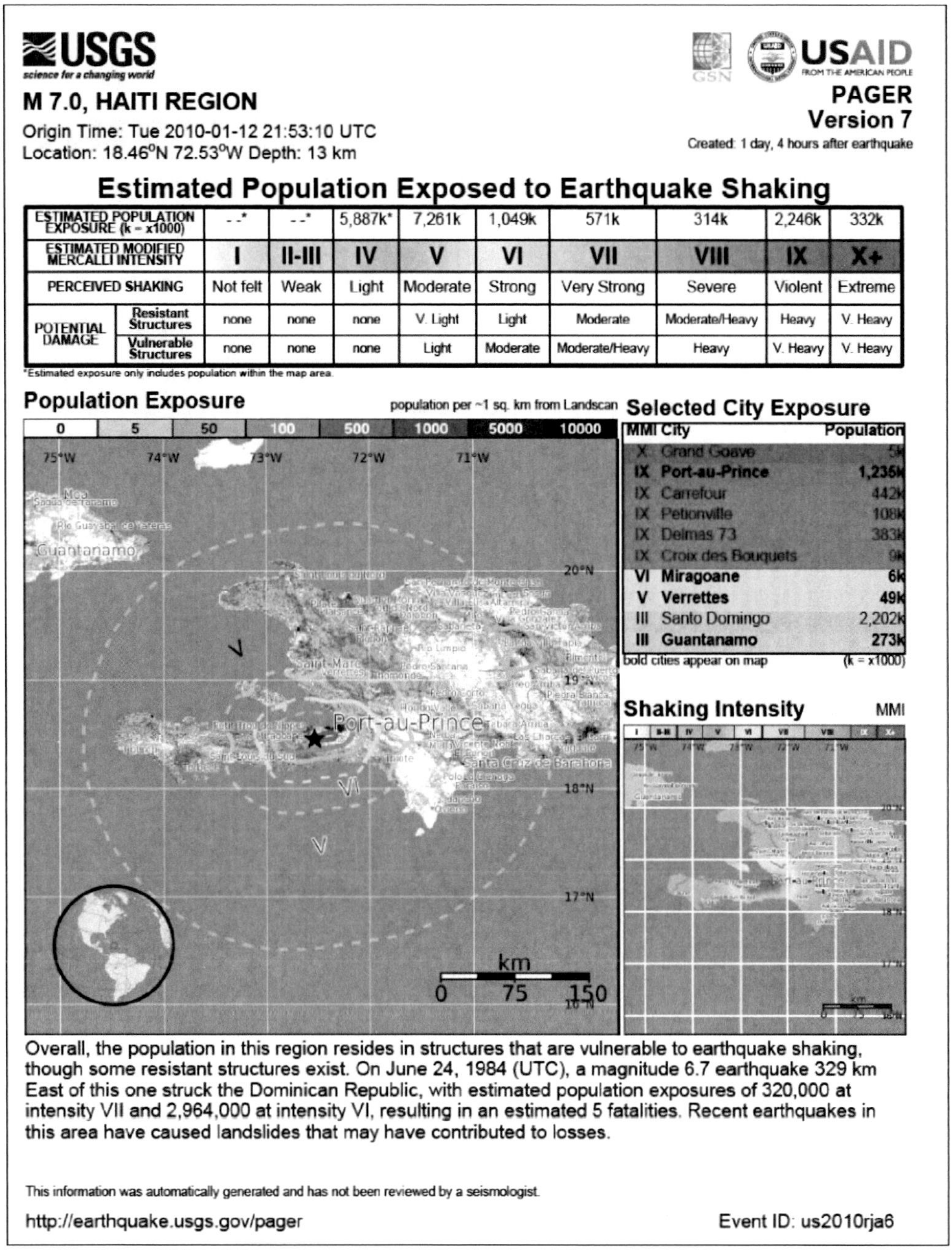

USGS
science for a changing world

USAID
FROM THE AMERICAN PEOPLE

M 7.0, HAITI REGION

PAGER
Version 7

Origin Time: Tue 2010-01-12 21:53:10 UTC
Location: 18.46°N 72.53°W Depth: 13 km

Created: 1 day, 4 hours after earthquake

Estimated Population Exposed to Earthquake Shaking

ESTIMATED POPULATION EXPOSURE (k = x1000)		--*	--*	5,887k*	7,261k	1,049k	571k	314k	2,246k	332k
ESTIMATED MODIFIED MERCALLI INTENSITY		I	II-III	IV	V	VI	VII	VIII	IX	X+
PERCEIVED SHAKING		Not felt	Weak	Light	Moderate	Strong	Very Strong	Severe	Violent	Extreme
POTENTIAL DAMAGE	Resistant Structures	none	none	none	V. Light	Light	Moderate	Moderate/Heavy	Heavy	V. Heavy
	Vulnerable Structures	none	none	none	Light	Moderate	Moderate/Heavy	Heavy	V. Heavy	V. Heavy

*Estimated exposure only includes population within the map area.

Population Exposure

population per ~1 sq. km from Landscan

0 | 5 | 50 | 100 | 500 | 1000 | 5000 | 10000

Selected City Exposure

MMI	City	Population
X	Grand Goave	5k
IX	**Port-au-Prince**	1,235k
IX	Carrefour	442k
IX	Petionville	108k
IX	Delmas 73	383k
IX	Croix des Bouquets	9k
VI	**Miragoane**	6k
V	**Verrettes**	49k
III	Santo Domingo	2,202k
III	**Guantanamo**	273k

bold cities appear on map (k = x1000)

Shaking Intensity MMI

I | II-III | IV | V | VI | VII | VIII | IX | X+

Overall, the population in this region resides in structures that are vulnerable to earthquake shaking, though some resistant structures exist. On June 24, 1984 (UTC), a magnitude 6.7 earthquake 329 km East of this one struck the Dominican Republic, with estimated population exposures of 320,000 at intensity VII and 2,964,000 at intensity VI, resulting in an estimated 5 fatalities. Recent earthquakes in this area have caused landslides that may have contributed to losses.

This information was automatically generated and has not been reviewed by a seismologist.

http://earthquake.usgs.gov/pager Event ID: us2010rja6

Source: USGS, http://earthquake.usgs.gov/earthquakes/pager/events/us/2010rja6/onepager.pdf.
Note: This is version 7 of the PAGER output, accessed on January 14, 2010.

Figure 5. Example of PAGER Output for the January 12, 2010, Magnitude 7.0 Haiti Earthquake

- physical damage to residential and commercial buildings, schools, critical facilities, and infrastructure;
- economic loss, including lost jobs, business interruptions, repair and reconstruction costs; and

- social impacts, including estimates of shelter requirements, displaced households, and number of people exposed to the disaster.

In addition to furnishing information as part of earthquake mitigation efforts, HAZUS-MH can also be used to support real-time emergency response activities by state and federal agencies after a disaster. Twenty-seven HAZUS-MH user groups—cooperative ventures among private, public, and academic organizations that use the HAZUS-MH software—have formed across the United States to help foster better-informed risk management for earthquakes and other natural hazards.[49] HAZUS-MH software was first released to the public in 1997 and the first user group, the Bay Area HAZUS-MH User Group, was formed the same year.

Pre-disaster Planning: HAZUS-MH

FEMA developed a methodology and software program called the Hazards U.S. Multi-Hazard (HAZUS-MH).[48] The program allows a user to estimate losses from damaging earthquakes, hurricane winds, and floods before a disaster occurs. The pre-disaster estimates could provide a basis for developing mitigation plans and policies, preparing for emergencies, and planning response and recovery. HAZUS-MH combines existing scientific knowledge about earthquakes (for example, ShakeMaps, described above), engineering information that includes data on how structures respond to shaking, and geographic information system (GIS) software to produce maps and display hazards data including economic loss estimates. The loss estimates produced by HAZUS-MH include

RESEARCH—UNDERSTANDING EARTHQUAKES

U.S. Geological Survey

Under NEHRP, the USGS has responsibility for conducting targeted research into improving the basic scientific understanding of earthquake processes. The current earthquake research program at the USGS covers six broad categories:[50]

- *Borehole geophysics and rock mechanics*: studies to understand heat flow, stress, fluid pressure, and the mechanical behavior of fault-zone materials at seismogenic[51] depths to yield improved models of the earthquake cycle;

- *Crustal deformation*: studies of the distortion or deformation of the earth's surface near active faults as a result of the motion of tectonic plates;

- *Earthquake geology and paleoseismology*: studies of the history, effects, and mechanics of earthquakes;

- *Earthquake hazards*: studies of where, why, when, and how earthquakes occur;

- *Regional and whole-earth structure*: studies using seismic waves from earthquakes and man-made sources to determine the structure of the planet ranging from the local scale, to the whole crust, mantle, and even the earth's core; and

- *Strong-motion seismology, site response, and ground motion*: studies of large-amplitude ground motions and the response of engineered structures to those motions using accelerometers.

National Science Foundation

NSF supports fundamental research into understanding the earth's dynamic crust. Through its Earth Sciences Division (part of the Geosciences Directorate), NSF distributes research grants and coordinates programs investigating the crustal processes that lead to earthquakes around the globe.[52] In 2003, NSF initiated a Major Research Equipment and Facilities Construction (MREFC) project called EarthScope that deploys instruments across the United States to study the structure and evolution of the North American Continent, and to investigate the physical processes that cause earthquakes and volcanic eruptions.[53] EarthScope is a multi-year project begun in 2003 that is funded by NSF and conducted in partnership with the USGS and NASA.

EarthScope instruments are intended to form a framework for broad, integrated studies of the four-dimensional (three spatial dimensions, plus time) structure of North America. The project is divided into three main programs:

- *The San Andreas Fault Observatory at Depth (SAFOD)*, a deep borehole observatory drilled through the San Andreas fault zone close to the hypocenter of the 1966 Parkfield, CA, magnitude 6 earthquake;

- *The Plate Boundary Observatory (PBO)*, a system of GPS arrays and strainmeters[54] that measure the active boundary zone between the Pacific and North American tectonic plates in the western United States; and

- *USA rray*, 400 transportable seismometers that will be deployed systematically across the United States on a uniform grid to provide a complete image of North America from continuous seismic measurements.

Through its Engineering Directorate, NSF funds the George E. Brown Jr. Network for Earthquake Engineering Simulation (NEES), a project intended to operate until 2014, aimed at understanding the effects of earthquakes on structures and materials.[55] To achieve the program's goal, the NEES facilities conduct experiments and computer simulations of how buildings, bridges, utilities, coastal regions, and materials behave during an earthquake.

CONCLUSION

At present earthquakes can be neither accurately predicted nor prevented, and in its 1990 reauthorization NEHRP shifted its program emphasis from prediction to hazard reduction. The program's focus has been on understanding the earthquake hazard and its risk to populations and infrastructure in the United States, developing effective measures to reduce earthquake hazards, and promoting the adoption of earthquake hazards reduction measures in vulnerable areas. In the 111th Congress, legislation introduced to reauthorize NEHRP, H.R. 3820, reemphasizes that approach but casts it in terms of hazard *mitigation* by stating that a major goal for the program should be "to reduce the loss of life and damage to communities and infrastructure through increasing the adoption of hazard mitigation measures." The bill further emphasizes the social aspects of mitigating earthquake hazards, calling for research to better understand institutional, social, behavioral, and economic factors that influence how risk mitigation is implemented, in addition to the traditional research into understanding how, why, and where earthquakes occur.

The emphasis on mitigation presented by H.R. 3820 reflects at least two fundamental challenges to increasing the nation's resiliency to earthquakes, and to most other major natural hazards such as hurricanes and major floods. The first is to assess whether social, behavioral, and economic factors can be understood in sufficient degree to devise strategies that influence behavior to mitigate risk posed by the hazard. Put simply, what motivates people and communities to adopt risk mitigation measures that address the potential hazard? A second challenge, which is more squarely an issue for Congress, which oversees and funds the federal earthquake program, is how to measure the effectiveness of NEHRP more quantitatively. It is inherently difficult to capture precisely, in terms of dollars saved or fatalities prevented, the effectiveness of mitigation measures taken before an earthquake occurs. A major earthquake in a populated urban area within the United States would cause damage, and a question becomes how much damage would be prevented by mitigation strategies underpinned by the NEHRP program.

A precise relationship between earthquake mitigation measures, NEHRP and other federal earthquake-related activities, and reduced losses from an actual earthquake may never be possible. However, as more accurate seismic hazard maps evolve, as understanding of the relationship between ground motion and building safety improves, and as new tools for issuing warnings and alerts such as ShakeMap and PAGER are devised, trends denoting the effectiveness of mitigation strategies and NEHRP activities may emerge more clearly. Without an ability to precisely predict earthquakes, Congress is likely to face an ongoing challenge in determining the most effective federal approach to increasing the nation's resilience to low-probability but high- impact natural hazards, such as major earthquakes.

Author Contact Information

Peter Folger
Specialist in Energy and Natural Resources Policy
pfolger@crs.loc.gov, 7-1517

End Notes

[1] ANSS is a nationwide network of seismographic stations operated by the USGS. GSN is a global network of stations coordinated by the Incorporated Institutions for Seismology. NEES is an NSF-funded project that consists of 15 experimental facilities and an IT infrastructure with a goal of mitigating earthquake damage by the use of improved materials, designs, construction techniques, and monitoring tools.

[2] U.S. House, Committee on Science, *National Earthquake Hazards Reduction Program Reauthorization Act of 2003*, H.Rept. 108-246 (Aug. 14, 2003), p. 13.

[3] Lifelines are essential utility and transportation systems.

[4] U.S. Congress, House Committee on Science and Technology, Subcommittee on Technology and Innovation, *Reauthorization of the National Earthquake Hazards Reduction Program: R&D for Resilient Communities*, testimony of Dr. Michael Lindell, 111th Cong., 1st sess., June 11, 2009.

[5] Ibid. Tornadoes could be considered similar to earthquakes as "rapid onset disasters" that provide little or no warning, but which elicit similar societal responses, according to Dr. Lindell.

[6] The current interagency coordinating committee for NEHRP submitted a strategic plan to Congress in October 2008. It is available at http://www.nehrp.gov/pdf/strategic_plan_2008.pdf.

[7] U.S. Geological Survey, Dept. of the Interior, *Earthquake Hazards—A National Threat*, Fact Sheet 2006-30 16, March 2006, http://pubs.usgs.gov/fs/2006/3016/2006-3016.pdf. During the period 1975-1995, only four states did not experience detectable earthquakes: Florida, Iowa, North Dakota, and Wisconsin. See USGS Earthquake Hazards Program, *Earthquake Facts*, at http://earthquake.usgs.gov/learn/facts.php.

[8] *Seismometers* are instruments that measure and record the size and force of seismic waves, essentially sound waves radiated from the earthquake as it ruptures. Seismometers generally consist of a mass attached to a fixed base. During an earthquake, the base moves and the mass does not, and the relative motion is commonly transformed into an electrical voltage that is recorded. A *seismograph* usually refers to the *seismometer* and the recording device, but the two terms are often used interchangeably.

[9] See USGS "Earthquakes Facts and Statistics" at http://neic.usgs.gov/neis/eqlists/eqstats.html#table_2.

[10] Estimates of earthquake-related fatalities vary and an exact tally of deaths and injuries is rare. For more information on the difficulties of counting earthquake-related deaths and injuries, see http://earthquake.usgs.gov/regional/world/ casualty_totals.php.

[11] U.S. Geological Survey, *Earthquakes with 1,000 or More Deaths Since 1900*, at http://earthquake.usgs.gov/ earthquakes/world/world_deaths.php. This estimate does not include fatalities from the February 27, 2010, magnitude 8.8 Chilean earthquake, which has resulted in widespread destruction but few fatalities compared to the Indonesian, Pakistan, and Haiti earthquakes.

[12] Insurance Information Institute, http://www.iii.org/media. Loss estimates are in 2005 dollars.

[13] Risk Management Solutions (RMS), Newark, CA, press release (Sept. 2, 2005), at http://www.rms.com/NewsPress/ PR_090205_HUKatrina.asp.

[14] FEMA Publication 366, *HAZUS MH Estimated Annualized Earthquake Losses for the United States* (April 2008), at http://www.fema.gov/library/viewRecord.do?id=3265. Hereafter referred to as FEMA 366.

[15] A. M. Best Company Inc., *2006 Annual Earthquake Study: $100 Billion of Insured Loss in 40 Seconds* (Oldwick, NJ: A.M. Best Company, 2006), p. 12. The A. M. Best report includes estimates from catastrophe-modeling companies of predicted damage from hypothetical earthquakes in Los Angeles, the Midwest, the Northeast, and Japan. The report cites an estimate by one such company, Risk Management Solutions (RMS), that a hypothetical 7.4 magnitude event along the Newport-Inglewood Fault near Los Angeles would cause $549 billion in total property damage. A hypothetical 6.5 magnitude earthquake along a fault between Philadelphia and New York City would produce $901 billion in total loss, according to an RMS estimate.

[16] Building inventory refers to four main inventory groups: (1) general building stock, (2) essential and high potential loss facilities, (3) transportation systems, and (4) utility systems (FEMA 366).

[17] FEMA 366, p. 37.

[18] Ibid., p. 32.

[19] FEMA 366, p. 32 and p. 36.

[20] Ibid., p. 36.

[21] Ibid., p. 10.

[22] The largest earthquakes in New York, New Jersey, and Massachusetts were, respectively: 1944, Massena, NY, magnitude 5.8, felt from Canada south to Maryland; 1783, New Jersey, magnitude 5.3, felt from New Hampshire to Pennsylvania; and 1755, Cape Ann and Boston, MA, intensity of VIII on the Modified Mercalli Scale, felt from Nova Scotia to Chesapeake Bay (USGS Earthquake Hazards Program).

[23] USGS Circular 1188, Table 3.

[24] In contrast to California, where earthquakes occur on the active margin of the North American tectonic plate, the New Madrid seismic zone is not on a plate boundary but may be related to old faults in the interior of the plate, marking a zone of tectonic weakness.

[25] Andrew Newman et al., "Slow Deformation and Lower Seismic Hazard in the New Madrid Seismic Zone," *Science*, v. 284 (April 23, 1999), pp. 619-62 1.
[26] Seth Stein, Joseph Tomasello, and Andrew Newman, "Should Memphis Build for California's Earthquakes?" *Eos*, v. 84, no. 19, (May 13, 2003), pp. 177, 184-185.
[27] Seth Stein, "Code Red: Earthquake Imminent?" *Earth*, vol. 54, no. 1 (January 2009), pp. 52-59.
[28] The GSN also monitors nuclear explosions.
[29] USGS Earthquake Hazards Program, at http://earthquake.usgs.gov/research/monitoring/anss/.
[30] USGS FY2010 Budget Justification, at http://www.usgs.gov/budget/2010/greenbook/ FY2010_USGS_Greenbook.pdf, p. I-10.
[31] Ibid., p. I-11.
[32] Email from William Leith, Advanced National Seismic System Coordinator, USGS, December 22, 2009.
[33] USGS FY20 10 Budget Justification, pp. T-32 and T-33.
[34] USGS FY20 10 Budget Justification, pp. T-32 and T-33.
[35] The number of stations necessary to generate a data-based ShakeMap depends on the urban area and geology, but roughly correspond to about half the number of planned stations per urban area, at a spacing of about 20 kilometers between stations. Personal communication, William Leith, Advanced National Seismic System Coordinator, USGS, January 11, 2010.
[36] Strong motion seismometers, or accelerometers, are special sensors that measure the acceleration of the ground during large (>6.0 magnitude) earthquakes.
[37] $9 million reflects the combined appropriations for USGS and NSF for FY2009. ARRA (P.L. 111-5) enabled approximately $10 million to be made available via the USGS and NSF for the GSN through FY2010. Email from William Leith, Advanced National Seismic System Coordinator, USGS, Dec. 21, 2009.
[38] IRIS is a university research consortium, primarily funded by NSF, that collects and distributes seismographic data.
[39] USGS Open-File Report 2004-1390, and California 24-hour Aftershock Forecast Map, at http://pasadena.wr.usgs.gov/step/.
[40] The California Integrated Seismic Network is the California region of ANSS; see http://www.cisn.org/.
[41] Earthquakes typically occur in clusters, in which the earthquake with the largest magnitude is called the main shock, events before the main shock are called foreshocks, and those after are called aftershocks. See also http://pasadena.wr.usgs.gov/step/aftershocks.html.
[42] The *epicenter* of an earthquake is the point on the earth's surface directly above the hypocenter. The *hypocenter* is the location beneath the earth's surface where the fault rupture begins.
[43] Stuart Simkin, NEIC, Golden, CO, telephone conversation, Nov. 4, 2006.
[44] In early 2006, the NEIC implemented an around-the-clock operation center and seismic event processing center in response to the Indonesian earthquake and resulting tsunami of December 2004. Funding to implement 24/7 operations was provided by P.L. 109-13.
[45] ShakeMaps for some areas outside the United States are also available. See http://earthquake.usgs.gov/eqcenter/shakemap/.
[46] See the USGS Earthquakes Hazards Program for more information, at http://earthquake.usgs.gov/earthquakes/pager/.
[47] See also USGS Fact Sheet 2007-3101 at http://pubs.usgs.gov/fs/2007/3101/.
[48] See http://www.fema.gov/plan/prevent/hazus/hz_overview.shtm.
[49] See http://www.hazus.org/.
[50] See http://earthquake.usgs.gov/research/.
[51] Seismogenic means capable of generating earthquakes.
[52] See http://www.nsf.gov/div/index.jsp?div=EAR.
[53] See http://www.earthscope.org/.
[54] A strainmeter is a tool used by seismologists to measure the motion of one point relative to another.
[55] A non-profit NEES consortium (NEESinc.) has operated the facilities for the 10-year operating lifespan at the following institutions: Cornell University; Lehigh University; Oregon State University; Rensselaer Polytechnical Institute; University of Buffalo-State University of New York; University of California-Berkeley; University of California-Davis; University of California-Los Angeles; University of California-San Diego; University of California-Santa Barbara; University of Colorado-Boulder; University of Illinois at Urbana-Champaign; University of Minnesota; University of Nevada-Reno; and University of Texas at Austin. See http://www.nees.org/.

In: The Big Shake Implications of a Major Earthquake in... ISBN: 978-1-61728-512-7
Editor: Christopher D. Monahan

Chapter 2

THE SHAKEOUT SCENARIO

Lucile M. Jones, Richard Bernknopf, Dale Cox, James Goltz, Kenneth Hudnut, Dennis Mileti, Suzanne Perry, Daniel Ponti, Keith Porter, Michael Reichle, Hope Seligson, Kimberley Shoaf, Jerry Treiman and Anne Wein

1.EXECUTIVE SUMMARY

Overview

This is the initial publication of the results of a cooperative project to examine the implications of a major earthquake in southern California. The study comprised eight counties: Imperial, Kern, Los Angeles, Orange, Riverside, San Bernardino, San Diego, and Ventura. Its results will be used as the basis of an emergency response and preparedness exercise, the Great Southern California ShakeOut, and for this purpose we defined our earthquake as occurring at 10:00 a.m. on November 13, 2008. As members of the southern California community use the ShakeOut Scenario to plan and execute the exercise, we anticipate discussion and feedback. This community input will be used to refine our assessment and will lead to a formal publication in early 2009.

Our goal in the ShakeOut Scenario is to identify the physical, social and economic consequences of a major earthquake in southern California and in so doing, enable the users of our results to identify what they can change now—*before* the earthquake—to avoid catastrophic impact after the inevitable earthquake occurs. To do so, we had to determine the physical damages (casualties and losses) caused by the earthquake and the impact of those damages on the region's social and economic systems. To do this, we needed to know about the earthquake ground shaking and fault rupture. So we first constructed an earthquake, taking all available earthquake research information, from trenching and exposed evidence of prehistoric earthquakes, to analysis of instrumental recordings of large earthquakes and the latest theory in earthquake source physics. We modeled a magnitude (*M*) 7.8 earthquake on the southern San Andreas Fault, a plausible event on the fault most likely to produce a major earthquake. This information was then fed forward into the rest of the ShakeOut Scenario (figure 1-1).

Earth Science in the ShakeOut Scenario

The Earthquake Source

The ShakeOut Scenario earthquake is a magnitude 7.8 earthquake on the southernmost 300 km (200 mi) of the San Andreas Fault, between the Salton Sea and Lake Hughes. The southern San Andreas Fault was identified in the most recent assessment of seismic risk as most likely source of a very large earthquake in California. A magnitude 7.8 is not the largest earthquake that the southern San Andreas Fault can produce, nor is the San Andreas the only fault to threaten the populated areas of southern California with very large earthquakes. However, those other faults have *recurrence intervals* (an estimate of the average time) between larger earthquakes that are considerably longer, measured in thousands of years. By contrast, the southern San Andreas Fault has generated earthquakes of ShakeOut size on average every 150 years—and on a portion of the fault that ruptures in the ShakeOut Scenario, the last earthquake happened more than 300 years ago. The extent of the fault rupture in this earthquake was determined from geologic characteristics, after considerable discussion among geologic experts. The most likely rupture initiation point is one of the endpoints of the fault. We started at the southern end of the San Andreas Fault, and ruptured the fault to the northwest. We assumed that the average amount of slip to be released anywhere along the fault would be the amount accumulated since the last event on that portion of the fault, ranging from 2 to 7 meters (6 to 23 ft). We then added a randomized variation of the average slip within each 30 km section of fault. The maximum amount of slip is at the southern end of the rupture near the Salton Sea, where it has been more than 300 years since the last earthquake.

Ground Motions

The sudden rupture of a fault produces shaking as one of its effects. This shaking moves the ground, and it is these ground motions that we feel and that cause most of the damage in an earthquake. We estimated these ground motions with physics-based computer simulations of the earthquake with computer systems developed by the Southern California Earthquake Center information technology research program.

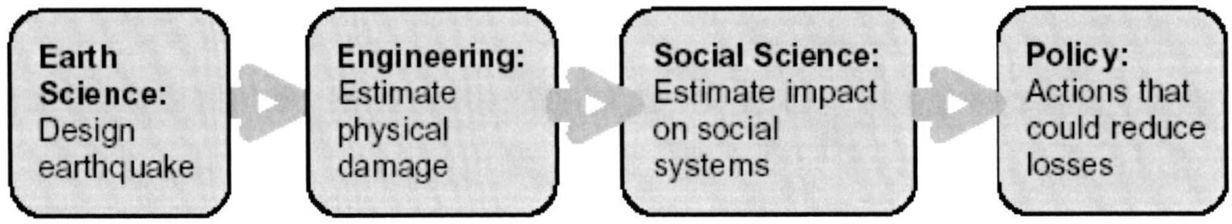

Figure 1-1. ShakeOut Scenario flow-chart.

For the past 30 years, before recent advances in information technology that have enabled scientists to obtain meaningful results from physics-based computer simulations, ground motion predictions have typically been made using **attenuation relations**, which forecast the expected shaking at a site from the magnitude and distance from the fault. However, in any earthquake there are pockets of shaking that are considerably higher or lower because of other factors that affect shaking, including site effects, directivity, and radiation pattern. Our physics-based simulations modeled all of these factors, primary and secondary, that affect ground shaking, using two inputs: (1) the ShakeOut kinematic rupture description and (2) a velocity model that describes the seismic characteristics of the southern California rocks through which the waves propagate. The results are shown to be consistent with the newest attenuation relations from the Next Generation Attenuation (NGA) relations.

We validated our modeling results through comparison of multiple methods, use of distinct velocity models, and comparison with empirically based attenuation relations. In all, four teams were engaged to make independent models of the ground motions. Several features of the ShakeOut earthquake ground motions are consistent across all the models including:

- Very strong shaking (approaching 3 m/sec) near the fault;
- Strong shaking with medium to long durations (20-45 sec) in the basins near the fault, including the Coachella, San Bernardino, and Antelope Valleys;
- Damaging shaking (at least 0.5 m/sec) over large areas (~10,000 km^2) of Los Angeles, San Bernardino, and Riverside counties;
- Pockets of very strong shaking (≥1 .5 m/sec) with long durations (45-60 sec) in areas of the San Gabriel Valley and East Los Angeles.

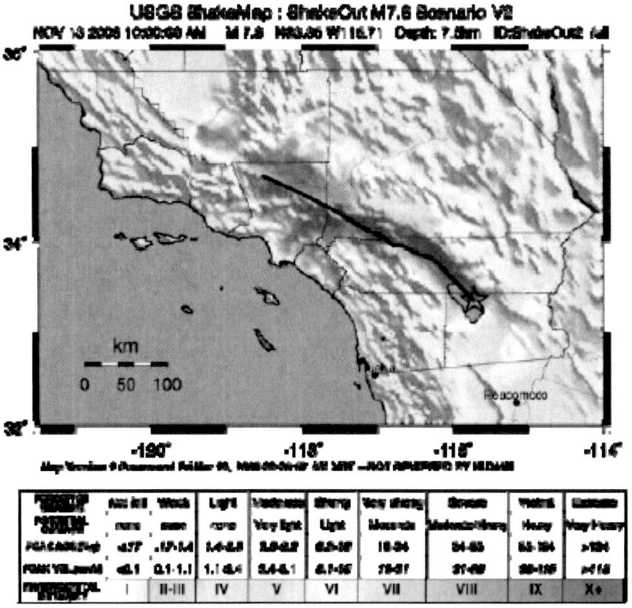

Figure 1-2. This "ShakeMap" is a representation of the shaking produced by the ShakeOut Scenario earthquake. The colors represent the Modified Mercalli Intensity with the warmer colors representing areas of greater damage.

Duration of strong shaking will be an important contributor to damage in any earthquake as large as the ShakeOut Scenario earthquake. Shaking lasts a long time because it takes about 100 seconds for a fault this long to rupture and because some of the waves get trapped and reverberate in sedimentary basins. In the ShakeOut Scenario earthquake (figure 1-2), the San Bernardino Valley is shaken extremely strongly but for a relatively short duration, as are Wrightwood and Palmdale, while the Coachella Valley has strong shaking with a long duration. Lower amplitude, but much longer duration ground motions occur in the Los Angeles and Ventura sedimentary basins.

To estimate damages from ShakeOut ground motion, the ShakeOut Scenario next calculated *ground motion parameters* used by engineers to estimate damage to structures. Ground motion parameters describe how the ground moves due to different measures of earthquake waves, and are needed because different kinds of structures are damaged by different kinds of waves. The ShakeOut Scenario created all the standard ground motion parameters: peak ground acceleration (PGA), peak ground velocity (PGV), Modified Mercalli Intensity (MMI), and spectral accelerations at 0.3, 1.0, and 3 seconds.

Fault Offsets

Fault offsets occur where the fault that moves in the earthquake is exposed at the Earth's surface. The ShakeOut fault rupture is on the San Andreas Fault and will be dominated by strike- slip, or horizontal displacement, causing structures and lifelines that straddle the fault to be sheared and offset as much as 9 meters (30 feet). Fortunately, there are few structures at risk of direct fault damage from the ShakeOut earthquake, due to the rural setting of the southern San Andreas Fault zone, and to the Alquist-Priolo Earthquake Fault Zoning Act of 1972, which prevents the construction of buildings used for human occupancy on the surface trace of active faults.

Damage from ShakeOut surface rupture is most serious where lifelines (roads, railroads, and utilities) cross the fault. Many of these crossings are concentrated within a few mountain passes and the disruption to these *lifeline corridors* has a major economic impact. Roads cross the fault at 966 places; the most critical damage occurs to Interstate 10 in the Coachella Valley and in San Gorgonio Pass, Interstate 15 in Cajon Pass, CA-14, CA-1 11, CA-62, Box Canyon Road, and Big Pines Highway. Other disrupted lifelines include fiber optic cables (90 crossings), petroleum and natural gas pipelines (39 crossings), railroads (21 crossings), aqueducts (32 crossings), and overhead electric power transmission lines (141 crossings).

Secondary Hazards

We investigated secondary hazards that can be triggered by large earthquakes in southern California including liquefaction, landslides, tsunamis, and seiches. All of these have caused significant additional damage in many big earthquakes, but only landslides and liquefaction will produce significant impacts in the ShakeOut Scenario. The ShakeOut Scenario earthquake will produce between 10,000 and 100,000 individual landslides, the vast majority of which will consist of rock falls, rock slides, rock avalanches, soil falls, disrupted soil slides and soil avalanches. Most of these will occur on steep slopes within the Transverse Ranges, primarily in the eastern San Gabriel Mountains. Conditions that can lead to liquefaction are potentially widespread in parts of the eight-county area impacted by the ShakeOut Scenario earthquake, particularly the Santa Clara River/Oxnard Plain areas of Ventura County, parts of the San Fernando and San Gabriel Valleys, portions of the coastal basin or flatland areas of Los Angeles and Orange Counties, the Santa Ana River corridor, the Imperial Valley, the southern Coachella Valley, and coastal areas of San Diego County. However, liquefaction requires both strong shaking and a high ground-water table. Strong ground motions from the ShakeOut Scenario earthquake mostly occur within the inland desert and mountain regions of southern California where ground water levels are typically low year-round. As a result, only the southern Coachella Valley will suffer significant liquefaction impacts in the ShakeOut Scenario earthquake, with localized liquefaction otherwise confined mostly to areas adjacent to perennial stream and river channels, such as in the upper Santa Ana and Santa Clara river basins. Because of the large distance from the earthquake to the coast, tsunamis are not a significant risk.

Aftershocks

Aftershocks are earthquakes and cause shaking and damage just like any other earthquake. Their additional shaking can damage weakened structures, necessitate evacuations, endanger rescue workers, and undo efforts to restore and rebuild. Based on experience in numerous earthquakes worldwide, after a *mainshock* earthquake as large as the ShakeOut Scenario earthquake, damaging aftershocks can occur for decades in a broad region around southern California, and any given region may experience more severe shaking from a close aftershock than from the original mainshock. Aftershock behavior in the aggregate can be well described by some simple, empirical laws, and these can be used to simulate sequences of aftershocks that realistically mimic actual aftershock sequences. For the ShakeOut Scenario, we generated ten random realizations of aftershocks for the first week following our mainshock. In reality, large, damaging aftershocks may occur months or years after the initial event.

We picked one of the simulations to be the aftershocks for the ShakeOut drills. This sequence includes two magnitude (*M*) 7 aftershocks. A *M*7.0 aftershock occurs 33 minutes after the mainshock, beginning at the southern end of the mainshock, near the Salton Sea, and rupturing south toward Mexico. It causes damage in Imperial and eastern San Diego Counties as well as in Mexicali, Mexico. A *M*7.2 event occurs 17 hours after mainshock on the Cucamonga Fault, rupturing along the front of the San Gabriel Mountains from Cajon Pass to Monrovia. The aftershocks in this sequence would cause substantial additional damage, but neither large aftershock has been evaluated in detail.

Engineering in the ShakeOut Scenario

The damage and impacts of the ShakeOut Scenario earthquake were estimated through a three-step process. First, FEMA's loss estimation program, HAZUS, was run using the physics- based ground motion model. For Los Angeles County, HAZUS used a refined database of structures created from tax assessor's data. For the other counties, this was not available and the default HAZUS database was used. In addition, HAZUS default mapping schemes (the relationships between basic inventory data and the assumed structural characteristics) were modified to reflect available information on unreinforced masonry buildings tabulated by the California Seismic Safety Commission, building density concentrations in urban core areas, and construction pattern changes over time throughout the eight counties. In the second step, expert opinion was collected through 13 special studies and 6 expert panels. Panels generally estimated impacts to public utilities, especially where multiple utility companies provide a public service such as water supply or electricity. Engineers and operators were invited to attend the half-day panel discussions, and were presented the results of prior Earth science studies (shaking, faulting, etc.), as well as damage to other interacting lifelines that had already been assessed. They were then asked to posit a realistic scenario of damage, service interruption, restoration, and to suggest promising mitigation options. To complement the panels, special studies were used for buildings and for lifelines when the panel process was impractical, such as private utilities or utilities (such as highways) where in-depth analysis was desired. In these cases, contributors were selected for their specialized expertise. They too were presented with all previously estimated Earth-science and relevant utility impacts, and asked to summarize assets exposed to damage, evidence of past seismic vulnerability, and to posit a realistic scenario of damage, loss of function, restoration, and promising mitigation measures. Crucial special studies were reviewed by panels of highly qualified experts. In the third step, the expert evaluations were merged with the HAZUS results to create the final estimates of probable damages.

The major losses for this earthquake fall into four categories: building damages, non-structural damages, damage to lifelines and infrastructure, and fire losses. Within each category, the analysis found types of losses that are well understood—that have been seen in previous earthquakes and the vulnerabilities recognized but not removed—and types of losses that had been less obvious – where the type of failure is only recently understood or the extent of the problem not yet fully recognized. The study also found numerous areas where mitigation conducted over the last few decades by state agencies, utilities and private owners, has greatly reduced the vulnerability. Because of these mitigation measures, the total financial impact of this earthquake is estimated to be "only" about $200 billion with approximately 1,800 fatalities. However, these are still big numbers

Buildings

Total losses to buildings are estimated at $33 billion. The two classes of older, known, poor performers--*unreinforced masonry* (where bricks or stone blocks with mortar form the bearing walls, called "URM") and *non-ductile reinforced concrete buildings*--pose the greatest risk to life safety. These types of buildings are no longer allowed to be built, but many of these buildings still exist and are not retrofitted. These types of buildings will be heavily damaged or destroyed near the fault, but in general will suffer less damage in the Los Angeles area. All URM buildings in the City of Los Angeles have been evaluated, and most have been strengthened to reduce loss of life. The strong shaking in Los Angeles will have very long periods (the waves will be big but slow) and these smaller buildings will in many cases ride out the shaking with less damage.

Woodframe construction generally fares well in earthquake shaking and woodframe buildings are less likely than other types of buildings to be damaged. However, because woodframe construction is so prevalent in California, substantial losses will still occur. Woodframe building damage is most likely:

- in older homes where the house is not bolted to the foundation or the cripple wall is not reinforced.
- in buildings with a "soft first story" – a large opening such as garage door or display windows on the first floor and without compensating reinforcement.
- in buildings where building codes were not rigorously followed--a condition difficult to recognized until after the earthquake.

Steel moment frame buildings built before 1994 were found to form cracks in their connections during the 1994 Northridge earthquake. Similar damage occurred in the 1995 Kobe earthquake and some buildings collapsed. Special study was conducted to analyze the behavior of steel frame high-rise buildings in the ground motions modeled for this earthquake. This event shows amplified long period motions caused by resonance in the sedimentary basins, particularly the very deep Los Angeles Basin. A special panel of structural engineers evaluated the analytical study and concluded "Given these ground motions, the collapse of some pre-1994 welded-steel moment-frame buildings is a credible scenario." Because this result comes from the long period ground motions, the area where this type of damage is possible is relatively large and includes much of the urbanized areas of Los Angeles, Orange, Riverside and San Bernardino Counties. It is impossible to determine how many and which buildings are the most susceptible without detailed structural analysis which is beyond the scope of this study. For the purposes of the ShakeOut emergency drills, we posit that 5 steel moment-frame high-rise buildings will collapse and that 10 more will be "red-tagged."

Non-structural and contents damage

Non-structural and contents damage is damage to the parts of a building other than what is holding it up, including interior walls, water pipes, air conditioning systems, and all moveable property such as electronics, and dishes. As building codes improve and buildings remain standing during earthquakes, the relative importance of non-structural damage increases. In recent earthquakes, the non-structural and contents losses have typically been comparable to the structural losses. Non-structural damages and mitigation have not been regulated in any way. Many of these losses are simple to prevent through securing contents and non-structural elements of the buildings. This is one of the most important ways that individuals can reduce the losses.

Utilities, Lifelines, and Infrastructure

California's investments in mitigation have paid off most obviously in increased robustness and resiliency of the region's lifelines. The retrofitting of highway bridges, conversion of ceramic insulators in the electric grid to polymers, and replacement of cast iron pipes mean that many utilities will be able to restore function much more quickly after the earthquake.

Significant vulnerabilities remain in the water conveyance system and in the lifelines that cross the San Andreas Fault. Pipes of concrete and iron are brittle and break in many places in an earthquake. The number of pipe breaks will be large enough that recreating the water system will be necessary in the hardest hit areas. Because this earthquake affects such a large area, there will not be enough pipe and connectors or trained manpower to repair all the breaks quickly. The worst hit areas may not have water in the taps for 6 months. This damage to the water system will also greatly increase the problems in fighting the fires that will follow the earthquake. The cost to repair water and sewer lines will be $1 billion.

The lifelines that cross the fault will all break when the fault moves. This will disrupt the movement of water, petroleum products, telecommunications, and general transportation. Repair of the lifelines will be slowed because the lifelines all cross the fault at just a few passes in the mountains and therefore interact with each other. For instance, repairing pipelines broken at Cajon Pass will require access that depends upon repair to Interstate 15. That in turn could be delayed if a wildfire starts after damage to the electric lines in the same location.

Many roads and highways will be impassable in the first few days after the earthquake because of debris on the roads, damage to bridges, and lack of power for the traffic signals. This will have a significant negative impact on the emergency response. Because of the major highway bridge retrofit program of the last 20 years, highway bridges are not expected to completely collapse, but some will not be passable. Many bridges on local roads have not been retrofitted and more damage is expected on those. The continuing impairment of the roads for months after the earthquake until everything can be repaired has a significant economic cost, estimated at $5 billion over one year.

Fire Following Earthquake

Southern California is unfortunately well situated for major fires to be generated following earthquakes. The number of ignitions that will create fires large enough to call the fire department can be extrapolated from previous earthquakes and depends upon the number of households at different levels of seismic shaking. This leads to an estimate of 1,600 ignitions of which 1,200 will be too large to be controlled by one fire engine company. In areas of dense woodframe construction, these fires if not controlled will grew quickly to involve tens or hundreds of city blocks. The fire risk is increased by the damage to the water distribution system and by the traffic gridlock that will result from the ShakeOut earthquake.

The final level of fire damage is difficult to assess because it depends upon several unpredictable factors, especially the degree to which fires spread when the fire protection services lose water and are overwhelmed. We use the minimum value from the fire estimates at $40 billion in damage to buildings and $25 billion in damage to building contents.

Social Science in the ShakeOut Scenario

The ShakeOut Scenario earthquake causes damage to the built environment that then ripples through and damages the social systems of the study region. This study has investigated the impacts of the earthquake on emergency services, human health, the regional economy, and trade operations from the Ports of Los Angeles and Long Beach.

Emergency Services

An emergency response matrix has been developed to help understand what the demands for emergency services will be like. Seventeen functions of emergency services are grouped into seven general classes of activities, including crisis information (public information and responder communications), search and rescue, victim services (shelter, provision of food and water and the management and distribution of donated goods and services), access management and law enforcement (control and security and traffic control), the staffing and functioning of emergency operations centers, fire suppression, medical emergency response, and service restoration, (repair of utilities, route recovery and debris removal). Research results and experience in past earthquakes have been analyzed to create this response matrix. Among the findings are that:

- 95% of rescues from downed buildings are carried out by fellow victims. Training ordinary citizens how to search safely could greatly reduce injuries.
- Many Emergency Operations Centers have not considered the impact of earthquakes on the contents of their Centers. Securing computers and desks and other non-structural mitigation activities would have large payoffs at low cost.

Mortality and Morbidity

Shaking in the ShakeOut Scenario earthquake will kill and injure many people, by causing buildings to collapse, creating falling debris and flying objects, and increasing traffic accidents when drivers lose control of automobiles. Many additional deaths and injuries will result in fires that follow the shaking. Estimating the total number of injuries and deaths is very uncertain particularly because the Scenario posits types of building failures that have not yet been observed. Because of strong life-safety building codes over the years, the ShakeOut Scenario estimates only approximately 1,800 deaths, of which about half occur because of the fires following the earthquake. There will also be about 750 people with very severe injuries who will require rapid, advanced medical care to survive. Approximately 50,000 people will have injuries that need emergency room care. The final mortality could increase if hospitals cannot function because of damage or if the transportation disruptions prevent people getting to emergency rooms.

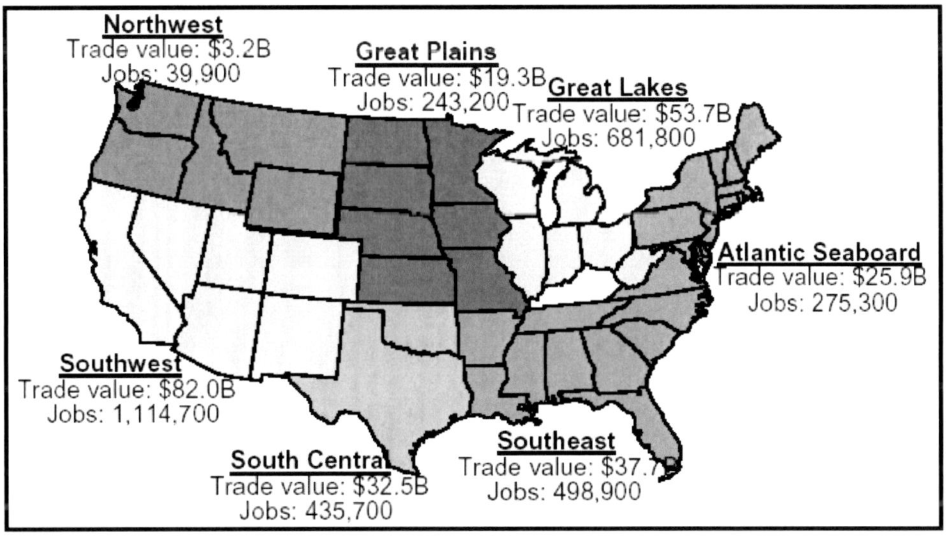

Source: BST Associates Trade Impact Report, 2007.

Figure 1-3. National Impact of San Pedro Ports.

Business Interruption

The economic impact of the earthquake is not limited to the structures and goods broken or burnt in the event. Much of the economic activity of the southern California region will be interrupted by the damage to structures and infrastructure. In particular, beyond their direct losses in stock (such as buildings, machines, and inventory), businesses will be unable to function because of loss of electricity, gas, water, and a transportation system. Some of the losses can be recaptured when the business resumes but the amount recaptured decreases with time as customers and suppliers find alternatives. Because the duration of outage is so long, the lack of water conveyance becomes the largest factor in business interruption losses for the ShakeOut earthquake, resulting in $50 billion in lost economic activity.

Movement of Goods

The ShakeOut Scenario earthquake will be far enough from both Los Angeles International Airport and the Ports of Los Angeles and Long Beach that the damage there will be minimal. This of course is not the case for many other possible earthquakes. The availability of these transportation resources is a significant asset in mobilizing the emergency response. Transportation from the Ports to the rest of the country is carried predominately by rail lines which will be rendered impassable by the fault offsets and ground motions. Significant economic disruption will result and the extent of the damage depends critically on how rapidly the railways and highways can be rebuilt.

The modeling estimates that the Ports will not function for the first 3 days after the event because of lack of electricity, general chaos, and the potential for slight damage to large structures such as cranes. For the next 2 weeks, the Ports will operate at 10% of capacity because there will be limited rail service and limited alternative transportation. They will gradually return to full capacity from 2 weeks to 2 months as rail service is reestablished and highways reopen. We estimate that 85% of the lost business will be recaptured but that 15% will be permanently lost to ship diversions, perished products, cancelled Far East shipments, and declined bookings.

Conclusions

The magnitude 7.8 ShakeOut earthquake is modeled to cause about 1,800 deaths and $213 billion of economic losses. These numbers are as low as they are because of aggressive retrofitting programs that have increased the seismic resistance of buildings, highways and lifelines, and economic resiliency. These numbers are as large as they are because much more retrofitting could still be done. The sources of the different losses are shown in Table 1-1.

The earthquake modeled here may never happen. Big earthquakes on the San Andreas Fault are inevitable, and by geologic standards extremely common, but probably will not be exactly like this one. The next very damaging earthquake could easily be on another fault. However, lessons learned from this particular event apply to many other events and could provide benefits in many possible future disasters.

Table 1-1. Total Regional Economic Impacts of Shake-Out (in billions of 2008 dollars)

Indicator	Total Impacts
Building Damage	$32.7
Related Content Damage	10.6
High-Rise Building Damage	2.2
Related Content Damage	0.7
Fire Damage	40.0
Related Content Damage	25.0
Highway Damage	0.4
Pipeline (water, sewer, gas) Damage	1.1
Sub-total Property Damage	112.7
Business Interruption	96.2
Relocation Costs	0.1
Traffic Delay Costs	4.3
Sub-total Additional Costs	4.4
Total	$213.3

The ShakeOut Scenario has identified five major areas of loss:

- Older buildings built to earlier standards.
- Non-structural elements and building contents that are generally unregulated.
- Infrastructure crossing the San Andreas Fault.
- Business interruption from damaged infrastructure, especially water systems.
- Fire following the earthquake.

The ShakeOut Scenario also found that previous efforts to reduce losses through mitigation before the event have been successful. There are dozens more actions and policies that could be undertaken at the individual and community levels to further reduce these losses. For instance, actions to improve the resiliency of our water delivery system would reduce the loss from business interruption, as well as reduce the risk of catastrophic conflagrations. At an individual and business level, actions to secure non-structural items in buildings and retrofitting of existing structures will greatly reduce individual risk. Planning and preparedness can improve personal and business resiliency.

Over the next 6 months, the ShakeOut Scenario will be used to prepare for future earthquakes and exercise in the Great Southern California ShakeOut in November 2008. This process will encourage public discussion of these risks and possible solutions. The risks can be analyzed and described by scientists but the solutions will come from southern California itself.

2. Introduction

Motivation

The U.S. Geological Survey (USGS) has recently initiated the Multi-Hazards Demonstration Project to demonstrate how hazards science can be used to improve a community's resiliency to natural disasters. To launch this project, earthquake and other hazard scientists held strategic planning workshops with stakeholders such as local officials and emergency response professionals. The workshops determined what information stakeholders find most useful from scientists and what additional information they need but might not have been getting. The top priority that emerged from these workshops was for disaster scenarios that could more fully support decision-making in planning and preparedness by detailing and quantifying anticipated consequences of natural disasters.

Hurricane Katrina was on the minds of all workshop participants. In Katrina we saw how a catastrophe can strain the fabric of society and lead to decades of economic disruption. Since Katrina, we have distinguished between a natural *disaster*--an inevitable event such as a hurricane, flood, wildfire, or earthquake--and a *catastrophe*, which occurs when a disaster disrupts a large region and the effects continue for decades. In southern California, the most likely source of a catastrophe is an earthquake so powerful that it causes widespread damage and consequently affects lives and livelihoods of all southern Californians. A catastrophe is a disaster that runs amok when a society is not prepared for the amount of disruption that occurs.

The ShakeOut Scenario was developed to meet the needs of stakeholders at the strategic planning sessions. It considers the impacts of a *M*7.8 earthquake on the southern San Andreas Fault, an earthquake selected because it is so probable. It is not the worst earthquake possible. Southern California has more than 300 faults capable of producing damaging earthquakes, and includes several faults capable of producing earthquakes with catastrophic consequences. Some of the earthquakes are much more likely than others to happen in the lifetime of a person or building. A full assessment of earthquake risk requires a probabilistic approach that accounts for all of the faults, earthquakes, and likelihoods. Instead, the ShakeOut Scenario considers the impact of a single event that is large enough and likely enough to create a catastrophe in our lifetimes. The ShakeOut Scenario is not predicting – and does not need to predict -whether this particular earthquake will actually ever happen. Examining the consequences and far-reaching impacts of one such event can help us prepare for other such events.

The ShakeOut Scenario was also developed to break through a common, dangerous misconception that goes something like this: My home/my business made it through the Northridge earthquake so I know what future earthquakes will be like and can rest assured I will make it through the next one, too. Natural disasters come in many sizes, and the disasters most likely to cause catastrophes are those large enough to have regional, long-term consequences. No Californians have experienced an earthquake like this except for survivors of the 1906 San Francisco earthquake.

The 1994, *M*6.7 Northridge earthquake is not an appropriate point of reference for a catastrophe, because the Northridge earthquake was simply not large enough to cause catastrophic devastation:

- Many buildings and other structures that were able to withstand the 7 to 15 seconds of shaking during the Northridge earthquake, will not withstand the nearly 2 minutes of shaking in an earthquake the size of that in the ShakeOut Scenario;
- Northridge was a local, not a regional disaster; even in the hardest-hit areas, one could drive five minutes and reach an area that was relatively unaffected; this will not be the case after an earthquake the size of the ShakeOut Scenario;
- After Northridge, most businesses were able to regroup fairly quickly; after a regional disaster, so many will struggle for such a long time that a much greater number will fail, creating a domino effect that hurts employees, customers, and surviving businesses;
- After Northridge, the Los Angeles area could turn to other southern California communities for mutual aid; after a regional disaster, those neighbors will need help too; mutual aid will be slower to arrive, coming from Arizona, Nevada, and northern California.

Objectives

The ShakeOut Scenario exists to support decision-makers in their efforts to make southern California a safer community. The most immediate users of the Scenario will be members of the emergency response community who are participating in the November 2008 Golden Guardian exercises. Other decision-makers include business owners, homeowners, employees, and tenants, as well as public officials, emergency responders, and planners. The ShakeOut Scenario analyzes how a large, regional earthquake will affect the social and economic systems that make southern California a desirable place, because an earthquake with similar kinds of impacts is an inevitable part of southern California's future. Thus, **appropriate uses of the ShakeOut Scenario** include:

- Urban planning;
- Emergency response training;
- School, business, and public earthquake drills;
- Prioritization of preparedness efforts;
- Understanding potential impacts on financial and social systems; and
- Identifying possible vulnerabilities of infrastructure, especially due to interactions among systems that are usually considered separately.

The ShakeOut Scenario has created as complete a description as possible of the regional, long-term impacts of a particular earthquake on the southern San Andreas Fault. It is not a probabilistic assessment of risk or cost-effectiveness nor is it a prediction that this particular earthquake will occur. This is only one of thousands of possible, damaging earthquakes that could hit southern California. Being spared in this event does not mean you are spared in other events. Thus, **inappropriate uses of the ShakeOut Scenario** include:

- Deciding where to live or work;
- Concluding you don't have an earthquake problem;

- Changing building codes; or
- Evaluating cost-effectiveness of mitigation.

Review Process

Early on we recognized that it was not practicable to expect any single reviewer to have expertise spanning the full range of ShakeOut Scenario components. So, in addition to having two, traditional reviewers of the entire document, throughout the project we brought in additional, topic- specific expert reviewers, who reviewed material as members of expert panels and invited guests at internal presentations, as well as by reading report sections or by less formal participation via email and telephone contact.

Contributors

The ShakeOut Scenario was created through a major collaborative effort involving more than 300 contributors. Our goal was to engage the full range of expertise needed to understand the complex interactions and to include experts and professionals from the public and private sectors, some who could share experience gained in previous earthquakes and others who understood the strengths and weaknesses of our systems. One challenge was to make collaborative use of this wide range of expertise, while integrating findings into a coherent result that could be delivered in time for the 2008 Golden Guardian planning meeting on May 5, 2008. Through this trial-by-fire process we have created a blueprint for future scenario efforts regarding earthquakes and other natural disasters. There is widespread recognition that now is the time to make such efforts.

To drive progress in the ShakeOut Scenario's varied endeavors, the USGS turned to partners with essential expertise. The California Geological Survey created the first earthquake scenarios over 20 years ago and provided many of their experts on California faults and geology. The Southern California Earthquake Center has assembled a state-of-the-science team of computer and earthquake scientists to perform and validate modeling of ground shaking. The Earthquake Engineering Research Institute has created scenarios for many events and their members in southern California understand the nature and vulnerabilities of our infrastructure. Social scientists affiliated with California's Office of Emergency Services, the Seismic Safety Commission, and UCLA contributed their understanding of casualties, disaster response and effective preparedness campaigns. Economists from the USGS and University of Southern California knew how to assess earthquake shocks to the region's economic health.

We assembled teams for different parts of the ShakeOut Scenario project led by Coordinators who enlisted and managed contributors. All are listed in the following pages of this report, and their specific contributions are identified in pertinent report sections.

Many of the studies that were conducted for the ShakeOut Scenario are available as reports on-line. For details go to http://urbanearth.usgs.gov/scenario08.

ShakeOut Scenario Coordinators

Chief Scientist:
  Lucile Jones, USGS
Project Manager:
  Dale Cox, USGS
Staff Scientist/Writer:
  Sue Perry, USGS
Earthquake Design:
  Kenneth Hudnut, USGS
Secondary Hazards:
  Daniel Ponti, USGS
  Michael Reichle, California Geological Survey
  Jerry Treiman, California Geological Survey
Physical Damages:
  Keith Porter, University of Colorado
HAZUS Loss Estimations:
  Hope Seligson, MMI Engineering
Emergency Response:
  Dennis Mileti, California Seismic Safety Commission
  James Goltz, Governor's Office of Emergency Services
Health and Safety:
  Kimberley Shoaf, University of California, Los Angeles
Economics:
  Anne Wein, USGS
  Richard Bernknopf, USGS

ShakeOut Scenario Contributors

ShakeOut Scenario Contributors

(@) = SoSAFE (Southern San Andreas Fault Evaluation)
(#) = SCEC ShakeOut Simulation Group

Brad Aagaard, USGS
Corey Abbott, USGS
Marty Adams, LA Dept. of Water and Power
Sinan Akciz, UC Irvine (@)
Roman Alarcon, Metrolink
Greg Albright, Caltrans
Patrick Allen, BNP Paribas
Richard Almanzan, Caltrans
Mariana Amatullo, Art Center College of Design
Naresh Amatya, Southern Calif. Association of Governments
John Amrhein, San Bernardino County Sheriff's Dept.

Michael Armstrong, Southern Calif. Association of Governments
D. J. "Joe" Andrews, USGS
Ramon Arrowsmith, Arizona State Univ. (@)
Arul Arulmoli, Earth Mechanics, Inc.
Donald Ballantyne, MMI Engineering
Frances Banerjee, Banerjee & Associates (retired LA Dept. of Transportation)
Larry Banks, Caltrans
George Barber, Metropolitan Water District
Eric Batman, LA County Public Works
Whitney Behr, USC (@)
Linda Bell, LA County Public Works
Denise Benson, County of San Bernardino
Mark Benthien, SCEC
Marc Bentojova, Sacramento Fire Dept.
Tracy Berge, Metrolink
Richard Bernknopf, USGS
Mike Bevis, Ohio State Univ.
Jacobo Bielak, Carnegie-Mellon Univ. (#)
Don Boland, Calif. Utilities Emergency Association
Adrian Borsa, USGS
Peggy Brutsche, American Red Cross
Bill Bryant, Calif. Geological Survey
Monica Buchanan, State Farm Insurance
Dai Bui, LA County Public Works
John Bwarie, Office of LA City Councilmember Greig Smith
William Byers, Burlington Northern & Santa Fe Railway, retired
Scott Callaghan, USC (#)
Larry Caminiti, Southern California Edison
Ken Campbell, EQECAT Inc.
Robert Castanon, LA County Metropolitan Transportation Authority
Robert Cavaglieri, NBC-Universal
Richard Champion, USGS
Ping Chang, Southern Calif. Association of Governments
Shane Chapman, Metropolitan Water District
Winston Chai, Metropolitan Water District
Mohan Char, Lim & Nascimento Engineering
Sungbin Cho, ImageCat, Inc. and Southern Calif. Association of Governments
Amit Chourasia, San Diego Supercomputer Center (#)
Mandy Chu, Caltrans
Bruce Clark, Calif. Seismic Safety Commission
Larry Collins, LA County Fire Dept.
Don Conner, Southern Calif. Edison
Joel Conte, UC San Diego
Pauline Cotelle, Université Paul Valéry, Montpellier III
Tom Cotter, LA County Public Works
Larry Cotrill, Port of Long Beach

Dale Cox, USGS
Ron Cox, City of Riverside
Yifeng Cui, San Diego Supercomputer Center (#)
Joseph Crisologo, Calif. State Dept. of Health
Luis Dalguer, San Diego State Univ. (#)
Julie Davenport, Wells Fargo Bank
Craig Davis, LA Dept. of Water and Power James Davis, Caltrans
Tim Dawson, USGS (@)
Steve Day, San Diego State Univ. (#)
Greg Deierlein, Stanford Univ.
Lori Dinitz, USGS
Barry Edelman, Zenith Insurance Company
Ronald Eguchi, ImageCat, Inc.
Geoff Ely, UC San Diego, San Diego Supercomputer Center (#)
Thomas Eng, Metropolitan Transit Authority
Raul Escandon, County of LA Public Works
Doug Failing, Caltrans
Karen Felzer, USGS
Yuri Fialko, UC San Diego
Ned Field, USGS
Eric Fielding, Jet Propulsion Laboratory
Laure Fossorier, Ecole Nationale des Travaux Publics de L'Etat
Dan Freeman, Caltrans
Gary Fuis, USGS
Tom Fumal, USGS (@)
Rick Gailing, Southern Calif. Gas Company
Kevin Garrity, LA Dept. of Water and Power
Leonard Gaydos, USGS
Eric Geist, USGS
Shubharoop Ghosh, ImageCat, Inc.
Katy Gibson, Metropolitan Water District
Fay Glass, Riverside Fire Dept.
Jim Goltz, Governor's Office of Emergency Services
Bob Gorham, Office of the State Fire Marshall
William Graf, URS Corporation
Rob Graves, URS Corporation (#)
Vladimir Grazer, Calif. Geological Survey
Michael Griffin, LA Dept. of Transportation
Hector Guerrero, LA County Metropolitan Transportation Authority
Ron Guss, Intermodal West, Inc., Calif. Trucking Association
Ron Hamburger, SGH, Inc.
John Hardcastle, Palm Springs Fire Dept.
Bruce Hamer, LA Dept. of Water and Power
Christopher Harris, Caltrans
Ruth Harris, USGS
Douglas Headrick, San Bernardino Valley Municipal Water District

Steve Hennessee, LA County Public Works
Caroline Hermans, USGS
Mike Herrera, Coachella Valley Water District
Lu Hersh, Alameda Corridor Engineering Team
Richard Hess, Hess Engineering Inc.
Dave Hester, USGS
Gill Hicks, Gill V. Hicks and Associates, Inc.
Chris Hill, Metropolitan Water District
Mark Hoffman, Infragard
Chris Holbrook, Office of Homeland Security
Richard Holden, Bureau of Labor Statistics
Wesley Hong, Southern Calif. Association of Governments
Hsi Hwa Hu, Southern Calif. Association of Governments
Guoxiong Huang, Southern Calif. Association of Governments
Robert Huddy, Southern Calif. Association of Governments
Ken Hudnut, USGS (@)
Joe Hull, Caltrans
Eugenia Hyung, USGS
Hasan Ikhrata, Southern Calif. Association of Governments
Tom Jasmin, LA County Metropolitan Transportation Authority
Mike Jenkinson, Caltrans
Chen Ji, UC Santa Barbara
Jeff Johnson, Caltrans
Ted Johnson, Water Replenishment District Of Southern Calif.
Laurie Johnson, Laurie Johnson Consulting
Martin Johnson, ABS Consulting
Lucy Jones, USGS
Tom Jordan, USC (#)
Gideon Juve, USC (#)
Timothy Kaiser, San Diego Supercomputer Center (#)
Robert Kamm, Los Angeles Unified School District
Erdem Karaca, Swiss Re Insurance
Michael Keever, Caltrans
Stephanie Kelly, USGS
Katherine Kendrick, USGS (@)
Payman Khalili-Tehrani, UCLA
Keith Killough, Southern Calif. Association of Governments
Richard Koshalek, Art Center College of Design
Swaminathan Krishnan, Caltech
Ryan Kuo, Southern Calif. Association of Governments
Sue Lai, Port of Los Angeles
Kwan Lam, Caltrans
Molly Latham, Southern California Edison
Homa Lee, USGS
Keith Lee, LA County Public Works
Robert Lee, Dept. of Homeland Security

Robert Leeper, USGS
Peter Lent, Riverside County Fire Dept.
K. Eric Leung, Long Beach Water Dept.
Steve Levy, Center for Continuing Study of the Calif. Economy
Marshall Lew, MACTEC Engineering
J Leyva, City of San Bernardino Public Services
Wan Lin, USGS
William Lloyd, City National Bank
Julio Lopez, Carnegie-Mellon Univ. (#)
Kate Long, Governor's Office of Emergency Services
Roberto Lugo, USGS
Ralph Loyd, Calif. Geological Survey
Lisa Grant Ludwig, UC Irvine (@)
Kai Lui, LA County Sanitation
Dave Luker, Desert Water Agency
Le Val Lund, LA Dept. of Water and Power
Paul Lundgren, Jet Propulsion Laboratory
Jeff Lusk, FEMA
Phil Maechling, USC (#)
Michael Mahoney, FEMA
Gerry Malais, LA City Fire Dept.
Pam Malam, USGS
Jim Malley, Degenkolb Engineering
Steve Maloney, Office of Homeland Security
Don Manning, LA City Fire Dept., retired
Kaolyn Mannino, Southern Calif. Edison
Peter Martin, USGS
Enrique Martinez, LA Dept. of Water and Power
Dawn Matthews, Cal National Bank
Jon Matti, USGS (@)
Darrell Maxey, Southern Calif. Regional Rail Authority
Richard McCarthy, California Seismic Safety Commission
Timothy McCrink, Calif. Geological Survey
Sally McGill, Calif. State Univ., San Bernardino (@)
Richard McKenna, Marine Exchange
Mike McManus, Caltrans
John Megreff, City of Torrance
Mike Mehrain, URS Corporation
Martha Merriam, Caltrans
Mutaz Mihyar, Dept. of Water and Power, Division of Safety of Dams
Dennis Mileti, Calif. Seismic Safety Commission
Javier Minjares, Calif. Association of Governments
Philip Mo, Southern California Edison
Reagan Moore, San Diego Supercomputer Center (#)
Phil Moores, Riverside Transit
Mike Morel, Metropolitan Water District

Mary Moreland, Riverside County Office of Emergency Management
Basam Muallem, Caltrans
K. Murthy, Metropolitan Transportation Authority
Matthew Muto, Caltech
William Nascimento, Lim & Nascimento Engineering
Charles Nestle, LA County Public Works
Craig Nicholson, UC Santa Barbara
Jerry Nickelsburg, UCLA
David O'Hallaron, Carnegie-Mellon Univ. (#)
David Okaya, USC (#)
Kim Olsen, San Diego State Univ. (#)
Gustavo Ortega, Caltrans
Dennis Ostrom, Southern Calif. Railroad
Frank Paczona, LBS Financial Credit Union
Louis Palm, Office of Homeland Security
Don Parker, Calif. Seismic Safety Commission
John Parrish, Calif. Geological Survey
Naresh Patel, Metrolink
Mark Pickett, Univ. of Toledo
Ines Pearce, Pearce Global Partners
Gilbert Pedroza, LA County Public Works
Michael Perovich, Caltrans
Sue Perry, USGS
David Phillips, UNAVCO, Inc.
Belle Philobosian, Univ. of Oregon (@)
Alen Pijuan, LA City Emergency Preparedness Dept.
Cliff Plumb, LA Dept. of Water and Power
Dan Ponti, USGS
Jack Popejoy, KFWB Radio
Keith Porter, Univ. of Colorado
Erik Pounders, USGS
Ren Powers, City National Bank
Cynthia Pridmore, Calif. Geological Survey
James Province, Caltrans
David Quirk, Metrolink
Vince Quitoriano, USGS
Frank Quon, Caltrans
David Raleigh, Ohio State Univ.
Leonardo Ramirez-Guzman, Carnegie-Mellon Univ. (#)
Sved Raza, Caltrans
Charles Real, Calif. Geological Survey
Eric Reichard, USGS
Michael Reichle, Calif. Geological Survey
Tim Richison, Calif. Earthquake Authority
Ralph Ricketson, Caltrans
Mike Ripley, NBC-Universal

Steve Robbins, Coachella Valley Water District
Marlene Roberts, Federal Deposit Insurance Corporation
Tom Rockwell, San Diego State Univ. (@)
Chris Rojahn, Applied Technology Council
Adam Rose, USC
Marco Ruano, Caltrans
Aram Sahakian, LA Dept. of Transportation
Richard Sanchez, Dept. of Water Resources State Water Project
David Sandwell, UC San Diego
Rosie Santilena, USGS
Victoria Saylor, Countrywide Bank
Charles Scawthorn, Kyoto Univ. / SPA Risk LLC
Katherine Scharer, Appalachian State Univ. (@)
Anshel Schiff, Precision Measurement Instruments
Merritt Schreiber, UC Los Angeles
David Schwartz, USGS (@) Caz Scislowicz, Caltech
Gordon Seitz, San Diego State Univ. (@)
Hope Seligson, MMI Engineering
Stephen Sellers, Governor's Office of Emergency Services
Tom Shantz, Caltrans Ben Sherrouse, USGS Kimberley Shoaf, UCLA
Richard Shapall, Metropolitan Water District
Michael Shulters, USGS
Kerry Sieh, Caltech (@)
Armand Silva, Caltrans
Michael Silva, Calif. Geological Survey
Mark Simons, Caltech
Michael Smith, Surface Transportation Board
Bridget Smith-Konter, UC San Diego
Bob Spears, LA Unified School District
Ellis Stanley, City of LA
Lisa Star, UCLA
Judith Steele, Judith Steele Consultant Services
Jonathan Stewart, UCLA
Phil Stolarski, Caltrans
David Strong, USGS
Mark Stuart, Dept. of Water Resources
Henry Stultz, Caltrans
Gary Sturdivan, East Valley Water District
Ricardo Taborda, Carnegie-Mellon University (#)
Ertugrul Taciroglu, UCLA
Alex Tang, L & T Engineering
Doug Thiessen, Port of Long Beach
Alan Thompson, Southern Calif. Association of Governments
Kevin Thompson, Caltrans
John Tinsley III, USGS
Ron Tognazzini, Tognazzini and Associates

Jerry Treiman, Calif. Geological Survey
Jeroen Tromp, Caltech
Loren Turner, Caltrans
John Urbanic – Pittsburgh Supercomputing Center (#)
Randy Van Gelder, San Bernardino Valley Municipal Water District
John Vaughn, Calif. State Employment Development Dept.
Margaret Vinci, Caltech
David Wald, USGS
Richard Walker, Metrolink
Zhongren Wang, Caltrans
Kathleen Webb, Office of Governor Schwarzenegger
Dan Wei, USC
Monique Weiland, Calif. Independent System Operator
Anne Wein, USGS
Deborah Weiser, USGS
Ray Weldon, Univ. of Oregon (@)
Frank Wen, Southern Calif. Association of Governments
Stu Werner, Seismic Systems & Engineering Consultants
Gedion Werrede, Caltrans
Troy Whitman, Southern California Edison
Pat Williams, San Diego State Univ. (@)
Chris Wills, Calif. Geological Survey
Rick Wilson, Calif. Geological Survey
Bill Winter, LA County Public Works
Liz Wojdak, Southern Calif. Association of Governments
Ray Wolfe, Caltrans
John Yang, Caltrans
Mark Yashinsky, Caltrans
Bill Yearsley, Univ. of Colorado, Bolder
Peter Yin, Port of Los Angeles
Doug Yule, Calif. State Univ., Northridge (@)
Jing Zhu, San Diego Supercomputer Center (#)
(@) = SoSAFE (Southern San Andreas Fault Evaluation)
(#) = SCEC ShakeOut Simulation Group

Funding for the HAZUS database enhancements and HAZUS runs was provided by the Governor's Office of Emergency Services.

The SCEC ShakeOut Simulation Group was funded by NSF grants EAR-0623704 and OCI-0749313.

3. Constructing the Scenario Event

A. Overview

Our goal in the ShakeOut Scenario is to identify the long-term social and economic consequences of an enormous earthquake in southern California and in so doing, enable the users of our results to identify what they can change now—relatively easily and *before* the earthquake - to avoid catastrophic impact *after* the inevitable earthquake occurs. Let's work backwards to put this into perspective. Our end users want to identify actions, including policy changes, that will minimize the social and economic consequences (blue boxes, upper right of figure 3-1) of an earthquake. To provide them with the information they need, we had to determine the physical damages (casualties and losses) caused by the earthquake. But before we could estimate physical damages, we needed to know about the earthquake ground shaking and fault rupture. So we had to construct an earthquake, and to do that we took all available earthquake research information, from trenching and exposed evidence of prehistoric earthquakes, to analysis of instrumental recordings of large earthquakes and the latest theory in earthquake source physics. We combined these elements to create a realistic "Big One"—a major earthquake on the San Andreas Fault—and then we simulated the shaking produced by this earthquake, using supercomputers and several alternative computer programs, as well as expert opinion and experience in real earthquakes, in order to test and validate the ground motions that went into estimating physical damages. This information then fed forward into the rest of the ShakeOut Scenario.

B. The Earthquake Source by Kenneth Hudnut, Brad Aagaard, Robert Graves, and Thomas Jordan

Contributing Authors: Jonathan Stewart, Lisa Star (also see Appendix B)

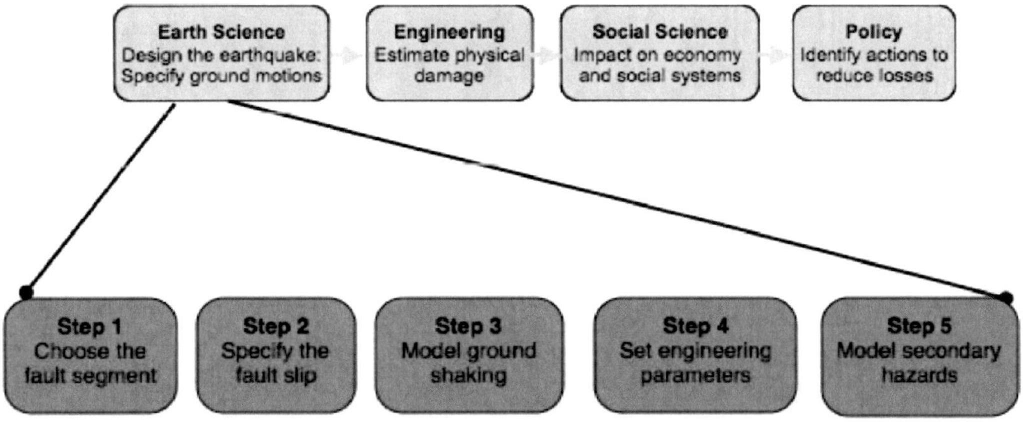

Figure 3-1. ShakeOut Scenario flow-chart

Step 1. Choose the fault segment

The earthquake used in the ShakeOut Scenario had to meet several distinct criteria. It had to be:

- scientifically plausible, in keeping with the latest scientific findings;
- large enough and close enough to population centers that it would have regional, long-term consequences; and
- likely enough that it would not be dismissed as a rare or extreme event.

We chose the southern San Andreas Fault as the source of the ShakeOut Scenario earthquake because of the short recurrence times between great earthquakes on that fault and because it is one of the best studied faults in the world, with a rich data set to inform our decisions. A magnitude 7.8 is not the largest earthquake that the southern San Andreas Fault can produce. Moreover, there are other faults that menace the populated areas of southern California and that will someday produce earthquakes as large as, or larger than. the event in this Scenario. However, the *recurrence intervals* (an estimate of the average time) between larger earthquakes on those faults are considerably longer, measured in thousands of years. By contrast, the southern San Andreas Fault has generated earthquakes of ShakeOut size every 150 years (on average, with actual times between earthquakes ranging from 45 to more than 300 years).

In the ShakeOut Scenario earthquake, the portions of the fault that rupture include sections of the San Andreas Fault that last broke about 1680 and 1812, as well as the southeastern part of the 1857 rupture. The 1680 and 1812 rupture sections are the most likely to rupture in a great earthquake, because they have gone the longest without an earthquake. The slip distributions and rupture speeds for those events are not well known, so we did not model a repeat of a relatively well-documented slip event, as has been done previously in modeling of the 1857 Fort Tejon and 1906 San Francisco earthquakes.

To define the large-scale features of the ShakeOut earthquake's slip distribution— the endpoints, magnitude, and overall rupture length—we used the best available geological slip rates for the San Andreas Fault, as well as paleoseismic evidence for the dates of the most recent earthquakes. We employed a simple earthquake recurrence model, and consensus on parameters and methods that was reached during expert discussions at multiple meetings and workshops. All features of the ShakeOut Scenario rupture were decided after considerable expert discussion. In particular, San Andreas Fault experts participated in two workshops, hosted by the Southern California Earthquake Center (SCEC) in November 2006 and January 2007, during which compilations of fault slip rates were combined with knowledge of the dates of the last event at different points along the fault, as indications of the amount of accumulated strain. At a fault parameter workshop that was held for the Working Group on California Earthquake Probabilities (WGCEP) in November 2006, discussion centered on the selection of a northwestern endpoint, as well as on rupture directivity.

The Scenario earthquake starts at the southeastern end of the San Andreas Fault, at Bombay Beach (figure 3-2). This southeastern portion of the fault has not ruptured since approximately 1680 (Sieh, 1986) and thus has accumulated strain far beyond that released in the average San Andreas event. Publications which hypothesize that rupture might initiate on the southernmost San Andreas Fault have tended to select Bombay Beach as the nucleation point. The other likely nucleation point is Parkfield, in central California, based on evidence from 1857 foreshocks, for example, Sieh, 1978; Agnew and Sieh, 1978; Meltzner and Wald, 1999). To the NW of Parkfield, the San Andreas Fault creeps. Southeast of Bombay Beach, the San Andreas also creeps, as it merges into the Brawley Seismic Zone. Hence, both Parkfield and Bombay Beach appear to be natural physical limits to seismic rupture. Both are also thought to be places where end-on loading of the San Andreas Fault occurs on an ongoing basis, and therefore they are likely places for events to nucleate (Stuart, 1986). It has been further hypothesized that a moderate earthquake on a cross-fault in the Brawley Seismic Zone could trigger a San Andreas rupture (Hudnut and others, 1989). Ultimately, Parkfield was ruled out as nucleation point for the ShakeOut Scenario earthquake because it lies at the northwestern terminus of the 1857 rupture; thus less strain has accumulated there.

Deciding how far to the northwest the ShakeOut Scenario rupture should extend in turn determines the magnitude and the likelihood of the ShakeOut event—a longer fault rupture produces a larger but less common earthquake. The ShakeOut earthquake ruptures to the northwest and stops at Lake Hughes, slightly southeast of the Cow Springs paleoseismic site. Because of the ShakeOut earthquake's size and relationship to urban areas, this event, should it occur, would have greater consequences than either a Coachella-only event (approx. *M*7.1, on only the southernmost section south of San Gorgonio Pass, figure 3-2) or an event like the SCEC TeraShake scenario (*M*7.7).

Figure 3-2. Map of California showing the extent of rupture—and thus size—of the last three earthquakes on the San Andreas Fault

Selection of the northwestern endpoint was based in part on spatial variation of slip in the 1857 earthquake (Sieh, 1978) and earlier events (Rush, 2005). Farther to the southeast, in the 1857 event the San Andreas Fault does not appear to have had slip as large as the 7 to 7.5 meters found at the Cow Springs site (Sieh, 1978), and evidence suggests that the amount of slip takes a sudden drop or sharply tapers to the southeast of the Cow Springs site (Sieh, 1978). Additional evidence from the Cow Springs site (Rust, 2005) finds large slip in three past events, the most recent of which is thought to be the 1857 earthquake. If slip in each of the last several events was large from Cow Springs to the NW, yet small to the SE, then more strain energy has been released to the NW, and more is still stored, awaiting release, to the SE. The precise southeastern terminus of the 1857 rupture is not known, and we do not know whether an earthquake like that in the ShakeOut Scenario would stop rupturing at the 1857 terminus. After considering all these inherent uncertainties, ShakeOut fault experts found it reasonable to terminate the ShakeOut rupture at Lake Hughes, slightly southeast of the Cow Springs site.

Extensive discussion also considered whether a rupture coming from either direction along the southern San Andreas could continue through the fault's complex structure at San Gorgonio Pass (figure 3-3 fence diagram). Despite concerns about this point, the majority held the view that a rupture initiating at Bombay Beach would plausibly continue through San Gorgonio Pass. For some, this view was substantiated by research within the dynamic rupture simulation group at SCEC, using a simplified fault model that is vertical, piecewise, and planar (Steve Day, SDSU, personal communication, 2008). Research using more detailed representations of the actual, complex fault surface may eventually provide fuller verification of the plausibility of through-going rupture at San Gorgonio Pass.

After the workshop, discussion of ShakeOut Scenario fault parameters continued informally and resumed during the first Southern San Andreas Fault Evaluation (SoSAFE) workshop. Discussion included evidence for along-strike variations in the dip of the San Andreas Fault, and additional research regarding the dip, which would be submitted for the next version of the SCEC Community Fault Model (CME). Given the short deadlines of the project, however, the ShakeOut Scenario used the present version of the SCEC CME.

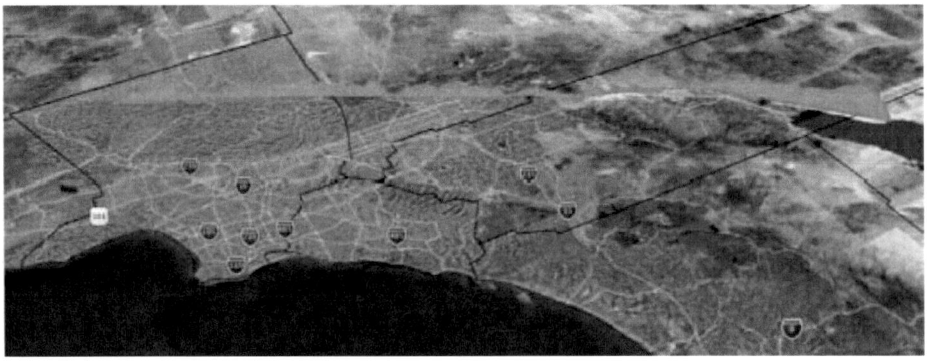

Figure 3-3. Slip along the San Andreas Fault, as modeled for the ShakeOut Scenario earthquake, is shown by the height of the red "fence" along the fault. Note that the maximum amount of slip is at the southern end of the rupture near the Salton Sea, where it has been more than 300 years since the last earthquake. Slip varies with position along the fault because of variations in slip rate and in time since the last earthquake. The diagram shows a further level of variability added as random variation to make a more realistic fault rupture.

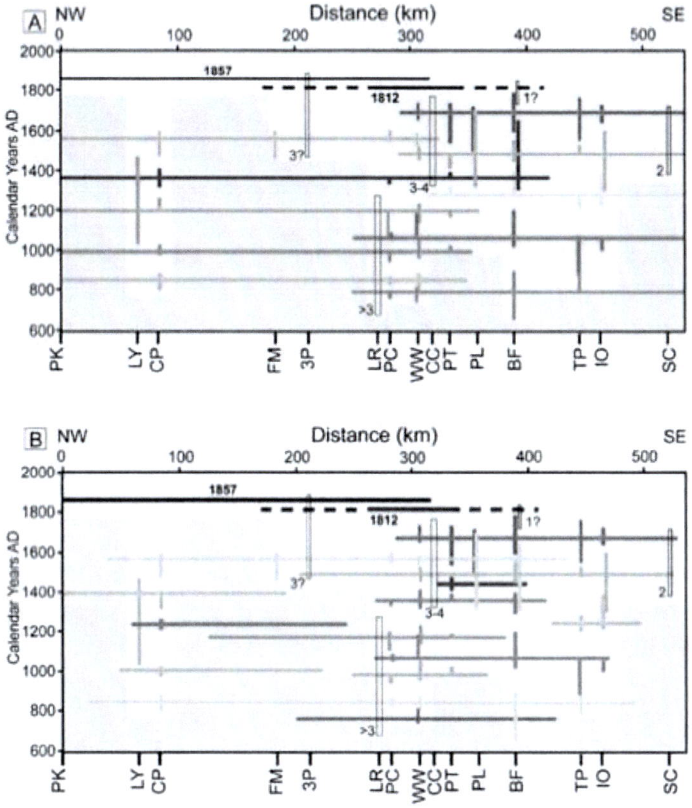

Figure 3-4. From Weldon and others (2005), the orange line at about A.D. 1460-1480 represents an event similar to the ShakeOut earthquake, with endpoints at Bombay Beach and Lake Hughes

Once decided, the endpoints of the ShakeOut earthquake defined an event remarkably similar to one proposed by Weldon and others (2005) at approximately A.D. 1480 (see orange line with bars in lower panel of their Figure 12, reprinted here as figure 3-4).

Step 2. Specify the fault slip

After the fault rupture was decided, we could define the slip along the fault. This was done at two scales. First, we defined the **static rupture description** (also called *a background slip distribution or average slip distribution*) for several portions of the fault, based on paleoseismic and geological data. This provides an estimate of accumulated slip along each portion of the fault. However, we know from study of past large earthquakes that these long sections of the fault will not rupture uniformly, and if we did model uniform rupture, we would create unrealistically large ground motions. Therefore, computer modeling was done to create a **kinematic rupture description** with a randomized variation of the average slip within each 30 km section of fault.

In one or more ways, our approach has departed from prior methods to simulate large earthquakes on the southernmost San Andreas Fault. For the 1857 Fort Tejon earthquake, enough is known of surface slip, endpoints, and magnitude to construct a simple rupture model and to estimate ground motions using attenuation relations (a method of estimating ground motion using data from past earthquakes, based on size and distance of an earthquake). However, insufficient evidence exists to reconstruct the slip distribution from Cajon Pass to the southeast. Another common modeling approach would be to project slip details of a roughly similar earthquake that occurred on another fault; for example, projecting the *M*7.9 Denali earthquake (for example, Eberhart-Phillips and others, 2003), onto the San Andreas Fault. This has been done by several investigators for the 1857 rupture zone as well as for the southernmost San Andreas (for example, Krishnan and others, 2006; Olsen and others, 2006). We opted instead to model the rupture that might occur on this fault, based on accumulated slip as determined from studies along the fault (figure 3-4).

The **static rupture description** was computed by assuming that the average amount of slip to be released in the ShakeOut Scenario earthquake would be the amount accumulated since the last event at study sites along each portion of the fault. This was calculated using the latest, best estimates of fault dip, slip rate, date of last earthquake, and seismogenic depth values from Wills and others (WGCEP App. A., 2008). This method is similar, but not identical, to the slip-predictable model of Shimazaki and Nakata (1980). The difference is that our calculations have a variable slip rate along the fault, whereas their models considered uniform event slip. Although it has become common-use terminology to describe our method as a "slip-predictable" construction, this is not strictly true to the original work. In our case, we took the time difference between the 2008 ShakeOut date and the date of the last event as the "open" time interval. We multiplied this by the slip rate from WGCEP App. A table, and thereby obtained the slip. As these parameters vary along-strike, accordingly so does the slip in our ShakeOut static rupture description. The resulting slip distribution and assumed parameters are given in Table 3-1, and the static rupture description is shown in figure 3-5.

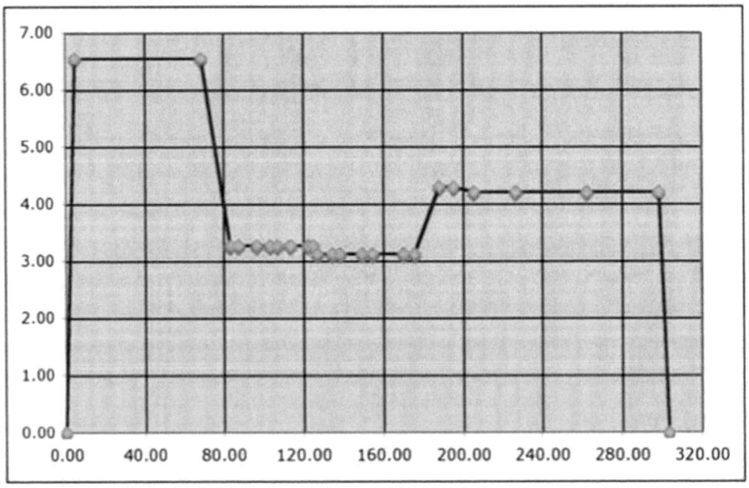

Figure 3-5. From Bombay Beach at zero on the x-axis to Lake Hughes at 305 km, slip (in meters) varies as shown for the ShakeOut average slip distribution model. After initiating at Bombay Beach and rapidly attaining large slip (and high rupture speed, not shown), the rupture then slips at a reduced level through San Gorgonio Pass, then increases again from Wrightwood to Palmdale, and finally ends at Lake Hughes

Table 3-1. Definition of the ShakeOut Static Rupture Description

	Latitudes	Longitudes	Depths (km)	Dip (deg)	Rate (mm/yr)	Yrs	Slip (m)	Length (km)	Section Boundary Points
1	34.698495	-118.508948	13.1	90	28±7	150	4.20	40.63	NW end: Lake Hughes
2	34.547849	-118.103936	13.1	90	28±7	150	4.20	35.90	
3	34.402927	-117.753579	13.1	90	28±7	150	4.20	21.10	
4	34.316300	-117.549000	13.1	90	28±7	150	4.20	10.39	
5	34.270900	-117.451000	12.8	90	22±6	195	4.29	7.12	Cajon Pass - Sect. Jct. Pt.
6	34.232843	-117.388692	12.8	90	22±6	195	4.29	12.43	
7	34.173137	-117.274161	12.8	90	16±3	195	3.12	5.44	Hwy. 18 - Sect. Jct. Pt.
8	34.150027	-117.222023	12.8	90	16±3	195	3.12	15.59	
9	34.092795	-117.067674	12.8	90	16±3	195	3.12	5.39	
10	34.073768	-117.013900	12.8	90	16±3	195	3.12	11.18	
11	34.033837	-116.902350	12.8	90	16±3	195	3.12	3.63	
12	34.011347	-116.873541	12.8	90	16±3	195	3.12	7.60	
13	33.959114	-116.819795	12.8	90	16±3	195	3.12	1.80	
14	33.953154	-116.801391	16.4	58	10±3	327	3.27	2.72	Millard Cyn. - Sect. Jct. Pt.
15	33.937411	-116.778598	16.4	58	10±3	327	3.27	8.61	
16	33.944163	-116.685809	16.4	58	10±3	327	3.27	6.47	
17	33.917569	-116.623871	16.4	58	10±3	327	3.27	3.75	
18	33.907018	-116.584856	16.4	58	10±3	327	3.27	6.76	
19	33.884664	-116.516889	16.4	58	10±3	327	3.27	9.35	
20	33.848123	-116.426527	16.4	58	10±3	327	3.27	4.05	
21	33.848518	-116.383007	16.4	58	10±3	327	3.27	14.37	
22	33.788250	-116.246290	11.1	90	20±3	327	6.54	69.22	Biskra Palms Oasis
23	33.350090	-115.711920							SE end: Bombay Beach

This very simple construction was used as the starting background slip in the kinematic rupture description described later, but was not otherwise used directly in any of our calculations of ground motions. The attenuation relations could not account for even this level of complexity in variable slip along-strike, and yet this was far too simplistic for use in the kinematic modeling. Real earthquakes tend not to exhibit stair- step slip distributions, and if forward-modeled, the coherence of such a rupture would produce singularly large and unphysical ground motions.

Our work to create a **kinematic rupture description** built upon the the recent experience of Aagaard and Graves in simulating the 1906 earthquake (Aagaard and others, in press). Their process of creating a fully detailed kinematic rupture description for the 1906 earthquake led to innovations that we used in creating the ShakeOut kinematic rupture description.

Beginning with the static rupture description and the relatively complex fault geometry available in the SCEC CFM-triangular element representation, several complexities were added to the source with the intent to make it more realistic. Instead of large rectangular patches with uniform slip, we wanted a rupture description compatible with the slip found in kinematic source inversions. As described in detail in Appendix A, transforming the static rupture description into the full kinematic rupture description involves several steps.

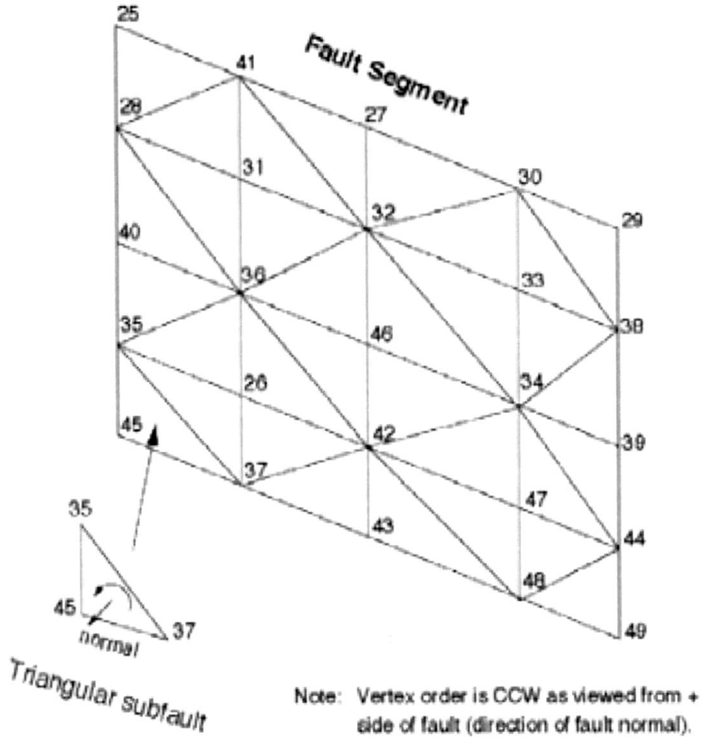

Figure 3-6. The SCEC Community Fault Model (CFM) exists both in the form of rectangular elements, and also triangular elements. This finer resolution, more smoothly varying surface model was used for our kinematic rupture description for the ShakeOut source

The kinematic rupture description includes shorter length scale variations in slip than those in the static rupture description. We add a random field with a wavenumber squared spectral falloff, a standard deviation of 2.0, and wavelengths less than 30 km to the background slip distribution. This results in maximum slip values about four times greater than the average slip.

Rake angles were randomized with a standard deviation of 10 degrees. Incorporating the temporal evolution of slip requires specifying both how slip occurs at a point and the progression of the rupture propagation. Brune' s far-field time function (Brune, 1970) defines the slip time history at each point on the fault with the peak slip rate related to the final slip by $V_{max} = 1.2\sqrt{D[m]}$. The local rupture speed correlates with slip using a piecewise linear variation. The maximum rupture speed of 1.4 Vs corresponds to regions of maximum slip (16 m), regions of average slip (4 m) have a rupture speed of 0.85 Vs, and regions with negligible slip have a rupture speed of 0.2 Vs. Additionally, the rupture speed is tapered by 50% over 3 km along both the top and bottom edges of the rupture (consistent with rupture propagating from regions of unstable sliding to stable sliding). Slip initiation times are determined from this rupture speed distribution by tracing the rupture front away from the hypocenter assuming locally circular wave fronts. The full kinematic rupture description projects the spatial and temporal evolution of slip onto the 3-D, non-planar fault geometry of the SCEC Community Fault Model (figure 3-6).

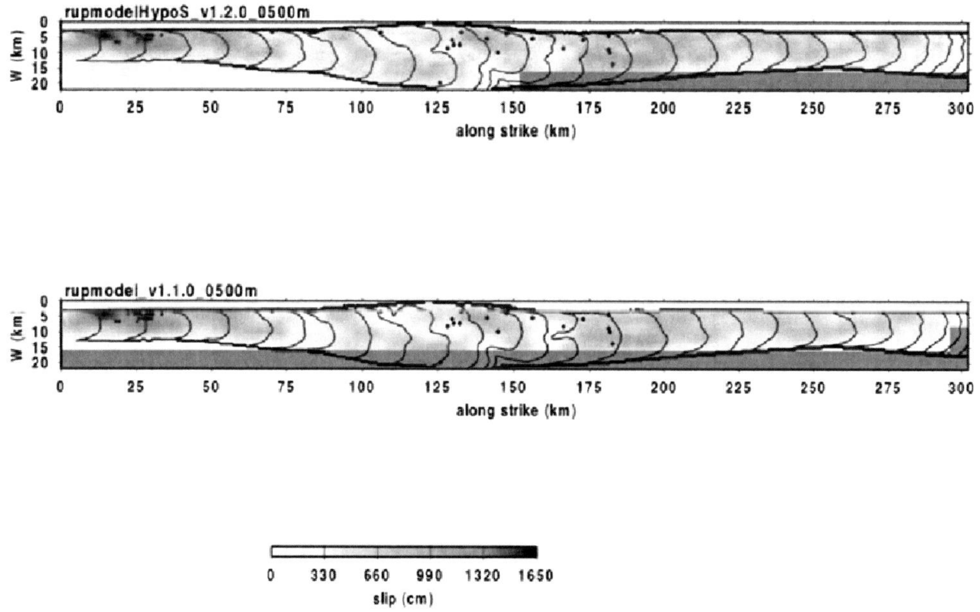

Figure 3-7. Cross-sectional view of the comparison between version 1.1 (bottom) and version 1.2 (top) of the ShakeOut kinematic rupture description.

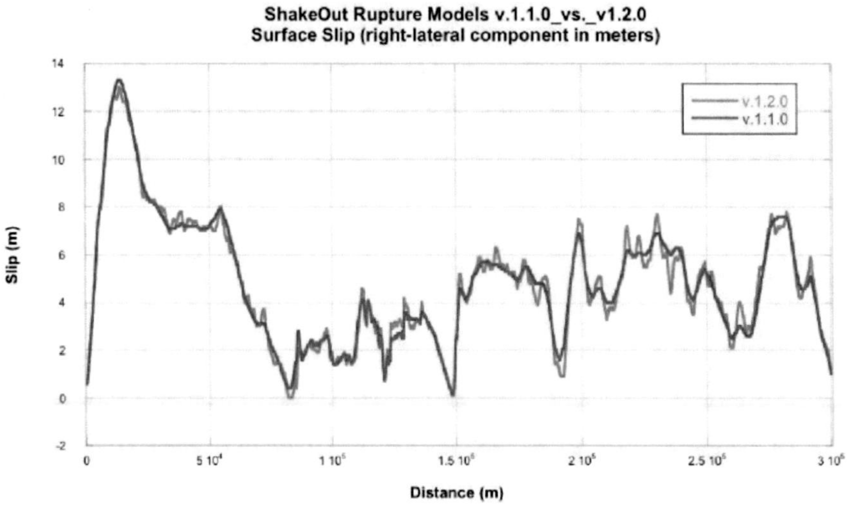

Figure 3-8. Comparison of slip in meters at the surface of the Earth, along-strike of the fault for both the version 1.1 (red line) and version 1.2 (blue line) kinematic rupture descriptions.

For each subfault, the following parameters were defined:

- Slip time (ime in seconds at which slip begins)
- Slip vector (slip vector in meters in 3-D coordinate system associated with displacement on east side of fault)
- Slip (slip vector in meters in along-strike, up dip, opening coordinate system)
- Slip rate (peak slip rate in meters)
- Rupture speed (rupture speed in meters/second)
- Strike dist (distance along-strike in meters from southeast end)
- Dip dist (distance down dip in meters from top end)
- Rise time (yime in seconds for 95% of the slip to occur)

Simulated ground shaking was calculated for two versions of the kinematic rupture description. The difference between version 1.1 and version 1.2 was in the amount of slip heterogeneity (that is, random variability) at short length scales, which was increased in version 1.2. Figure 3-7 shows a cross-sectional comparison of the two versions, and figure 3-8 compares surficial slip along-strike for both versions. Where critical lifeline infrastructure crosses the fault, these seemingly minor differences in slip were significant in some cases.

Step 3. Model ground shaking

The sudden slip of one side of the fault past the other, described in the last sections, produces shaking as one of its effects. This shaking moves the ground, and it is these ground motions that we feel and that cause most of the damage in an earthquake. Thus, accurate estimates of damage depend first and foremost on a realistic description of the ground motions. The goal of this aspect of the ShakeOut Scenario effort was to predict what ground motions would arrive at sites around southern California to shake the buildings, roads, pipelines, and other structures that are needed by our society.

Table 3-2. Summary of ShakeOut Rupture Description

Fault Segment
• Magnitude 7.8
• Unilateral rupture from southeast to northwest
• SE endpoint (Bombay Beach): 33.35009, -115.71192
• NW endpoint (Lake Hughes): 34.698495, -118.508948
Static Rupture Description
• 23 points along-strike, from SCEC Community Fault Model—rectangular surface representation
• Slip recurrence model to construct slip distribution along-strike
• Slip rates, dips, and depths for all sections of the San Andreas from the WGCEP; used Appendix A. by Wills, Weldon and Bryant, March 1, 2007—draft version
Kinematic Rupture Description
• Uses SCEC CFM – triangulated surface representation
• Convolves a 30-km wavelength random slip function with the static rupture description.
• Applies scaling criteria to the slip distribution to generate the rise time and rupture speed.
• Computes contours showing the rupture front at one-second intervals.

For this part of the ShakeOut we turned to seismologists, who can study ground shaking at sites in order to understand motion on a fault and the structure of the Earth. For many years, seismologists have understood how to apply the basic physics of waves to calculate the waves produced by a fault as it ruptures, and they have estimated how the waves and thus the ground shaking will change as the waves move away from the fault through different types of rock. The techniques to do these estimations are not difficult or new and have been validated through repeated experiments. What is new is that information technology has advanced to the point that seismologists can now use supercomputers to address the complications that develop when waves travel through the complex geologic structures that underlie southern California. The ShakeOut Scenario was fortunate to be able to take advantage of major advances in the application of information technology to seismology made within the SCEC information technology research program.

Ground motions depend on three first-order effects (which will affect shaking at every site) and several secondary effects (which will affect shaking at some sites). The first factor is **magnitude**—a bigger earthquake produces more energy, which means more energy arrives at any site. The magnitude depends on both the area of the fault that moves and the amount of slip. Each point on the fault radiates energy proportional to the amount of slip at that point. The second factor is **distance from the fault** because the shaking attenuates as it travels through the crust. The third factor is **soil conditions**—the characteristics of the soil or rock at a particular location affect the amplitude and duration of the shaking at that site. The secondary factors include **directivity**, in which ground motions are focused in the direction of rupture propagation and are diffused at the 180- degree-opposite direction) and **radiation pattern**, variations in energy distribution that depend on the orientation of the fault that is rupturing.

Our physics-based simulations can model all of the factors, primary and secondary, that affect ground shaking, using two inputs: (1) the rupture model we created in Step 2 and (2) a *velocity model* that describes the seismic characteristics of the southern California rocks through which the waves propagate. The computer codes then run on supercomputers to use physics-based simulation algorithms that model how the waves propagate, scatter, attenuate and resonate through the different types of rock and sediment. We validated our modeling results through comparison of multiple methods, use of distinct velocity models, and comparison with empirically-based attenuation relations. In all, four teams were engaged to use different computer codes and modeling algorithms to make independent models of the ground motions using the same input rupture model and two distinct velocity models.

Three separate, collaborative groups within the SCEC Community Modeling Environment modeled the ShakeOut earthquake ground motions. Initially, the kinematic rupture description was announced through e-mail to all potentially interested SCEC modelers (SCEC ShakeOut Simulation Group, in preparation), and shared in standard format from an anonymous ftp site. The initial release (version 1.1) later was refined in version 1.2, which used a rougher slip distribution and less coherent rupture front. Version 1.2 also provided two alternative rupture models, discussed in more detail below. In one of the alternative rupture models, the rupture initiates at the northwest end and propagates to the southeast; in the other, rupture nucleates near San Gorgonio Pass and propagates *bilaterally* (in two directions) to the northwest and southeast. Some simulation groups had already used version 1.1 and have not since re-run their models with version 1.2. One of the simulation groups only ran version 1.2, then also ran the alternative scenarios. As a result, we have the best basis for comparing three groups' results using the version 1.1 model, but for assessing rupture directivity effects and differences among modeling groups, we have the best basis for comparisons using version 1.2.

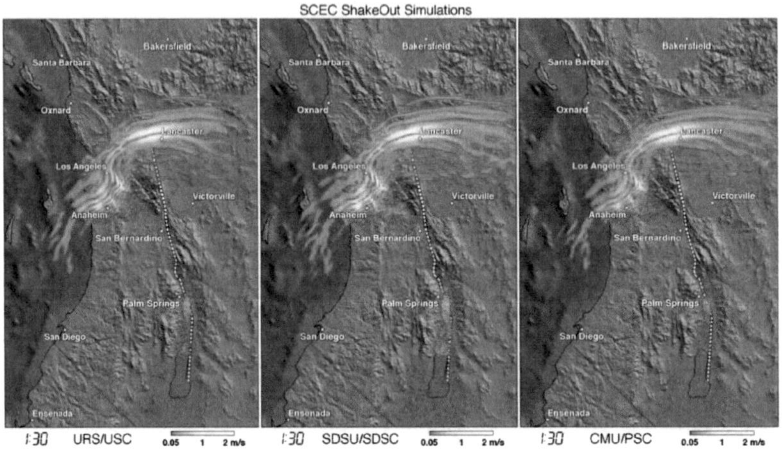

Figure 3-9. To illustrate the comparison among results of the three simulations conducted by the SCEC ShakeOut Simulation Group (funded by the National Science Foundation), this snapshot was taken at 1 minute, 30 seconds after rupture initiation. The color scale is linear, and the map scale is common to all panels. In general, the positions of the main P- wave and S-wave arrivals are similar, as are the patterns of the largest-amplitude— hence most damaging—shaking. Left panel result is from URS, Inc. and University of Southern California. Center panel result is from San Diego State University, San Diego Supercomputer Center. Right panel is from Carnegie-Mellon, TeraGrid Pittsburgh Supercomputing Center. (Image courtesy of Geoff Ely, SDSU/UCSD/SDSC.)

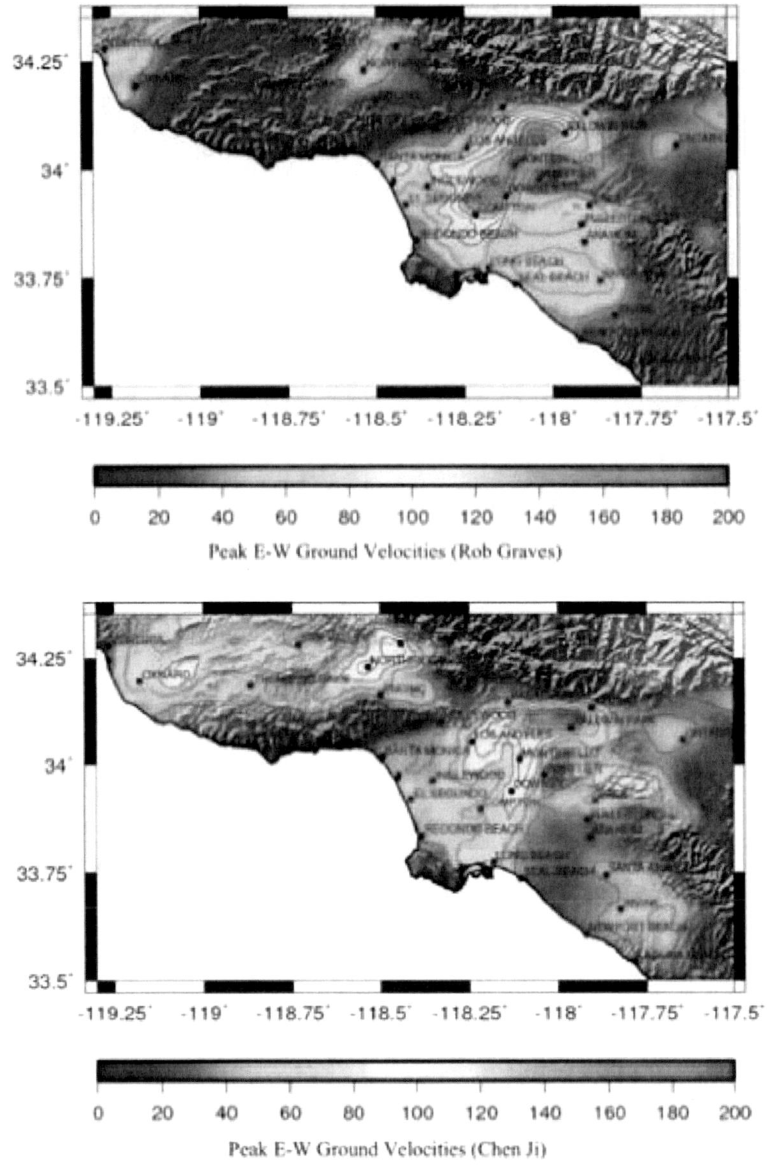

Figure 3-10. Upper panel shows the E-W component of Graves's result, and lower panel shows the E-W component of Ji's result for comparison, using the same color scale. It is important to ShakeOut damage estimates to note that the results have comparable amplitudes in many areas with tall buildings.

In general, agreement seen among the simulation methods is reasonably good, as illustrated by the Carnegie-Mellon group on their SCEC 2007 poster (Taborda and others, 2007) and in figure 3-9. In ongoing research, the SCEC collaboration is actively pursuing more rigorous cross-comparisons among the modeling results so that discrepancies may be identified and understood (SCEC ShakeOut Simulation Group, in preparation); and is also pursuing full dynamic rupture simulations to examine the effects of source complexity on the ground motions from ShakeOut-type events (Day and others, in preparation).

An additional set of simulations, performed by Prof. Chen Ji of University of California, Santa Barbara, used a significantly different approach and a different velocity model and further validated results obtained by the SCEC group. Because the velocity structure used by Ji is more complex, it was expected that results might differ greatly from those of the other groups. His initial results emphasize differences in the amplitude and pattern of ground motions through the Los Angeles area, as shown in figure 3-10. However, although significant differences exist in some places, they are on the order of less than a factor of two, lending a much greater degree of certainty to the ShakeOut modeling results, given the consistency of the overall results of all forward simulations despite the wide range of approaches taken to this problem. It is important to ShakeOut damage estimates that the results in figure 3-10 have comparable amplitudes in many areas with tall buildings.

Several features are consistent across all the models, including the following:

- Very strong shaking (approaching 3 m/sec) near the fault;
- Strong shaking with moderate to long durations (20-45 sec) in the basins near the fault, including the Coachella, San Bernardino and Antelope Valleys;
- Damaging shaking (at least 0.5 m/sec) over large areas (~10,000 km2) of Los Angeles, San Bernardino, and Riverside Counties;
- Pockets of very strong shaking (≥1 .5 m/sec) with long durations (45-60 sec) in areas of the San Gabriel Valley and East Los Angeles.

The first three features could be called "mundane" – these are the standard motions seen in all great earthquakes. The last feature, the pockets of very strong shaking in Los Angeles, are the type of variability that has caused damage—sometimes far from the fault—in many previous large earthquakes, including Mexico City in the 1985 Michoacan earthquake, Santa Monica in the 1994 Northridge earthquake, and Santa Cruz or Watsonville in the 1989 Loma Prieta earthquake.

The details of these features depend on which model is used, but some version of them shows up in each model. Variability in the pattern of shaking results from both the focusing of energy towards Los Angeles (see discussion of rupture directivity below), and the amplification and resonance of energy in the sediments of the San Gabriel and Los Angeles basins, as is typical of sedimentary basins.

For the past thirty years, before the recent advances in information technology, ground motion predictions have typically been made using **attenuation relations**, and these have become standard tools that engineers use in order to forecast the expected shaking at a site. Attenuation relations are based on statistical evaluations of empirical data from a database of past, recorded earthquakes. They generally do a poor job estimating ground motions in an earthquake like the ShakeOut earthquake, for several reasons:

- Scientists have only been collecting empirical data for a very short amount of geologic time and so we lack needed data, particularly for the largest earthquakes, which occur more rarely than smaller events.
- The relations cannot account for the physics of the fault rupture or of wave propagation and instead predict the mean value of ground shaking expected at a site as a function of magnitude, distance from the fault, and site conditions. The actual recorded ground motion could be substantially higher or lower.

- The relations can lead to a very uniform distribution of ground motions, which is potentially unrealistic when then used to model one particular earthquake. A real earthquake has significant spatial variability from directivity and radiation patterns as well as propagation effects, and the physics-based modeling is better able to identify the places that are surprisingly spared as well as the places where the shaking is more intense and continues for longer than expected.

In real earthquakes, most damage is concentrated in the areas of highest shaking. Damage is not a simple linear function of shaking level but rather often demonstrates a strongly non-linear acceleration toward failure at the highest shaking levels. Pockets of strong shaking that also undergo a long duration of high shaking are particularly prone to damage, and attenuation relations (unlike the physics-based simulation of the ShakeOut Scenario) cannot identify areas that undergo a long duration of strong shaking.

Because of the prevalence of attenuation relations in engineering, we have compared the results of the synthetic ground motion models to the ground motions predicted by the latest attenuation relations from the Next Generation Attenuation (NGA) project.

Figure 3-11 shows the values of one of the ShakeOut Scenario models compared to the predictions of NGA models. The comparisons are made for several engineering parameters, including peak ground acceleration, and 1-second and 3-second spectral acceleration. These show that the physics-based models create a wide distribution of values at any distance from the fault (as is seen in real earthquakes), with a mean value that appears adequately close to the NGA predicted value. When we look more closely at the comparison between the two, we find that there is a trend with period that is plausible because it falls within one standard deviation.

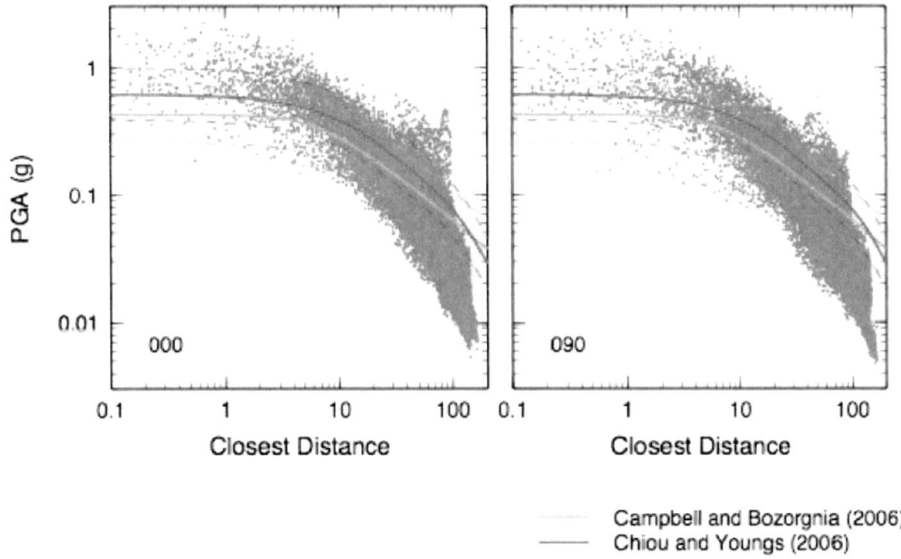

Figure 3-11. Comparison of next-generation attenuation (NGA) relation (Campbell and Bozorgnia, 2008 in green, and Chiou and Youngs, 2006 in red) with physics-based simulations (shown as grey points are values computed by Rob Graves, using source version 1.1) of the ground motions for the ShakeOut Scenario earthquake.

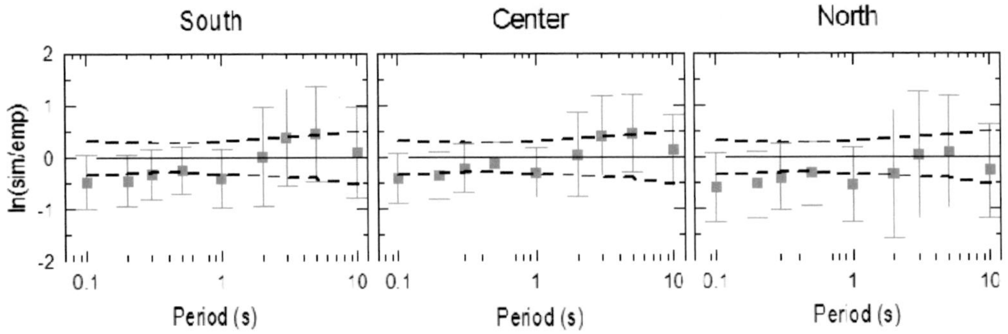

Figure 3-12. Average of residuals (red squares) between simulations and empirical predictions from the model of Campbell and Bozorgnia (2008) as a function of period for three hypocenter cases. Positive values indicate the simulations predict larger motions than the empirical model; negative values indicate smaller simulated values compared to the empirical model. The error bars indicate +/- 1 sigma for the residuals. The heavy dashed lines plot the +/- 1 sigma level of the inter-event term from the empirical model.

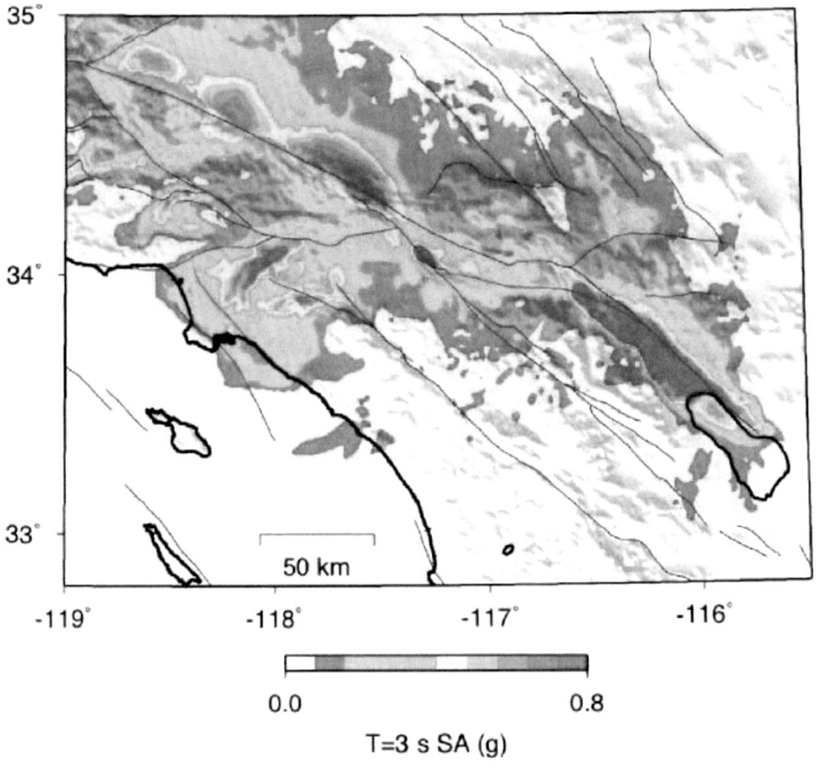

Figure 3-13. Ground motions throughout southern California from the ShakeOut event are large, especially along the San Andreas Fault and in deep basins such as the Coachella Valley and Los Angeles Basin. Energy that is focused through the basin in San Bernardino passes into East Los Angeles and near the downtown Los Angeles area. Shown here is the spectral acceleration (SA) at a period of 3 seconds.

Figure 3-12 shows the mean of the residuals of the Graves model with the Campbell and Bozorgnia (2008) NGA relation for three possible hypocenter locations. The figure also shows the event terms, which express the average offset of any single earthquake's motions from those predicted by the attenuation relation. At short periods the synthetic ground motions are generally low compared to the attenuation relation, while at long periods they are generally high. This was true when we looked at comparisons with other attenuation relations as well. However, because the residuals generally fall within one attenuation relation event term standard deviation, we would not consider the physics- based ground motion simulation scenarios to be unrealistic. Based on these comparative analyses we conclude that these ground motions are plausible, displaying the mean values and variability that we see in real events.

A discrepancy was found between the synthetic models and the attenuation relation in the attenuation of ground motions with distance. We found that the computer simulation generally gives larger ground motion close in to the fault, but attenuates faster with distance than predicted by attenuation relations. The disparity between the attenuation rates varies with period and may explain some of the bias compared to the event terms. We corrected for this discrepancy and found indication that the correction was valid during a preliminary examination to compare the differences between the attenuation relation and the synthetic model predictions for site conditions. Once the discrepancy for distance had been taken into account, trends with basin depth and shearwave-velocity for residuals of the computer model and the Campbell and Bozorgnia NGA relation are negligible.

As seen in figure 3-13, the ShakeOut earthquake produces large ground motions throughout much of southern California. Motions along the fault are especially strong, as are bands of strong shaking that radiate outward along the axes of sedimentary basins that happen to be elongated in the direction that energy radiates from the fault.

Duration of strong shaking will be an important contributor to damage in any earthquake as large as the ShakeOut Scenario earthquake, and is due to the length of time that it takes for such a long fault to rupture and the reverberation of waves trapped in sedimentary basins. In the ShakeOut Scenario earthquake, the San Bernardino Valley is shaken extremely strongly but for a relatively short duration, as are Wrightwood and Palmdale. The Coachella Valley is strongly shaken for a longer duration. Lower amplitude, but much longer duration ground motions (Table 3-3) are seen in the Los Angeles Basin. Notably, a band of strong ground motions arcs from East Los Angeles towards (but not quite reaching) Torrance.

Rupture directivity occurs when energy is focused in the direction that a fault is rupturing, and it can greatly increase ground shaking. We examined the contribution of directivity on the ShakeOut ground motions and determined that rupture from the southeast to the northwest produces the strongest directivity effect within southern California. This finding correlates with SCEC Terashake (Olsen and others, 2006) simulation results. Unfortunately, then, the San Andreas Fault rupture that is most in keeping with accumulated slip—the rupture that begins at Bombay Beach and ruptures to the northwest—would increase shaking in many of the heavily populated valleys of southern California, because of rupture directivity as well as the amplification of shaking in sediments that fill those valleys.

To examine the contribution of rupture directivity we created synthetic ground motion waveforms for the ShakeOut and two alternative kinematic rupture models. In one alternative (the "Central" alternative), the earthquake rupture begins near San Gorgonio Pass and spreads *bilaterally* (in two directions) to the northwest and southeast. In the other alternative (the

"North" alternative), the rupture starts in the northwest at Lake Hughes and ruptures southeast to terminate at Bombay Beach. Table 3-4 specifies the rupture initiation points for the ShakeOut earthquake ("South") and both alternatives. Figure 3-14 compares the waveforms. The "Central," bilateral rupture produces a similar directivity effect yet significantly smaller ground motions in Los Angeles. The "North" alternative shows little or no directivity effects. However, all three alternatives would still produce extended duration of shaking as well as strong ground motions in long-period waves that are potentially very damaging, even without directivity.

Step 4. Set engineering parameters

Using the SCEC simulation done by Graves, the ShakeOut Scenario then calculated additional engineering parameters. The physics modeling was used to create a synthetic seismogram of ground motion at each of 25,500 grid points across southern California, with spacing of 2 km x 2 km. These seismograms were processed in standard ways to calculate various *ground motion parameters* used by engineers to estimate damage to structures. Ground motion parameters are descriptions of how the ground moves as a result of different measures of earthquake waves, because different kinds of structures are damaged by different kinds of waves. Ground motion parameters used in the ShakeOut Scenario are peak ground acceleration (PGA), peak ground velocity (PGV), Modified Mercalli Intensity (MMI), and spectral accelerations at 0.3, 1.0, and 3.0 seconds (figure 3-15).

Table 3-3. Onset Times and Durations of Strong and Very Strong Shaking

Location	Seconds after start of earthquake that strong shaking begins at this location	Seconds after start of earthquake that strong shaking ends at this location	Duration of very strong shaking
Palm Springs	25	60	35 sec
San Bernardino	45	75	30 sec
Los Angeles (downtown)	70	125	55 sec
Orange County	70	105	35 sec
Santa Monica	85	150	65 sec
Palmdale	75	90	15 sec
Ventura	105	160	55 sec

Table 3-4. Kinematic Rupture Initiation Points

Rupture Alternative	Longitude	Latitude	Depth (km)
South (ShakeOut)	-115.7068	33.3451	7.6
Central	-118.2900	34.6169	8.1
North	-116.7419	34.0445	15.1

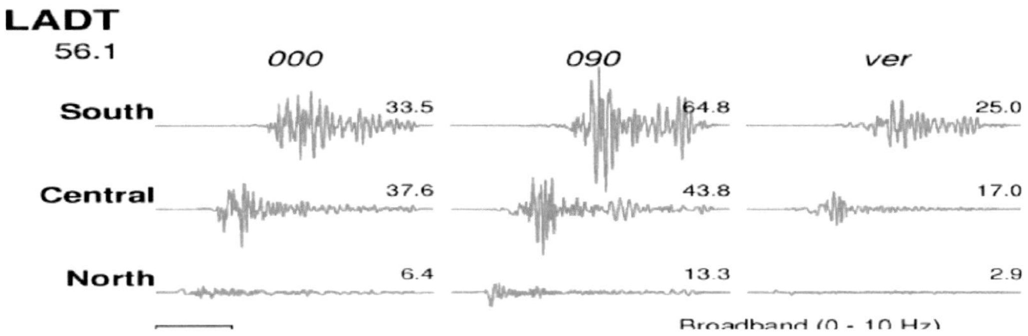

Figure 3-14. Comparison of ground motion waveforms for the three different cases of rupture directivity as recorded in Los Angeles. The uppermost panel, labeled South, is the ShakeOut rupture with rupture beginning at the southeast end of the fault at Bombay Beach. The middle panel, labeled Central, is for the bilateral case with rupture initiating near San Gorgonio Pass. The lowermost panel, labeled North, is for the case in which rupture begins at the northwestern end near Lake Hughes and propagates towards the southeast.

These parameters describe different aspects of the ground motion. Each will matter more in some locations than in others, and to some structures more than others. PGA is the most commonly used, but it does not capture all the information needed to estimate structural damage due to an earthquake as large as that in the ShakeOut Scenario, because it measures only the force applied at one moment in time and thus does not reflect the increased duration of strong shaking in the largest earthquakes. PGV is also an instantaneous measure, but the peak velocity increases when force is applied for a longer time and therefore PGV does a better job at indicating the impact of the long duration of very large earthquakes.

We also plotted Modified Mercalli Intensities (MMI), because this is a way of looking at damage that is familiar to many in the engineering and emergency management communities. We estimate the MMI values from the relationships between instrumental recordings of PGA and PGV developed by ShakeMap (Wald and others, 1999). At lower levels of shaking, the intensity is determined from PGA. At higher levels of shaking, above MMI VII, the dependency is on PGV. That relation is

$$MMI = 3.47*\log10(PGV) + 2.35$$

Wald and others (1999) found PGV to be a better predictor of higher levels of damage because it better reflects the impact of longer-period accelerations, which when applied for a longer time will result in larger ground motions. However, the PGV-MMI relationship was developed using data from smaller earthquakes (up to magnitude 7) and thus does not include the effect of the particularly long durations observed in the sedimentary basins in earthquakes as large as the ShakeOut Scenario event.

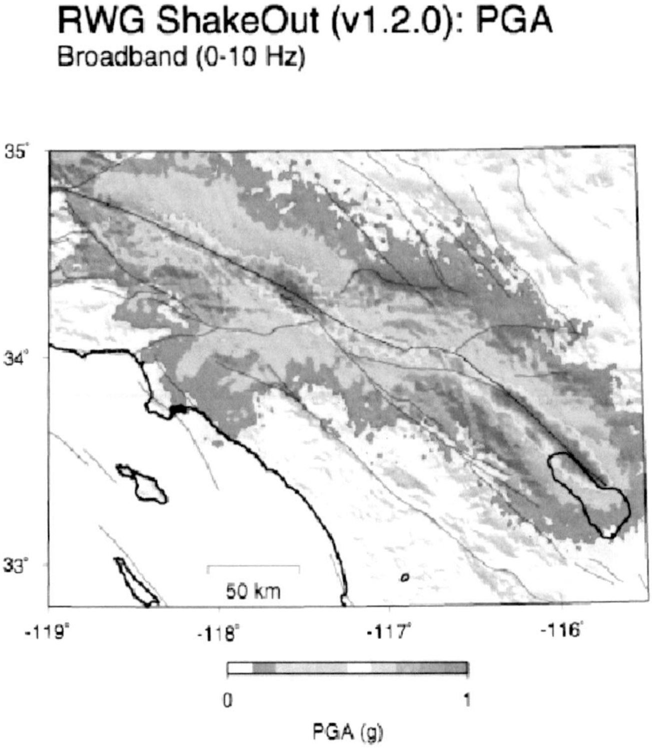

A

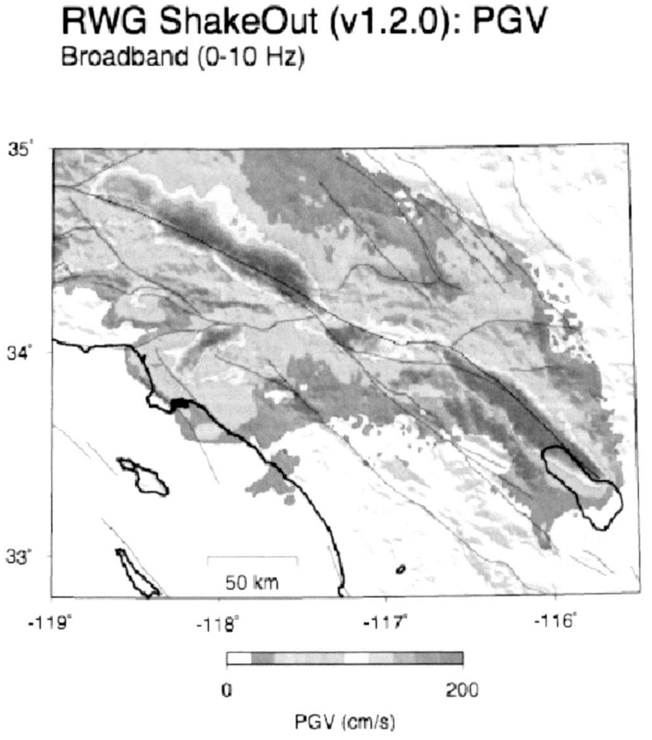

B

Figure 3-15. Continued on next page.

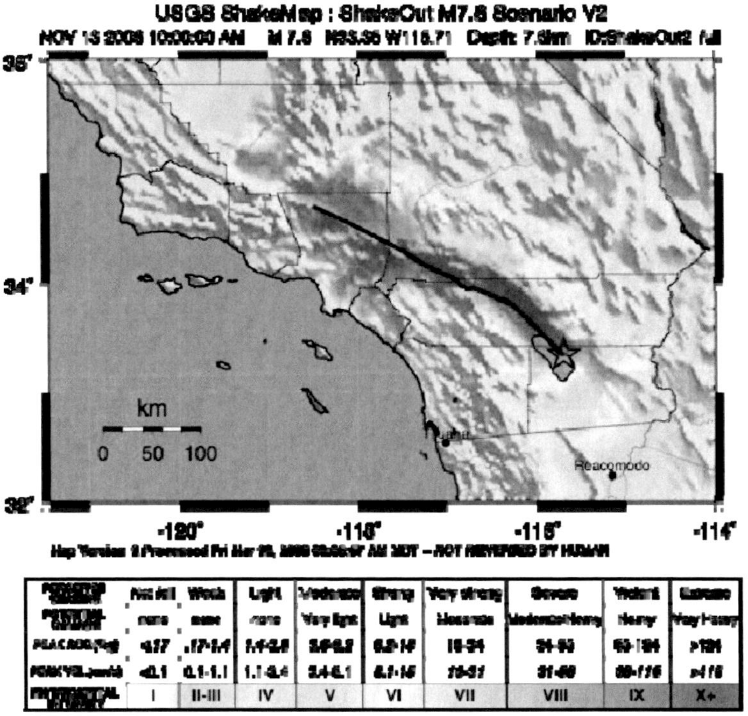

C

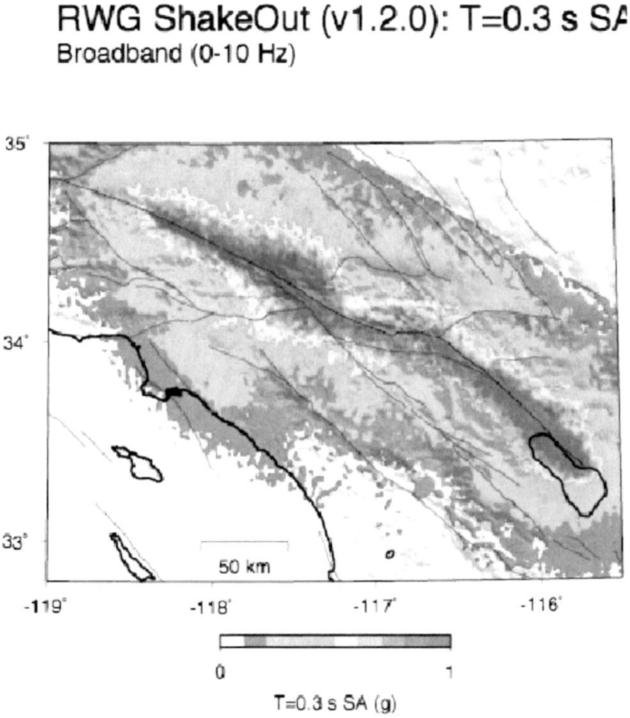

D

Figure 3-15. Continued on next page.

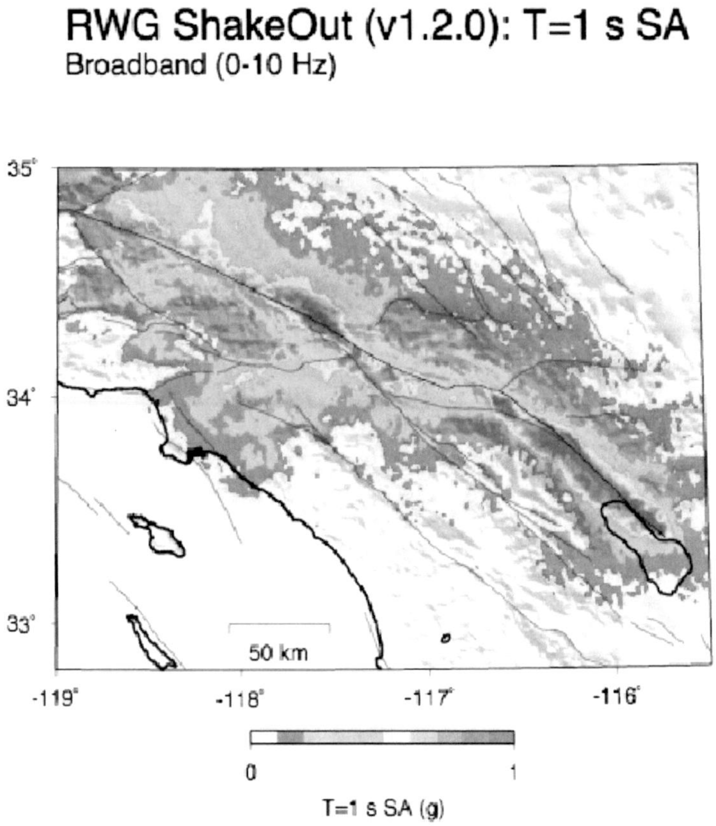

E

Figure 3-15. Engineering parameters of the ShakeOut Scenario earthquake: A) Peak Ground Acceleration (PGA). B) Peak Ground Velocity (PGV). C) Modified Mercalli Intensity. D-E) Spectral Acceleration (SA) at 0.3 and 1.0 seconds. Spectral Acceleration at 3.0 seconds is shown earlier, in figure 3-13.

C. Ground Deformation by Daniel J. Ponti, John C. Tinsley III, Jerome A. Treiman and Hope Seligson

Overview

Earthquakes of moderate to large magnitude commonly produce permanent deformation of the ground surface. The deformation may occur as displacements across planar fractures and narrow deformation zones, or as mass movement of earth materials; the sizes of these displacements range from as little as a millimeter across in the case of small cracks, to tens of kilometers in the case of rock avalanches and soil flows. Common displacement features include open cracks and fissures, various combinations of horizontal and vertical dislocations across surface fractures or zones of shearing, and buckling or heaving of the ground surface. Ground deformation features produced by an earthquake are highly localized and affect a small region when compared to the area affected by shaking. Nevertheless, even small amounts of ground displacement can be devastating to structures and buried utility systems and may produce significant casualties. Therefore, where ground deformation occurs, the impacts can significantly increase losses and damages from those produced by shaking alone.

Ground displacements from an earthquake results from two different kinds of mechanisms. **Tectonic deformation** produces direct movement along the earthquake fault, and this displacement can reach the surface as the fault rupture propagates from depth. Even where the fault rupture does not reach all the way to the surface, faulting to shallow depths can cause strain concentrations that result in fissures or buckling of the ground surface. Tectonic deformation is highly localized along the surface fault trace or along the surface projection of the fault. Fault rupture that breaks through to the surface is commonly referred to as *primary surface faulting* and is the principal type of tectonic ground deformation that the ShakeOut Scenario earthquake will produce.

The second type of earthquake-induced permanent ground deformation is **ground failure**, a secondary effect of earthquake ground motions that occurs where shaking is sufficiently strong to cause masses of earth material to move under the influence of gravitational forces as well as inertial forces from the earthquake shaking. The two principal kinds of earthquake ground failure mechanisms are landsliding and liquefaction. Landslides occur when ground motion forces are sufficiently strong to overcome the shear strength of the surficial materials. *Liquefaction* occurs where strong ground motions produce a rise in pore-water pressures that in turn causes granular material to briefly lose strength and liquefy. This can lead to settlement and a special type of earthquake-induced landslide known as a lateral spread. The likelihood that an earthquake-induced ground failure will occur at any given location depends on the intensity of ground shaking and the overall susceptibility of near-surface materials at that location.

Figure 3-16. The surface rupture of the Denali Fault in Alaska is shown about 140 km east of the epicenter of the *M*7.9 earthquake of November 3, 2002. Although the trace is not perfectly straight, note the narrow width of the rupture itself across much of the countryside; this attribute is often characteristic of strike-slip faults such as the San Andreas Fault. We anticipate that the width of the surface rupture of the southern San Andreas Fault will be narrow over much of its length, much as this photo depicts. Photograph by David P. Schwartz, U.S. Geological Survey.

The ShakeOut Scenario earthquake is sufficiently large that ground deformation can be expected to occur throughout the eight-county ShakeOut study region. Effects include significant *primary surface faulting* along the trace of the San Andreas Fault between the Salton Sea and Lake Hughes, *landslides* within the San Gabriel and San Bernardino Mountains and other areas with steep terrain, and liquefaction-induced lateral spreads and settlement in basins and river valleys where susceptible conditions prevail. In this section we present a regional assessment of the surface faulting hazard, and the probability of liquefaction and landsliding occurrence throughout the region from this event, and we discuss the likely impacts of these failures on the built environment. General discussion on the origin and types of ground deformation features, and examples of surface faulting, liquefaction and landslide features that have been observed from past earthquakes are presented in Appendix C. In addition, Chapter 4B and Appendices E-G provide more a detailed analysis of ground deformation impacts along several major lifeline corridors, highlighting one of the major consequences of ShakeOut ground deformation.

Expected Deformation Due to Primary Surface Faulting

Primary surface faulting ranks among the more visually impressive effects of many moderate and large earthquakes (figure 3-16), and like many of the physical characteristics of the Earth's surface, is largely a result of plate tectonics, where a brittle upper crust rides atop a ductile and deforming lower crust and mantle. Tectonic ground deformation has created California's principal landscape elements, including the Transverse Ranges, the Great Valley, and the Salton Trough. The centerpiece is the San Andreas Fault, an active tectonic plate boundary that extends from the Salton Sea in southern California to the Mendocino Escarpment in northern California.

Although the San Andreas Fault hasn't ruptured in southern California since 1857, careful observations and studies of other San Andreas Fault earthquakes and on similar fault ruptures elsewhere in the world have provided numerous examples of what the surface rupture of a great southern San Andreas Fault earthquake will look like. The San Andreas Fault is a right-lateral, strike-slip fault, which means that displacement across the fault surface is dominantly horizontal, with one side of the fault zone moving to the right relative to the other side. Where faulting occurs along a single strand or trace (as shown for the Denali Fault in figure 3-16), the displacement is typically concentrated in a zone that is rarely more than a few meters wide, even if the displacement across the fault is large. Where the fault-surface geometry changes at depth, or where fault ruptures traverse areas underlain by relatively thick unconsolidated deposits or encounter a contrast in materials along their trend, the pattern of surface fractures that develops may become significantly more complex. Wide zones that contain multiple fault traces or splays, uplifted areas (pressure ridges), and depressions (graben and sag ponds) are typical expressions of these types of variations and complexities. Where fault slip at the surface is low (less than 1 meter), rupture is often represented by short, discontinuous, en echelon fissures that define a zone of surface faulting a few meters wide. Such features are expected to occur near the ends of the ShakeOut rupture and on secondary strands within the fault zone that carry only a portion of the total slip.

Large surface-rupture displacements can cause considerable damage to structures that are built across the fault. However, the number of structures affected by surface faulting is extremely small compared to the number of buildings that are damaged from strong ground motion across a region. Specific effects of fault rupture to structures or buried utilities depend on both the amount of fault slip and the orientation of the fault trace relative to the manmade features. Depending on this relationship, man-made structures may be stretched or shortened by the faulting, in addition to being sheared. Examples of the variety of fault displacements that have been observed from past earthquakes and the kinds of impacts that can occur are described in more detail in Appendix C.

In addition to primary surface rupture on the causative earthquake fault, a few millimeters of surface displacement may occur on other faults within a region of strong ground shaking. Such displacements are commonly referred to as *triggered slip* and have been reported in southern California following the 1986 Palm Springs, 1992 Landers and 1999 Hector Mine earthquakes (Williams and others, 1988; Bodin and others, 1994; Rymer and others, 2002). The process controlling triggered slip is poorly understood, but it is generally thought to be a result of near-surface strain release resulting from transient stress changes produced by earthquake ground motions. Given the size of the ShakeOut earthquake and evidence for triggered slip that has occurred in other southern California earthquakes, it appears likely that triggered slip will occur on nearby faults; we do not anticipate any significant damage from triggered slip, however, and we have not evaluated this phenomenon for the ShakeOut study.

Regional Assessment of Primary Surface Rupture from the ShakeOut Earthquake

Methodology

The ShakeOut earthquake rupture model described in Chapter 3B explicitly defines rupture on the San Andreas Fault for the Scenario event. As described previously, a background (static) rupture distribution was created to constrain a more complex, kinematic rupture description overlain on the SCEC triangular-element Community Fault Model. For the purpose of estimating primary surface rupture, we used the kinematic rupture model, v. 1.2, and took the total slip on the fault computed for each of the uppermost triangular facets of the fault surface, which represents the average slip computed for the top 500 m of the fault. Each of the slip values was assigned to the location of the centroid of its corresponding fault facet, yielding a modeled slip value every 500 m along the surface of the fault rupture zone.

To predict how the modeled fault slip will impact the ground surface, it was necessary to develop a method for translating the predicted slip points to mapped traces of the San Andreas Fault where the actual ground displacements are likely to occur. The fault traces selected for the ShakeOut surface rupture were traced from best available large- to medium-scale maps, primarily the most recent version of the California Geological Survey (CGS) Quaternary fault database (Bryant, 2005), with additional adjustments based on mapping from Barrows and others (1985), Clark (1984), Matti and others (unpublished), Morton and Miller (2003), Perez and Bryant (2007), Smith (1979), Treiman (1994), Weldon (1986) and Yule and Sieh (2003). At a few locations, additional fault-trace mapping was performed using limited interpretation of aerial photo and LiDAR remote sensing.

In the simplest cases, where the San Andreas Fault is expressed at the surface as a single trace, modeled slip was assigned to 500 m-long segments of the fault by projecting slip orthogonally to the fault trace from the slip points provided by the kinematic rupture model. However, along much of its length, the San Andreas Fault is not expressed as a single fault trace, but by a zone of active traces; variation in complexity is generally due to complexities in the geometry of the fault plane at depth, as well as the nature of the rock or soil at the surface. Where multiple traces exist, expert judgment was used, incorporating both familiarity with the San Andreas Fault and observations of slip distributions in similar strike-slip earthquakes on other faults, to apportion the modeled fault slip among the mapped fault strands. Models for distributing fault slip among mapped strands of the San Andreas Fault were constructed first by CGS in four focus areas along principal lifeline corridors (Chapter 4B) and extended, where practicable, to mapped strands throughout the entire rupture zone. These assigned slip values are reasonable for defining effects specifically from the ShakeOut earthquake, but should not be treated as an actual prediction of fault slip at any particular location, nor should these values be taken as the maximum slip expected at any particular location for any other San Andreas Fault earthquake.

Even where a single fault trace is mapped, observations of surface faulting in past events have shown that not all displacement occurs across a single plane, but can be distributed across a zone from less than one to hundreds of meters wide, although the majority of slip is usually concentrated across a single fault plane surface. To account for the possibility that some fault disruption may occur away from mapped traces, and also to account for potential map inaccuracies in the locations of the fault traces, a 40-meterwide "buffer" zone, centered on the mapped fault strand, was assigned to each individual trace. For the purpose of our analyses, we assume that all of the fault displacement will occur within this 40-meter-wide zone centered on a mapped fault strand. Treating fault slip in this way has the added benefit of providing some guidance about the potential maximum length over which lifelines may be disrupted where they cross the fault, especially where lifelines cross the fault at highly oblique angles. Figure 3-17 gives an example of the association among the mapped fault strands, the modeled buffer zones, slip model points, and crossing lifelines (lifelines that cross the fault), as evaluated for this study.

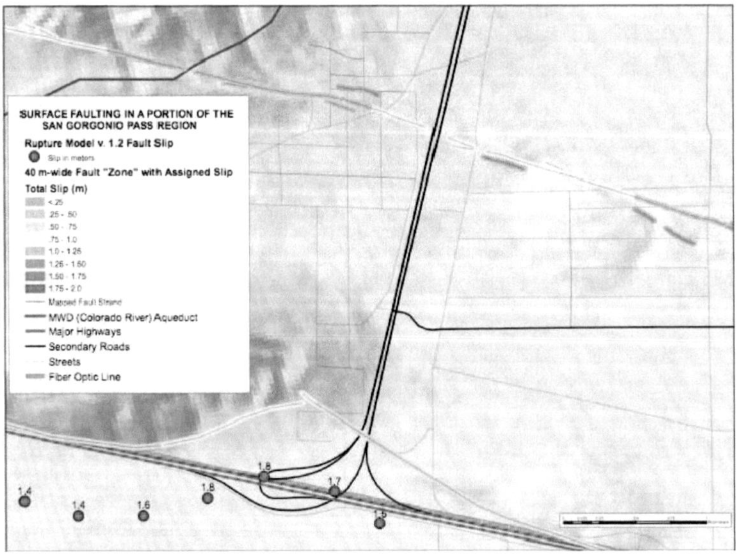

Figure 3-17. Map of a portion of the San Gorgonio Pass area, showing mapped fault strands (thin dark grey lines), the 40-m-wide modeled fault buffer zones color-coded by slip magnitude, fault slip from the v. 1.2.0 kinematic rupture model (red dots), and several crossing lifelines. Note how the Interstate-10 freeway (double red lines, bottom of figure) and the fiber optic line (blue) run nearly parallel to the southernmost fault trace. Here, fault disruption impacts these lifelines over a long distance.

Surface Rupture Impacts

Fault displacement of more than a few centimeters can have devastating effect on structures. Overall, ShakeOut fault rupture will be dominated by strike-slip, or horizontal displacement, causing structures and lifelines that straddle the fault to be sheared laterally. In some areas, especially in the San Gorgonio Pass area, small components of vertical displacement are likely to disrupt road grades, drainage systems, and any structures on the fault. Fortunately, there are few structures at risk of direct fault damage from the ShakeOut earthquake, because of the rural setting of the southern San Andreas Fault zone, and to the Alquist-Priolo Earthquake Fault Zoning Act of 1972, which prevents the construction of buildings used for human occupancy on the surface trace of active faults. Some older buildings in the town of Wrightwood (which straddles the fault), on the outskirts of Palmdale, and at a few other locations will likely suffer severe damage from surface rupture; but overall loss to structures due to surface rupture will be minimal, relative to losses from ground shaking.

Damage from ShakeOut surface rupture is most serious where lifelines (roads, railroads, and utilities) cross the fault. Many of these crossings are concentrated within a few mountain passes, and the effects of ground deformation to lifelines in these "corridors," along with the economic impacts of these disruptions, are treated in great detail later in this report. We summarize here regional statistics with respect to all lifeline fault crossings. Details of disruptions to lifelines within several key corridors are given in Chapter 4B.

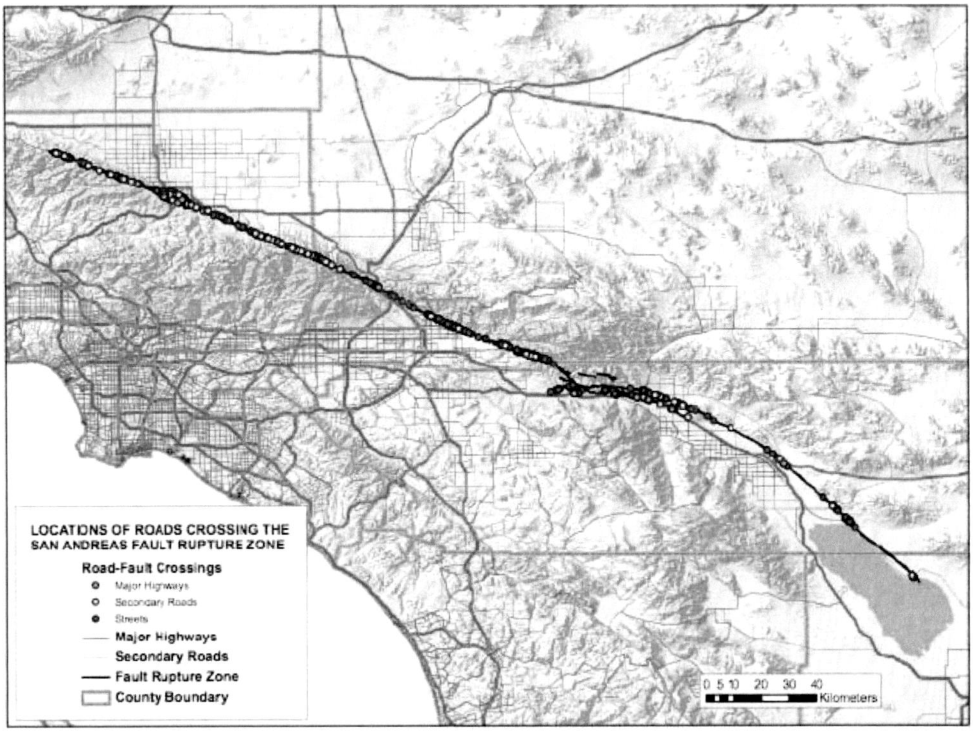

Figure 3-18. Map showing locations (dots) where roads cross the ShakeOut Scenario fault rupture. Roads cross strands within the fault zone 966 times; road displacements range from 2 cm to 8.3 meters. Road crossing details are given in Appendix D, Table 1

Roadways—Major highways, secondary roads, and surface streets intersect fault strands within the ShakeOut Scenario fault rupture zone at 966 locations (figure 3-18; Appendix D, Table 1). Road displacements range from 2 cm to 8.3 meters. Most critical of these crossings are the Antelope Valley Freeway (CA-14), which is displaced a maximum of 2.95 m, Interstate 15 at Cajon Pass (maximum displacement 2.38 m), the Interstate 10 at San Gorgonio Pass (maximum displacement 0.7 m), and Interstate 10 in the Coachella Valley (maximum displacement 6.7 m). Important secondary highways affected by surface rupture include CA-111 near Niland (3.9 m displacement), CA-62 near Desert Hot Springs (0.71 m displacement), Box Canyon Road near Desert Center (6.6 m displacement), and Big Pines Highway near Valyermo (7.6 m displacement), among others.

Railways and Transit Lines—Railroads cross the ShakeOut rupture zone 21 times, with displacements from 4 cm to 8.3 meters (figure 3-19; Appendix D, Table 2). Railroad crossings are confined to the Palmdale area, Cajon Pass, San Gorgonio Pass, and near the Salton Sea. The largest displacement occurs to the Union Pacific tracks near the Salton Sea (8.3 m). In addition to the main rail lines, the MetroLink Antelope Valley commuter line is disrupted as much as 3.12 m at the fault crossing near Palmdale.

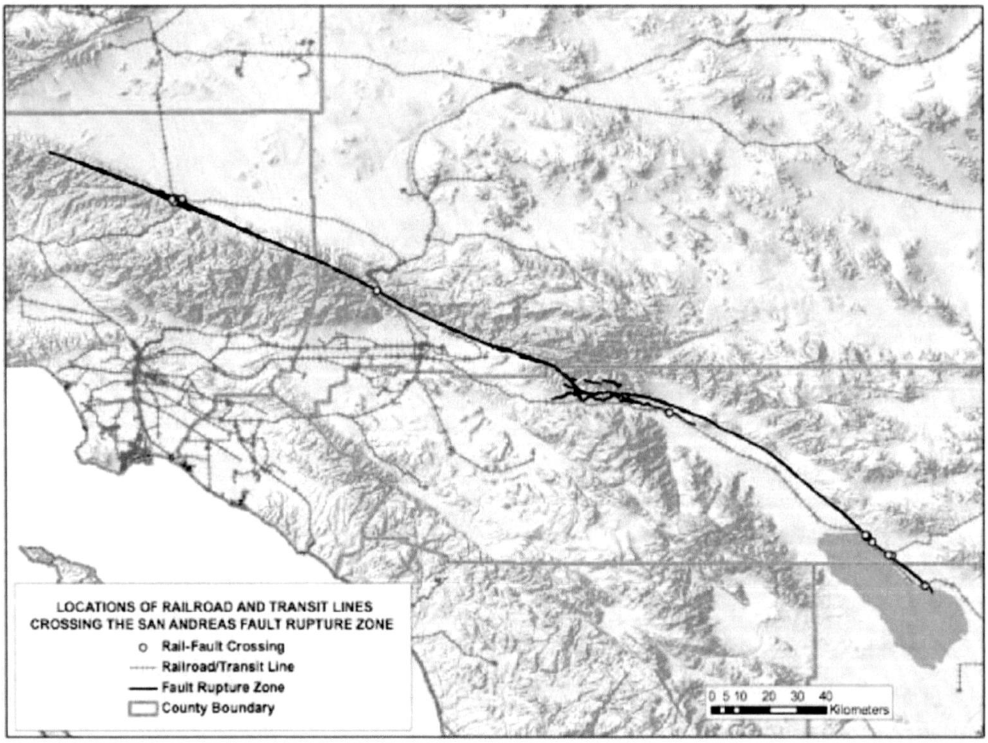

Figure 3-19. Map showing locations (dots) where railroad and transit lines cross the ShakeOut Scenario fault rupture. There are 21 rail crossings of strands within the fault zone (some dots represent multiple strands); modeled displacements range from 4 cm to 8.3 m. Details of the crossings are given in Appendix D, Table 2.

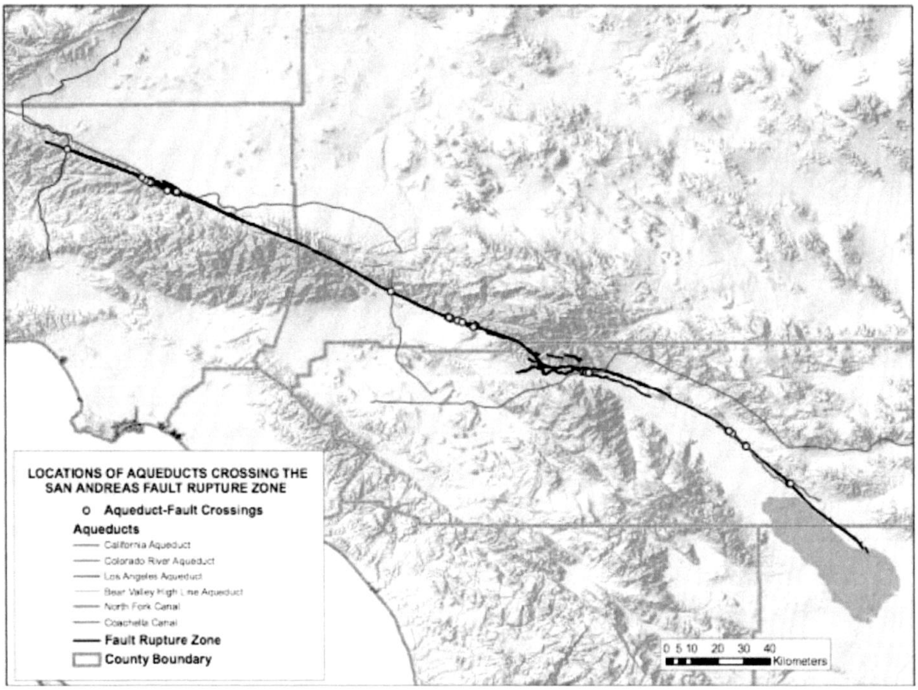

Figure 3-20. Map showing 32 locations (dots) where major water supply aqueducts (colored lines) cross rupturing strands within the ShakeOut Scenario fault rupture zone (some dots represent multiple strands). Modeled displacements range from 4 cm to 8 m. Details of the crossings are given in Appendix D, Table 3.

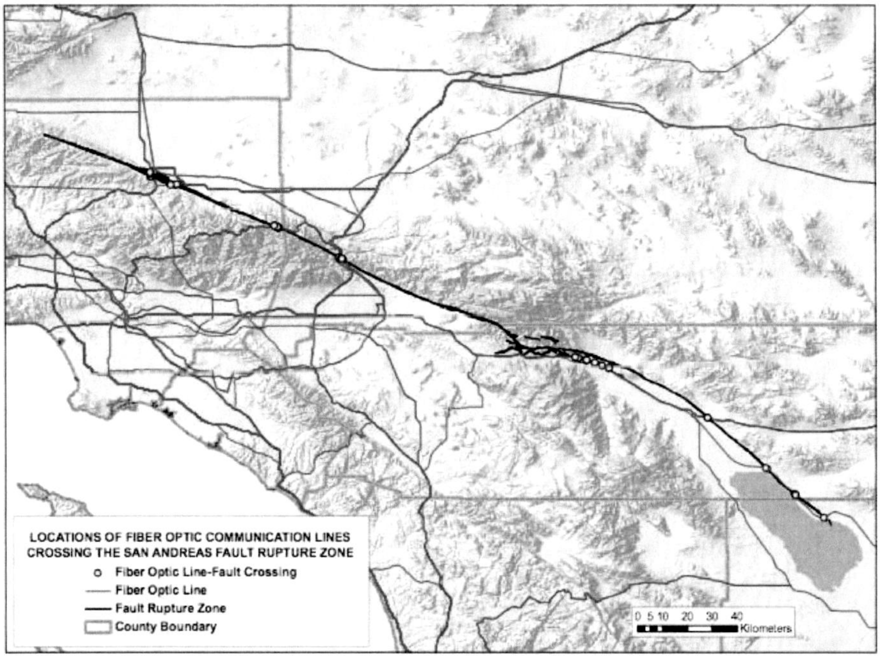

Figure 3-21. Map showing locations (dots) where major fiber optic lines cross the ShakeOut Scenario fault rupture. There are 90 fiber optic line crossings of strands within the rupture zone; modeled displacements range from 7 cm to 11.16 m. Details of the crossings are given in Appendix D, Table 4.

Aqueducts—Several major water-supply aqueducts serving urban southern California cross the ShakeOut fault rupture; these include the Los Angeles Aqueduct, the California Aqueduct, the MWD Colorado River Aqueduct, and the Coachella Canal. These aqueducts cross the ShakeOut Scenario rupture zone at 32 locations and offsets range from 4 cm to 8 m; the largest displacement occurs on the Coachella Canal near the town of Coachella (fig 3-20; Appendix D, Table 3).

Fiber Optic Communication Lines—Based on a database of fiber optic trunk lines, fiber optic lines cross the ShakeOut Scenario rupture zone 90 times, with displacements ranging from 7 cm to 11.16 meters (figure 3-21; Appendix D, Table 4). These lines are concentrated along the four principal lifeline corridors (Palmdale, Cajon Pass, San Gorgonio Pass, Coachella Valley), but a few crossings occur at Valyermo and along the eastern shore of the Salton Sea.

Oil and Gas Pipelines—Major petroleum and natural gas pipelines also cross the ShakeOut Scenario rupture zone through the major lifeline corridors at 39 locations. Displacements range from 2 cm to 8.26 meters, with the largest displacement occurring near the Salton Sea (figure 3-22; Appendix D, Table 5). In general, pipelines can best withstand fault displacements when deformation places the pipeline in tension rather than compression or shear. Based on their orientations relative to the fault zone, pipelines in the Palmdale, San Gorgonio Pass, and Coachella Valley areas would likely undergo both shearing and tension, whereas in the Cajon Pass region, most pipelines would likely experience both shearing and compression as a result of the ShakeOut earthquake.

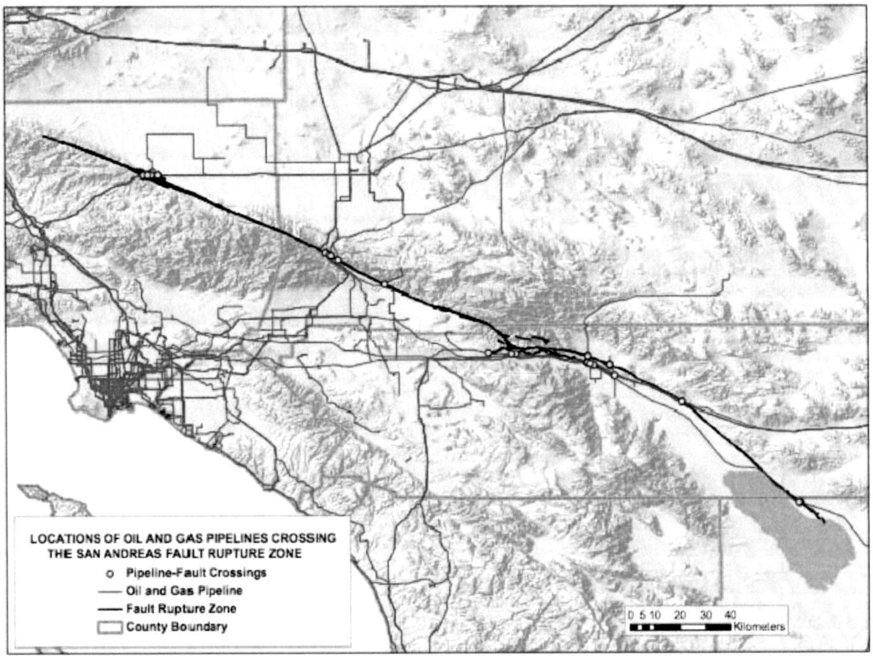

Figure 3-22. Map showing 39 locations (dots) where oil and gas pipelines cross the ShakeOut Scenario fault rupture, primarily in lifeline corridors. Modeled displacements range from 2 cm to 8.26 m. Details of the crossings are given in Appendix D, Table 5.

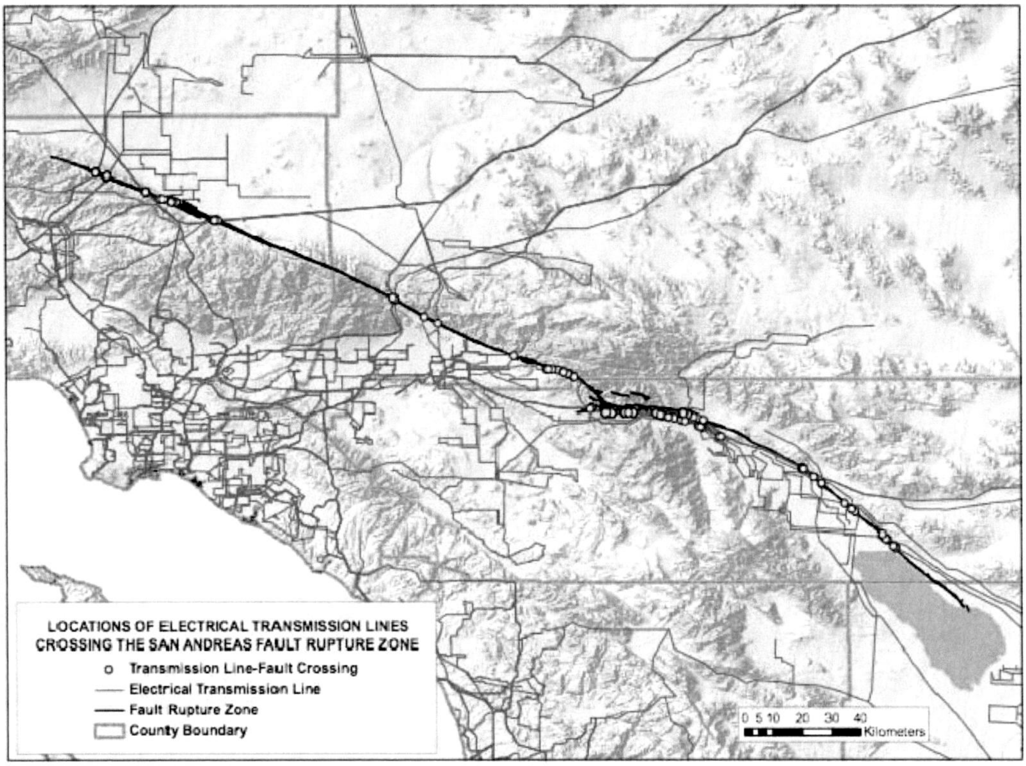

Figure 3-23. Map showing the 141 locations (dots) where electric power transmission lines cross fault strands in the ShakeOut Scenario fault rupture. Modeled displacements range from less than 2 cm to 7.2 m. Details of the crossings are given in Appendix D, Table 6.

Electric Power Transmission Lines—Overhead electric power transmission lines cross traces of the ShakeOut Scenario fault rupture at 141 locations. Many lines cross the fault within the four principal lifeline corridors, but others cross the fault elsewhere (figure 3-23; Appendix D, Table 6). Fault displacement at these crossings ranges from less than 2 centimeters to 7.2 meters. It is likely that many of these fault crossings will not result in significant damage to the transmission lines, because the cables are able to accommodate significant slip. Where lines cross the fault with a more easterly trend than that of the fault zone, power lines will tighten when placed in tension; this could damage towers and sever cables. Conversely, where lines cross the fault zone with a more northerly trend than the fault has, the distance between towers will be shortened, thus relaxing the cables. While this may require repair, it is unlikely that lines will be severed in these instances. Fault rupture will only damage towers where they straddle a rupturing fault strand; we do not have sufficient information on the locations of towers to estimate the likelihood of such damage; however, there is a far greater probability of tower damage from shaking than from fault rupture.

Expected Deformation Due to Liquefaction

During liquefaction, formerly solid ground is transformed temporarily to a softened or liquefied state that can no longer support the built environment. Effects of liquefaction commonly are observed following moderate to great earthquakes throughout the world and can produce significant damage (figure 3-24) over and beyond what might be expected from ground shaking alone. The occurrence of liquefaction during a specific earthquake is restricted chiefly to certain geologic and hydrologic settings that experience relatively high levels of ground shaking. In general, areas susceptible to liquefaction are underlain by water-saturated, cohesionless granular sediment within less than 50 feet of the ground surface. These conditions are potentially widespread in parts of the eight- county area affected by the ShakeOut Scenario earthquake, particularly the Santa Clara River /Oxnard Plain areas of Ventura County, parts of the San Fernando and San Gabriel Valleys, portions of the coastal basin or flatland areas of Los Angeles and Orange Counties, the Santa Ana River corridor, Imperial Valley, the southern Coachella Valley, and coastal areas of San Diego County.

Four types of ground failure commonly result from liquefaction. These are: 1) lateral spread, 2) ground oscillation, 3) loss of bearing strength, and 4) flow failure. Flow failures, or soil flows, are restricted to slopes of greater than 3°, whereas the other failure types typically occur on level ground or gentle slopes of less than 3°. Descriptions of these types of liquefaction failures, and examples from past earthquakes, can be found in Appendix C. Ground deformation due to lateral spreading, and settlements from lateral spreading, ground oscillation, and loss of bearing strength are all expected to occur within portions of the eight-county study region as a result of the ShakeOut Scenario event.

Figure 3-24. Photo of the Moss Landing Marine Laboratory facility, Monterey County, CA, destroyed as a result of about 1.2m of displacement due to liquefaction-induced lateral spreading caused by the *M*6.9 1989 Loma Prieta earthquake. Ground motions at this site were sufficiently low that the structure would likely have survived the earthquake with minimal damage had the lateral spreading not occurred. Photograph by John Tinsley, U. S. Geological Survey.

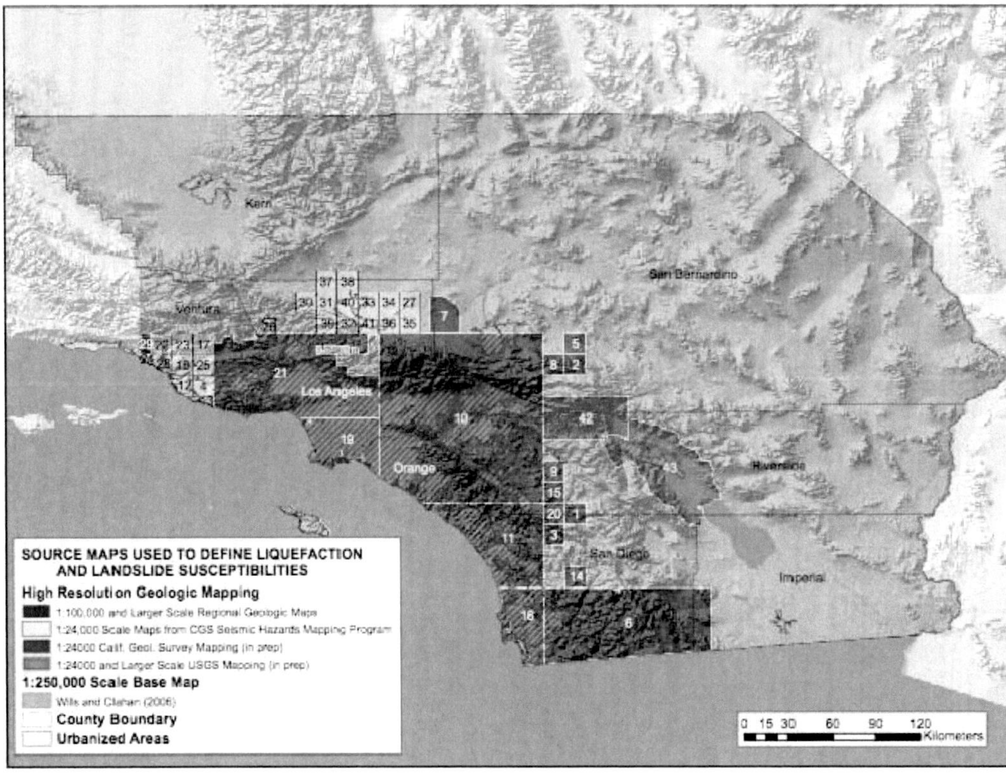

Figure 3-25. Index map of source digital geologic maps used to define liquefaction and landslide susceptibilities within the eight-county study area. Medium- and large-scale maps are used to supplant the base map derived from Wills and Clahan (2006) to provide greater detail in the urbanized areas (yellow stripes). Maps shown are: 1 - Aguanga 7.5' quadrangle (Tan and Kennedy, 2003); 2 - Big Bear City 7.5' quadrangle (Miller and Cossette, 2004); 3 - Boucher Hill 7.5' quadrangle (Kennedy, 2006); 4 - Camarillo 7.5' quadrangle (Tan and others, 2004a); 5 - Cougar Buttes 7.5' quadrangle (Powell and Matti, 2004); 6 - El Cajon 30' x 60' quadrangle (Todd and others, 2004); 7 - El Mirage Lake area (Miller and Bedford, 2000); 8 - Fawnskin 7.5' quadrangle (Miller and Matti, 1998); 9 - Hemet 7.5' quadrangle (Morton and Matti, 2005); 10 - San Bernardino and Santa Ana 30' x 60' quadrangles (Morton and Miller, 2006); 11 - Oceanside 30' x 60' quadrangle (Kennedy and Tan, 2005b); 12 - Oxnard 7.5' quadrangle (Clahan, 2003); 13 - Point Mugu 7.5' quadrangle (Tan and Clahan, 2003); 14 - Ramona 7.5' quadrangle (Todd and others, 2006); 15 - Sage 7.5' quadrangle (Morton and Kennedy, 2005); 16 - San Diego 30' x 60' quadrangle (Kennedy and Tan, 2005a); 17 - Santa Paula Peak 7.5' quadrangle (Tan and Irvine, 2005b); 18 - Saticoy 7.5' quadrangle (Tan and others, 2004c); 19 - Long Beach 30' x 60' quadrangle (Saucedo and others, 2003); 20 - Vail Lake 7.5' quadrangle (Kennedy, 2003); 21 - Los Angeles 30' x 60' quadrangle (Wills, in prep); 22 - Matilija 7.5' quadrangle (Tan and Jones, 2006); 23 - Ojai 7.5' quadrangle (Tan and Irvine, 2005a); 24 - Pitas Point 7.5' quadrangle (Tan and others, 2003a); 25 - Santa Paula 7.5' quadrangle (Tan and others, 2004b); 26 - Whitaker Peak 7.5' quadrangle (CGS, 2003g); 27 - Hi Vista 7.5' quadrangle (CGS, 2003a); 28 - Ventura 7.5' quadrangle (Tan and others, 2003b); 29 - White Ledge Peak 7.5' quadrangle (Tan and Clahan, 2004); 30 - Lake Hughes 7.5' quadrangle (CGS, 2003b); 31 - Del Sur 7.5' quadrangle (CGS, 2005b); 32 - Ritter Ridge 7.5' quadrangle (CGS, 2003e); 33 - Lancaster East 7.5' quadrangle (CGS, 2005d); 34 - Alpine Buttes 7.5' quadrangle (CGS, 2005a); 35 - Lovejoy Buttes 7.5' quadrangle (CGS, 2004); 36 - Littlerock 7.5' quadrangle (CGS, 2003c); 37 - Little Buttes 7.5' quadrangle (CGS, 2005c); 38 - Rosamond 7.5' quadrangle (CGS, 2005f); 39 - Sleepy Valley 7.5' quadrangle (CGS, 2003f); 40 - Lancaster West 7.5' quadrangle (CGS, 2005e); 41 - Palmdale 7.5' quadrangle (CGS, 2003d); 42 - San Gorgonio Pass Area (Matti, in prep); 43 – Coachella Valley photointerpretation (from 2005 NAIP Imagery, this study).

Regional Assessment of Liquefaction from the ShakeOut Earthquake

Methodology

There are a number of methods in the literature for evaluating liquefaction potential (for example, Seed and Idriss, 1971; Seed and others, 1983; Roth and Kavazanjian, 1984). More recently, the California Geological Survey (CGS) drew on these and other works to designate seismic hazard zones in accordance with the Seismic Hazards Mapping Act of 1990. These zones show where liquefaction and seismically induced landsliding are more likely to occur, and must be investigated before construction of structures. CGS also provides guidelines for evaluating and mitigating these seismic hazards (California Geological Survey, 1997). For each area designated as a seismic hazard zone, CGS assembles data on the geologic and hydrologic factors that can lead to liquefaction or seismically induced landsliding. These data have been incorporated and supplemented to estimate the hazards for the ShakeOut Scenario. Liquefaction hazard evaluation is an evolving science, and new approaches such as that devised by Holzer and others (2006) continue to advance the evaluation of liquefaction susceptibility and potential.

For the ShakeOut Scenario earthquake, we have relied on a generalized geological analysis for evaluating liquefaction hazard that was originally proposed by Youd and others (1975) and Youd and Perkins (1978), but which has been updated to include liquefaction ground failures in clay-bearing sediments (Seed and others, 1983; Tinsley and others, 1985). Conceptually, this approach requires the development of a liquefaction susceptibility map, which classifies surficial deposits based on the likelihood that they would fail via liquefaction, assuming that the susceptible materials are saturated and that ground motions are sufficient. This map is then "intersected" with maps of the depth to ground water and the ShakeOut Scenario ground motions to produce a liquefaction potential map that provides the probability of liquefaction occurrence and estimates of maximum ground displacement due to lateral spreading and settlement. This method is fairly well suited for this large regional study, because the required input parameters can be estimated from regional-scale maps. Other, more recent and sophisticated approaches generally require that more detailed geotechnical parameters and subsurface conditions be known. Given the paucity of detailed geotechnical data that are available and compiled regionwide, more detailed analyses are neither justifiable nor practical. While our regional approach can provide reasonable estimates of the likelihood of liquefaction occurrence, developing predictions of specific effects of liquefaction requires detailed, site-specific geologic and geotechnical data. Such evaluations were performed for this Scenario within the focus areas centered on the principal lifeline corridors (Chapter 4B), but not for the region as a whole.

Liquefaction Susceptibility

To construct a comprehensive, regionwide liquefaction susceptibility map, we digitally compiled published and unpublished geologic maps from 43 separate sources that represent the most recent and best available digital geologic mapping of the region (figure 3-25). In addition, we developed a liquefaction susceptibility map for the Coachella Valley area based on photo interpretation using ca. 2005 NAIP imagery (1 m pixel resolution), in order to provide more detail for this critical region. Because digital large- to medium-scale maps do not cover the entire eight-county study area, we used the 1:250,000 scale materials map from

Wills and Clahan (2006) to provide the base data for the entire region. This map is highly generalized in that the most potentially susceptible deposits are lumped into only a couple of map units. This map was therefore supplanted by larger scale and more detailed mapping in the most urbanized areas to provide better detail where liquefaction would be most likely to affect the built environment. These maps include 1:100,000-scale (generally compiled from 1:24,000-scale line work) regional geologic maps published by USGS and CGS as part of the Southern California Areal Mapping effort, as well as 1:24,000-scale maps compiled principally to support the State's Seismic Hazard Mapping Program, as well as other ongoing mapping efforts.

Table 3-5. Liquefaction Susceptibility Classification Scheme for Sedimentary Deposits. (modified from Youd and Perkins, 1978)

Depositional Environment	Modern (~<500 years)	Susceptibility Class (by age of deposit)		
		Holocene (~< 10 ka)	Pleistocene (~10 ka – 2 Ma)	Pre-Pleistocene (> 2 Ma)
River channel	Very High	High	Low	Very Low
Flood Plain	High	Moderate	Low	Very Low
Alluvial fan and plain	Moderate	Low	Low	Very Low
Marine terraces and plain	---	Low	Very Low	Very Low
Delta and fan-delta	High	Moderate	Low	Very Low
Lacustrine and playa	High	Moderate	Low	Very Low
Colluvium	High	Moderate	Low	Very Low
Talus	Low	Low	Very Low	Very Low
Dunes	High	Moderate	Low	Very Low
Loess	High	High	High	?
Glacial till	Low	Low	Very Low	Very Low
Tuff	Low	Low	Very Low	Very Low
Tephra	High	High	?	?
Residual soils	Low	Low	Very Low	Very Low
Sebka	High	Moderate	Low	Very Low
Coastal delta	Very High	High	Low	Very Low
Estuarine	High	Moderate	Low	Very Low
Hi-wave-energy beach	Moderate	Low	Very Low	Very Low
Low-wave energy beach	High	Moderate	Low	Very Low
Lagoonal	High	Moderate	Low	Very Low
Fore shore	High	Moderate	Low	Very Low
Uncompacted Fill	Very High	---	---	---
Compacted fill	Low	---	---	---

Following this compilation, the various geologic map units were assigned to one of five susceptibility classes, generally following the classification system presented by Youd and Perkins (1978). Rocks of Tertiary age and older are considered for this analysis to present no liquefaction hazard and are therefore assigned a susceptibility of "none". Younger deposits are classified based on their relative ages and inferred depositional environments as shown in Table 3-5.

Although the process of classifying geologic map units into corresponding susceptibility classes appears straightforward, in practice the translation can be ambiguous, primarily because many geologic map units do not adequately differentiate among the various depositional environments that are critical to inferring susceptibility, nor do many maps provide information on the average grain size of a deposit, which could, for example, influence the classification of river-channel or beach deposits. Many geologic maps lump all or many Holocene continental deposits into a single map unit (typically Qal), which generally includes stream channel, alluvial fan, and floodplain deposits, each of which have very different liquefaction susceptibilities. In other cases, such as in the Morton and Miller (2006) map of the San Bernardino and Santa Ana 30' x 60' quadrangles, the grain size character of facies within an alluvial fan or flood plain deposit is differentiated and therefore allows for identifying the more susceptible portions of these deposits. However, even these excellent maps still do not provide information on facies thickness, which would substantially improve the classification assignments. Where classification was ambiguous, we tended to classify conservatively. For example, the generalized Qal unit of Wills and Clahan (2006), which includes Holocene sediment deposited in multiple environments, was generally placed into the "Moderate" class, even though the unit overall likely contains significant areas of low-susceptibility alluvial fan deposits. Likewise, all modern river channel deposits, even if they may contain substantial cobbles and gravel (and are therefore likely not to liquefy), were assigned to the "Very High" class. Similarly, most maps do not differentiate between engineered and uncompacted fill (although engineered fills might be reasonably inferred for dam and highway construction); in general, deposits identified as artificial fill were assigned to the "Very High" class. In cases where mapping or independent geomorphic analysis allowed for differentiating more and less susceptible facies within a deposit of similar age and depositional environment, we would reflect this in our classifications by assigning the more susceptible facies to a higher susceptibility class than would normally be justified for an undifferentiated deposit.

Following classification, any differences that may have been present across map boundaries were reconciled using expert judgment, in order not to produce any major map boundary discontinuities. Minor shifts in the location of unit boundaries between maps were generally not reconciled, however, nor were adjustments typically made across maps where artificial boundaries occurred because of differences in the detail in the mapping (such as would occur typically at the boundaries between the Wills and Clahan (2006) map and maps published at larger scale). The resultant liquefaction susceptibility map for the eight-county study region is shown in figure 3-26. Areas of highest susceptibility occur in the Coachella and Imperial Valleys in Riverside and Imperial Counties, in the San Bernardino area, locally along the coast, and within and adjacent to modern drainages.

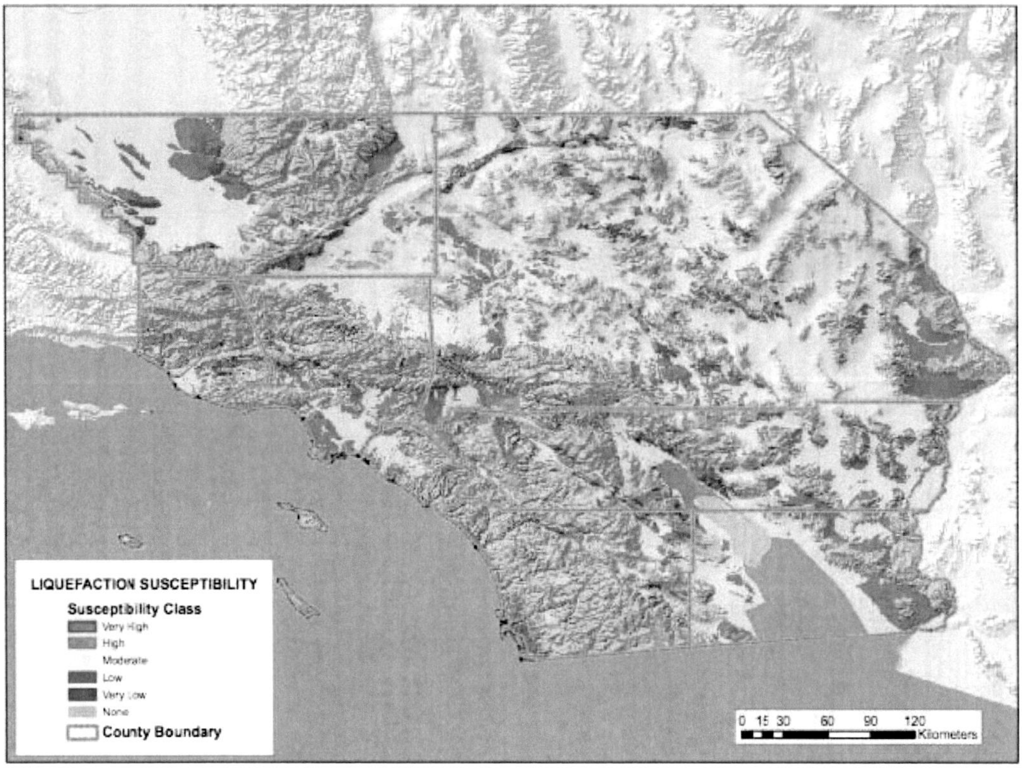

Figure 3-26. Map of liquefaction susceptibility showing relative likelihood for liquefaction failures to occur, given saturated conditions and ground motions sufficient to produce failure. Bedrock areas (grey) are assumed to pose no liquefaction hazard regardless of ground water conditions or shaking levels.

Depth to Ground Water

Various contour maps of depth to ground water have been published for portions of the eight-county study region (for example, Tinsley and others, 1985; various California Geological Survey Seismic Hazard Zone reports), but these were generally considered not to be directly applicable for use in this scenario. In addition to the fact that existing maps do not cover the entire study region, many published maps are quite old and do not reflect more recent water-use practices, which have caused water levels to rise in some regions (for example, the Los Angeles Basin) or fall dramatically in others (for example, the Antelope Valley) within the last decade. Other maps, such as those produced to support the State's Seismic Hazard Mapping Program, reflect historic high water levels, which may not be reflective of water levels that are likely to exist on November 13, 2008, the proposed date of the ShakeOut Scenario earthquake. For this study we therefore created a "Fall" water level map, which is constrained primarily by water level measurements collected within the last ten years during the months of October through December. Water level measurements were compiled from three principal sources:

- The USGS National Water Information System (NWIS). Water levels were compiled from observation wells with screens identified to be located within the upper 100 ft, or, if screen depths were not specified, from wells drilled no deeper than 100 ft.

Included in this collection are water level measurements collected as part of the Mohave River project and not yet included in the full NWIS database. These measurements are presumed to reflect the unconfined water table, although some may be from confined aquifer systems.

- The California State Water Resources Control Board Leaking Underground Fuel Tank program, which provides water level data from shallow observation wells online at: http://www.geotracker.swrcb.ca.gov/.
- Water level measurements compiled from shallow geotechnical borings and other sources by CGS in support of the Seismic Hazards Mapping Program (written communication, C. Real, 2007).

The result of this compilation effort yielded nearly 500,000 individual water level measurements for the eight-county study region. These results were then culled to eliminate measurements taken during unusual conditions (for example, while a nearby well was pumping), and the measurements were then combined at each location to derive minimum, maximum, and average water levels for individual well sites. These measurements were then grouped into three measurement periods for use in developing the ground water contours. The three datasets include (a) water levels measured during the months of October, November and December between 1997 and 2007, (b) water levels measured during the months of August, September, January and February between 1997 and 2007, and (c) water levels measured during the months of October, November and December prior to 1997. The resultant compiled dataset provides data at over 37,000 localities (figure 3-27). In general, abundant data exist within the urbanized coastal basins, but are generally lacking within the rural desert areas.

Using these data, ground water contours were constructed by hand within basin deposits (for example, regions where the liquefaction susceptibility map indicated a potential for liquefaction to occur) using as constraints the minimum water levels recorded at each site. In constructing the contours, priority was given to measurements taken in the fall (October—December) within the last 10 years. Where these data are lacking, measurements taken in the late summer and early winter months were added to the analyses, and if these data too were lacking for a region, water level measurements taken in the fall prior to 1997 were used. No water levels were inferred for bedrock areas where there is no liquefaction hazard. In general, most water level measurements are located in regions where liquefaction susceptibilities are moderate or higher. The paucity of water level measurement in regions of low or very low susceptibilities (generally Pleistocene or older deposits) in part reflect the fact that these deposits have typically undergone erosion, uplift, and diagenetic changes and generally do not contain shallow ground water. Unless data existed to indicate that shallow water levels (< 50 ft depth) were present in low-susceptibility deposits (which was the case locally within the coastal basins), these regions were assigned a default depth to ground water of 51 ft. Also, narrow drainages within mountainous regions that contain young stream channel and floodplain deposits, but where water measurements were absent, are assigned default depths to ground water of 14 ft, or depths equivalent to water levels measured in nearby drainages. The resultant contours and water level assignments were then checked against published water level maps, where applicable, to ensure that the contours were generally consistent in form with earlier compilations. The resultant map (figure 3-28) provides depth to ground water estimates for the region at 4-ft intervals for the uppermost 24 ft below ground surface, and at 10-ft intervals between 30 and 50 ft. Although there are considerable uncertainties, in large part due to the sporadic distribution of available data, these levels may be considered a reasonable, albeit somewhat conservative, estimate of depth to ground water that would be present at the time of the ShakeOut Scenario event. The map (figure 3-28) shows that relatively shallow ground water is prevalent in the vicinity of the Salton Sea, along the coast, and in parts of the coastal basins of Orange County, southern Los Angeles County, and the Oxnard Plain in Ventura County.

Input Ground Motions

The final requirement for estimating liquefaction probabilities for the ShakeOut Scenario event is an estimate of peak ground accelerations (PGA) across the study area. Geometric mean PGA's, spaced at 2-km intervals, were computed from N-S and E-W components derived from the broadband ground motion simulation (Graves and others, 2008), using the ShakeOut v. 1.2 kinematic rupture model. This model, however, does not cover the entire eight-county study region, so ground motions for the remaining area were derived using OpenSHA software, the ShakeOut Scenario earthquake parameters, and the next-generation attenuation (NGA) model of Campbell and Bozorgnia (2008). The two models were resolved onto equivalent grids with 0.02 degree cell spacing, and then stitched together to produce a single ground motion grid (figure 3-29). A similar process was utilized to derive the grids for peak ground velocity (PGV), and 0.3-second and 1.0-second spectral accelerations that are used in the HAZUS loss model described elsewhere in this report.

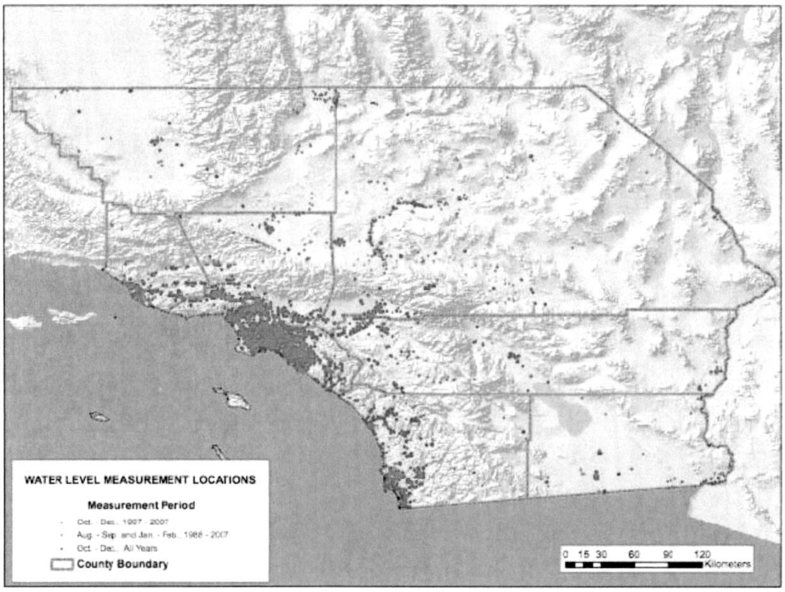

Figure 3-27. Locations of water level measurements used to construct the depth-to-ground-water map used in this study, distinguished by measurement period. More than 37,000 measurement localities were used to constrain ground water contours. In general, water level measurements are lacking in the rural areas.

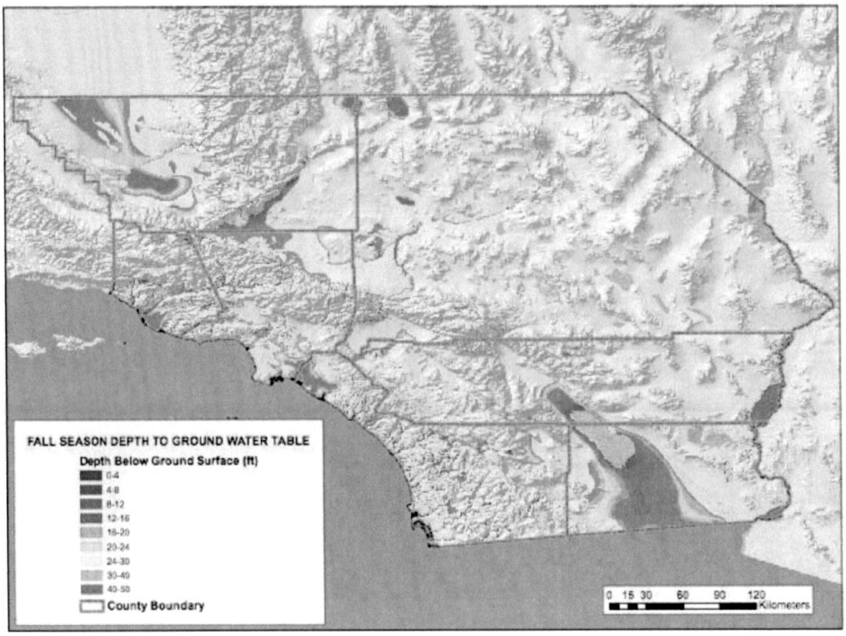

Figure 3-28. Map of depth below ground surface to the water table estimated for the fall season, utilizing water level measurements shown in figure 3-27 as constraints. Uncolored areas are regions where the water table is at depths greater than 50 ft, or in bedrock areas where there is no susceptibility to liquefaction.

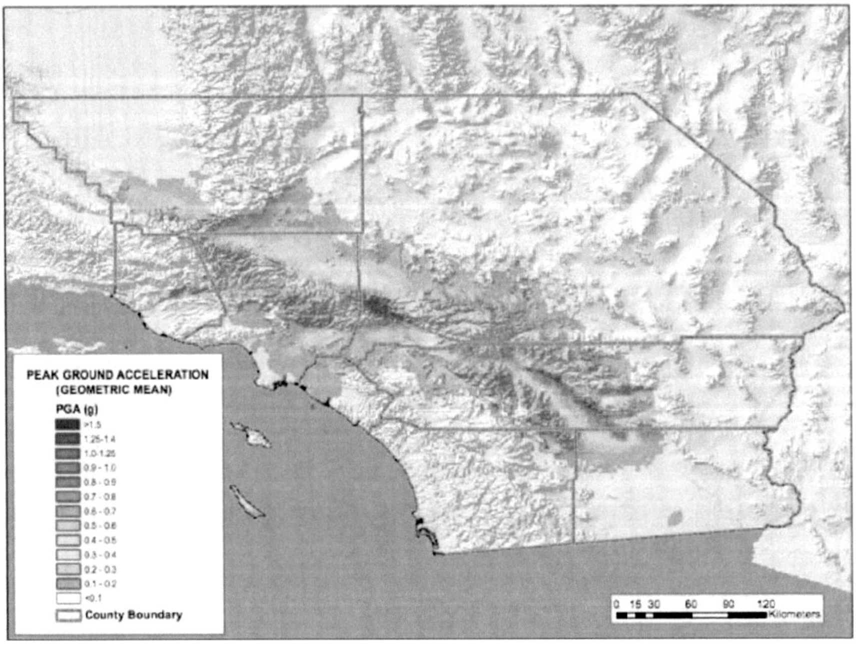

Figure 3-29. Map of geometric mean peak ground acceleration (PGA) computed for the ShakeOut Scenario event, derived as described in the text.

Regional Liquefaction Probabilities and Peak Ground Deformation

We use the procedure described in the HAZUS MR3 Technical Manual (FEMA, 2003, Chapter 4) to estimate regional liquefaction probabilities and peak ground deformation from the susceptibility, ground water, and PGA inputs described previously. The method for computing liquefaction probabilities derives from the procedures presented by Seed and Idriss (1982), Seed and others (1985), and the National Research Council (1985) that define relationships between PGA and liquefaction probability based on empirical observations and include statistical modeling of the empirical catalog by Liao and others (1988). Spatial probabilities are computed using the default HAZUS map unit proportions assigned to the various susceptibility classes, but we defined a cutoff and assigned a liquefaction probability of 0 to all areas where depths to ground water exceeded 50 ft (40 ft in low susceptibility regions). The HAZUS methodology can theoretically derive liquefaction probabilities > 0 where saturation depths are much greater than 50 ft, given sufficiently high ground motions, but in practical terms, if deep liquefaction does in fact occur, there is little evidence of it producing significant surface displacements (California Geological Survey, 1997).

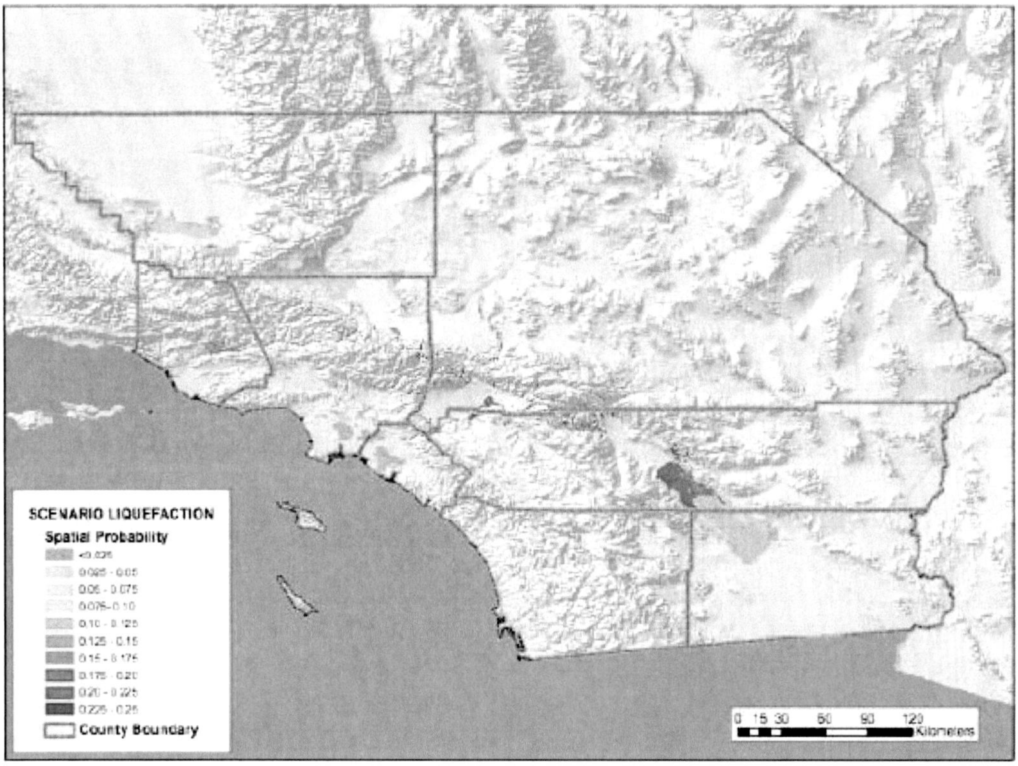

Figure 3-30. Map showing the probability of liquefaction occurrence from the Scenario earthquake, using the computational methods described in the HAZUS MR3 Technical Manual (FEMA, 2003). Highest probability of liquefaction (red) occurs in the southern Coachella Valley, along the Santa Ana River corridor west of San Bernardino, and locally along active stream channels in regions of relatively high ground motion.

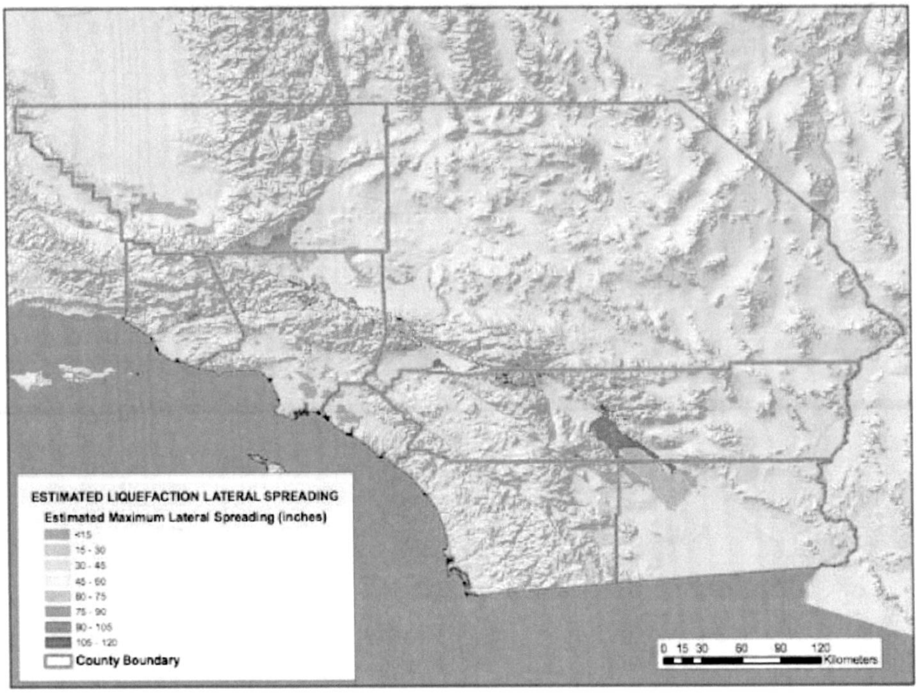

A

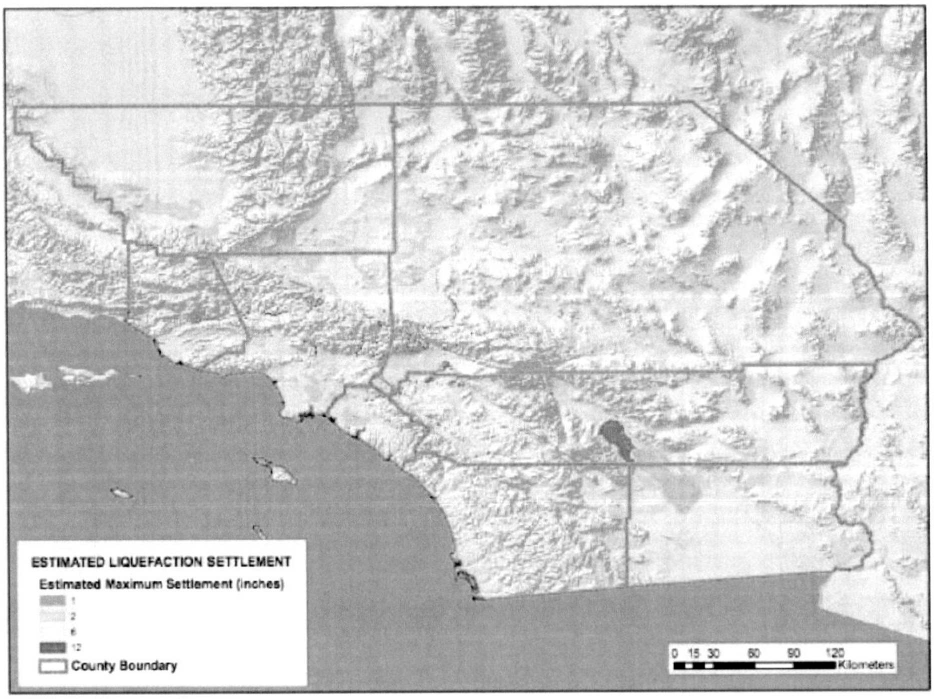

B

Figure 3-31. Maps showing estimated peak ground deformation resulting from liquefaction: A) peak ground deformation due to lateral spreading. B) peak ground deformation due to settlement. The southern Coachella Valley (north of the Salton Sea) is clearly the region where largest liquefaction-related displacements will occur during the ShakeOut earthquake.

The method HAZUS uses for estimating peak ground deformation due to lateral spreading for each location with liquefaction probability >0 was developed by combining the Liquefaction Severity Index (LSI) relationship presented by Youd and Perkins (1987) with the ground motion attenuation relationship developed by Sadigh, et. al. (1986) as tabulated in Joyner and Boore (1988). Computed settlement displacements, on the other hand, are assigned to various liquefaction susceptibility classes independent of ground shaking, following a relationship presented by Tokimatsu and Seed (1987) that indicates strong correlation between settlement due to volumetric strain and soil relative density, which is associated with susceptibility. Maps showing estimated liquefaction probability and expected peak ground displacements due to settlement and lateral spreading are presented in figure 3-30 and figure 3-31.

The results of this analysis show that in the ShakeOut Scenario earthquake, the southern Coachella Valley north of the Salton Sea is the region with the highest likelihood of experiencing liquefaction-related failures with large displacements. Other regions with relatively high-likelihood for liquefaction failure include the Santa Ana River corridor between Prado Dam and Colton, along other active river and stream channels within the Transverse Ranges, and extending along parts of the Santa Clara River in Ventura County.

As a "reality check" on these predictions, a qualitative evaluation of the liquefaction probabilities computed in this fashion, compared against mapped failures from the 1994 Northridge earthquake, was performed using Northridge ShakeMap ground motions and the identical susceptibility and water level maps used for the ShakeOut Scenario. Results of this comparison showed excellent agreement between predicted regions of high liquefaction probability and the mapped occurrences of liquefaction and soft-clay failures reported from the Northridge earthquake, although the spatial extent of observed failures fell considerably short of that predicted in the Scenario. This is probably a result of several factors: (1) The reported failures may not represent the full inventory of failures produced by the Northridge earthquake because small areas of failure and very small displacements would likely be missed by investigators or otherwise not interpreted to be due to liquefaction. (2) As discussed previously, the susceptibility classes we assigned to geologic map units were intentionally conservative. 3) The default probability factors we used to quantify the proportion of a geologic map unit deemed susceptible to liquefaction within a particular susceptibility class may be too high. The provenance and tectonic setting that dominates much of the southern California region is such that deposits would possibly be less susceptible to liquefaction than equivalent types of sediment deposited elsewhere in less tectonically active regions. We therefore suggest that the spatial liquefaction probabilities presented for the ShakeOut Scenario may be high by a factor of 2 or more, and that areas with liquefaction probabilities of 5% or less would very likely not experience significant liquefaction failure from the Scenario event.

Expected Deformation Due to Landsliding

Earthquakes greater than *M*4 commonly trigger landslides in susceptible materials. Earthquake ground motions produce landslides on hill slopes when the inertial forces produced by shaking overcome the static forces that hold a mass of earth material in place on the hillside. Aside from the steepness of the slope and the intensity of ground shaking, a primary control on landslide occurrence is the strength of the earth material, which can be reduced by the presence of ground water. Earth material strength is dependent on the type of rock or soil, the degree of cementing, weathering, and fracturing, whether the material has been weakened by previous landsliding, and the orientation of any bedding surfaces.

The number of landslides triggered and the geographic area affected by landslides generally scale with earthquake magnitude. Keefer (1984) has observed that earthquakes of comparable magnitude to the ShakeOut Scenario event have produced between 10,000 and 100,000 individual landslides and that seismically induced landslides can be grouped into three major categories: Category 1—disrupted slides and falls (80% relative abundance, figure 3-32), Category 2—coherent slides (12% relative abundance); and Category 3—lateral spreads and flows (8% of total abundance). Additional information on the characteristics of expected ShakeOut Scenario landslides and examples from past earthquakes are given in Appendix C. Overall, we anticipate that the numbers and types of landslide features that will be produced by the Scenario earthquake will fall within the parameters described by Keefer (1984). That is, the ShakeOut Scenario event will likely produce between 10,000 and 100,000 landslides, and the vast majority of them will be Category 1 slides.

Regional Assessment of Landsliding from the ShakeOut Earthquake

Methodology

The amount of ground shaking needed to initiate downslope movement of a slide mass (called the critical acceleration) is dependent upon the strength of the surficial geologic materials (which includes the occurrence and orientation of joints and fractures), the steepness of the slope, the ground water conditions, and the type of landslide generated. Critical inputs into a regional landslide hazard analysis, therefore, are a reliable landslide susceptibility map that classifies the relative strength of a given geologic unit, and a slope map. Using these data, relationships can be applied, like those of Wilson and Keefer (1985), to estimate, for any given susceptibility class and slope angle, the critical acceleration necessary to trigger failure. This information can then be compared against an earthquake's expected ground motion at a site to determine whether failure is likely. This is a generalized approach that provides estimates of the relative likelihood of landslide occurrence at a regional scale, but does not predict specific landslide behavior. More rigorous approaches require site-specific evaluations that are not practical to perform at the regional level; however, such analyses have been performed locally within the focus areas surrounding the major lifeline corridors so that impacts to lifelines from landsliding within these critical localities can be better evaluated (see Chapter 4B).

Landslide Susceptibility

For the regional analyses, we have chosen to categorize the mapped geologic units into landslide susceptibility classes according to the approach presented by Wilson and Keefer (1985). In this approach, 10 landslide susceptibility classes are defined based on various combinations of three geologic groupings and slope. The geologic group classifications are derived from the same base geologic mapping that was compiled to produce the liquefaction susceptibility map (figure 3-25), plus an additional landslide inventory map (CGS, unpublished) that provides more comprehensive data on landslides for parts of eastern Ventura County and western Los Angeles County. Geologic map units were classified into one of three groups as follows:

- Group A: Strongly Cemented Rocks (crystalline rocks and well-cemented sandstones);
- Group B: Weakly Cemented Rocks and Soils (sandy soils and poorly cemented sandstone);
- Group C: Argillaceous Rocks (shales, clayey soil, existing landslides, poorly compacted fills).

We encountered many of the same issues and ambiguities when assigning landslide susceptibility classifications to the geologic map units as we did when assigning liquefaction susceptibility classes. As with the liquefaction effort, where such ambiguities existed, we chose to err by being conservative and generally assigned units of mixed lithologies to the more susceptible geologic group included within that unit.

Slope information used for the analysis came from a slope angle map that was constructed for this study by the USGS EROS Data Center, using existing 10-m horizontal-resolution digital elevation models and processing techniques developed for the Elevation Derivatives for National Applications (EDNA) database. We then grouped the elevation grid into eight slope classifications (figure 3-33). Slope data were then intersected with the geologic groupings to generate the Wilson and Keefer (1985) susceptibility units shown in Table 3-6.

Figure 3-32. Photo of a Category 1 rock fall along the Scotia Bluffs adjacent to the Eel River near Fortuna, California. The triggering earthquake was the Fortuna, California, 7 June 1975, Ml 5.2 event. Note the disaggregated nature of the failed material, the steep slope, and the orange tractor (at red arrow) for scale. (Photograph by David Keefer, USGS, 1975.)

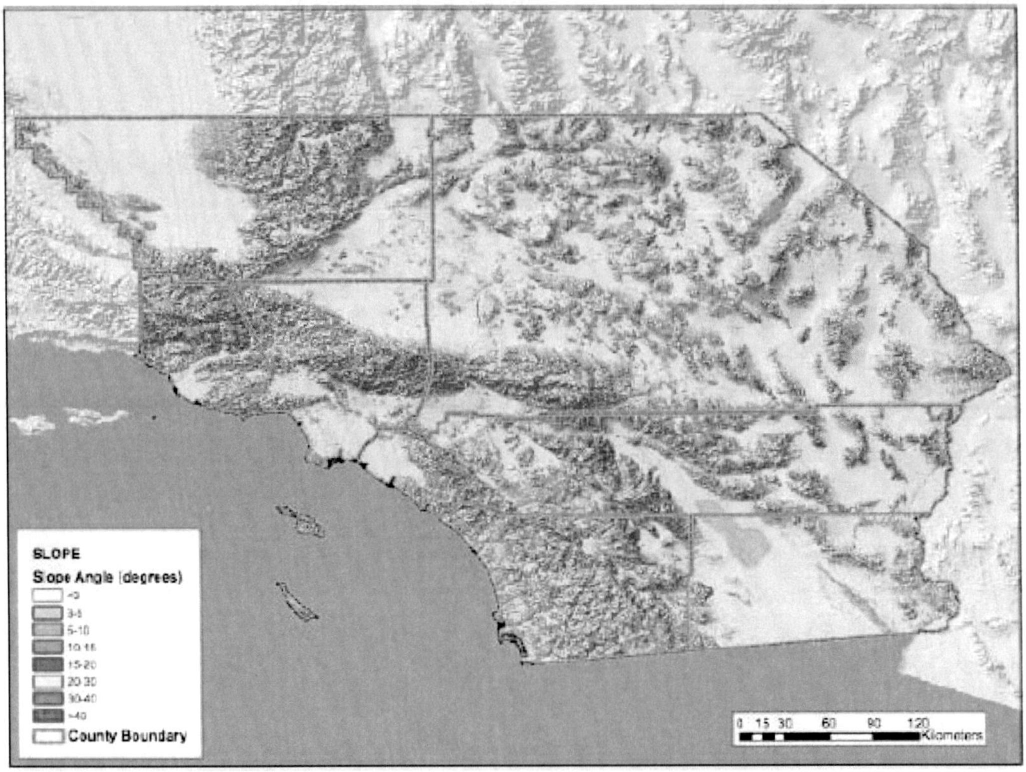

Figure 3-33. Map showing slope angles for the eight-county ShakeOut study area, derived from USGS 10 m-resolution digital elevation models and processed according to EDNA standards. Colors show slope angle in degrees classified into the 8 slope categories used for computing landslide susceptibility classes.

Table 3-6. Landslide Susceptibility of Geologic Groups According to Slope and Ground Water Conditions. (Modified from Wilson and Keefer (1984) and Keefer, personal Communication, 2007)

Geologic Group	Slope Angle (degrees) Dry Conditions (ground water below level of sliding)						
	0-5	5-10	10-15	15-20	20-30	30-40	>40
Group A	None	None	None	I	II	IV	VI
Group B	None	None	III	IV	V	VI	VII
Group C	None	V	IV	VII	IX	IX	IX
	Slope Angle (degrees) Wet Conditions (ground water at ground surface)						
	0-3	3-10	10-15	15-20	20-30	30-40	>40
Group A	None	None	III	VI	VII	VIII	VIII
Group B	None	V	VIII	IX	IX	IX	IX
Group C	None	VII	IX	X	X	X	X

From Table 3-6, it is apparent that the landslide susceptibility classification is highly dependent on local ground water conditions. For the purposes of the ShakeOut Scenario, we chose the Dry Conditions classification given that the ShakeOut event occurs in November, at the end of the summer and fall dry season.

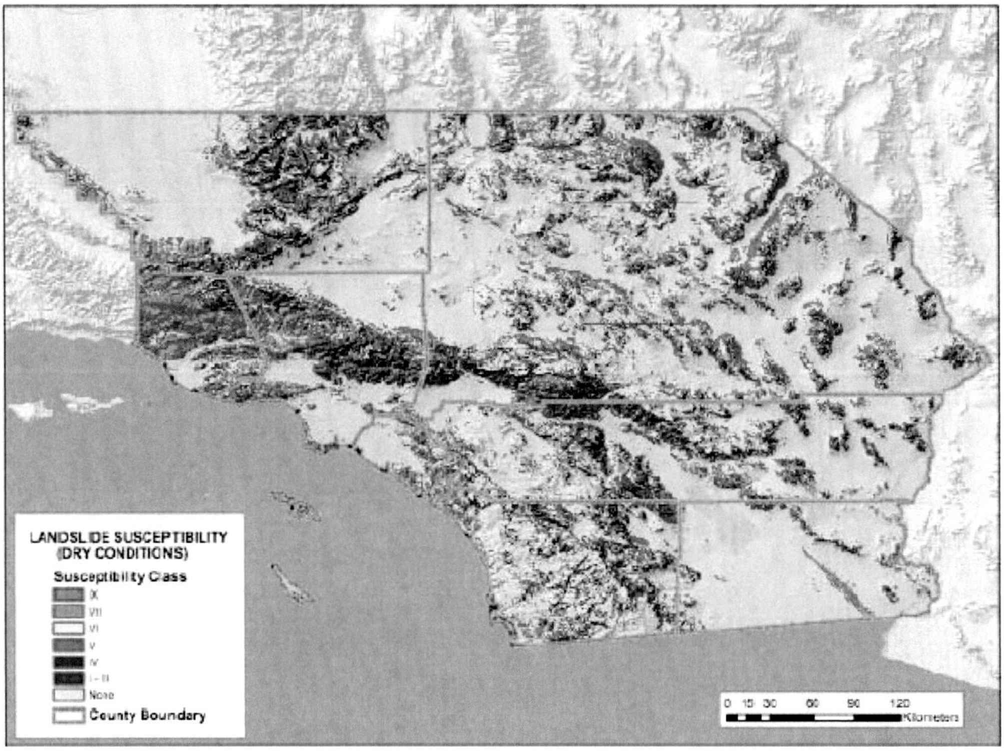

Figure 3-34. Landslide susceptibility map, derived using the procedure defined by Wilson and Keefer (1985), for the eight-county southern California region. Highest landslide susceptibilities exist within the Transverse Ranges of Ventura County and westernmost Los Angeles County, with smaller areas of high susceptibility in the eastern San Gabriel Mountains, Puente Hills, and northern Santa Monica Mountains.

The resultant landslide susceptibility map (figure 3-34) reveals that the highest landslide susceptibilities exist within the Transverse Ranges of Ventura County and westernmost Los Angeles County, with smaller areas of high susceptibility in the eastern San Gabriel Mountains, Puente Hills, and northern Santa Monica Mountains; smaller pockets of high susceptibility areas also occur elsewhere around the region.

Regional Landslide Probabilities and Peak Ground Deformation

As with liquefaction, we use the procedure described in the HAZUS MR3 Technical Manual (FEMA, 2003, Chapter 4) to estimate the probability of landslide occurrence and peak ground deformation for the Scenario event (figs. 3-35 and 3-36). This approach uses the static landslide susceptibility map (figure 3-34) and PGA inputs described previously (figure 3-29), along with relationships defined by Wilson and Keefer (1985), to relate the critical acceleration needed for failure to susceptibility class and slope angle. Because the derived critical accelerations generally apply only to the most susceptible portions of a geologic group, correction factors suggested from mapping by Wieczorek and others (1985) are applied to derive the proportion of a susceptibility unit that is likely to fail when the critical acceleration has been exceeded at a site. Maximum slope displacements are computed using the results of Makdisi and Seed (1978), who showed that displacement increases as the ratio of the induced peak ground acceleration within a slide mass to the critical acceleration increases; because of the nature of the analysis, estimated displacements are really only applicable to coherent slides (soil slumps or block glides), as opposed to rock slides or falls that break apart once movement is initiated.

Our regional analysis of landslide probability from the Scenario event suggests that the region of most concern with respect to landslide impacts is in the easternmost San Gabriel Mountains near Cajon Pass, where both estimated failure probabilities and peak displacements are high. Several other areas close to the fault rupture zone have similarly high hazard potential. As with the liquefaction results, we compared the resultant probability maps against more detailed local-scale predictions of landslide occurrence that were performed within the lifeline corridor focus areas (see Chapter 4B and Appendix E-G, and also compared a similar model constructed to compare against reported landslides from the Northridge earthquake. Again, qualitative agreement was generally quite good between the regional predictions and reported occurrences of landslides (for the Northridge comparison), and to predicted landslide occurrence in the lifeline focus areas. As with the liquefaction results, the regional analysis appears to over- predict the spatial probability of landslides when compared against reported observations from Northridge by perhaps a factor of 2 or more, with the overestimate even greater in low probability (<10%) zones. Estimated landslide displacements computed for the regional analysis also were consistently higher by a factor of 2 as compared to displacements derived for landslides evaluated in the focus areas.

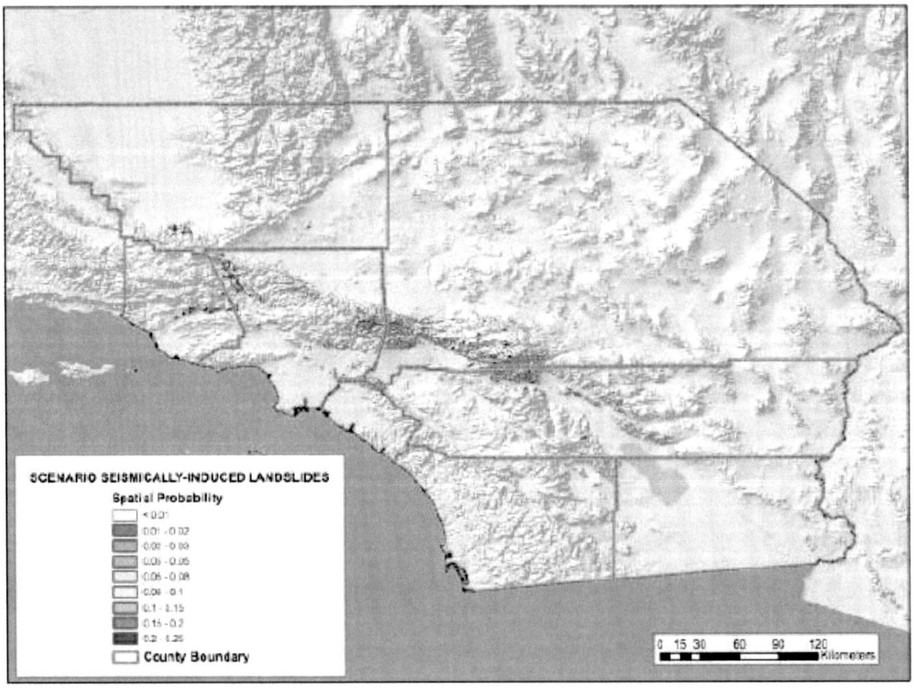

Figure 3-35. Map showing probability of landslide occurrence for the ShakeOut Scenario earthquake. Largest regions of high landslide probability are inferred for the easternmost San Gabriel Mountains, the western San Gabriel Mountains near the Interstate-5 corridor, and on steep slopes adjacent to the Santa Clara River in Ventura County.

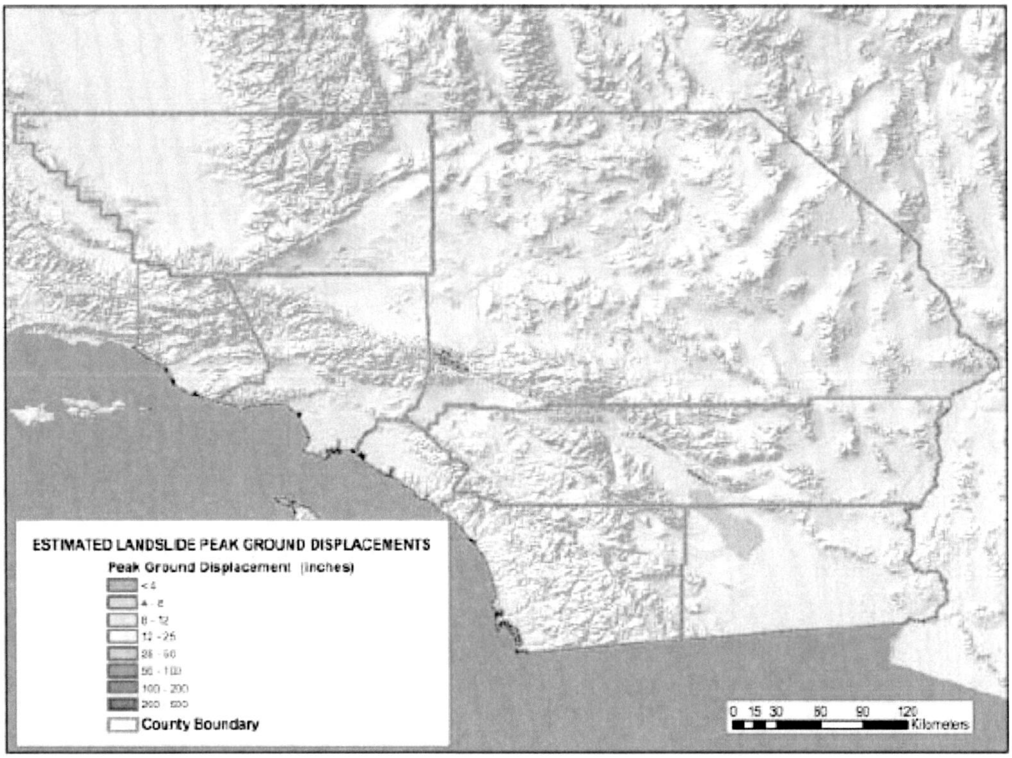

Figure 3-36. Map of estimated peak ground displacements inferred for landslides produced by the ShakeOut Scenario earthquake. Highest displacements are constrained to locations close to the San Andreas Fault rupture zone, especially in the vicinity of Cajon Pass.

One area of disagreement between the regional analysis and those performed in the focus areas—although consistent with the tendency for the regional analysis to over- predict failure—is in the Interstate 5/Pyramid Lake area, where local-scale evaluations of several large coherent landslides indicate that ShakeOut Scenario ground motions at those sites were too low to trigger failure. In contrast, many nearby slopes in the vicinity were identified in the regional analysis as having a high probability for landslide failure, albeit predicted displacements are quite small (a few inches or less). We interpret this to suggest that while the large landslide masses near Interstate 5 will not fail in the Scenario, there is a possibility for Catcgory 1 failures (for example, rock and soil falls) to occur in this region, and the ShakeOut Scenario posits rockfalls that close Interstate 5 for one day.

Evaluating Loss Estimates from Permanent Ground Deformation

A goal of the overall ShakeOut Scenario effort is to provide as reasoned and detailed a description of a large earthquake on the southern San Andreas Fault as possible, not only to aid emergency planners, but to provide the best inputs for estimating the direct and indirect economic losses that would result from an event of this magnitude. The detailed fault displacement estimates provided here, and the ground failure probability maps we have developed ultimately contribute to estimates of direct losses from the earthquake. The HAZUS earthquake model is being used for the ShakeOut Scenario to estimate losses from ground shaking for most building stock, and the ground failure maps have been constructed to feed into the HAZUS application. However, use of HAZUS for estimating building loss due to ground failure has not been thoroughly tested, primarily because ground failure susceptibility or probability maps are rarely available for input into the HAZUS loss model. We have found that translating the probability maps into losses using the HAZUS application has yielded some unexpected results, which we discuss below.

Because the southern San Andreas Fault trace is in a rural setting, and also because of legislation that controls development in the fault zone, we do not anticipate any significant loss to structures directly as a result of fault offset. We do, however, anticipate significant impact to lifeline function as a result of surface faulting; those losses are handled outside of the HAZUS loss model and are discussed elsewhere in this report. Below we discuss HAZUS results for ground failure due to liquefaction and landsliding.

Methodology

With regard to losses due to ground failure, we have shown that liquefaction and landslide effects are generally localized and dependent on specific geologic conditions. One of the challenges in working with the HAZUS earthquake model is that it computes regional building damage at the census-tract level and expects input values that are relevant to the tract as a whole. The default manner in which HAZUS obtains this information for a tract is to select the information from input maps at the location of the tract's centroid. In regions where census tracts are small in area, such as in urbanized regions, and where the geology, slope, ground water, and ground motions do not vary substantially, this approach works well. However, census tracts are fairly large close to the southern San Andreas Fault, and in these regions both the geology and ground motions from the ShakeOut Scenario vary over short distances; furthermore, building exposure is typically not evenly distributed within a tract, especially in the more rural parts of southern California. Therefore, if HAZUS is allowed to select input hazard probabilities at tract centroids from existing detailed probability maps, the values obtained may not be representative of the tract as a whole. Where ground failure effects are highly localized, tracts with significant ground failure may be missed or a large hazard value may be selected that is only representative of a small portion of the tract. To provide HAZUS with ground failure probabilities and ground displacements that should be more representative of average conditions for the tract, we have devised an exposure-weighting approach for computing tract level ground failure parameters as follows:

- The replacement value for all building occupancy classes and contents within a census block are added together and divided by the sum total of building and content value for the entire tract, to produce an "exposure weighting" for each block.
- For each census block, ground failure probabilities and peak displacements are computed as an area-weighted average within the block, to account for cases where a single census block may contain more than one value of ground failure hazard. In the case of peak displacements (PGD), the area-weighted average only includes portions of the block that contain non-zero displacements.
- The area-weighted values for each block are then multiplied by the block's exposure weighting, and then these values are summed across all of the blocks within the census tract to obtain the final result for the tract.

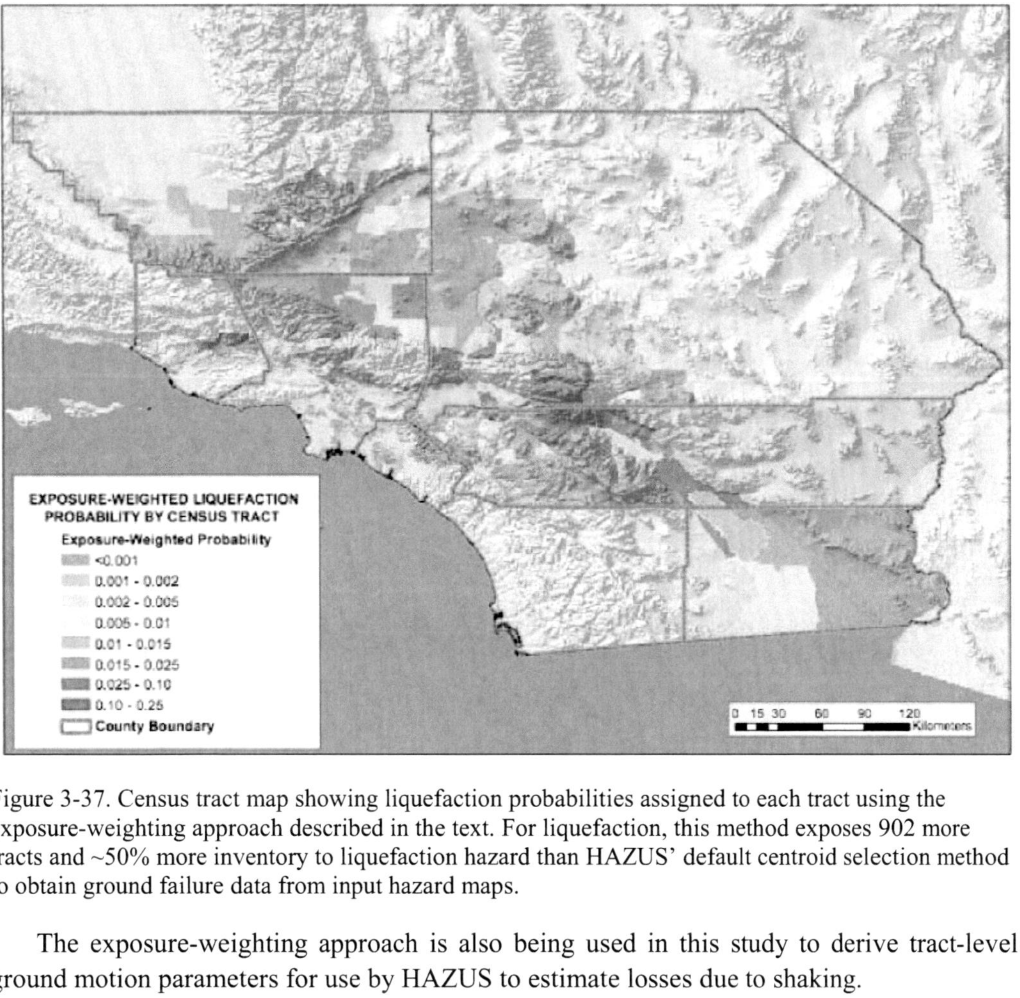

Figure 3-37. Census tract map showing liquefaction probabilities assigned to each tract using the exposure-weighting approach described in the text. For liquefaction, this method exposes 902 more tracts and ~50% more inventory to liquefaction hazard than HAZUS' default centroid selection method to obtain ground failure data from input hazard maps.

The exposure-weighting approach is also being used in this study to derive tract-level ground motion parameters for use by HAZUS to estimate losses due to shaking.

The resultant ground failure maps for input to HAZUS are therefore census tract maps, with the exposure-weighted probabilities and displacements assigned to each tract (figures 3-37 and 3-38). To compare the two approaches: With exposure weighting, 1,406 tracts (out of 4,147 total) are exposed to liquefaction hazard, with an average probability of 0.015; total inventory exposed to liquefaction amounts to approximately $11.9 billion. Using the default HAZUS centroid approach, only 504 tracts are exposed to liquefaction, but with an average probability of 0.04, and a total exposed value of just under $8 billion. Therefore, using exposure weighting, 50% more inventory is exposed to liquefaction hazard—a result of the fact that exposure weighting picks up localized hazard zones that would otherwise be missed with the centroid selection approach

Inferred Losses due to Ground Failure

HAZUS assumes that any ground failure impact to a structure causes either extensive or complete damage and therefore acts to alter the damage state distribution within a tract relative to what results from shaking alone. The amount of inventory exposed to ground failure is simply the product of the total inventory times the failure probability, and the likelihood for this exposed inventory to move into an extensive or complete damage state is controlled by a fragility function (FEMA, 2003, p. 5-63) that relates displacement to the probability of reaching an extensive or complete damage state.

Adding liquefaction and landslide hazard into the HAZUS analysis for the ShakeOut Scenario results in an increase in total direct economic losses of 26.8% over the losses obtained from shaking alone. This increase is apparently dominated by damage due to liquefaction; landslide damages alone result in only a 3.6% increase in losses over shaking. The minimal impact of landsliding relative to liquefaction appears due to the fact that most damaging landslides are confined to mountainous regions that contain few buildings.

We believe however, that the computed losses due to ground failure (amounting to ~$ 1 3B over shaking alone) are unreasonably conservative. Factors that could contribute to this apparently errant estimate are:

- As discussed previously, comparison of failure probabilities to actual failures (in the case of the Northridge earthquake) suggests that our probability maps may be overestimating the extent of failures by a factor of 2. Although this would certainly have some impact on the overall losses, whether this issue is a major contributor is unclear, since damage is mostly controlled by peak displacement, not just the likelihood of failure. Based on limited comparison, it appears that our displacement estimates, especially for liquefaction, are in agreement with other approaches.

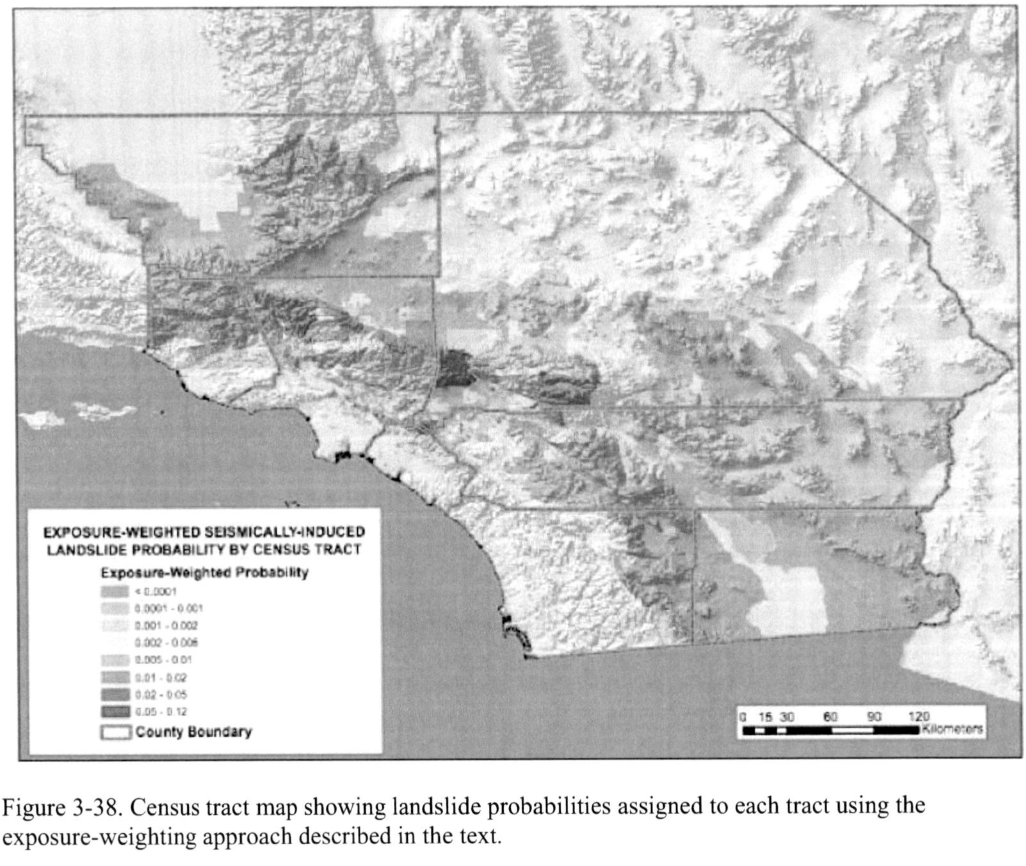

Figure 3-38. Census tract map showing landslide probabilities assigned to each tract using the exposure-weighting approach described in the text.

- The HAZUS loss methodology presumes that all inventory exposed to ground failure is subject to the maximum estimated displacement. In our view, this is an overly conservative assumption. Unless failure occurs as a soil flow or disrupted slide, structures sited on coherent failures are not likely to experience differential displacements equal to the maximum predicted unless they straddle a block boundary. In other words, for many structures subject to ground failure, the damage they incur will likely be less than the HAZUS model predicts because the earth beneath these structures essentially moves as a single mass, thus producing significantly less differential displacement beneath the structure than the maximum predicted.
- The exposure-weighting method we use to derive tract-level probabilities may be contributing to this overestimate in the losses by picking up small hazard zones that are not likely to impact the built environment, but do get factored into the tract losses. Many zones of high liquefaction or landslide potential occur in limited areas within modern river channels or associated with steep slopes. Most of these areas are not developed, and therefore failures at these locations will likely have little impact on the built environment. The exposure-weighting approach used here to compute tract- level failure probabilities would likely pick up these zones (if there is exposure elsewhere in the census block), whereas the default HAZUS centroid selection method might not.
- We examined the distribution of damage states for a few selected tracts relative to the damage state distribution for the same tracts exposed to shaking only, and it appears that nearly all affected inventory is completely destroyed once subjected to ground failure, independent of the inferred PGD. Whether this is an intended effect of HAZUS' loss estimator is unclear; this aspect of the HAZUS loss estimation approach is currently under review by the HAZUS developers. Clearly, however, this effect can easily lead to overly conservative loss estimates.

Until these issues are examined in closer detail and resolved, we do not believe that we can reliably report HAZUS-derived economic losses due to ground failures for the ShakeOut Scenario. However, we can use the loss estimates to help identify, in a relative sense, where the largest ground failure impacts are likely to occur. A map of relative liquefaction losses by census tract (figure 3-39) shows that the Coachella Valley region near the town of Indio will likely be hard-hit as a result of liquefaction-related ground failure; both expected failure probabilities and displacements are inferred to be high in this region. Much of this area is agricultural (figure 3-40), but a small part of Indio may be affected. Beyond structure loss in this region, liquefaction related lateral-spreads and settlement would likely severely impact field drains and irrigation systems.

A small area near Vernon in the Los Angeles Basin is also inferred to suffer significant liquefaction losses, but this is probably due more to high exposure value in this tract rather than a significant liquefaction risk. Our model computes only a 2% probability for liquefaction occurrence, and inferred displacements are small. Given that our probability estimates may be high, we question the notion that this area will suffer significant liquefaction losses.

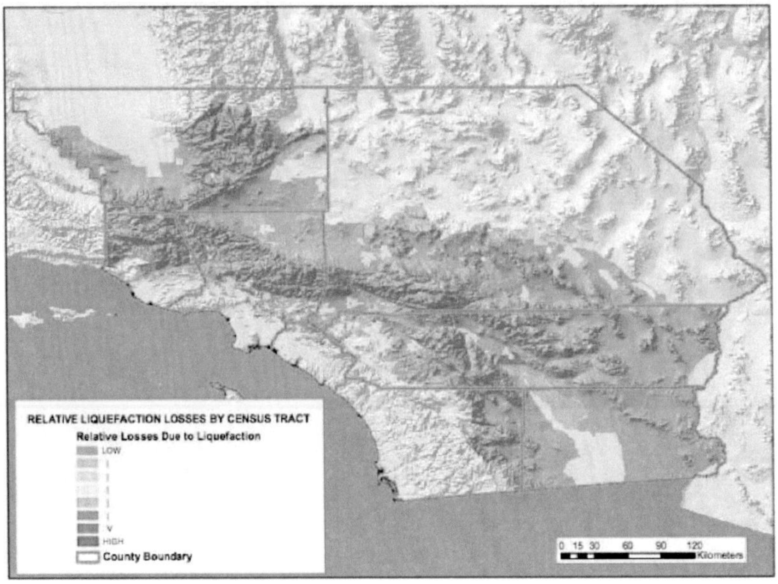

Figure 3-39. Map showing relative losses due to liquefaction (by census tract). Significant losses from liquefaction may occur in the Coachella Valley area near Indio, where both inferred probabilities and inferred displacements are high. A census tract near Vernon in the northern Los Angeles Basin also indicates high losses from liquefaction. However, the probability for liquefaction occurrence there is only 2% and displacements are small. Given that our computed probabilities may be overestimates, we feel there is considerable uncertainty as to whether Vernon would indeed suffer significant economic loss due to liquefaction from the ShakeOut event.

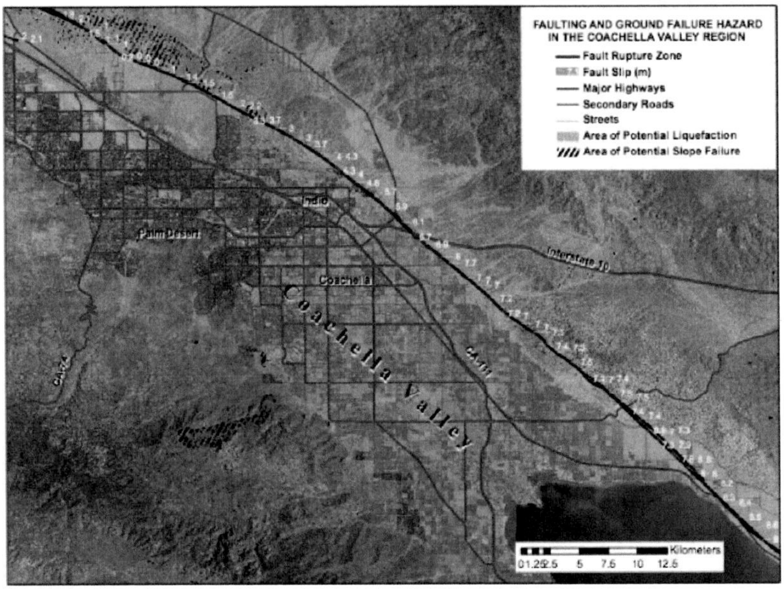

Figure 3-40. Region of high liquefaction potential overlain onto ca. 2005 NAIP imagery of the Coachella Valley area near Indio. Most of the region within the zone of high liquefaction potential is agricultural, and lateral spreads and settlement could cause extensive damage to field drains and irrigation systems. Black line is the San Andreas Fault rupture, with inferred surface slip (in meters) in yellow.

Significant economic losses to structures due to landslides appear most likely to occur in the eastern San Gabriel Mountains between Wrightwood and Cajon Pass, in the Chino Hills, and in the San Jose Hills near San Dimas (figure 3-41). Near Wrightwood, existing landslides are mapped near the San Andreas Fault (figure 3-42) that may likely reactivate under strong shaking with some potential impact to buildings, although the largest concern is likely within the Cajon Pass lifeline corridor itself. Both landslide probabilities and predicted displacement for the Chino Hills and San Jose Hills areas are small, so the potential for significant economic loss due to landslides in these areas are somewhat uncertain, although both areas are urbanized and underlain by susceptible geology.

In conclusion, while ground failure is likely to occur over a broad area of southern California, significant economic losses due to ground failure damaging buildings are predicted in only a few areas. The most significant of these is in the Coachella Valley between the Salton Sea and Indio. Significant damage to drains and irrigation systems in this area from liquefaction lateral spreading and settlement appears likely given the prevalence of susceptible material in this region; liquefaction damage to structures near Indio may occur as well.

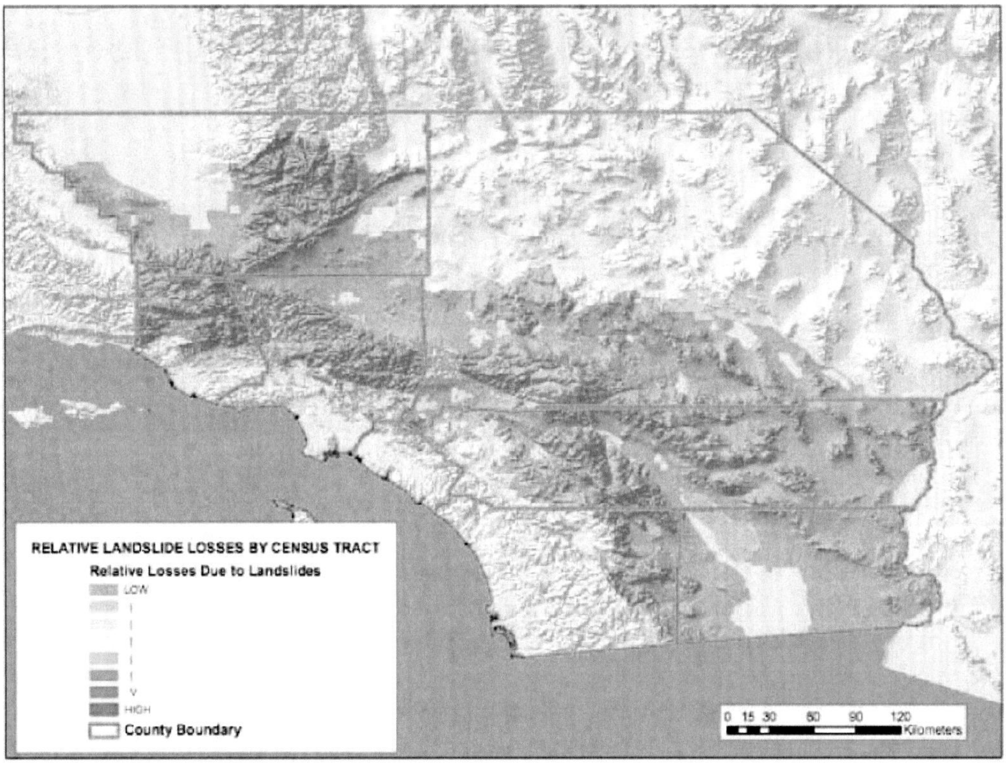

Figure 3-41. Map showing relative landslide losses (by census tract). Significant losses from slope failure are inferred near Wrightwood in the eastern San Gabriel Mountains (yellow), in the Chino Hills (red, near the Orange County border) and in the San Jose Hills near San Dimas. Inferred displacements at the latter two locations are small.

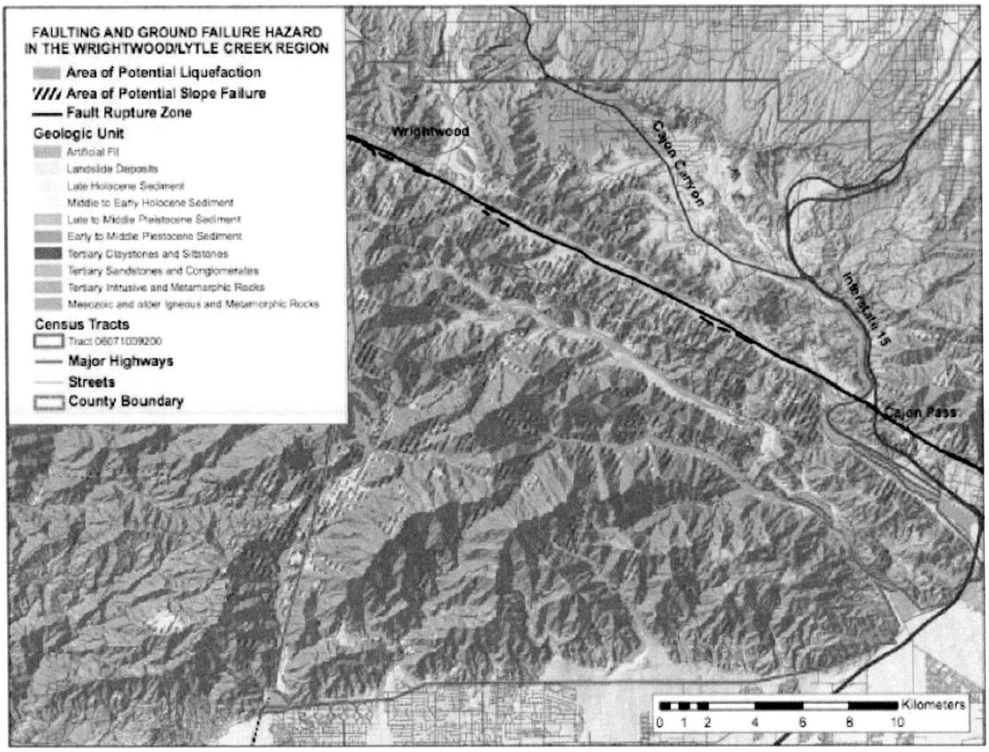

Figure 3-42. Map of the eastern San Gabriel Mountains between Wrightwood and Cajon Pass, where significant losses due to ground failure from the Scenario earthquake are inferred; geology is generalized from Morton and Miller (2006). Numerous landslides (yellow) exist just south of the San Andreas Fault (black line) and may reactivate during the Scenario earthquake causing damage to nearby structures. Analysis of hazards within the lifeline corridor at Cajon Pass is detailed in Chapter 4B and Appendix E.

D. Aftershocks by Karen Felzer

Large aftershocks pose significant hazards for years after a large mainshock. Aftershocks are earthquakes and can cause shaking and damage just like any other earthquake. Moreover, additional shaking can damage weakened structures, necessitate evacuations, endanger rescue workers, and undo efforts to restore and rebuild. Based on experience in numerous earthquakes worldwide, after a mainshock earthquake as large as the one in this scenario, damaging aftershocks can occur for decades in a broad region, and any given area may experience more severe shaking from a close aftershock than from the original mainshock.

Why aftershocks occur is not fully understood, and there is no scientific consensus on the physics that underlies aftershock triggering. Nonetheless, aftershock behavior in the aggregate can be well described by some simple, empirical laws, and these can be used to simulate sequences of aftershocks that realistically mimic actual aftershock sequences. Such simulations are particularly realistic if every generated aftershock is allowed to generate its own aftershocks, a technique generally known as the ETAS (Epidemic Type Aftershock Sequences, Ogata, 1998) model. Felzer (2008) uses one variation of the ETAS model to generate ten random realizations of aftershocks for the first week following our Scenario's mainshock. In reality, although aftershock rate diminishes with time, the distribution of aftershock magnitudes does not change (Lomnitz 1966), so a large, damaging aftershock may occur months or years after the initial event. Felzer simulates one week because that is the extent of the ShakeOut for response planning and because this will be the most intense period of aftershock activity. The long-term aftershock risk, however, should not be overlooked.

Felzer (in press) first initializes the mainshock rupture planes, and then simulates a set of primary, or direct, aftershocks produced by the mainshock, with smallest size *M*2.5, over a duration of seven days. A set of aftershocks of these aftershocks is then generated, and sets of aftershocks of those aftershocks, and so on, until no new earthquakes are produced within the 7-day time period. Throughout the simulation, the smallest magnitude earthquake that is allowed to trigger its own aftershocks is *M*2.5. Earthquakes smaller than this certainly exist in the real system and produce their own aftershocks, but calculation time increases exponentially as the minimum simulation magnitude decreases. At *M*2.5 we find a minimum magnitude small enough to make the simulations realistic while keeping the calculations tractable.

We use an application of the ETAS model based on Felzer and others (2002), with the addition of spatial components. As explained in Felzer and others (2002) the simulation starts with statistical distributions—cumulative probability density functions that describe the rate of aftershocks at different times and locations. These are then translated into discrete aftershock times, locations, and magnitudes for each simulated catalog by use of the inverse Poissonan function and a random number generator. See Felzer (in press) for additional details.

Table 3-7 summarizes parameters of damaging aftershocks from all ten simulations (figure 3-43). Like real aftershock sequences of earthquakes, these sequences are quite varied. Any of these could be reasonably expected to follow this *M*7.8 mainshock. See Felzer (in press) for additional simulation parameters. For the purposes of the ShakeOut exercises in November, 2008, one of the simulated aftershock sequences, number 10, was chosen to be used during the drill. It is one of the most damaging simulations with a magnitude 7.2 that begins near the San Andreas Fault at Cajon Pass, but causes rupture on the Cucamonga Fault. The total rupture area of this event extends about 50 km from Lytle Creek west to near Monrovia. This event would cause substantial further damage throughout the San Gabriel Valley, perhaps increasing the financial losses and deaths by 20-30%.

E. Tsunamis by Homa Lee and Eric Geist

The ShakeOut Scenario earthquake has the potential to produce submarine landslides along the basin slopes of the southern California continental borderland. Recent experience has shown that large, rapidly moving submarine landslides can induce significant localized tsunamis that can strike the coastline, causing damage to property and loss of life (Tappin and others, 2003, Lee and others, 2007, Goff and others, 2006).

Within the offshore, southern California continental borderland, two large submarine landslide complexes have so far been identified that are thought to have generated tsunamis during the Holocene (past 10,000 years, Lee and others, submitted). Both of these complexes show repeated episodes of failure, and they are likely to fail again at some time in the future.

The two complexes are the Goleta slide near Santa Barbara (Fisher and others, 2005) and the Palos Verdes debris avalanche (Locat and others, 2004). According to Lee and others (2004), the last major failure of the Goleta slide was 5,500 years ago and the last major failure of the Palos Verdes debris avalanche was 3,500 to 7,500 years ago. Of these two major failure complexes, the Palos Verdes deposit is about 90 km from the nearest point of fault rupture for the Scenario earthquake, whereas the Goleta slide is about 150 km away. If the ShakeOut Scenario earthquake were to cause a major submarine slope failure, the most likely event would be a recurrence of failure within the Palos Verdes complex, because it is much closer than the Goleta complex and consequently will experience stronger ground motions.

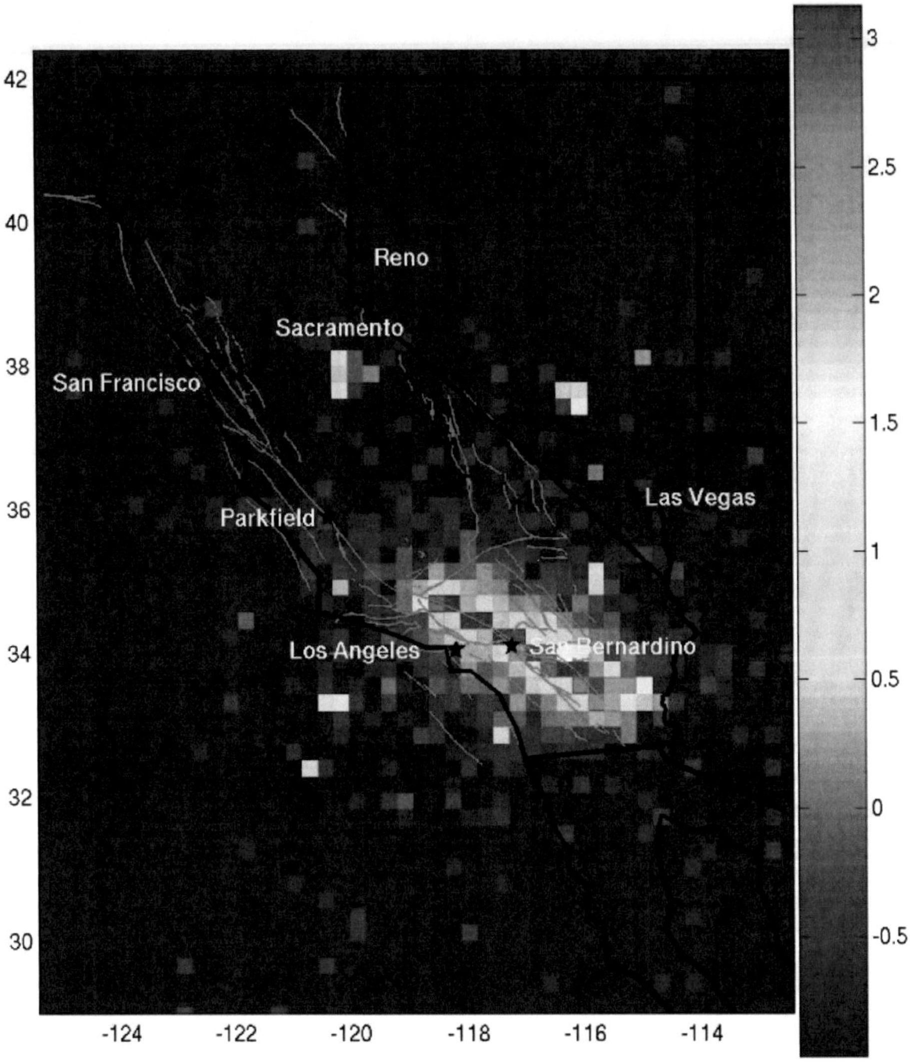

Figure 3-43. This map shows the likely spatial distribution of aftershocks to the ShakeOut Scenario. The colors represent the number of $M \geq 2.5$ aftershocks that occur in the first week after the ShakeOut mainshock in 25 x 25 km squares, averaged over all 10 simulations with hot colors indicating the most aftershocks. The numbers are given by a logarithmic scale so that the full range of aftershock rates can be seen without saturating the color scale. Bright spot of activity at great distances from the fault are places where a large aftershock occurred in a single simulation and produced its own aftershocks.

Table 3-7. Parameters and Narrative Descriptions of Simulated, Damaging Aftershocks. [Day 1 is the Day of the Mainshock. Any Earthquake of $M \geq 5.5$ is Capable of Causing Damage. Earthquakes Smaller Than M 5.5 May Damage Already-Weakened Structures]

After-Shock Simulation Number	Largest After shock (M)	Day of largest event	Number of events $M \geq 5.5$	Areas and communities most affected	Notes about largest aftershock
1	6.95	4	20	Sacramento, Modesto, Mariposa	Occurs far to the north—east of Sacramento. It impedes recovery efforts conducted from the State capital and endangers the levies, creating a serious new problem.
2	6.87	1	9	San Bernardino, Crestline, Rialto, Fontana	This rupture occurs in the mountains, just fourteen minutes after the mainshock, and may worsen landslides.
3	7.09	3	14	Palmdale, Lancaster	Nearly three days after the mainshock, occurs right at the edge of Palmdale and Lancaster.
4	6.39	1	9	Lancaster	Epicenter is about 30 km from Lancaster. Also, a $M5.57$ aftershock occurs in the mountains above Palm Springs and may prove more damaging.
5	6.75	1	21	Wrightwood	A number of moderate-size aftershocks, close to communities, pose additional threats to Palm Desert Country, Palm Desert, Rancho Mirage, Palm Springs, Desert Hot Springs, Mentone, Highland, Lancaster, and Palmdale.
6	6.73	1	10	San Bernardino, Yucaipa	Largest aftershock has an epicenter 10 km from Yucaipa. Of additional concern is a M5.52 outside of Mentone, near the Seven Oaks Dam.
7	7.71	4	30	Palm Springs, Palm Desert, Rancho Mirage, Coachella, Thermal, Mecca, Imperial Valley, Brawley, El Centro	An aftershock almost as large as the mainshock ruptures from Palm Springs south, running close to many communities. Significant effects may be expected throughout southern California, particularly in Imperial and San Diego Counties.
8	6.48	1	13	Little Rock, Lancaster, Hwy 14	Epicenter is directly under Lancaster, near the intersection of Avenue L and 20th.
9	7.28	2	24	Little Rock, Palmdale, Lancaster	Widespread damage would result from this rupture that passes Palmdale and Lancaster, and is less than 50 km from Santa Clarita, Burbank, and Pasadena, This simulation also generates a $M6.61$ in central Nevada.
10	7.22	1	23	Redlands, foothill communities from the Inland Empire to the San Gabriel Valley, Imperial and eastern San Diego Counties	Two $M \geq 7$ aftershocks, with serious conesequences. One ruptures west through the Inland Empire. The other ruptures south toward Niland.

Locat and others (2004) performed a geotechnical analysis of the most recent failure of the Palos Verdes debris avalanche and determined that a pseudo-static seismic acceleration of 0.3 to 0.4 g was probably necessary to cause the event. The pseudo-static seismic acceleration corresponds to steady horizontal force that, if applied constantly, would produce a failure. Actual earthquake loading is dynamic, so such a pseudo-static acceleration does not correspond to the peak ground acceleration (PGA) needed to cause failure. The critical PGA would typically be higher. Ten Brink and others (submitted) estimated that the PGA associated with slope failure would be about 1.9 times the critical pseudo-static acceleration, or 0.6 to 0.75 g for the case of the Palos Verdes debris avalanche. According to models for the ShakeOut Scenario earthquake, such high values of PGA extend to only about 10 km from the fault rupture, a distance much shorter than the 90-km span between the Palos Verdes feature and the nearest point of fault rupture. Accordingly, the probability is extremely low that the ShakeOut Scenario earthquake would cause a major submarine landslide that would in turn cause a damaging tsunami. However, the probability is considerably higher for a variety of other earthquakes that menace southern California.

F. Seiches by Homa Lee and Eric Geist

The ShakeOut Scenario earthquake has the potential to produce seiches affecting coastal facilities and inland bodies of water. Seiches are long-period, water-level oscillations within closed or open basins that can be excited by a number of different phenomena, including earthquake ground motions (for example McGarr, 1965). Like all waves, the oscillations may have different periods, that is, the length of time over which the wave repeats its motion; and every body of water will be of size and shape to enhance resonance in waves of certain periods. These harmonic periods of the oscillation are dependent on the dimensions and geometry of the water basin. Free oscillations excited by, for example, coseismic tilting of the water level, decay primarily as a function of bottom friction and topographic irregularities of the basin. Because of the shallow water depth (average, 9 m) and large area of the Salton Sea, free oscillations are expected to decay more rapidly compared to those in relatively deeper lakes. Differences in water levels (±6 cm) between two Salton Sea hydrologic stations have been ascribed to meteorological seiches with a period of approximately 1.5 hr (Wilson and Wood, 1980). This approximately corresponds to the second mode using a simplified basin geometry (Dean and Dalrymple, 1991) and thus we would expect long-period seismic waves of similar 1.5 hour periods to be most likely to resonate in the Salton Sea. However, rupture directivity sends most of the energy in the ShakeOut earthquake away from the Salton Sea.

Seiches may be more of a problem in the coastal regions of the Los Angeles sedimentary basin, where seismic surface waves from the ShakeOut earthquake are higher in amplitude and longer in duration. Although seismic surface wave periods are considerably shorter than seiche periods, the longer surface wave train will make it more possible that seiches will be set up, compared to the Salton Sea, where earthquake rupture initiates. As an example, the period of the fundamental mode for San Pedro Bay is 59.6 min. and that for the outer basin of Los Angeles Harbor is 6.5 min. (Wilson, 1972).

Seiche amplitudes are more difficult to estimate than seiche resonant periods because of interference patterns and non-linear hydrodynamic interactions. That is, as the various waves interact, some will grow in size and some will shrink. As a possible analog to the ShakeOut Scenario earthquake, the M=7.9 Denali, Alaska earthquake generated a destructive seiche in Lake Union, Washington, where seismic surface waves were amplified by the Seattle sedimentary basin (Barberopoulou and others, 2004). Observed seiche amplitudes were approximately 15-30 cm. Most of the damage caused by seiches from this event, however, apparently arose from the horizontal motion of these long- period waves.

4. PHYSICAL DAMAGES BY KEITH PORTER

A. Overview of Physical Damages

The damage impacts of the ShakeOut Scenario earthquake were estimated through a three step process. First, FEMA's loss estimation program, HAZUS, was run using the physics-based ground motion model. For Los Angeles County, HAZUS used a refined database of structures created from tax assessor's data. For the other counties (Imperial, Kern, Orange, Riverside, San Bernardino, San Diego, and Ventura), this was not available and the default HAZUS database was used, with improvements to the structural characterization. In the second step, expert opinion was collected through 13 special studies and 6 expert panels. Panels generally estimated impacts to public utilities, especially where multiple utility companies provide a public service such as water supply or electricity. Engineers and operators were invited to attend the half-day panel discussions, and they were presented with results of prior earth science studies (shaking, faulting, etc.) as well as damage to other interacting lifelines that had already been assessed. They were then asked to posit a realistic scenario of damage, service interruption, restoration, and to suggest promising mitigation options. To complement the panels, special studies were used for buildings and for lifelines where the panel process was impractical, such as private utilities or utilities (such as highways) where in-depth analysis was desired. In these cases, contributors were selected for their specialized expertise. They too were presented with all previously estimated earth-science and relevant utility impacts, and they were asked to summarize assets exposed to damage, evidence of past seismic vulnerability, and to posit a realistic scenario of damage, loss of function, restoration, and promising mitigation measures. Crucial special studies were reviewed by panels of highly qualified experts. In the third step, the expert evaluations were merged with the HAZUS results to create the final estimates of probable damages.

The major losses for this earthquake fall into four categories: building damages, non-structural damages, damage to lifelines and infrastructure, and fire losses. Within each category, the analysis found types of losses that are well understood—that have been seen in previous earthquakes and the vulnerabilities recognized but not removed— and types of losses that had been less obvious, where the type of failure is only recently understood or the extent of the problem is not yet fully recognized. The study also found numerous areas where mitigation conducted over the last few decades by state agencies, utilities and private owners, has greatly reduced the vulnerability. Because of these mitigation measures, the total financial impact of this earthquake is estimated to be "only" about $200 billion with approximately 1,800 fatalities. These are still big numbers.

The fault offset causes extensive damage to lifelines that cross the rupture: two interstate highways (I-15 and I-10) cross the fault and lose lanes as a result of the offset. Several oil and natural gas pipelines are also ruptured by the fault offset, as are several rail lines, aqueducts, and potable water pipelines. Shaking-related damage to highway bridges renders most freeways in Los Angeles, San Bernardino, and Riverside counties impassible at several locations, with some damages taking as long as 5-7 months to repair.

Electric power is lost throughout the study area immediately, and it is restored to 90% of those capable of receiving it within 3 days. Pipeline damage causes the loss of piped drinking water in much of the most strongly shaken areas (with MMI VIII+ shaking) for a week or more. Telecommunications are severely impacted as a result of heightened demand after the earthquake, and to a limited extent because of damage to telephone switching facilities and fiber-optic cables. Between 100,000 and 200,000 addresses lose phone and Internet service for between 2 and 5 days.

The earthquake causes the ignition of 1,600 fires. Owing in part to the loss of piped water for firefighting, 200 million sq ft of residential and commercial property valued at $40-100 billion is burnt. This is in addition to shaking-related property and direct income losses valued at approximately $60 billion. Five pre-Northridge highrise steel moment-frame buildings completely collapse, with approximately 5,000 people inside. Approximately 50 low- and midrise older reinforced concrete moment-frame buildings are hypothesized to collapse, most partially, as opposed to complete pancake- style collapse. These involve 800 people in completely collapsed concrete buildings and 7,000 in partially collapsed ones. Approximately 900 unreinforced masonry buildings are irreparably damaged.

None of the region's dams ruptures. Three experience damage serious enough to cause their operators to immediately draw down the reservoir behind the dams and call for evacuation of downstream communities. The fault ruptures the Palmdale Reservoir, which combined with damage to the California Aqueduct, results in flooding in some residential areas of Palmdale. Damage to sewer pipelines and equipment at wastewater treatment plants causes sewage spills at 50 to 100 locations throughout the study region, with untreated sewage flowing into nearby creeks, subsequently requiring cleanup.

The Ports of Los Angeles and Long Beach experience minor damage, but port operations are nonetheless hindered for 2 weeks or so by the aforementioned damage to the rail and highway network, by which most land shipments enter and leave the ports. The loss of electric power causes dozens to hundreds of elevator occupants to be trapped for several hours or more in elevators between floors, requiring first responders to extract them. The tripping of seismic switches requires large numbers of elevators to be inspected by elevator mechanics before being put back into service, resulting in long restoration times for elevators throughout the study region. Of hospital buildings in Los Angeles, Orange, Riverside, and San Bernardino Counties, it is posited for ShakeOut Scenario planning purposes that over 60% of the buildings are nonfunctional and suffer irreparable damage. Over 85% of the buildings suffer significant non-structural damage. Damage to four hospitals in San Bernardino County is particularly severe, impacting almost 40% of licensed beds in the county.

Many of the studies that were conducted for the ShakeOut Scenario are available as reports on-line. For details go to http://urbanearth.usgs.gov/scenario08.

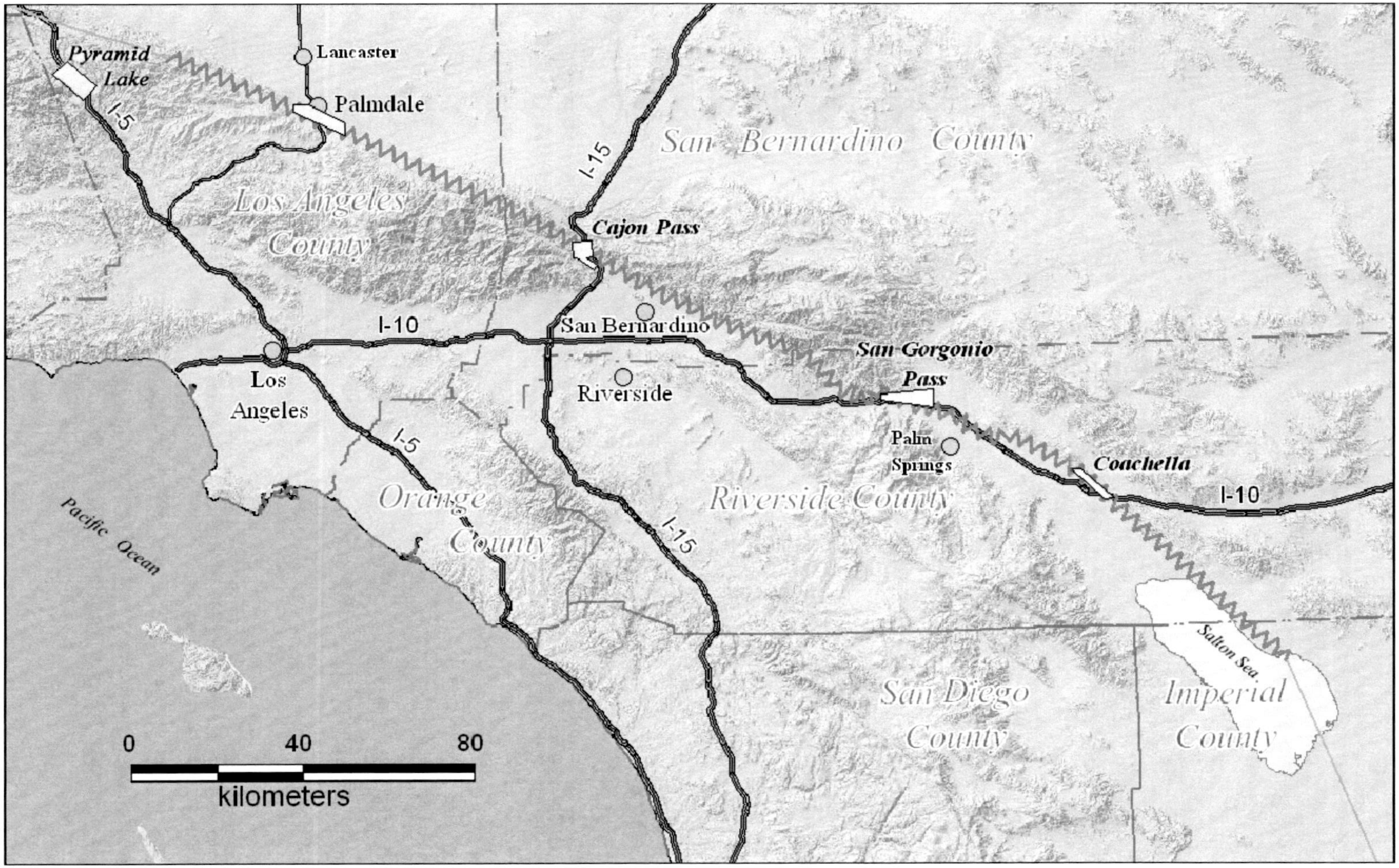

Figure 4-1. Principal lifeline corridor focus areas. Red line enotes rupture of southern San Andreas Fault in the ShakeOut Scenario.

B. Ground Deformation Impacts to Lifeline Corridors by Jerome Treiman and others

This excerpts a study by Jerome A. Treiman, Charles R. Real, Rick I. Wilson, Michael A. Silva, Cynthia L. Pridmore, Timothy P. McCrink, Ralph C. Loyd, and Michael S. Reichle, California Geological Survey. See Appendix E for full study.

Overview of the Issues

Lifelines are the veins and arteries of the Los Angeles region. Pipelines (oil, gas, and water), aqueducts, canals, highways, railway lines, fiber-optic communication cables, and power transmission lines move energy, goods, information, and people within, into and out of southern California. These lifelines are critical to the economy of the area, indeed to the economy of the nation. As a result of the unique southern California geography, lifelines going to the north and the east tend to funnel through only a few topographic passes. Steinbrugge and others (1987), referred to these areas as "lifeline corridors." At each of these passes, where the lifelines are concentrated in a small area, they must cross the southern San Andreas Fault. Because of the importance of these corridors, we give them special attention in the ShakeOut Scenario. The focus areas examined are Interstate 5 near Pyramid Lake (Tejon Pass), Palmdale/Highway 14 (Soledad Pass), Cajon Pass, San Gorgonio Pass, and the Coachella Valley near Indio. In each area, we examined three earthquake-related hazards and their effects on the lifelines: fault rupture, earthquake-induced landslides and liquefaction. The locations of the areas considered are shown in figure 4-1. This chapter summarizes the analyses of Treiman (2008), Wilson and others (2008), and Real and others (2008). More detail on the analyses and results can be found in those reports.

Damage Estimates for the ShakeOut Scenario

Ground deformations from the first stages of this study were used to posit damages to lifelines that cross the fault within the lifeline corridors, and these are described in the lifelines sections below. In some cases, additional work has been done since that time and some additional damages are posited, which are also listed here.

Interstate 5/Pyramid Lake Focus Area

Surface Fault Rupture—Fault rupture from the scenario earthquake does not extend as far west as Interstate 5. Therefore, there would not be damage to lifelines due to surface fault rupture in this corridor.

Landslides—The Interstate 5 corridor was examined from Gorman to Pyramid Lake for potential earthquake-induced landslide hazard. The priority targets identified for analysis are a large landslide complex and an adjacent 60-meter-high, slide-prone road- cut near the Vista Del Lago Visitor Center (California Department of Water Resources facility) above Pyramid Lake, approximately 25 kilometers from the western end of the scenario fault rupture. This area has produced numerous recent landslides activated during winter rains that have impacted the highway and various utility lifelines that cross the area. Based on the results of v.1.1.0 of the ShakeOut ground-shaking analysis by Graves and others (2007; see Chapter 3), it was concluded that ground motions calculated for the scenario event are too low to trigger significant slope instability.

Palmdale/Highway 14 Focus Area

Surface Fault Rupture—Surface displacement within the Palmdale/Highway 14 focus area is proposed to occur primarily on the main strand of the San Andreas Fault, but as much as 20% of the rupture may be distributed to several adjacent parallel fault strands, namely the Littlerock, Cemetery and Nadeau faults (figure 4-2). Fault trace locations are based on detailed field mapping and aerial photo interpretation by Barrows and others (1985).

Slip distributions between the fault traces are a judgment call based on experience and familiarity with the San Andreas Fault and observation of slip distribution in strike- slip earthquakes on other faults. Intense shearing and deformation adjacent to the main trace are expected to cause some broader zones (200+ meters) of distributed shear, most importantly at the two aqueduct crossings and at Lake Palmdale dam. Lateral fault offset at these locations is proposed in the range of 3.0-3.5 meters. Highway 14 could be offset 2.8 meters at the main trace of the San Andreas Fault and about a third of a meter at each of two other fault crossings. Natural gas pipelines, railway lines and fiber-optic communication cables could experience comparable offsets at several fault lines. Power lines, where they cross the fault traces, may be affected by extension.

Landslides—Within this focus area, the only significant slopes that could have impacts on lifelines are a pair of 20-meter-tall road-cuts that face each other along Highway 14 between the main and northern strands of the San Andreas Fault Zone. Based on an analysis performed by the California Geological Survey (CGS) for the Seismic Hazard Zone map of the area, these road-cut slopes are considered to have a high hazard potential for landslides (Wilson and others, 2003; Department of Conservation(DOC)-CGS, 2003). During the scenario earthquake, an estimated 1,000 cubic-meters of material could be deposited on the highway.

Liquefaction—In the Palmdale area, several lifelines could be affected by liquefaction-induced lateral spreading; these include Highway 14, buried fiber-optic communication cables, petroleum pipelines, railway lines, and the California Aqueduct. The liquefaction potential for the Palmdale area was reviewed in earlier studies by CGS (Mattison and Barrows, 2003; Pridmore, 2003); and more recently by DOC-CGS (2003a,b).

Cajon Pass Focus Area

The Cajon Pass focus area is approximately 20 square kilometers in size and is located along the Interstate 15 corridor centered about 39 kilometers north of the City of San Bernardino. The focus area was also extended to the south of the fault rupture area to encompass steep unstable slopes above the railroad rights of way. The impacts to lifelines from a similar scenario earthquake for the Cajon Pass area were previously evaluated and summarized in reports published by the Federal Emergency Management Agency (FEMA, 1992a; 1992b).

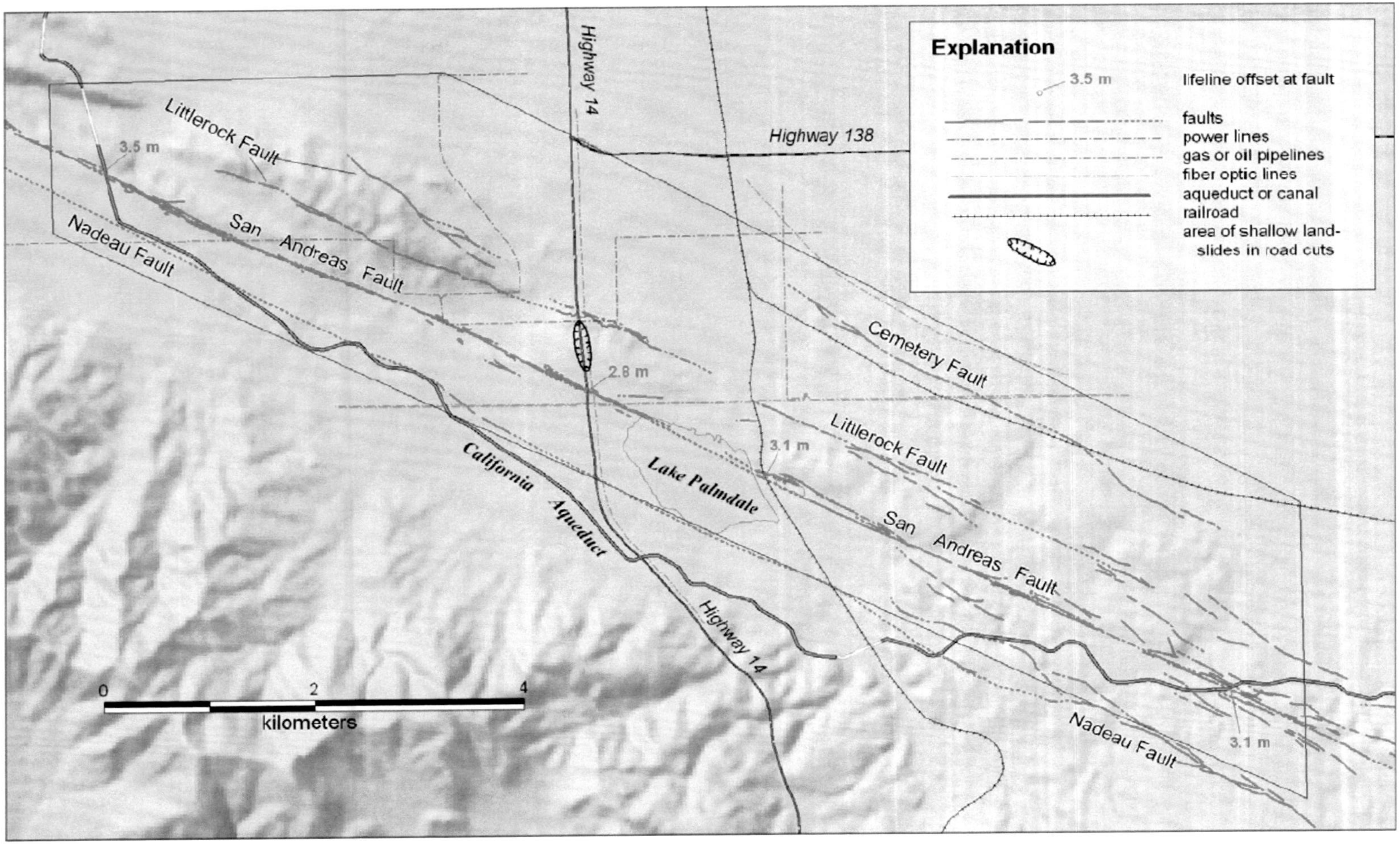

Figure 4-2 . Hazards to lifelines in the Palmdale/Highway 14 focus area.

Surface Fault Rupture—Surface displacement within the Cajon Pass focus area is modeled to occur only along the principal trace of the San Andreas Fault, including a few narrow stepovers (figure 4-3). Fault trace locations are modified from mapping by Weldon (1986); they have been relocated to match fault-geomorphic features observed in aerial imagery and LiDAR data. Ground rupture is proposed to offset rail lines as much as 4.0 meters; I-15 by 2.8 meters; and a number of gas, oil, communications and power lines by as much as 5.3 meters.

Landslides—Five moderate to deep-seated landslides, identified in figure 4-3 (LS-1 through LS-5), were selected for evaluation based on their proximity to specific lifelines. A detailed analysis was made to provide an estimate of landslide movement during the scenario earthquake (Wilson and others, 2008). If the analysis indicates that earthquake- induced displacements are relatively large (greater than one meter), there is a higher chance that the landslide, or portions of it, could disaggregate and change behavior from a coherent sliding block to a flow-type failure. As a consequence, rather than landslide displacements of tens to hundreds of centimeters expected for a coherent sliding block, a flow failure initiates where displacements can be an order of magnitude greater, that is, on the order of tens of meters. Such catastrophic failure is posited for landslides LS- 1, LS-2 and LS-4, which could involve from one- to 15-million cubic meters of rock and soil. Landslides LS-3 and LS-5 are partially buttressed by fill and would have lesser displacement but could still have an impact.

Aside from the five major landslides, CGS has identified at least six areas where lifelines cross slopes with high hazard potential for shallow landslides to occur during a large earthquake. The areas are shown as areas A through F in figure 4-3. These are areas that are likely to cause the most significant impact to transportation and utility lifelines in the Cajon Pass focus area.

Probably the most significant impact of the slope failures will be to transportation routes. The large artificial fill prisms constructed for I-15 in area A will likely have significant seismic compression/settlement and possibly landslide failures during a very large earthquake. There is evidence from numerous earthquakes that seismic compression and/or vertical settlement occurs on large highway fill prisms such as the one through Cajon Pass. Seismically-induced settlement could displace the highway as much as several meters vertically. In addition, considering the long duration of high ground motions expected, a worst-case scenario could include a large portion of the highway fill prism failing catastrophically in a westerly direction and sections of both the north and south bound lanes being displaced significantly. Road cuts in area E could fail along zones of weakness and adverse structure within the bedrock. Overall, the total amount of debris that could be shed on I-15 could be about 73,000 cubic meters. At area C, the old Route 66 roadbed (important access for lifeline repair) could be displaced or blocked by bedrock failures below and road cut failures above.

Rail lines could be displaced at several locations, significantly so at LS-1 and LS- 2; the tracks may also be buried under millions of cubic meters of debris at LS-1, LS-2, and LS-4. Several adjacent sections of railroad tracks in areas D and F may be covered by tens of thousands of cubic meters of additional debris.

Shallow and deeper landslides in this scenario could cause failure of power transmission line towers along five separate rights-of-way. A 36 inch gas line could be displaced at up to four locations. Other buried petroleum product lines and fiber-optic communication lines are expected to be displaced by landslides and/or buried by debris at multiple locations.

Liquefaction—The current channel of Cajon Creek is flanked by a series of four elevated terraces formed as the mountain range uplifted and Cajon Creek eroded downward. As terrace elevation and age increase, the deposits are less susceptible to liquefaction because depth to saturation increases and the deposits become more indurated. Therefore, the lowest terrace, which lies in the Cajon Wash, is considered the most susceptible, while the highest terraces are considered unlikely to liquefy. It is assumed for purposes of the scenario that the lower two terraces are susceptible to liquefaction lateral spreading, so our deformation analysis is focused there.

Seven sites were identified from aerial imagery as potential liquefaction sites near co-located lifelines (figure 4-3). Of the seven sites, field inspection revealed four that have a high potential for liquefaction-induced lateral spread. Because these sites are less than two kilometers from the primary scenario rupture of a *M*7.8 event, it is assumed that the high amplitude and long duration of shaking will build pore pressures so rapidly that liquefaction will occur in the heterogeneous channel and terrace deposits.

Estimates of horizontal displacement are as much as 10 meters, affecting railway lines, highway, petroleum and gas pipelines, and fiber-optic communication cables.

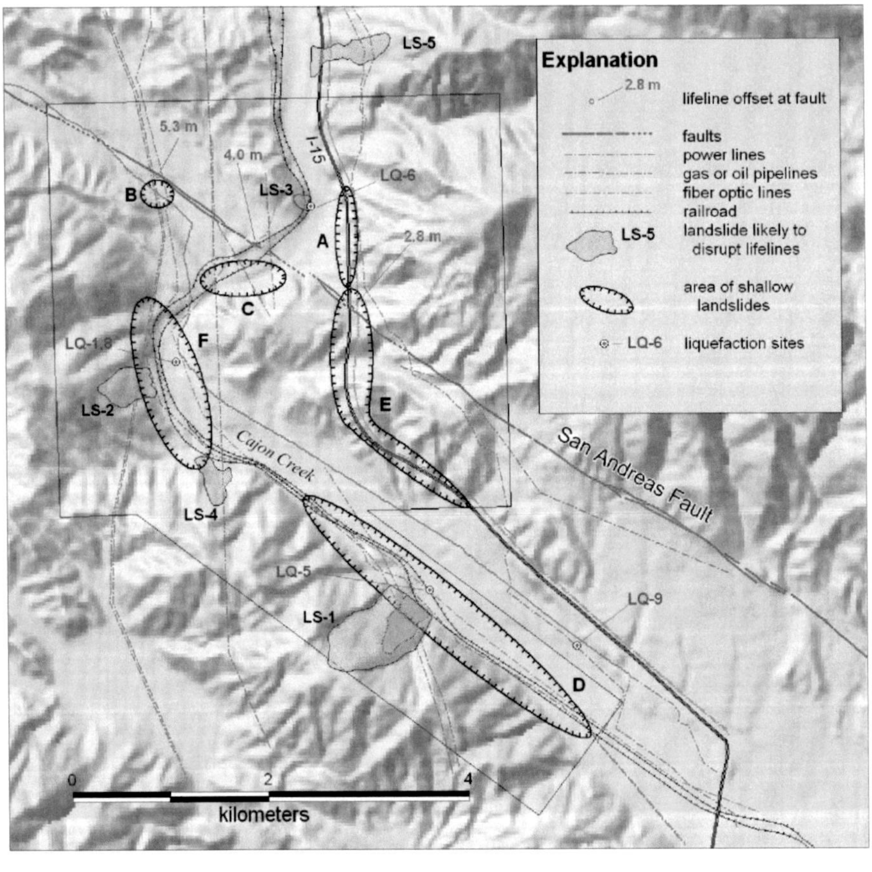

Figure 4-3. Hazards to lifelines in the Cajon Pass area.

San Gorgonio Pass Focus Area

Surface Fault Rupture—The San Gorgonio Pass focus area includes a broad stepover from the Banning Fault to the Garnet Hill and San Gorgonio Pass faults. Based on geomorphic expression, the Banning Fault is the primary fault trace to the east but slip appears to diminish westward into the San Gorgonio Pass. Meanwhile, the Garnet Hill Fault, poorly expressed to the east, becomes prominent to the west and merges into the active San Gorgonio Pass Fault Zone. Surface slip (1.4-4.6 meters) is modeled to transfer gradually between the Banning Fault and the Garnet Hill-San Gorgonio Pass faults. Fault trace locations in this focus area (figure 4-4) are based on aerial photo interpretation and field mapping (See Fault Evaluation Reports by Smith, 1979; Treiman, 1994; and sources therein; and Yule and Sieh (2003).) It is anticipated that in the scenario event the Metropolitan Water District aqueduct will be disrupted by strikeslip displacement at four locations, including the tunnel section in the west-central part of the focus area. Proposed offsets vary from 0.46 to 1.28 meters. Highway 62 will be broken in two localities with offset up to one meter. The Garnet Hill Fault roughly parallels a section of I-10 and may cause breakage along as much as a kilometer of pavement, as well as affecting co-located fiber-optic communication lines. Overhead power lines will be affected in different parts of the focus area by extensional and compressional displacements.

Landslides—Figure 4-4 shows the two specific areas that were analyzed for potential slope failure in the San Gorgonio Pass area (Wilson et al, 2008): 1) a 60-metertall, steep slope, named LS-Bluff, and 2) a large "landslump" mapped by Proctor (1968), modified and shown as LS-Slump.

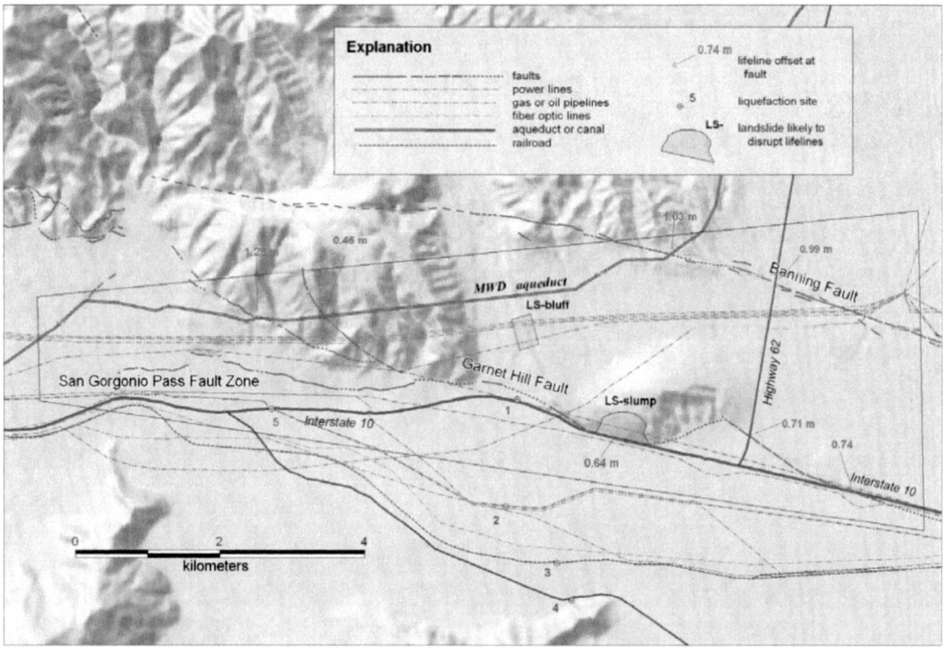

Figure 4-4. Hazards to lifelines in the San Gorgonio Pass area.

The LS-Bluff locality sits below several overhead power-line towers. Catastrophic failure is not considered likely, although slight (0.5 meter) westward displacement of the towers is possible. The LS-Slump feature, including a 60-meter slope directly above I-10, may fail onto the interstate. In doing so it would bury co-located fiber-optic communication lines and probably disrupt overhead power transmission lines.

Liquefaction—Five sites were evaluated in the San Gorgonio Pass area as potential liquefaction sites near co-located lifelines (figure 4-4). Two of the sites are along I-10 where fiber-optic cables and the interstate highway share right-of-way (sites 1 and 5), and are proposed to experience from 4 to 8 meters of displacement. The three other sites evaluated are located further south where high-pressure gas pipelines (site 2), rail lines (site 3) and Highway 111 (site 4) are located. Site 2 might experience 4 to 6 meters of displacement and sites 3 and 4 might have 3 to 5 meters. Site 4 is located on the San Gorgonio River, and site 5 is located on Cottonwood Creek.

Coachella Valley Focus Area

Surface Fault Rupture—The Coachella focus area is characterized by a single fault trace with some minor extensional stepovers (figure 4-5). Slip values shown represent the net surface slip; co-seismic slip at the surface is projected to be 60% of these values with afterslip continuing for weeks to months as strain propagates through the deep alluvium in this area.

Previous fault trace mapping by Clark (1984) has been relocated based on interpretation of vintage aerial photos and careful registration of the images within a GIS database as well as from local interpretation and mapping by consultants (Miles Kenney, personal communication, 2007).

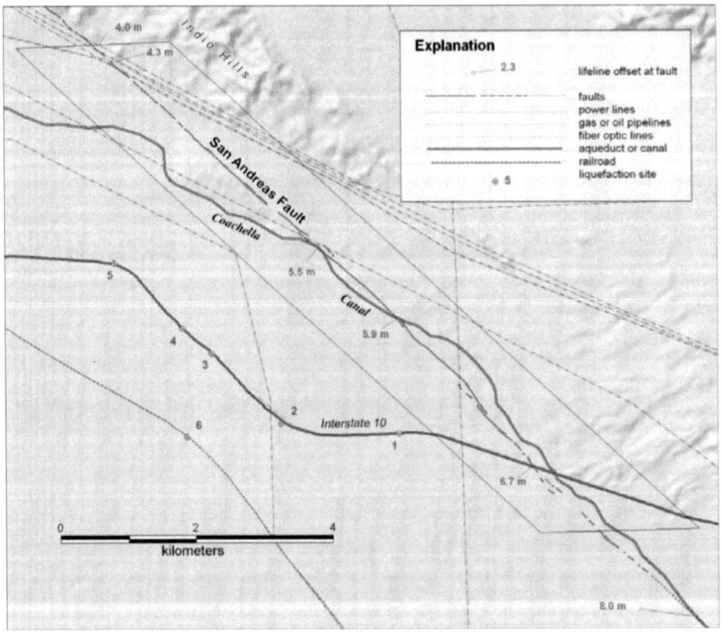

Figure 4-5. Hazards to lifelines in the Coachella area.

Several lifelines, including Interstate 10 and the Coachella Canal, may be affected immediately by 2.2 to 4.8 meters of offset along this rupture segment. The scenario should anticipate afterslip amounting to an additional 1.8 to 3.2 meters along this section of the fault, which may interfere with reconstruction efforts. Total cumulative slip is shown on figure 4-5.

Liquefaction—Six sites were evaluated for liquefaction displacements (figure 4-5). Five of these (sites 1-5) are along Interstate 10 where fiber-optic cables and the interstate highway share right-of-way. Site 6 evaluated the rail lines near Avenue 46 and Highway 111/86. Among the six sites, sites 2 through 5 were evaluated for lateral spread deformation with respect to the free face of the nearby, channelized Whitewater River and are proposed to experience 2 meters of horizontal displacement. Two other sites (sites 1 and 6) were evaluated for lateral spreading using the ground slope parameters of the Bardet and others (2002) model. Site 1 might have up to 3 meters horizontal displacement and site 6 would have 1 meter of displacement.

Conclusions

The ShakeOut Scenario earthquake has significant impacts on the transit of people, information, goods, and services through four key lifeline corridors in southern California. Aqueducts and pipelines that convey water and fuel products, fiber-optic communication lines, railroads, and highways will be severed directly by fault rupture, displaced by liquefaction-related ground failure, and displaced and/or buried by landslides. Overhead power transmission lines will fail where towers collapse or shift dramatically due to landsliding; other failures may occur where fault displacement stretches overhead lines beyond design tolerances. All of these impacts are likely in the Cajon Pass focus area with lesser, but still significant, combinations of these hazards in the other focus areas.

As an example of the widespread impacts to just one lifeline category, railroad tracks in Cajon Pass will be displaced three to four meters at the San Andreas Fault, but also displaced elsewhere in the Pass by landsliding, with the right-of-way also being buried by millions of cubic meters of landslide debris from several large landslides and an undetermined number of smaller slope failures. Rail traffic will also be disrupted at Palmdale due to fault displacement of at least three meters affecting the tracks and the adjacent dam. Liquefaction in San Gorgonio Pass and near Coachella may also imperil rail lines. Although not in one of our focused study areas, it should be noted that nearly 3.5 kilometers of rail line adjacent to the Salton Sea so closely parallels the scenario rupture that it too, may see extensive failure.

Other lifelines will similarly experience multiple failures throughout these vital corridors.

C. HAZUS by Keith Porter and Hope Seligson

HAZUS-Multi-Hazards (HAZUS) is FEMA's loss-estimation tool for emergency planning. Developed in the 1990s and early 2000s for FEMA by the National Institute of Building Sciences and a large number of technical experts in a wide variety of disciplines, HAZUS produces societal-level estimates of human and economic consequences of earthquakes, hurricanes, and floods. To understand how HAZUS' earthquake module was used and enhanced for the present scenario, it is helpful to understand generally how HAZUS works.

Assets exposed to loss—HAZUS uses economic and population census data, as well as commercial economic data, to estimate the number of people exposed to various hazards, on a geographic basis. It also encodes engineering judgment in a number of matrices to estimate the quantities and structural characteristics of the built environment (such as buildings, bridges, and pipelines) that are exposed to damage and loss. From these estimates it derives estimates of loss and damage, with quantities that are generally aggregated at the level of census tracts, as opposed to individual buildings.

Earth science hazards—To estimate the shaking at any given site in any given fault rupture, HAZUS uses built-in earthquake fault maps and estimates of how frequently each fault experiences an earthquake of a certain magnitude, along with probabilistic relationships between magnitude and distance from the fault rupture to any given site. By assuming a particular fault rupture occurs (such as the ShakeOut Scenario earthquake), HAZUS can estimate the shaking experienced by any given portion of the built environment, such as the shaking experienced by all high-rise, steel-moment-frame buildings built during a particular era and standing in a particular census tract of downtown Los Angeles. HAZUS also estimates liquefaction effects, using ground shaking, soil maps, and assumptions about depth to groundwater.

Damage and loss—HAZUS uses an analytical, structural engineering methodology to estimate the probabilistic physical damage to each such portion of the built environment, such as the fraction of the high-rise, steel-moment-frame buildings that can be expected to collapse in the scenario earthquake, or the fraction of unreinforced masonry buildings that might experience slight structural damage. Given probabilistic physical damage, HAZUS then estimates property losses by using built-in estimates of repair costs as a function of structure type, damage state, and occupancy. In addition to property loss due to earthquake shaking, HAZUS also estimates business interruption losses, homeless caseload, casualties, fire following earthquake, losses from liquefaction and other ground-failures, and other secondary sources of loss. It does not estimate insurance loss.

Enhancing HAZUS default data—The power and scope of HAZUS are increased by the ability to augment default databases and supplement or enhance certain of its methodologies. For the ShakeOut Scenario, HAZUS was supplemented in several ways. First, maps of shaking intensity developed for the Scenario project were substituted for the maps that HAZUS would have produced for the fault rupture. HAZUS treats shaking intensity with four parameters - peak ground acceleration (PGA), 5%-damped elastic spectral acceleration response at 0.3-second and 1-second periods (Sa[0.3 sec, 5%] and Sa[1.0 sec, 5%]), and peak ground velocity (PGV) - so the Scenario generated maps of these parameters at the census-tract level, weighted to account for population distribution at the census-block level, and adjusted to account for shaking amplification produced by soil near the ground surface. Landslide maps developed by the California Geological Survey were also imported to HAZUS, although this proved to have little effect on HAZUS-estimated loss.

Southern California structural engineering experts provided local judgment to enhance HAZUS' default assumptions about the distribution of structure types. We improved the structural characterization for all counties, and used additional inventory improvements in LA County. In a previous project for the Governor's Office of Emergency Services, tax-assessor data for Los Angeles County were used to replace HAZUS' default estimates of the quantities of taxable property. Similar tax-assessor data for other Southern California counties could not be acquired in time for this first phase of the ShakeOut Scenario. However, for use in the ShakeOut Scenario, exposure values (both building and content) for the Los Angeles County data were updated for consistency with the latest HAZUS building and content valuation models. Comparison of the updated inventory data to the HAZUS default demonstrates the importance of this type of update; for commercial and industrial occupancies, the HAZUS default data has 40% less building square footage than the database derived from assessor's data. The use of improved inventory data in Los Angeles County makes the Los Angeles damage and loss estimates more reliable. For other heavily impacted counties, San Bernardino & Riverside, use of the HAZUS default inventory and exposure data most likely leads to an underestimate of the exposure and associated loss. The losses estimated in Riverside and San Bernardino Counties should therefore be considered lower bound estimates and with inventory enhancement, they might increase on the order of 20 – 30%.

Because HAZUS was intended to facilitate natural hazard risk assessments on a consistent basis nationwide, the software is provided with a significant amount of default data to allow the user to run a simplified (HAZUS Level 1) analysis "straight from the box", without input of any additional data. By virtue of the fact that the default databases must be assembled at a national level, the data may be, in some cases, incomplete (having been collected by other agencies for different purposes) or out of date (some data are no longer available in the public domain due to security and other concerns). While Level 1 analyses are useful for gauging the approximate magnitude of potential impacts, the accuracy of results can be enhanced by the input of "user-supplied" data, resulting in a HAZUS Level 2, and potentially Level 3 analysis. The HAZUS enhancements conducted for the ShakeOut scenario, including generation of custom ground motion and other secondary hazard inputs, incorporation of significant building inventory database enhancements, and economic and population parameter adjustment (tested using a calibration exercise for the 1994 Northridge earthquake), result in a HAZUS earthquake loss assessment that would be considered "Level 3".

With these enhancements, HAZUS generated estimates of damage and loss (Seligson, 2007). Section E of this chapter provides HAZUS and supplemented values used in the ShakeOut Scenario.

D. Supplemental Studies

To examine certain important topic areas of physical damage in greater depth than HAZUS can achieve, or to check or improve HAZUS estimates of physical damage, 19 additional studies of physical impacts were performed for this Scenario. These took the form of modest studies by individuals or small groups, as well as discussions among experts. Other supplemental studies were also conducted outside HAZUS that involved emergency response and casualties; they are described in chapters 5 and 6 of this report. The supplemental studies regarding physical impacts are listed below, and described in detail in the sections that immediately follow.

Many of the studies that were conducted for the ShakeOut Scenario are available as reports on-line. For details go to http://urbanearth.usgs.gov/scenario08.

Individual or group studies:

- Unreinforced masonry buildings
- Mid- and High-rise Pre-Northridge welded-steel moment-frame buildings
- Non-ductile reinforced concrete moment-frame buildings
- Woodframe buildings
- Ports of Los Angeles and Long Beach
- Elevators
- Hospitals
- Highways
- Rail
- Telecommunications
- Natural gas and liquid fuels pipelines
- Fire following earthquake
- Hazardous materials

Expert panels:

- Dams, reservoirs, and aqueducts
- Wastewater and debris disposal
- Mass transit
- Surface streets
- Water supply
- Electric power

The supplemental studies were typically limited to a few days of effort and relied to varying degrees on expert judgment. In several cases, the special studies built upon other work by the authors, and so represent more than a few days' effort. Most notably these include steel-frame buildings, non-ductile concrete buildings, fire following earthquake, and highways. The expert panels were generally limited to a four hour discussion among engineers and operators of local utilities, who were briefed on the shaking and other earth-science impacts of the Scenario, and then asked to judge a realistic damage outcome of the earthquake. In each study or panel, participants also identified useful mitigation measures that might greatly reduce either the damage or the consequences of damage. When all damage estimations for the Scenario were completed, panel members and authors were invited to participate in a symposium to hear and discuss the findings.

i. Key Building Types

Unreinforced Masonry (URM) Buildings

Study by Richard L. Hess, S.E.,*Hess Engineering Inc.*

"Impacts of a M7.8 Southern San Andreas Earthquake on Unreinforced Masonry (URM) Buildings"

Overview of the Issues

Unreinforce masonry (URM) buildings have performed poorly in earthquakes worldwide and that they are dangerous in earthquakes is indisputable. Due to their inherent brittleness, lack of tensile strength, and lack of ductility, URM buildings are prone to collapse even in earthquakes of moderate size. Recognizing the considerable seismic vulnerability of URM buildings, California prohibited the construction of them after the 1933 Long Beach earthquake, and in 1986 passed legislation to mandate communities at high seismic risk to identify and attempt to mitigate URM building risk. Although the URM building mitigation rate in southern California approaches 90% in some counties, thousands of URM buildings remain, and partial retrofits have only partially reduced the hazard in many buildings. Moreover, URM retrofit is designed to protect life safety, not building functionality; thus a successful retrofit will still result in a building that can no longer function, creating significant long-term societal and economic impacts.

Experience with URM buildings in the 1987 Whittier, 1989 Loma Prieta, and 1994 Northridge earthquakes is relevant to future performance where level of shaking is concerned. An individual building may have damage levels above or below the average depending on its structural characteristics and the local ground motion. For example, if it has features other than being rectangular with continuous floor and roof diaphragms, additional damage will occur unless a retrofit engineer and contractor correctly address these irregularities.

We do not have experience with the performance of southern California's URM buildings during the duration of strong shaking that will occur in the ShakeOut Scenario earthquake. However the prospects are poor. Durations of shaking in the Northridge and Loma Prieta earthquakes were seven seconds, with some places suffering continued shaking for up to fifteen seconds. The ShakeOut Scenario earthquake will instead produce strong shaking that lasts 30-50 seconds, with continued shaking for as much as 100 seconds. After the Northridge event, some URM buildings that had not collapsed showed displaced supports that appeared ready to collapse after just one or two more ground oscillations.

Table 4-1 shows the status of URM buildings in the city of Los Angeles after the 1994 Northridge earthquake, from a report for the City of Los Angeles Task Force on Building Damage. This tabulation shows that, in this much smaller earthquake, 2.5% of retrofitted buildings suffered damage that was more than 10% of their replacement cost, and 0.3% suffered damage that exceeded 50%. For non-retrofitted buildings, 10.5% were damaged more than 10%, and 7% had damage that exceeded 50% of replacement cost. This demonstrates the value of the retrofit program but also that retrofitting URM buildings does not eliminate risk. Building configuration and the quality of the retrofit evaluation, design and construction make a substantial difference in the degree of improvement.

Table 4-1. Ground Motion Intensity and URM Building Damage in City of Los Angeles During 1994 Northridge Earthquake

				URM Status				*Unstrengthened*			*Strengthened*		
Council District	*Location (approx.)*	*MMI*	*Total URM Buildings*	*Prior Demolition*	*Exempt*	*Divison 88 Strengthened*	*Non-Division 88 Unstrengthened*	*< 10% Damage*	*< 50% Damage*	*> 50% Damage*	*< 10% Damage*	*< 50% Damage*	*> 50% Damage*
1	[25,-15]	VII	906	151	15	654	86	10	5	2	94	9	0
2	[20,5]	VII	31	3	1	23	4	0	0	0	2	0	0
3	[-5,0]	VIII	54	7	1	45	1	0	0	0	8	7	2
4	[20,-10]	VII	644	81	7	490	66	15	9	2	73	8	2
5	[10,-10]	VII-IX	220	20	9	183	8	1	2	1	7	3	1
6	[10,-25]	VI-VII	265	34	12	178	41	2	2	1	9	1	0
7	[10,10]	VII-VIII	23	5	0	11	7	5	0	1	1	2	2
8	[20,-25]	V-VII	613	105	20	383	105	4	1	3	22	7	0
9	[25,-20]	VII	3013	735	41	1935	302	23	11	6	74	27	1
10	[15,-20]	VII-VIII	607	51	11	466	79	12	11	3	80	15	0
11	[5,-10]	VII	125	18	0	94	13	1	4	2	13	3	2
12	[0,10]	VIII-IX	3	1	0	2	0	0	0	0	0	1	0
13	[25,-10]	VII	858	101	14	654	89	17	7	5	90	9	8
14	[30,-15]	VI-VII	560	78	12	378	92	15	12	5	18	6	0
15	[25,-45]	V	320	60	4	213	43	5	2	1	0	0	0
			8242	1450	147	5709	936	110	66	32	491	128	18

Damage Estimates for the ShakeOut Scenario

Within approximately 30 km of the fault, 900 (90% of the) URM buildings are damaged beyond economical repair, especially multistory, non-retrofitted buildings. An additional 200 buildings that have been retrofitted and 300 that have not be retrofitted experience damage costing at least 10% of their replacement cost, causing them to be rendered unsafe to enter or occupy during repair, and most likely causing them to be replaced rather than repaired. It is estimated that, on average, 40 people occupy each building, either for residential or commercial purposes, so the damage displaces approximately 60,000 residents or other building occupants. It is further estimated that to replace these buildings costs between $1 and $2 billion. Deaths and injuries associated with damage to URM buildings are estimated elsewhere in this report.

Expert-Recommended Mitigation

Properly retrofitting URM buildings saves lives. The minimum required procedures include:

- Remove parapets and ornamentation above the roofline or brace these items to the roof.
- Anchor the exterior and interior URM walls to the roof and floor framing.
- Check the height/thickness ratios of the URM walls to verify their out-of-plane stability. Brace wall if required.
- Develop horizontal diaphragms at each wall-bracing level. Verify adequacy of these diaphragms to control the relative dynamic displacement of the center of the diaphragm span.
- Develop adequate in-plane strength of URM walls and other elements that control interstory displacements.

High-Rise Pre-Northridge Welded-Steel Moment-Frame (PNWSMF) Buildings

Study by Swaminathan Krishhan and Matthew Muto, *California Institute of Technology*
"ShakeOut 2008: Tall Steel Moment-Frame Building Response"

Review Panelists:
Greg Deierlein, *Stanford University*
Ronald Hamburger, *SGH Inc.*
Jim Malley, *Degenkolb Engineers*

Overview of the Issues

Welded-steel, moment-frame buildings are a relatively new building type, common in mid- and high-rise buildings. This structural style has been expected to perform well during earthquake shaking, because steel is a ductile material that will deform in shaking but not fail catastrophically. However, after the Northridge earthquake, significant damage was discovered where welds developed cracks and some of those cracks spread into certain steel building components. This is brittle rather than ductile behavior and it may be that the chemical changes caused by the welding process changed the way the steel behaves. The following year in Kobe, Japan, steel buildings constructed to a different building code collapsed after exhibiting similarly brittle behavior. Since that time, building codes have been changed and mitigation procedures have been developed for existing buildings, but these changes have not been applied to most of these buildings in California.

An earthquake as large as the ShakeOut earthquake has not occurred near a metropolitan area since these kinds of buildings have become common. We do not have the experience of many earthquakes to inform our understanding regarding the performance of these buildings. We hired teams to consider the performance analytically, and estimate the effect of waves generated during an earthquake of this size on these kinds of pre-Northridge, welded-steel, moment-frame (PNWSMF) buildings.

Major steel building damage has only been observed in three earthquakes, the 1985 Mexico City earthquake, the 1994 Northridge earthquake, and the 1995 Kobe earthquake (see figs. 4-6 and 4-7), although isolated cases of steel building damage has been observed in other instances. These observations, described in some detail in Krishnan (2007), provide pointers to a spectrum of damage possibilities to southern Californian steel buildings due to the ShakeOut Scenario earthquake.

In the 1994 Northridge earthquake, the specifics of the fault rupture sent most of the energy away from the heavily populated Los Angeles metropolitan region into the mountains north of San Fernando Valley (Wald and others 1996). Nonetheless this earthquake revealed serious weaknesses in welded-steel moment-frame buildings in the greater Los Angeles region. Welded-steel moment-frames were previously considered to be the most ductile of all the structural systems recommended by the building codes for seismic resistance. Ductile structural systems are capable of withstanding large inelastic deformations without significant degradation in strength, and remain stable without collapsing under strong ground motion. Following the Northridge earthquake, a number of steel moment frame buildings were found to have experienced brittle fractures in welded beam-to-column connections. Such nonductile behavior occurred in buildings of varied heights, from one to 26 stories (FEMA, 2000d), with some of them as old as 30 years and others being erected at the time of the earthquake. The buildings were distributed throughout the region with some sites experiencing only moderate levels of ground shaking. Typically, fractures initiated at the root of the full penetration weld connecting the beam bottom flange to the column flange. Once initiated, the fractures progressed along a number of different paths, some progressing completely through the thickness of the weld, some developing into a crack in the column flange with the column flange separating from the rest of the column, some progressing further into the column web and across the panel zone, and finally in some instances the fractures cracked the entire column cross-section. Such fractures can lead to a significant loss of stiffness and strength in the moment frames and can dramatically lower the capacity of the structure to resist collapse.

The magnitude 6.9, 1995 Kobe earthquake occurred exactly one year following the Northridge earthquake. Kobe had many welded-steel moment-frame buildings ranging from low-rise buildings constructed in the 1950s and 1960s to modern high-rise structures constructed within the preceding 10 years. While the design and construction of these buildings are significantly different from that in the US, the Japanese code prevalent in 1994 required 20-story steel moment frame buildings to be designed for more than 2.5 times the force levels prescribed by the 1994 Uniform Building Code (Hall 1997), and the extent of damage observed in steel structures point to the possibility of a set of unfortunate factors leading to disastrous consequences. Out of 630 modern steel buildings in the heavily shaken area, the Building Research Institute determined that approximately one-third experienced no significant damage, one-third relatively minor damage, and the remaining third severe damage, including partial or total collapses of approximately half of these buildings (FEMA 2000b; FEMA 2000d).

Damage Estimates for the ShakeOut Scenario

In the study area there are possibly several hundred high-rise buildings of the kind that experienced unexpected damage to the moment-frame connections in the 1994 Northridge earthquake. For the ShakeOut Scenario, an analytical study was carried out, examining the likely performance of PNWSMF buildings. The study results were reviewed by a special panel of structural engineers, which carried out a small additional analysis, and reached agreement that red-tagging of buildings of this type is likely, and that collapse is possible. The panel concluded that "Given these ground motions, the collapse of some pre-1994 welded-steel moment-frame buildings is a credible scenario." The Caltech team recommended assuming 8 collapses for emergency planning purposes, wheras the panel considered 1 or more collapses to be realistic. Because collapses of these building types result from the long-period ground motions, the area where this type of damage is possible is relatively large and includes much of the urbanized areas of Los Angeles, Orange, Riverside and San Bernardino Counties. It is impossible to determine how many and which buildings are the most susceptible without detailed structural analysis which is beyond the scope of this study. Thus for the ShakeOut Scenario emergency response drill it is posited that 5 high-rise, PNWSMF buildings completely collapse (3 in Los Angeles, 1 near Costa Mesa, and 1 in San Bernardino). The Caltech team judged that twice as many PNWSMF buildings would be red-tagged without collapse; and twice as many again would be yellow-tagged, with visible damage requiring temporary building closure. Thus for the Scenario it is posited that 10 are red- tagged without collapse, and 20 are yellow-tagged. Some of the remainder of the stock of these buildings may experience damage requiring repair, but remain operational during the repairs.

Figure 4-6. Collapsed story of steel moment-frame building, Kobe, Japan, 1995.

Figure 4-7. Continued on next page.

Figure 4-7. Damage in Mexico City, September 19, 1985, from an earthquake some 200 km distant. (A) Collapse of a 21-story building on top of a 14-story building in the Pino Suarez complex. Three towers are left standing with extensive damage to the two 21-story towers. Source: Bob Reitherman; (B) Failure in the welds connecting the flanges and webs of built-up box column possibly led to local flange-buckling of the column and eventual collapse. Source: Bob Reitherman. (C) Debris of tower collapse spilled on to the streets. Source: Mehmet Celebi, USGS. (D) This 21-story steel building remained standing, but was leaning six feet out of plumb at the roof level due to yielding that resulted in a permanent interstory drift (Source: Jim Beck, Caltech). Such permanent tilting would most likely render the vertical transportation system (elevators) unusable due to misalignment. In addition, doors to staircase shafts may be jammed due to deformation of the frames, possibly hampering evacuation efforts.

A typical high-rise might measure 350,000 square feet in gross area, and be occupied by 1,000 people at 10 a.m. on a Thursday, so it is posited that 5,000 people are in completely collapsed PNWSMF buildings, 10,000 are in red-tagged, non-collapsed PNWSMF structures, and 20,000 are in yellow-tagged PNWSMF structures. Number of stories, occupancy class, and number of occupants have also been posited for each hypothetical collapsed building.

For emergency planning purposes, each selected collapse location is within an area of real, pre-1994, high-rise buildings (of unknown construction), and also within a region estimated by the Caltech study to be most likely to produce collapses in PNWSMF construction due to the ShakeOut Scenario earthquake. However, the collapsed buildings are entirely hypothetical. To emphasize this point they are sited in parking lots, baseball fields, and other locations currently without buildings. Their possible proximity to real, nearby high-rise buildings does not imply anything about the likely performance of the real buildings.

Some of the designers of the ShakeOut emergency planning exercises in November may find it desirable to move the collapse locations to real, nearby clusters of high-rise buildings, in order to better confront problems such as limited access and collapse onto adjacent structures. However, the clusters of real high-rise buildings tend to be outside the regions with highest collapse probability due to this particular, ShakeOut Scenario earthquake, so moving the scenario collapses would make them less defensible as a realistic outcome of the ShakeOut earthquake.

Deaths and injuries associated with damage to PNWSMF buildings are described in Chapter 6 of this report.

Expert-Recommended Mitigation

Research is needed to improve modeling of building damage generally and to better understand situations where behavior of welded-steel moment-frame buildings changes from ductile to brittle and seismic performance consequently worsens.

Building-specific modeling might assess collapse risk and thus better determine the type, extent, and cost of needed retrofit. However neither the collapse modeling nor the retrofit are realistic investments for most PNWSMF building owners. Retrofitting may only become an economically justifiable expenditure if public perception changes, such that substantially more rent could be charged for a more earthquake-resistant building.

Non-Ductile Reinforced-Concrete Moment-Frame (NDRC) Buildings

Study by Ertugrul Taciroglu and Payman Khalili-Tehrani, UCLA

"*M7*.8 Southern San Andreas Fault Earthquake Scenario: Non-ductile Reinforced Concrete Building Stock"

Overview of the Issues

After URM buildings, the type of building most likely to kill people during an earthquake is a non-ductile, reinforced-concrete moment-frame (NDRC) building. The rate at which earthquakes cause collapse of these buildings varies due to variation in building code and code enforcement. The rate at which they kill people doesn't vary much, however, because regardless of building code, concrete coming down on people kills them. Whether there will be partial or total collapse of a NDRC building may dramatically affect the casualty rate but either way, the building will be a complete economic loss.

Thanks to building code changes that began with the 1976 Uniform Building Code, NDRC buildings are no longer built in California. However, this was the most common commercial building type (including schools) in the 1950s and 1960s and thus NDRC buildings are exceedingly widespread. That fact, plus the lack of legislation like the URM law to require mitigation, arguably makes the NDRC building an even more deadly building type than URM buildings. Certain Los Angeles City Council members are thus trying to enact NDRC building mitigation legislation.

In past earthquakes, NDRC buildings have suffered strikingly poor performance due to deficiencies in design, detailing, and sometimes construction. A reinforced concrete structural system has very limited capacity to absorb and dissipate the destructive energy of strong ground shaking beyond its limited elastic range, and hence, is extremely vulnerable to collapse. Reinforced concrete construction began in the United States at the turn of the twentieth century. Seismic design of reinforced concrete structures was at its infancy in 1950s, and lacked critical improvements to detailing requirements until the 1970s.

The watershed event that changed the building codes, eliminated construction of non-ductile reinforced concrete buildings and commenced design of ductile reinforced concrete design was the 1971 San Fernando Earthquake, particularly the devastating destruction at the recently completely Olive View Hospital (figure 4-8). The Olive View Hospital was rebuilt and became a testament to improved design for seismic resistance when it suffered equally strong shaking in the 1994 Northridge earthquake, yet little structural damage.

Damage Estimates for the ShakeOut Scenario

Structural engineers performed sophisticated engineering analyses of a hypothetical, non-ductile reinforced-concrete moment-frame building for the George E. Brown Network for Earthquake Engineering Simulation. For the ShakeOut Scenario, they offer a realistic range of possible outcomes, and from these outcomes the following ShakeOut damages are derived. It is posited that five NDRC buildings in the study area experience complete, pancake-style collapse, 45 experience partial collapse, and 100 are red-tagged but do not collapse. Because the ShakeOut Scenario earthquake occurs at 10 a.m. on a typical Thursday morning, perhaps 750 people are in buildings with complete collapse, 6,750 are in buildings with partial collapse, and 10,000 to 20,000 people are in heavily damaged buildings that are subsequently red-tagged, meaning they are unsafe to enter and possible complete economic losses.

Deaths and injuries associated with damage to NDRC buildings are described in Chapter 6 in this report.

Figure 4-8. Second story of psychiatric unit, Olive View Hospital, 1971, pancaked over first story. This and other damage at Olive View Hospital emphasized the poor seismic performance of NDRC construction.

Expert-Recommended Mitigation

Because of the widespread and serious problems with seismic performance in NDRC buildings, retrofit is recommended. However, because some NDRC building construction types perform better in earthquakes than others, it is not possible to make the blanket assertion (as can be said with URM buildings) that all NDRC buildings will perform poorly. This makes it more difficult to legislate required rehabilitation of these buildings because the action needed for the public good is not as clear as with URM buildings.

It could be beneficial for society to experiment with tougher laws that have phased requirements, as with the Field Act for public schools. For the Field Act, brief and thus relatively inexpensive work by a structural engineer determines whether further investigation and possible mitigation are needed, but many buildings are found to be sound and eliminated from additional requirements.

Protecting ourselves from buildings that perform poorly in earthquakes is as much a social issue as an engineering one, because often it is neither simple nor inexpensive to locate, identify, and rehabilitate problem buildings. Engineering research—and the results of damaging earthquakes—reveal seismic design flaws, and lead to improvements in building codes, which in turn lead to new buildings that perform better in earthquakes. However, we are still left with existing buildings that may perform poorly, and because building inventories are incomplete, the locations of all suspect buildings are not readily identified. Addressing the seismic performance of existing buildings requires resolve, persistence, and prioritization of resources by community members and their officials. This report offers no new or magic solutions to this pervasive and important problem, but emphasizes its urgency and also emphasizes the fact that numerous, varied communities have made significant progress despite the myriad challenges (For more information see FEMA 154, FEMA 17).

Woodframe Buildings

Study by William Graf, *URS Corporation*

"Damage to Wood-Framed Buildings"

Overview of the Issues

Woodframe structures are generally very safe in earthquakes because wood is light and flexible and the buildings tend to be smaller. However, some aspects of construction do create life-safety risk and some can create large financial vulnerabilities. In particular, if the structure is not properly bolted to the foundation it may slide off during relatively low levels of shaking, becoming a total financial loss. Foundation bolts were mandated in 1935 but not adopted by every jurisdiction until 1960. After foundation bolting became routine, the 1971 Sylmar earthquake exposed the vulnerability of the cripple wall, the short wall that forms a crawl space between house and foundation. The cripple wall may simply be composed of 2 inch by 4 inch wood pieces that fall like dominos when pushed by earthquake ground motion. Addition of plywood sheathing resolves this problem.

Woodframe structures include apartments and condos as well as single family homes and in some eras, designs featured a *soft* story, lacking the structural integrity to resist sideways earthquake motion and keep upper floors standing. This type of construction is widespread in condos and apartment units with tuck-under parking. The majority of deaths in the Northridge earthquake resulted from collapse of an apartment building with a soft first story.

Damage Estimates for the ShakeOut Scenario

Rather than revise damage estimates, this study was used to inform and improve the HAZUS analysis of woodframe buildings.

Expert-Recommended Mitigation

- For woodframe structures, foundation bolting is the single most important protection from seismic damage. Addition of plywood sheathing to cripple walls can be done with minimal cost and expertise and greatly reduces overall financial losses of individuals and our society.
- Not all tuck-under parking performs poorly; however, owners and condo associations should consider getting professional evaluation of their structure and mitigating as necessary.

ii. Critical Facilities

Ports of Los Angeles and Long Beach
Panelists:

Ron Guss. CEO, *Intermodal West, Inc., California Trucking Association*
Richard McKenna, *Captain, Marine Exchange*
Arul Arulmoli, Principal, *Earth Mechanics, Inc.*
Alan Thompson, Planning and Policy, *Southern California Association of Governments*
Michael Armstrong, *Government and Public Affairs, Southern California Association of Governments*
K. Murthy, Assistant Chief, *Construction, Metropolitan Transport Authority*
Lu Hersh, Technical Services Manager, *Alameda Corridor Engineering Team*
Sue Lai, Senior Transportation Engineering, *Port of Los Angeles*
Doug Thiessen, Managing Director Engineering, *Port of Long Beach*
Peter Yin, Senior Structural Engineer, *Port of Los Angeles*
Larry Cotrill, Master Planning, *Port of Long Beahb*
Gill Hicks, President, *Gill V. Hicks and Associates, Inc.*

Overview of the Issues

The Ports of Los Angeles and Long Beach handle 44% of imports to the United States and 80% of the container traffic. From these ports, goods are transported to the rest of the country—and most of the goods must cross the San Andreas Fault to leave southern California. The ShakeOut Scenario earthquake is sufficiently far away from the ports that little direct damage will result. However, there will still be business disruption that will have an impact on the U.S. economy.

Other earthquakes in southern California may cause direct as well as indirect damage to the ports. The cranes that load and unload containers respond to longer-period seismic waves, generated by large earthquakes, and liquefaction in soils is also a possibility.

Damage Estimates for the ShakeOut Scenario

Based on the regional liquefaction analysis, the likelihood of liquefaction occurring at the Ports of Long Beach and Los Angeles is low, with a probability of less than 7%. Lateral spread displacements are anticipated to be only a few inches (less than fifteen centimeters), although localized settlement up to twelve inches (30 cm) is possible. Assuming that some liquefaction will occur locally, we performed a review of a liquefaction study of the Port of Los Angeles, and shaking intensities were compared with readily available crane-damageability data, to arrive at the following estimate of damage: Localized liquefaction damage to some piers modestly reduces port capacity, and slight damage to cranes occurs, but the damage is repaired quickly. Damages to the rail lines across the fault and to the highway network (both described elsewhere) impose the principal limits on the cargo capacity of the ports.

Expert-Recommended Mitigation

Considering the impact of other earthquakes on the complex activities of the ports is beyond the scope of this study, but recommended, given the importance of the ports to the U.S. economy.

Hospitals

Study by Mark Pickett, *University of Toledo*

"Assessing the Impacts of a *M7*.8 Southern San Andreas Fault Earthquake on Hospitals"

Overview of the Issues

After a disaster, functioning hospitals are critical. However, in earthquakes worldwide, hospitals have lost functionality for critical time periods and have had to evacuate existing patients, either due to structural damage or as a result of non-structural damage that shuts down essential systems. All too commonly, generators and electrical systems are damaged from water that floods from broken pipes or toppling reservoirs.

Hospitals are built where people are, and thus there are many hospitals in older urban areas, with older buildings constructed based on older, less earthquake-resistant building codes. The state has tried to address this issue through a bill passed in the aftermath of the Northridge earthquake, that placed more stringent structural and nonstructural standards on hospitals than any other buildings in California. The bill had two goals: 1) structural retrofitting of old hospitals (recognizing that changing building codes only improves new buildings); and 2) addressing the non-structural systems whose damage will affect functionality.

The law has proved controversial because the required actions are very expensive at a time when the health care system is already in financial trouble. In addition, the law assumed that retrofitting an existing building would be cheaper than tearing down and constructing new buildings, but this has not been the case. Thus many hospitals have sought extensions of a 2008 deadline for retrofitting with an eye toward a 2030 deadline to get rid of old buildings completely. Another effort has been to limit hospital mitigation to the oldest buildings in the highest risk areas. Unfortunately, this effort did not limit mitigation sufficiently to improve the economic burden, as these highest-risk buildings stand throughout the urban centers of the Los Angeles and San Francisco Bay areas.

In the 1994 magnitude 6.7 **Northridge CA earthquake**, several hospitals lost functionality due to structural and/or non-structural damage (Pickett, 1995):

- The Olive View Medical Center, Sylmar, had partially collapsed during the 1971 San Fernando earthquake. The rebuilt, steel frame/steel shear plate wall structure performed well, despite very high peak ground acceleration (PGA) over 0.8g. Emergency power was intermittently lost due to equipment failures. However, there was significant water damage due to interior water lines rupturing after punching through non-structural walls, fire sprinkler heads shearing, and chill water lines in HVAC ducts rupturing. Because of these problems, 300 patients were evacuated and health care functions were performed in the parking lot for about 30 hours.
- At the St. John's Hospital, Santa Monica, non-ductile reinforced concrete frame buildings built in 1942 and 1954 failed and interior water lines ruptured due to failure of non-structural walls. Electrical fires occurred as a result of electrical grounds. The hospital evacuated approximately 195 patients. The facility was completely closed for three months, and it was not fully functional for six months.
- At the Veterans Administration Medical Center, Sepulveda, the earthquake caused a site free-field PGA of 0.94g and four buildings constructed in the period from 1952-1955 suffered major structural damage due to pounding at the intersections of wings of buildings and at seismic joints. Interior water lines ruptured at these locations. Other interior water lines also ruptured due to failure of non-structural walls. Emergency power was lost as a result of electrical grounds caused by these water line failures and by emergency generator batteries that toppled over. The four buildings were without electrical power for 48-60 hours. Approximately 330 patients were evacuated.
- At the Holy Cross Medical Center (constructed in 1976; never seismically retrofitted), water mains in the surrounding area ruptured. Interior water lines ruptured due to failure of non-structural walls and rupture of heating water coils in HVAC ducts. All functions of the facility were interrupted, which required the evacuation of approximately 50 patients. The facility was not fully functional for three weeks. Emergency power was lost in portions of the facility due to circuit breakers opening automatically after water caused electrical grounds. One patient died due to loss of power to the respirator.
- In the area surrounding the Granada Hills Community Hospital, water mains ruptured. The facility suffered interior damage when rooftop tanks moved in their saddles, ruptured their piping and spilled their contents. This, in turn, caused the loss of some functionality, resulting in over 1,000 patients being treated in a parking lot. The facility could not accept trauma patients for two days.
- At the Los Angeles County Medical Center, where the earthquake caused a site free-field PGA of 0.49g, both the Psychiatric Hospital and the Pediatric Hospital suffered significant damage to non-ductile reinforced concrete columns. Both facilities were constructed within the years 1952 to 1954 and had never undergone any seismic retrofit. The Psychiatric Hospital also suffered water damage when three rooftop tanks moved in their saddles, ruptured their piping, and spilled their contents. The Pediatric Hospital evacuated 67 patients. The Psychiatric Hospital evacuated approximately 100 patients. Both facilities were red-tagged for over 15 months.

- The University of Southern California "University Hospital" building was located across the street from the Los Angeles County Medical Center. A steel frame, base-isolated building, built in 1991, it remained functional and had less than $500 worth of damage.

In the 1999 **Izmit, Turkey** earthquake, several hospital buildings were located at sites subjected to PGA of 0.21g to 0.23g (Pickett, 2003). Little damage and no disruption of function occurred at the newer buildings, which were built after 1973 with ductile reinforced-concrete structural frames. In buildings built within the years 1938 to 1939, with non-ductile reinforced concrete frames or unreinforced masonry, there was significant wall and column damage that necessitated the evacuation of patients. Three hospital buildings were located at sites subjected to PGA of 0.4*g*. At these sites, wall and column damage occurred and patients were evacuated from two buildings constructed with non-ductile reinforced-concrete frames; however there was little damage and no disruption of function at a ductile reinforced-concrete frame hospital building.

In the 2007 **Pisco, Peru** earthquake (Pickett, 2007), the Felix Alva Hospital in Ica (PGA = 0.33g, Johansson) suffered damage to interior water and wastewater lines. This caused a 75% loss of function and the evacuation of 60-80 patients. In Pisco (MM VII- VIII, Tavera, Johans son), two hospitals lost all functions in their unreinforced masonry buildings, while several ductile reinforced concrete buildings in these facilities remained undamaged but lost functionality due to loss of electrical power.

In Hurricane Katrina there were damages that also pose concerns in an earthquake like the ShakeOut earthquake. Water supply and waste water elimination were major problems and one hospital, with no structural or power problems, was forced to shut down all functions, due to loss of water. That hospital subsequently made plans to dig wells. Another hospital was unable to eliminate its wastewater and the internally generated wastewater eventually flooded the vault area that contained the main electrical panels. Consequently, emergency power had to be shut down, and for several days the facility was without power, except for a few hand-carried portable generators.

The Office of Statewide Health Planning and Development (OSPHD) lists 516 hospitals in the document entitled "California Hospitals Licensed as of 6/30/2007." The Structural Performance Category (SPC) and Non-structural Performance Category (NPC) of 436 hospitals were reported to OSPHD in the "Summary of Hospital Seismic Performance Ratings," April 2001. In the 10 county region of the report by Pickett for the ShakeOut, there are 164 General and Acute Care (GAC) hospitals that provide licensed Emergency Medical Services (standby, basic or comprehensive). The 10 counties covered by this study include Imperial, Kern, Los Angeles, Orange, Riverside, San Bernardino, San Diego, San Luis Obispo, Santa Barbara, and Ventura. The Santa Barbara and Ventura County hospitals were not considered in the ShakeOut Scenario.

For the eight counties of the ShakeOut Scenario, plus San Luis Obispo and Santa Barbara counties, Table 4-2 lists the total number of hospital buildings reported in the various SPC and NPC ratings, and their total licensed beds and GAC licensed beds. A rating of SPC = 1 means that the building poses "a significant risk of collapse and a danger to the public after a strong earthquake." A rating of SPC = 3 signifies that the building "may not be repairable or functional following strong ground motion." A rating of SPC = 5 means that the building "is reasonably capable of providing services to the public following strong ground motion." A rating of NPC = 2 signifies that the building "is expected to suffer significant non-structural damage in a strong earthquake." A rating of NPC = 3 means that "non-structural systems are adequately braced in critical areas of the hospital" and "the hospital should be able to provide basic emergency medical care following a strong earthquake." A rating of NPC = 5 means that "contents are braced" so that the "hospital building should be able to function" and that the building has "on-site fuel" and "water and wastewater tanks, sufficient for 72 hours of emergency operations." A rating of zero means that no rating was reported for that building.

Table 4-2. Hospital Building Ratings and Licensed Beds in the Region of Study

Counties	Total Reported Buildings	SPC Ratings						NPC Ratings						Total Licensed Beds	GAC Licensed Beds
		0	1	2	3	4	5	0	1	2	3	4	5		
Imperial	22	0	9	2	0	7	4	0	16	0	0	6	0	221	221
Kern	63	1	19	8	0	28	7	1	33	13	2	13	1	1636	1253
Los Angeles - North	451	1	241	22	62	84	41	7	363	64	7	8	2	20665	17748
Los Angeles - South	214	0	95	18	29	41	31	0	161	41	6	6	0	8927	7310
Orange	181	14	61	9	15	63	19	14	139	23	0	4	1	6843	5699
Riverside	80	0	29	3	15	24	9	1	59	15	2	1	2	3203	2764
San Bernardino	135	1	56	9	19	26	24	1	104	18	2	10	0	4015	3446
San Diego	158	15	55	15	18	44	11	15	134	8	0	1	0	6790	5289
San Luis Obispo	25	0	5	1	4	10	5	0	12	10	1	2	0	554	498
Santa Barbara	45	0	16	6	2	13	8	0	28	10	0	7	0	1211	743
Ventura	48	0	18	2	9	13	6	0	32	8	3	4	1	1510	1120
10 County Total =	1422	32	604	95	173	353	165	39	1081	210	23	62	7	55575	46091
Statewide Total =	2507	35	975	211	291	672	323	49	1807	430	63	143	15	90136	73684
% of Statewide Total =	57%	91%	62%	45%	59%	53%	51%	80%	60%	49%	37%	43%	47%	62%	63%

Table 4-3. Peak Ground Acceleration (PGA) at the Sites of 164 GAC Hospitals in the Region of Study

Counties	Number of GAC Hospitals PGA<0.1	Number of GAC Hospitals 0.1•PGA<0.2	Number of GAC Hospitals 0.2•PGA<0.3	Number of GAC Hospitals 0.3•PGA<0.4	Number of GAC Hospitals 0.4•PGA<0.5	Number of GAC Hospitals 0.5•PGA<0.6	Number of GAC Hospitals 0.6•PGA<0.7
Imperial	2	0	0	0	0	0	0
Kern	9	1	0	0	0	0	0
Los Angeles	17	32	10	3	2	0	0
Orange	5	23	2	0	0	0	0
Riverside	3	4	3	1	0	0	0
San Bernardino	0	3	6	2	1	2	1
San Diego	17	1	0	0	0	0	0
San Luis Obispo	3	0	0	0	0	0	0
Santa Barbara	4	0	0	0	0	0	0
Ventura	6	1	0	0	0	0	0
Total =	66	65	21	6	3	2	1
% of Total	40%	40%	13%	4%	2%	1%	0.6%

Based on ground shaking maps developed for the postulated M7.8 southern San Andreas Fault earthquake, the peak ground accelerations (PGA) at the locations of the 164 GAC hospitals exposed to loss in the 10- county region are presented in Table 2.

Damage Estimates for the ShakeOut Scenario

The Scenario posits damages for general and acute care (GAC) hospitals providing licensed emergency medical services (standby, basic, or comprehensive). The eight county ShakeOut study area includes 1422 buildings at 157 GAC hospitals. Approximately 20% of the GAC hospitals, located in four counties (Los Angeles, Orange, Riverside, and San Bernardino), are subjected to damaging shaking, in excess of 0.2g peak ground acceleration. In two of these counties, Los Angeles and San Bernardino, there are hospitals located at sites with shaking in excess of 0.4g from the ShakeOut Scenario earthquake.

Of hospital buildings in Los Angeles, Orange, Riverside, and San Bernardino counties, over 60% of the buildings are nonfunctional and suffer irreparable damage. Over 85% of the buildings suffer significant non-structural damage. Hypothetical damage is particularly severe at two hospitals in Los Angeles County and four hospitals in San Bernardino County. See Table 4-4 for patient service statistics at these facilities. Most buildings in these hospitals suffer structural damage such as failures of non-ductile beam- column joints, pounding at wings and seismic joints, and wall failures, leading to leaks or ruptures of interior water and wastewater piping systems, and creating a likelihood of red-tagging. That is, the buildings are unsafe to enter and occupy, per building safety inspectors using the ATC-20 assessment methodology.

At these six hospitals most of the buildings experience extensive non-structural damage:

- Some roof-mounted HVAC cooling towers topple because of anchorage failure.
- Some roof-mounted saddle tanks move in their saddles, rupturing their inlet/outlet piping and losing their contents.
- Water from these sources causes grounds in the buildings' electrical distribution systems. To avoid fires caused by these grounds, circuit breakers trip open, either manually or automatically.
- Consequently, the facilities are without electrical power and critical systems and equipment are inoperable.
- Some of the hospitals have portable, hand-carried generators, but they lack sufficient quantity to power enough systems and equipment to regain normal functionality.
- Thus, electrical grounds caused by water-line ruptures cause near total loss of functionality for these facilities.

Table 4-4. Beds at Severely Damaged Hospitals

County	Total of severely damaged hospitals	Total licensed beds/ (% of county total)	Total GAC licensed beds(% county total)
Los Angeles	2	448 (2%)	370 (1%)
San Bernardino	4	1,479 (37%)	1,304 (38%)

Other non-structural damage impedes functionality when:

- Medical gas piping ruptures due to interior, non-structural wall failures;•
- Patient care functions are interrupted due to toppling of unrestrained patient record shelves and unanchored nurses' stations;
- Unanchored equipment falls from shelves, desks, and counters; and
- Unrestrained equipment topples, including medical gas and water filtration cylinders.

There is also significant impact from damage to utilities:

- External electric power is lost immediately and is restored in the subsequent 24-72 hours or so.
- The external supply of potable water is lost for several days.
- The earthquake causes immediate interruption of commercial telecommunications:
    - Land-lines briefly fail due to some combination of usage that saturates capacity, equipment or structural damage at central offices, and breaks in copper and fiber lines.
    - Cellular systems fail due to some combination of saturation, tower failure, and other equipment damage.
    - Some facility PBX (Private Business Exchange) systems are functional while UPS (uninterruptible power supply) batteries maintain power, but because of the water and electrical grounding problems, power distribution fails, and thus so do to PBX systems, as well as wall-mounted INTERCOM systems.
    - Consequently, interior and exterior communications can only be maintained by radio-telephones.

Expert-Recommended Mitigation

Structural—To ensure continuation of a building's function, additional seismic detailing is required for strength and ductility, such as in the use of ductile beam-column joints, shear walls, and possibly, based isolation.

Non-structural—Water has been the single cause of most major non-structural problems. Hospitals should identify vulnerable lines and mitigate as appropriate with:

- installation of flexible and/or telescoping connections; and
- creation of relative motion gaps in wall penetrations.

Shelter in place—Per Office of Statewide Health Planning and Development (2001), to meet the needs of patients and staff, hospitals must have 72 hours worth of batteries and storage tanks adequate to store water and wastewater for 72 hours.

Elevators

Study by: Anshel Schiff, *Precision Measurement Instruments*

"Assessing the Impacts of a *M7*.8 Southern San Andreas Fault earthquake: elevators"

Overview of the Issues

Overall, being trapped in an elevator during the ShakeOut Scenario earthquake, or unable to use one afterwards, may be very stressful for the limited number of people affected, but will not have widespread consequences. Most of the exceptions will concern non-functioning elevators in hospitals, and other rescue operations that are delayed while first responders rescue people out of elevators. In some cases, trapped but healthy people may become seriously injured because they try to free themselves. Also, slow restoration of elevator function in critical facilities, especially hospitals, is a life-safety issue, because of the need to transport injured patients between floors.

Damage Estimates for the ShakeOut Scenario

Some physical damage to elevators does occur but is not the significant problem. Loss of electric power traps dozens to hundreds of elevator occupants for several hours or more, requiring first responders to remove them from the elevators. The tripping of seismic switches requires large numbers of elevators to be inspected by elevator mechanics before being put back into service and causes long restoration times.

Expert-Recommended Mitigation

- In hospitals and other facilities where disruption of service is a life-safety issue, consider implementing proposed but unadopted guidelines for restoration of service after triggering of seismic switches, without inspection.
- Educate building engineers (and possibly others) so that they can help elevator mechanics and firefighters to safely extricate trapped people. Do not, however, widely disseminate the procedures because the knowledge can be dangerous.
- Augment building code to require key parts to be tested and qualified.
- Add P-wave sensors so that elevators will stop and open doors at nearest floor when strong shaking is detected.
- Because of strong competition, elevator companies sustain few personnel. To meet post-earthquake demand for inspection and restoration, establish mutual aid among companies in and out of area.

Dams, Reservoirs, and Aqueducts

Panelists:

Mutaz Mihyar, Supervising Engineer, *Department of Water Resources Division of Safety of Dams*

George Barber, *Metropolitan Water District*

Charles Nestle, Geotechnical Engineer, *County of Los Angeles Public Works*

Raul Escandon, *County of Los Angeles Public Works*

Linda Bell, Geotechnical Engineer, *County of Los Angeles Public Works*

Chris Hill, Senior Engineer, *Metropolitan Water District*

Mike Morel, Operations Manager, *Metropolitan Water District*

Katy Gibson, Manager, Emergency Operations, *Metropolitan Water District*

Eric Reichard, *USGS Califonia Water Science Center*

Craig Davis, Manager, Los Angeles Department of Water and Power

Overview of the Issues

Most of southern California's water is imported and must cross the San Andreas Fault to reach us. Based on previous earthquake experience, and anticipating disruption from future earthquakes, water companies store months of water on the near side of the fault and have thought in detail about the types of repairs that will be needed after an earthquake on the San Andreas Fault. Although most lay people tend to worry the most about this part of the water delivery system, this is an area of infrastructure where retrofitting and planning are in place.

Because dam inundation causes severe and widespread death and damage, dams are built to higher standards than other structures and California's Division of Safety of Dams exists to monitor dam safety. Still, over the years, around the world, dam failures have been some of the most catastrophic consequences of earthquakes. When a dam fails, the outflow of water sweeps away structures in a way no engineering can resist. When a dam is damaged to the extent that evacuation is required, the evacuation is complicated by earthquake damage to roads and communications, and the effort competes with other pressing, post-earthquake response needs.

After the San Francisco earthquake, California's second most deadly natural disaster was southern California's 1928 San Francisquito Dam failure, even though it occurred in a relatively unpopulated era and area. This failure was not caused by an earthquake but emphasizes the destructive power of dam failures. In 1971, the Sylmar earthquake caused near failure of the Van Norman Dam and required the evacuation of 50,000 people. The same earthquake caused severe damage to the Pacoima Dam, which has sat unused since.

As is true with other structures, standards for dam construction have been improved with lessons learned from earthquake damage, including the 1971 earthquake. There are many dams in California that were constructed before the current standards were enacted. The ShakeOut Scenario assembled a panel of experts to take all this into account when positing damages due to the ShakeOut earthquake.

Damage Estimates for the ShakeOut Scenario

Due to strong ground shaking, on the order of 30 dams within approximately 15 miles of the fault experience damage serious enough to cause safety concerns. Of these, three that are nearly full have transverse cracking with muddy water emerging at the toe, and therefore may require emergency discharge and possible evacuation of populated areas below the dam. None of the dams ruptures. (For emergency exercise purposes in November, in collaboration with county agencies, locations of three real dams will be identified, but their selection will not indicate real vulnerability.)

Fault offset ruptures the California Aqueduct at two places near Palmdale. At one location, the rupture causes water to run east along the fault across the Antelope Valley Freeway near Lakeview, into the Palmdale Reservoir, and thence into the neighborhood to the east. At the second location, the rupture floods scrublands to the north. In addition, the fault rupture offsets the levee at the east end of Palmdale Reservoir by approximately 3 m, allowing water to flow along the arroyo toward Cemetery Road, flooding nearby neighborhoods. The fault also crosses the Coachella Canal at three places, and rupture by offset seems inevitable at these locations, with resulting flooding of the scrublands to the south and west of the canal. Particular locations of these damages have been identified and are available in a Google Earth formatted file. The fault also ruptures the Colorado River Aqueduct at several locations and a tunnel portion of the Los Angeles Aqueduct at one location. In the tunnel, the ShakeOut Scenario fault offset of 3.5 m greatly reduces flow to the City of Los Angeles. It is posited that 6,000 feet of roof and sidewall block fail in the tunnel starting 11,000 feet from the North Portal. In addition, the first 5,000 feet experience failures of the roof and tunnel lining. An additional 2,000 feet of roof failures block the tunnel at about 19,500 feet from the North Portal in an area of particularly weathered and cracked rock. As a result, water backs up into the Fairmont Reservoir at the North Portal until both barrels of the aqueduct are shut down. That is likely to overflow the relatively small capacity of the Fairmont Reservoir and flood a portion of the surrounding area. The hydrostatic head of the ground water and added depth of water at the North Portal provide some small flow through the tunnel that will back up behind the 2,000 feet roof failure further downstream. That will in turn seep through the debris and make repairs difficult and dangerous for miners entering from the south. Repairs, if undertaken, are posited to take between 80 and 800 days, once the debris is removed in the first 5,000 feet. It might take six months after the earthquake before tunnel excavation operations begin. One alternative to repairing the tunnel would be to abandon it and build a surface route to deliver this water to the City, although there are serious challenges to this approach.

Expert-Recommended Mitigation

Use the Scenario:

- to evaluate how agencies will interact and improve coordination among them;
- to practice emergency plans and determine whether existing protocols will work efficiently and effectively;
- to educate and engage owners of the small-capacity dams, and canals with high storage capacity that are not under the jurisdiction of California Division of Safety of Dams. (These are similar to structures that breached in Hawaii.)
- to rethink and improve:
  - warnings, and communication with the hazards community;
  - blowoff;
  - prepare blowoff inundation maps.
  - the evacuation decision-making process;
  - channels between release point and detention area; and
  - evacuation routes by using Geographic Information Systems rather than pdf files.

Wastewater and Debris Disposal

Panelists:
Randy Van Gelder, San Bernardino Valley Water
Joseph Crisologo, California State Department of Health
Kai Lui, Los Angeles County Sanitation

Overview of the Issues

In the ShakeOut study area are some 22 million people and that many people generate a lot of sewage. Most of it passes through brittle pipes that are vulnerable to earthquake shaking. Pipes that do not break completely may crack and wastewater may then contaminate nearby drinking water pipes as well as ground water reservoirs, at a time when disruption to aqueducts means more ground water will be drawn. It may be some time before such contamination is recognized. In addition to these sometimes subtle pipe damages, pipe breaks and above-ground damage to facilities will create sewage spills, each of which dramatically increases health risks.

Large earthquakes create enormous amounts of debris, measured in millions of tons. Debris must be cleared or removed before roads and buildings can resume functioning or be reconstructed. Transportation and disposal of debris each pose significant complications.

A panel discussion addressed the impacts of the ShakeOut Scenario on wastewater treatment and debris disposal in the 24 Sanitation Districts of Los Angeles County (SDLAC). We extrapolated from that discussion to the entire study area

Damage Estimates for the ShakeOut Scenario

It is posited that damage to sewer pipelines and equipment at **wastewater treatment plants** throughout the study area results in five to ten million gallons per hour of untreated sewage spilling onto streets in 50 to 100 locations throughout the study region. Although sanitation districts attempt to relieve flow by routing untreated sewage directly to the ocean through dedicated pipelines, most or all water treatment plants are forced to dump untreated, raw sewage into nearby creeks (which flow by gravity to the ocean), as happened in New Orleans after Hurricane Katrina. Dumping sewage into creeks causes temporary biohazard near the wastewater treament plants. Sanitation districts advise people not to go near these locations, and to stay away from open standing bodies of water. Dumping untreated sewage into creeks requires cleanup, involving the dumping of bleach in the areas of spills.

In response to the damage, sanitation districts ask people to minimize the use of toilets. Repairs proceed in a prioritized fashion, while the system is in use. The sanitation districts seek mutual aid for several weeks after the earthquake. Even with mutual aid, it takes several weeks to repair enough of the system so that all sewage flows directly to the ocean rather than via creeks, two to three months before all sewage is treated in a safe way, and five to six months to achieve normal sewage treatment. Ultimately some of the damaged pipelines, equipment, and electronics take up to five years to repair.

Active **landfills** accept debris as do some closed landfills, which maintain required personnel. Reopening closed landfills first requires a declared state disaster or a presidential directive. Before debris can be received, personnel at landfills receive baseline immunizations and protective equipment, and FEMA and state authorities review and approve the debris-handling plan. Operations run round-the-clock, and SDLAC alone can accept between 10,000 and 20,000 tons per day of debris. Tonnage varies depending on waivers to allow it to take construction debris; there would be no separating of recyclables, concrete, asbestos, or other materials.

Expert-Recommended Mitigation

- Use the Scenario to improve planning and training:
  - Review feasibility of current personnel plan: in case of natural disaster such as tsunami or earthquake, the plan expects to replace more than 50% of staff with contractors.
  - Establish priorities for repair and incident response.
  - Establish guidelines for restricted use while system is being repaired.
  - Practice emergency response exercises.
  - Train employees in updated response plan.
  - Train employees to document expenses, to avoid problem with unreimbursed expenses that was faced after the Northridge earthquake.
- Stockpile supplies and equipment including:
  - fuel supplies and secondary support generators;
  - food and facilities to shelter employees in place; and
  - radio repeaters and stand-alone antennas for base-station radios at district operating centers.
- Advocate seismic evaluation of critical buildings, equipment, and pipelines.

iii. Non-Structural Damage

In destructive earthquakes worldwide, non-structural and content-related damages in buildings have long been recognized as persistent, important sources of casualties and losses, including loss of function. Non-structural elements account for at least 75% of a building's cost, and losses attributable to contents alone are commonly estimated as at least one third of the total earthquake losses (FEMA, 1994). Of the $18.5 billion in building damage due to the 1994 Northridge earthquake, Kircher (1993) attributed 50% to non-structural damage.

Here are just a few examples of common types of problems (FEMA, 1994):

- In homes, schools, and businesses, seemingly innocent elements such as falling light fixtures or toppling furniture can cause life-threatening injuries, and broken glass is responsible for the majority of less severe injuries.
- Due to breakage of materials and unconstrained containers, toxic chemicals (such as cleansers) and substances (such as asbestos) are readily released.
- Businesses may only need a few minutes to right fallen file cabinets but a week to re-organize spilled contents; unsecured computers and electronics equipment are particularly prone to irreparable damage.
- In the 1987, Whittier Narrows earthquake, a decorative concrete panel fell and killed a pedestrian.
- In the 1993 Guam earthquake, non-structural partitions at a hotel blocked and jammed doors, trapping guests.
- Non-structural damage during the 1994 Northridge earthquake closed 10 essential hospital facilities. Other hospitals with little or no structural damage were inoperable due to water damage. At more than a dozen, water leaks occurred when fire sprinkler, chilled-water or other pipelines broke, and water could not be immediately

shut off. At one facility, the emergency generator was disabled when its cooling water line broke. At another facility, water in circuit breakers shut down a respirator, causing the death of a patient. Pickett (2008) cites many other examples of earthquake-induced water damage to hospitals. Similar experience exists for museums, libraries, and schools.

For building owners and occupants, mitigation of non-structural damage is often, by far, the simplest, most cost-effective way to prepare for and reduce earthquake impact. Numerous resources exist to help with this mitigation and many can be found through a web search on the phrase "non-structural earthquake damage".

iv. Transportation

Highways

Study by Stuart Werner, Sungbin Cho, and Ronald Eguchi, *Seismic Systems and Engineering Consultants* "Analysis of Risks to Southern California Highway System Due to *M*7.8 Earthquake Along Southern San Andreas Fault"

Panelists, Meeting Attendees, and Contributors:
Basam Muallem, Caltrans
Michael Keever, Caltrans
Phil Stolarski, Caltrans
Ralph Ricketson, Caltrans
Kevin Thompson, Caltrans
Henry Stultz, Caltrans
Gustavo Ortega, Caltrans
Michael Perovich, Caltrans
Greg Albright, Caltrans
Marco Ruano, Caltrans
Mike McManus, Caltrans
Joe Hull, Caltrans
Zhongren Wang, Caltrans
Mike Jenkinson, Caltrans
Dan Freeman, Caltrans
Frank Quon, Caltrans
Loren Turner, Caltrans
Richard Almanzan, Caltrans
Armand Silva, Caltrans,
Doug Failing, Caltrans
James Province, Caltrans
John Yang, Caltrans
Mark Yashinsky, Caltrans
Gedion Werrede, Caltrans
Sved Raza, Caltrans
Christopher Harris, Caltrans

Tom Shantz, Caltrans
Martha Merriam, Engineering Geologist, Caltrans
Larry Banks, Caltrans
Keith Killough, Southern California Association of Governments
Mandy Chu, Caltrans
Ron Eguchi, ImageCat
Stu Werner, Seismic Systems and Engineering Consultants
Sungbin Cho, ImageCat
Judith Steele, Judith Steele Consultants
Bill Yearsley, University of Colorado, Boulder
William Nascimento, Lim & Nascimento Engineering
Mohan Char, Lim and Nascimento Engineering
Kwan Lam, Caltrans
Rick Wilson, California Geological Survey
Jerry Treiman, California Geological Survey
Joel Conte, University of San Diego
Frances Banerjee, Banerjee & Associates
Kevin Thompson, Caltrans
James Davis, Caltrans

To posit damages for the ShakeOut Scenario, a team of geologists evaluated damages to the Interstate and California highways that may result from surface rupture, landslides, ground failure, and liquefaction. A team of structural and risk engineers performed a computer simulation to screen for ground-motion-induced damage to over 6,000 California highway bridges and to estimate highway-bridge repair times. The highway-bridge simulation analysis was reviewed by a panel of experts and five critical zones of bridge damages had additional evaluation by Caltrans. Experts at Caltrans reviewed a sample of bridge design plans and the damage results for bridges within the critical damage zones, and conducted independent analyses at 13 bridges most likely to suffer serious damage. The damages posited here reflect the results after this vetting process.

Overview of the Issues

The impact of disruption to the transportation system covers the full range of issues in earthquake response and recovery. Immediate disruption occurs when roads are offset during fault rupture, when roads are affected by earthquake-induced ground deformation such as a landslide or settlement, when structural debris falls into roads, and when bridges, including overpasses, go down. The southern California road system comprises hundreds of thousands of bridges – not because there are that many rivers to cross, but because roads need to cross other structures. Any time a road is not poured literally on the ground it is considered a bridge.

There are many ways a road can become impassable. Some problems, like debris, can be quickly removed. However repairs due to fault offsets or downed bridges can disrupt traffic flow for many months. In an earthquake the size of that in the ShakeOut Scenario, fault offset and other ground deformations can be significant, and so separate pieces of the study address these issues in detail (see Section B, this chapter). In addition, it is extremely likely that aftershocks will cause additional ground deformation such as landslides and ground rupture, causing some locations to require multiple rounds of repairs.

The California Department of Transportation (Caltrans) has devoted substantial effort and money to reduce the potential for disruption to the California highway system. In their most recent effort, begun after the 1989 Loma Prieta earthquake, they have invested over $6 billion to mitigate the types of damages seen in that earthquake. Nonetheless it is not cost effective to build to withstand every possible pocket of high shaking that will occur in a large earthquake (particularly because the precise locations of those pockets can rarely be anticipated). Moreover, earthquakes cause damage due to many factors, some of which are rarely seen. Some factors that imperil the safety of the bridges will escape engineers' best efforts to predict because they have not been seen in the short amount of geologic time that humans have been scrutinizing earthquake effects on infrastructure. Such factors can only be recognized after they occur in some earthquake.

Impacts of damage to the transportation system ripple out to all else the ShakeOut Scenario is considering. In the short term, transportation system disruption will affect response because it will slow travel times for fire trucks, ambulances, police cars and other emergency vehicles. In the intermediate term, as repairs are done to bridges and deformed roads, employees and customers across the region are forced to spend more time in traffic, resulting in regional economic cost. In both the intermediate and longer term, the disruption of movement of goods impedes the activities of the ports, causing economic disruption of national importance.

Damage Estimates for the ShakeOut Scenario

All highways, secondary highways and streets that cross the fault suffer surface rupture in the range of 0.2-8.3 meters (about 0.6-27 feet). Inspection of the major highway fault crossings reveals surface rupture as follows:

- **California Highway 14:** across multiple fault strands, up to 3 meter slip along as much as 320 meters of highway),
- **Interstate 15 (I-15):** up to 2.4 meter slip along 180 meters of highway,
- **Interstate 10 (I-10), San Gorgonio Pass:** up to 0.7 meter slip along as much as 1,660 meters of highway,
- **California Highway 62:** up to 0.94 meter slip along as much as 56 meters of highway, and
- **Interstate 10 (I-10), Coachella Valley:** up to 6.7 meter slip along as much as 140 meters of highway.

Damage from fault surface rupture takes two or more months to repair. Amount of fault displacement, and therefore direct damage to the roadway, is most severe in the Coachella Valley. At San Gorgonio Pass, although fault displacement is relatively small, repair times are increased because significantly more roadway is damaged by faulting that is nearly parallel to the highway.

Multiple landslides along the I-15 in Cajon Pass dump up to 75,000 cubic meters of debris on the highway. North of the fault on I-15, the large highway fill prisms suffer significant amounts of seismically induced settlement, with cracks displacing the highway vertically as much as several meters, affecting all lanes. In addition, some large slopes on the fill prisms experience landslide-type failures. The settlement and slides greatly inhibit access and highway repair. The old U.S. Highway 66 through Cajon Pass is not available as an alternate route due to fires caused by rupturing pipeline, erosional undercutting, and debris. On the I-10 through San Gorgonio Pass, due to a large slump on the steep slopes or reactivation of a large landslide mass north of the highway, 45,000 to 1.5 million cubic meters of debris cover the highway.

Potential liquefaction locations have been identified along the:

- I-10, San Gorgonio Pass: 2 sites with 4-8 meter displacements;
- California Highway 111, San Gorgonio Pass: 1 site with 3-5 meter displacements; and
- I-10, Coachella Valley: 5 sites, each with 2-3 meter displacements.

The most concentrated highway bridge damage is estimated to occur in San Bernardino. Highway segments affected by bridge damage are located in

- Palmdale along the California Highway 14;
- Cajon Pass and San Bernardino along the I-10, I-15, I-215, California Highways 30 and 138;
- San Gorgonio and Palm Springs along the California Highway 62 and I-10;
- Coachella Valley and Indio along the I-10, California Highways 86, and 111; and
- the vicinity of Baldwin Park along the I-2 10, I-10, I-605, and California Highway 60.

It will take up to seven months to restore highway segments affected by bridge damage, fault offsets, landslides and liquefaction, and bridge rebuilding is the critical factor in reopening a highway segment. Irreparable bridge damage will take 5-7 months to rebuild, and one month to open the roads beneath the bridges. Extensive damage and moderate damage will take weeks and days to repair, respectively. Highway rebuild and repair times have been developed in consultation with Caltrans, but without consideration of the effects of co-located lifeline damages, bonus incentive programs for early completion, additional damage due to large aftershocks, or parallel demands for and management of contracts.

Caltrans estimates that, based on their retrofitting, there will be no casualties because of failure of their highways or bridges during the ShakeOut Scenario earthquake.

Most of the regional traffic occurs within the study area, rather than through it. Post-earthquake traffic analysis for the region posits:

- during the first three days, a 21% regional trip time increase and a 3% decrease in the number of trips,

- two to four months later, an 8% trip time increase and a 0.3% decrease in number of trips

Maps of postulated highway surface slip, landslides, ground failure and liquefaction, and bridge damage zones are available in Chapter 3.C, Chapter 4.B and Appendices E-G. Cost estimates of highway repair and replacement, and trip delays are presented in Chapter 7.

Mass Transit and Surface Streets

Mass Transit Panelists:
Tracy Berge, *Metrolink, Safety/Security*
David Quirk, *Metrolink/Engineering*
Frances Banerjee, *Banerjee & Associates (retired Los Angeles Department of Transportation (LADOT)*
Aram Sahakian, *LADOT, Transportation Engineering*
Scott Richardson, *Riverside Transit*
Tom Jasmin, *Los Angeles County Metropolitan Transportation Authority (LACMTA), bus/rail*
Michael Griffin, *LADOT*
Keith Killough, *Foothill Transit (San Gabriel Valley)*
Judith Steele, *Judith Steele Consultants*
Robert Castanon, *LACMTA, bus*
Hector Guerrero, *LACMTA, rail*

Surface Streets Panelists:
Aran Sahakian, *LADOT*
Gilbert Pedroza, *LA County Public Works*
Dai Bui, *Los Angeles County Public Works*
Tom Cotter, *Los Angeles County Public Works*
Raul Escandon, *Los Angeles County Public Works*
Ralph Ricketson, *Caltrans*
Judith Steele, *Judith Steele Consultants*
James Faber, *Lam Civil Engineering*
J. Leyva, *City of San Bernardino Public Services*

Damage Estimates for the ShakeOut Scenario

Roadway damage and debris. Damage assessment takes two weeks citywide in Los Angeles, with more than 1,000 inspectors required, borrowed from other agencies. Clearing debris on roadways is a substantial task: block walls, fencing, and collapsed buildings require months of prioritized debris removal. It takes one to two weeks to clear most major arterials in the cities of Los Angeles, Riverside, San Bernardino, and elsewhere in the affected region.

Traffic signals. Power outage causes loss of operation to more than 4,000 traffic signals in the city of Los Angeles alone; and perhaps 12,000 to 15,000 throughout the region. Only a limited number of traffic officers are available to direct traffic; these officers are deployed to major arterials within perhaps four days of the earthquake. In the city of Los Angeles, approximately 3,500 traffic signals are connected via fiber-optic cable to the Automated Traffic Surveillance and Control Center (ATSCC). Fiber is damaged to some degree, so that once power is restored, 350 intersections still require officers to direct traffic for two to four weeks, until the fiber can be repaired. Moderate damage to mechanical, electrical, and plumbing (MEP) equipment and computers at the ATSCC also requires two weeks to repair, with critical repairs performed within seven days. Los Angeles Department of Transportation (LA DOT) does not expect many downed traffic light poles, probably fewer than the available supply of replacements, which could in any event be cannibalized from elsewhere. Riverside already suffers from extensive gridlock, and loss of power to traffic signals makes the situation worse.

Metrolink. Under normal conditions on a Thursday in November, Metrolink has 44,000 boardings, perhaps 10-20% of whom have no alternate means of getting to work, or home again if they have already arrived at 10 a.m. *Damages:* The Metrolink system shuts down for several hours until rail inspection is completed, but no Metrolink trains are derailed. Damage to the highway network and to pipes at the fuel storage facility impacts fuel availability and delivery, so Metrolink operates on a reduced schedule for two to four weeks after the earthquake. Track damage occurs to a 7,000-ft tunnel near Palmdale between Santa Clarita and Los Angeles. Moderate damage occurs to nonstructural components at the Metrolink dispatch facility in Pomona, but it is repaired within two weeks and does not significantly impact operations.

Riverside buses. Riverside provides approximately 23,000 boardings on a typical weekday, and 10,000 on weekend days. Of these, 85% are transit-dependent. Few are handicapped; these passengers rely on Dial-a-Ride. Dispatch relies on Nextel walkie- talkies, with no backup and no priority service. *Damages:* Bus service in Riverside and Hemet is impacted by communication problems and damage to natural gas pipelines. Immediate post-earthquake overload to the Nextel system eliminates communications between buses and the Riverside and Hemet operations centers. Because buses rely entirely on compressed natural (CNG), with no storage or backup, buses stop at the end of the day, and operations stops until Southern California Gas sufficiently repairs the transmission and distribution system pipes in the streets, perhaps one to two weeks after the earthquake.

Metropolitan Transportation Authority (MTA, Los Angeles County). MTA operates 11 bus divisions, 4 rail yards (one each in Long Beach and Hawthorne, two in Los Angeles), 1 underground subway (the Red Line), and 3 largely above-ground light rail lines (the Blue, Green, and Gold Lines). MTA operates 2300 buses running on CNG with little or no storage capacity, and somewhat fewer than 200 diesel buses. Support vehicles use diesel or nonleaded gas. MTA provides approximately 1 million average weekday boardings on its bus and rail system, and 760,000 on weekends. A large fraction (70-75%) is transit-dependent, without alternative transportation, and 10-15% are handicapped, such as blind and wheelchair-bound passengers. *Damages:* Trains stop with the loss of power, and it takes several hours to rescue all passengers and personnel from subways and aerial structures; the rescue is performed without police assistance. Damage inspections take 2 days. It is observed that several rail viaduct and aerial station columns are moderately to severely damaged, especially on the Blue

Line, where damage requires months to repair and renders the Blue Line inoperative from I-105 south. Liquefaction- related damage at the Compton Rail Control Center (which lacks a backup) causes loss of automated train control for at least one week after power is restored, so trains are temporarily run by line-of-sight until the signal system is restored and damaged fiber- optics are re-routed. MTA trains run on reduced service because of limited power due to damaged substations. Whereas on a normal November weekday trains run approximately every five minutes, for the next two to three weeks trains run approximately every twenty minutes. Buses also run at reduced schedule for several weeks after the earthquake.

LADOT buses. Approximately 65% of the fleet runs on propane, and the rest run on gasoline. Fuel is delivered via truck, with yards throughout the city as far south as San Pedro. Contractors run their own communications system that includes Nextel and 2-way radio. *Damages:* Immediate post-earthquake overload of the Nextel commercial radio network causes a loss of communication with LADOT buses. LADOT operations facilities suffer moderate damage to computer and other equipment, but these do not significantly impact operations. The most serious impact to LADOT bus operations comes from personnel disappearing for about one day to check on the safety of their families.

Expert-Recommended Mitigation

- Conduct needed investigations:
    - Check that equipment such as generators are secured, and if not, secure them.
    - At Metrolink Central Maintenance Facility, study liquefaction potential and determine whether mechanical, electrical and plumbing equipment there has been seismically evaluated.
- Establish contingency plans for situations that can be anticipated:
    - Develop procedures to rescue passengers without police assistance.
    - Develop procedures to remove structures from track.
    - Pre-identify NS and EW arterials most usable for detour routes.
    - Identify critical infrastructure (like hospitals) that must remain accessible and plan alternate routes.
    - Identify critical intersections to prioritize repairs and assign officers to do intersection control.
    - Decide traffic limitation rules such as truck traffic at night, or odd/even rules.
    - Define priorities for bridge repairs.
    - Define priorities for route restoration, such as those near emergency service facilities or in places where people can be trapped such as the Hollywood Hills.
- Expand advance planning:
    - Pre-plan with U.S. Army Corps of Engineers and others for repairs:
        - Arrange backup communications.
        - Determine internal staff who can be relied upon for repairs.
        - Expand mutual aid agreements.
    - Enact legislative program modifications that allow the suspension or extension or route reports, updates, and other mandated agency reporting requirements during emergencies.

  - Recognize that volunteers will be important to help with post-event damage surveys, perhaps create training or annotated forms for them?
- Consider funding and purchases:
  - Adding 2,070 adaptive controllers ($30,000/intersection) to 2,000 intersections would facilitate getting the roadway network operating quickly.
  - Buy new radios that are interoperable with police and fire.
  - Get generators with adequate fuel supplies.
  - Treat and fund traffic managers as first responders.
  - Acquire mobile command centers and portable emergency decontamination units.
- Other jurisdictions might consider emulating City of Los Angeles:
  - in effort to assess all bridges and tunnels for seismic resistance, and mitigate those that need mitigation;
  - in implementing multi-agency plans like the City's Emergency Operations Plan and its annexes;
  - by meeting needs documented in the City's emergency traffic plans, including acquisition of changeable message signs, both portable and permanent; and
  - by creating lists of pre-approved contractors and creating an emergency operations ordinance to bypass ordinary contracting constraints.

Rail

Study by William Byers, *P.E., retired*
"Impacts of a *M7*.8 Southern San Andreas Earthquake on the Railway Network"

Damage Estimates for the ShakeOut Scenario

Service Interruption—All lines in Southern California have service interruptions of at least four to eight hours until inspections establish their safety. However the Union Pacific (UP) line north to San Jose, because of its location, does not have any damage beyond possible signal malfunctions due to overturned relays. These do not pose a safety issue but do slow train movement through blocks governed by affected signals and across grade crossings with inoperative crossing protection. They are quickly repaired, within two or three hours after inspection is completed. The dispatching center at San Bernardino becomes temporarily inoperable, affecting train movement and thus near-term resumed operations, but not the immediate safety of trains. Radio communication with trains is uninterrupted.

Track Inspection: Locomotives are radio-equipped and all trains in the area are notified of the earthquake within minutes of its occurrence by the train dispatcher and/or the crews of trains that feel the earthquake. They then either reduce their speed to a restricted speed level or stop. In either case, they avoid running on damaged track. Once the epicenter location and earthquake magnitude are determined, trains within a to-bedetermined distance from the epicenter are stopped until the track, bridges, and signals are inspected and found safe. In moderate earthquakes the distance is 100 miles from the epicenter, but for an earthquake of this size, because several hundred kilometers of fault have ruptured, a more appropriate stopping distance should consider miles from the fault. Inspections are conducted from the track, using vehicles that can operate on both track and highway. Where the track is impassable, the vehicle is operated on the right-of-way maintenance road. Inspections are conducted from both ends of the affected area and, possibly, from intermediate points. If the track on the line is passable, inspections are completed within six to eight hours after the earthquake. Once a portion of the track is found to be safe, it is used to move stopped trains out of the affected area.

Track Damage—In this Scenario, rail lines through Cajon Pass and near the Salton Sea are severely damaged at the fault crossing, suffering bent rails and damaged alignment, and consequent loss of service that affects over half of the Los Angeles area rail traffic. Due to the topography, establishment of useable alignments through this area is relatively time-consuming. The amount of equipment that can be brought in is limited by congestion, and by earthquake-induced ground deformation affecting both tracks and the roads. Experience in past earthquakes suggests that tracks should be in service on the new alignment in no more than a week, provided the effort is not delayed by continuing ground deformation due to large aftershocks.

The rail line through the Coachella Valley runs close to the fault for several kilometers, crossing the rupture some eighteen times. It is posited that the repairs to this segment will be performed within two weeks, with several crews performing repairs in parallel, after which time rail operations resume through the Coachella Valley. However, there is limited equipment to grout railbed at the fault crossings, so completing the grouting takes up to two more weeks. During those two weeks, trains operate at reduced speeds, perhaps ten miles per hour, at locations that have not yet been grouted.

There are natural gas and petroleum product pipelines more-or-less parallel to the rail lines through Cajon Pass. In the ShakeOut Scenario these pipelines rupture, and although the rupture locations are far from the tracks, the resulting fires delay both inspections and repairs of railroad damage at Cajon Pass; hence the durations of inspections and repairs must be estimated from after the time when the fires are extinguished.

Derailments—Based on timetable speed restrictions for westbound trains and practical speed on the ascending grade for eastbound trains, along with experience in the derailment of trains in earthquakes worldwide, the ShakeOut Scenario posits that four trains are within an area of PGA > 0.5*g*, and that as many as three trains are derailed, including one in the Coachella Valley. One of the derailments, between Keenbrook and Devore, produces a hazmat release of twenty gallons of benzene, due to a broken pipe on the safety valve of an overturned tank car. No evacuation is posited here because when such a derailment and hazmat release occurred in reality, a railroad employee plugged the broken pipe to stop the leak, and no evacuation was required. Most derailments are now cleared by contractors. One of the primary contractors specializing in this type of work has branches in Fontana and Stockton and is able to bring equipment in from both sides of the fault. After the control of fires, typical times for clearing derailments are less than 24 hours.

Earthquake-induced Landslides—A ShakOut Scenario landslide at one or more locations in the Cajon Pass closes one or more of the Burlington Northern Santa Fe (BNSF) and Union Pacific (UP) tracks. Slide debris shed from the adjacent steep mountain blocks several portions of the tracks. In addition, movement of one landslide undercuts and displaces one of the tracks for several meters, causing disruption similar to that of fault rupture and similarly taking 1 to 2 weeks for repair. After equipment reaches the site, clearing of debris at the different locations will take on the order of hours, while repair of the track affected by landslides will take on the order of a week or two, at the same time as repairs performed at the fault offset.

Track System Replacement: There are twenty steel bridges consisting of 47 spans with a combined length of 2,227 feet on the part of the BNSF line subject to potentially damaging ground motions. (Parallel bridges on adjacent tracks are counted as separate bridges.) One of the piers is damaged similarly to a pier at the Southern Pacific's (SP)'s Pajaro River bridge in the 1906 earthquake: a corbel is shifted but does not collapse, and the line at this point tolerates operation at very low speed until the spans can be supported on falsework. Near Cajon Pass, there are two prestressed concrete trestles on steel piles crossing Cajon Creek with lengths of 180 feet and 277 feet, consisting of 28-feet, 30-feet and 33-feet spans. A realistic (upper-bound) estimation is that a few spans drop off their bents. Repairs consist of temporary trestles in the affected spans, which are constructed within 72 hours after the earthquake, followed by installation of replacement spans after service is restored. Between Verdemont and San Bernardino, there is a reinforced concrete underpass with spans of 36 feet and 44 feet under the two tracks. It is entirely possible that there is no bridge damage that would prevent continued operation of the railroad. In the worst case, repairs and reconstruction necessary to allow normal use of all bridges are completed in less than two weeks, with one BNSF track available in less than one week.

Expert-Recommended Mitigation

Anticipate post-event needs:

- Advance arrangement of waivers for permits, environmental impact reports, or other governmental requirements could hasten temporary repairs.
- Expect reduced operations that create conflicts between the needs of commuters and efficient movement of freight; develop arrangements to accommodate such needs.
- Recognize that helicopters are valuable tools for rapid post-earthquake reconnaissance, to identify areas that most quickly need detailed, on-track inspections; plan for their use in inspections and in rescue of stranded train crews.

Secure Equipment: Unnecessary loss of operability, particularly at the San Bernardino dispatch center, could occur due to damage to essential equipment. Communications, dispatching, and emergency power generation equipment should be physically secured to reduce the risk of earthquake damage.

v. Lifelines

Overview of the Issues

We call them *lifelines* because our lives require them: water, power, fuel, internet, and telephones. In addition, the lifelines depend on one another, and when one lifeline becomes unavailable that can slow the restoration and recovery of others. As just a few of many examples:

- Telecommunications equipment requires air conditioning, which in turn requires water and either electricity or gas for generators;
- Many lifelines are co-located and without coordination, repairs of one may interfere with repairs of others;
- Many lifelines are co-located and damage to one may increase damage to others; and
- Emergency response and utility restoration require the ability to communicate.

In general, lifeline providers are fully aware of the significance of their contribution to disaster damage, restoration, and recovery and many have taken important steps to resiliency. In many cases the larger providers, because they span wider regions, have more experience with previous earthquakes and have more mitigtion measures in place.

Telecommunications

Study by Alex Tang, P.E., *L & T Engineering*

"Assessing the Impacts of a *M7*.8 Southern San Andreas Fault Earthquake on Telecommunications"

Damage Estimates for the ShakeOut Scenario

The following damage estimate is based primarily on population and shaking intensity, rather than precise numbers and locations of telecommunication nodes, as such information is publicly unavailable for security reasons. Without precise locations, emergency responders cannot plan ahead in terms of access to particular facilities, but they can use estimates to plan resource needs.

Damage is concentrated in areas of high population density (where telecommunications assets tend to be concentrated) that experience high PGA (generally 0.4g or greater) or high MMI (generally VIII to IX or greater). Service impacts extend far beyond the heavy damage area, however, as a result of saturation, i.e., demand for telecommunications service suddenly surging after the earthquake in excess of the system's capacity. All services (voice and Internet) are impacted in these areas. For example, where the cables serving a community are disconnected, both the voice and Internet services are unavailable for that community. As summarized in Tables 4-5, 4-6, and 4-7, fourteen such breaks are posited and there are six regions where physical damage is likely to be concentrated. Table 4-6 indicates the approximate quantity of facilities in each region.

Table 4-5. Regions of Hypothetical Physical Damage to Telecommunications Facilities

Region	Location
1	Within approximately 30 km of the Ontario Airport.
2	San Dimas to South El Monte.
3	Eastern San Bernardino Valley (Colton, San Bernardino, Highland, Redlands)
4	Coachella Valley
5	Palmdale and Lancaster
6	Eastern San Gabriel Valley (Arcadia, Baldwin Park. West Covina), to east LA (Downey, Lynwood Rosemead, Southgate, Huntington Park)

Table 4-6. Approximate Quantities of Telecommunications Facilities

Region	CO/MTX/Remotes	Cell sites	911 PSAP centers
1	15-20	40+	5-8
2	10-15	25+	3-4
3	15-20	40+	8-10
4	20-25	40+	8-10
5	6-8	30+	4-6
6	20-25	50+	8-10

Structural damage to buildings has limited effect on telecommunications operations in the ShakeOut Scenario earthquake. The principal concern is red-tagging: damage to beams, columns, braces, or walls significant enough to cause building inspectors to post the building as unsafe to enter or occupy. Lesser structural damage might be costly for the building owner to repair and might hinder building access, but these costs are likely to be modest compared with overall regional economic loss (less than $100 M in total). Three kinds of non-structural impacts are more likely to cause service interruption:

- **Loss of commercial electric power**—This impacts cell service. Cell sites have backup power adequate for about three hours. A loss of commercial electric power therefore causes a loss of cell service from about three hours after the earthquake until power is restored or an emergency generator is brought to the cell site (often within one day).

- **Fiber-optic cable damage**—Fiber-optic cables are commonly co-located on bridges and overpasses. Bridge collapse or substantial offset can sever fiber-optic lines.

- **Equipment damage in Central Office/MTX/Remote buildings**—There are many modes of equipment damage that can render one of these facilities inoperative. Cable conveyance systems (overhead racks and trays) sway and can fall, sometimes on equipment below. Telephone switches are essentially computers that generate heat and must be continuously cooled to operate yet heating, ventilation, and air conditioning (HVAC) equipment commonly becomes dismounted from vibration isolators. Cards in telephone switching equipment can rattle loose. Suspended ceilings can collapse, potentially dropping light fixtures on equipment below and contaminating equipment with dust from broken tiles. Raised access floors can collapse if not properly braced. Emergency generators and their support equipment (tanks, pumps, and starter motors) can be damaged and fail to start. Battery racks and their supporting equipment, which provide uninterruptible power until emergency generators start, can be damaged and fail to operate. In cases where the batteries are lead-acid type, acid spills can occur. A variety of other mechanical, electrical, and plumbing equipment can be damaged from shaking and other causes, and result in operational failure. Table 4-7 summarizes the damage posited for facilities in the ShakeOut Scenario.

Companies that operate these facilities are typically well prepared to deal with equipment damage, and in the past have been able to fully return to normal within three days of an earthquake. For example, central offices (COs) with lead-acid batteries have neutralizing chemicals on site. Service providers have mobile COs: trailers filled with the necessary equipment that can quickly quick hook up to damaged facilities, along with power units to restore minimum service when an area is out of service. In many cases, these units provide free phone service for the victims. These mobile units are equipped with microwave or satellite dishes to connect to other COs in the network.

Many service providers also have agreements with manufacturers to fill emergency orders within a day. Mutual aid among the service providers in a state of emergency is an unwritten agreement. Spare parts and components are stored in all facilities. More expensive spares are stored in a few secured locations. For cell sites without power, service technicians can deliver small power generators from a central location within a region. Large electric power generators can be delivered within a day. Because cable cuts are a common occurrence in the industry, techniques and tools are well developed to locate the cut quickly from a CO where the transport equipment is located. The equipment will identify the link ID and the exact location of the cut or trouble point. In the case of multiple cuts in different locations, the technicians can identify the critical links and set priority to repair them first. When multiple cable cuts occur, out-of-state service teams will be called in to help, although access in a large earthquake such as the ShakeOut Scenario earthquake may be problematic, as described elsewhere in this Scenario. The Federal government operates a program called Telecommunications Service Priority (TSP), aimed at providing priority installation and restoration of the telecommunications lines that are considered critical in times of emergency.

Table 4-7. Hypothetical Damage to Telecommunications Facilities

Region	Red tags	Serious equipment damage	Fiber cable breaks
1	1-2	same 1-2 facilities	1 location: I-15 viaduct at Lytle Creek
2	0	2-5 facilities	3 bridges over San Gabriel River
3	2-3	same 2-3, plus 2 more	5 locations from Colton to Devils Canyon
4	2-3	same 2-3, plus 2 more	2 locations near I-10 fault crossing
5	0	2-5 facilities	3 overpasses in Palmdale and Littlerock
6	0	0-2 facilities	2 overpasses near Azusa and El Monte

Table 4-8. Hypothetical Telecommunication Service Impacts

Region	No service	Normal service	Extent of no service
1	1 day	2 days	10,000-20,000 addresses NE of I-10/I-210
2	1-2 days	3 days	20,000-50,000 addresses
3	2-4 days	4-5 days	40,000-50,000 addresses
4	2-4 days	4-5 days	40,000-50,000 addresses
5	1-2 days	3 days	20,000-50,000 addresses
6	0-1 day	2 days	0-10,000 addresses

Based on the hypothetical damage summarized above, and considering the service providers' ability to restore damage, Table 4-8 summarizes service impacts associated with physical damage to telecommunications facilities and power failure at cell sites.

Expert-Recommended Mitigation

Companies that operate these facilities are typically well prepared to deal with equipment damage. The real need is a standard procedure adhered to by all service providers regarding access control and network control with both LECs and IECs in a disaster site.

Natural Gas and Liquid Fuels Pipelines

Study by Donald Ballantyne, *MMI Engineering*

"M7.8 Southern San Andreas Fault Earthquake Scenario: Oil and Gas Pipelines"

Additional input by Rick Gailing, Principal Engineer, Southern California Gas Company

Damage Estimates for the ShakeOut Scenario

Worldwide, buried pipelines have been damaged by earthquake shaking and by permanent ground deformation, which can include fault rupture, earthquake-induced landslides, and liquefaction with associated lateral spreading or settlement. All three hazards were evaluated in special studies for this Scenario (see Chapter 4.B and Appendices E, F, and G). Damages posited for the ShakeOut Scenario earthquake are as follows. A landslide in the hills east of Whittier causes a product line to rupture, releasing jet fuel. At the Colton Receiving Station, earthquake shaking causes heavy damage to tanks and equipement in the tank farm. In addition, at Cajon Pass, Palm Springs and Palmdale, natural gas, gasoline, and diesel pipelines rupture. Based on the special study of the lifeline corridors, these areas are also susceptible to liquefaction-related settlement and spreading. However, additional study has determined that the water table is typically low enough that these locations will avoid liquefaction damage in this earthquake. Details of damage and repair times follow.

Two Southern California Gas transmission pipelines at Cajon Pass rupture at the fault where the offset causes 6.2 m of compression in the pipe. There is a subsequent explosion, possibly as a result of arcing from current within the pipeline, a current induced by overhead power transmission lines that cross over the pipelines 0.75 miles down a dry riverbed hill from the fault rupture. ShakeOut Scenario experts debated the most likely reason for an explosion, and whether the overhead power lines might deenergize and the current dissipate before the rupture reached the Cajon Pass. However, because there are numerous ways it could occur, an explosion is posited. The explosion results in a crater. The CalNev 14-inch product pipeline crosses the fault at approximately the same location, and also ruptures as a result of significant ground displacement. At the time of the earthquake, the CalNev pipeline is transporting gasoline, which adds to the fire. The fire reaches the overhead power lines, causing them to break.

Block valves on the Southern California Gas Company pipeline spaced at roughly 10-mile intervals near the Cajon Pass fault crossing are equipped with pneumatically- actuated, automatic line break controls that isolate the damaged pipeline within minutes. Operations personnel view the affected pipelines by helicopter shortly after the earthquake. However, delayed by the fire, highway traffic, and roadway damage, operations personnel reach the site several hours after the earthquake to examine the ruptures and plan repair efforts.

The 20-inch Kinder Morgan product line is ruptured at two points near Palm Springs and at one near the Salton Sea where the pipeline is aligned roughly parallel to the fault. When the earthquake occurs, the pipeline is placed into tension at two points (with 1.5 meters of offset at a point just north of Palm Springs and 11.7 meters near the Salton Sea), and the pipeline is placed in compression at a second point near Palm Springs, where the fault offset is 2.6 meters. The pipeline at that moment is carrying diesel fuel, which floods the surrounding area. Shutoff valves and topography limit the release of product to 200,000 gallons, which is discharged into the natural terrain drainage. A smaller natural gas distribution pipeline is located in the same right-of-way, and is also broken by the fault rupture. The Kinder Morgan Pipeline fails at several other locations due to ground failure (faulting and landslides) along the 60-km alignment near the fault trace. The repairs take up to six months, partly because of environmental assessment, cleanup, and permitting needs.

In Palmdale, two 30-inch natural gas transmission lines cross the fault multiple times and consequently rupture due to fault offset. They blow high-pressure gas into the air. First responders quickly evacuate the area and an automatic line break control is able to keep the gas from igniting until a Southern California Gas Company crew arrives. A landslide in the hills east of Whittier shears the 20-inch Nogales pipeline, releasing jet fuel. 100,000 gallons of product are discharged before the line can be shut in. The jet fuel flows into a local drainage. The Colton Receiving Station is subjected to 0.4 g shaking. The receiving station is a node for distribution of gasoline, diesel and jet fuel. The facility also controls flow of jet fuel to March Air Force Base. Some pipes break at the pipeline- tank interface, because of such causes as poorly- or un-anchored tanks, above-ground pipe flexure or displacements, or elephant foot deformation of tanks. Fuel discharges into the retaining dikes, but is not ignited. Southern California Gas Company has small- diameter transmission and distribution piping throughout the impacted area. In 1993, the company completed its replacement of all vulnerable, cast-iron pipe in the distribution system with more earthquake-resistant, steel and polyethylene pipe; but because of building damage, landslides, and other causes, they still suffer approximately 200 pipeline failures, primarily at fittings and transitions.

Most repairs to restore essential system operations and service to customers in a prioritized and strategic manner are made within two weeks. The remainder follow in days to weeks. The demand for natural gas drops significantly because of business interruption caused by damage to consumers' facilities, transportation impacts, and other earthquake impacts. Because of damage to distribution lines caused by building damage and ground failure, 50% of gas customers within MMI VIII+ and in areas with landslide and liquefaction damage (MMI X) are without gas for up to three weeks, mostly due to customer shut-off, and 5% of restorable customers are without gas for between three weeks and two months.

Expert-Recommended Mitigation

Successful mitigation of buried pipelines varies with the site:

- Old (pre-1930) pipe with oxy-acetylene welded joints should be replaced.
- Seismic-resistant design can accommodate fault crossings.
- Where practicable, when laying new pipe, areas prone to landslides, liquefaction, and lateral spreading should be avoided.
- Construction techniques can also improve seismic performance. For example, avoiding the use of "anchors" such as valves or sharp bends will help to distribute stresses, and thus reduce damage from concentration of stresses.
- To limit the consequences of pipe failure, implement an automated control system for quick shutdown, and construct redundant pipelines in independent alignments so that if one fails, the other may remain intact.

Water Supply

Panelists:

Ted Johnson, Chief Hydrologist, *Water Replenishment District of Southern California*

Chris Hill, *Metropolitan Water District*

Steve Robbins, *Chief Manager and General Manager, Coachella Valley Water District*

Mike Herrera, Operations, *Coachella Valley Water District*

Shane Chapman, Assistant Manager, Water Systems, *Metropolitan Water District*

Richard Shapall, Engineering, *Metropolitan Water District*

Marty Adams, Director, Water Resources, *Los Angeles Department of Water and Power*

K. Eric Leung, Director of Water Resources, *Long Beach Water Department*

Le Val Lund, Engineer (retired), *Los Angeles Department of Water and Power*

Randy Van Gelder, General Manager, *San Bernardino Valley Municipal Water District*

Douglas Headrick, Deputy General Manager, *San Bernardino Valley Municipal Water District*

Rick Sanchez, Division Chief, *Department Water Resources State Water Project*

Mark Stuart, District Chief, Southern Region, *Department of Water Resources Central and West Coast Basins*

Gary Sturdivan, *East Valley Water*

Eric Reichard, *U.S. Geological Survey California Water Science Center*

Peter Martin, *U.S. Geological Survey California Water Science Center*

Michael Shulters, Director, *U.S. Geological Survey California Water Science Center*

Mike Morel, Operations Manager, *Metropolitan Water District*

Craig Davis, Manager, *Los Angeles Department of Water and Power*

Damage Estimates for the ShakeOut Scenario

Communities within about 10 miles of the fault and small communities in isolated regions have pipeline damage so severe as to impair piped water supply for up to six months, with repairs proceeding in some prioritized fashion. Perhaps 5% of customers in small regions throughout Los Angeles, Riverside, and San Bernardino Counties have pipeline damage requiring between one week and two months to repair, before piped water supply is available. Throughout the region shaken with MMI VIII or greater, 50% of customers are without water supply for up to one week, owing to loss of power to pumps, and damage to pumps, tanks, and other equipment. Damage to water supply pipe contributes to the potential for post-earthquake conflagration, especially in the cities of San Bernardino and Riverside, as discussed in the Fire Following Earthquake section of this chapter. Mutual aid is problematic, because of limited resources to direct repair crews. There is damage to aqueducts that cross the fault (as discussed in Chapter 3.C), but there is water storage on the south side of the fault; assuming a 25% reduction in demand, there is adequate domestic water supply for six months or more, by which time much of the aqueduct damage has been repaired. The aqueduct damage impacts some heavy industrial users such as Anheuser Busch, while others are able to draw from their own groundwater. Thus water supply is probably adequate for all but heavy industrial and agricultural uses. Due to system damage, reaching end users with piped water supply is the graver issue. With widespread damage to water conveyance systems, the effort to find and repair the numerous individual leaks in many places is so slow and expensive a process that it is cheaper and faster replace the entire system.

Expert-Recommended Mitigation

- Use the ShakeOut Scenario for planning, training, and alliance-building:
  - Establish communication plans to help employees know whether their families are okay.
  - Work out the significant management aspects of mutual aid, which is hindered by lack of local knowledge, training, paperwork, and management issues.
  - Work out resource limitation: need technical people to make drawings and to direct repair crews.
  - Sustain employee cross-training and training between systems.
  - Bring Emergency Operations Centers together to cross-train and exercise emergency plans.
  - Promote participation in the Golden Guardian emergency response exercises.
  - In the ShakeOut Scenario, 50% of pipelines in some areas would have no pressure because of liquefied soil. Map liquefaction potential to inform providers about where to build in valves and redundancy.
  - Following the model of Earthquake Country Alliance, establish a working group or forum to identify and solve water supply system problems.
- If regulatory and water quality issues can be resolved, consider filling up groundwater basins for earthquake recovery purposes. Caveats:
  - would need to provide onsite power for pumps;
  - would increase peril of fire with boil-water orders.
- Consider Purchasing:
  - Stockpiles of clamps; stockpiling pipe would be too expensive.
  - Portable treatment systems and back-up supplies.

Electric Power

Panelists:

Enrique Martinez, *Los Angeles Department of Water and Power*

Philip Mo, *Southern California Edison*

Mike Morel, *Metropolitan Water District*
Don Conner, *Southern California Edison*
Troy Whitman, *Southern California Edison*
Kevin Garrity, *Los Angeles Department of Water and Power*
Bruce Hamer, *Los Angeles Department of Water & Power*
Ron Cox, *City of Riverside*
Dennis Ostrom, *Southern California RR*

Additional Input: Monique Weiland, Lead Contingency Planner, California Independent System Operator

Overview of the Issue

Restoration of electric service is contingent upon the following 3 factors: 1) The area must be safe for restoration. Electricity will not be restored in areas still experiencing leaks of gas, fuel, and other flammable gases. 2) Access to utility poles is available. Large utility vehicles need to get to the poles in order to provide repairs. And, 3) Inventory must be available to replace damaged equipment.

Damage Estimates for the ShakeOut Scenario

High-tension transmission towers are likely to collapse due to landslides and strong ground shaking near the fault, especially at the Cajon Pass. Generation plants may be taken offline for damage inspection, including the San Onofre Nuclear Generating Station, and some may be offline for weeks. Damage to transformers on overhead poles causes localized power loss. It may take two years or so to repair all damage. Mutual aid will be invoked by agreements. Assistance typically arrives anywhere from 1 to 7 days after request but may be complicated now by differences in processes and equipment, and by the large number of entities requesting assistance.

Los Angeles, San Bernardino, and Riverside Counties immediately lose all electric power. Gas pipeline damage reduces the ability to produce power within the affected areas of those counties. Within 24 hours, repairs restore 30-50% of service; within three to ten days, 75-90% of those capable of receiving power have service restored; and in one to four months virtually all power is restored. In Ventura, Orange, and Imperial counties, power is also immediately lost, but 75% of service is restored within 12-24 hours, and 90% of service is restored within two days. In Kern and San Diego counties, 90% of power is restored within 24 hours.

Los Angeles Department of Water and Power temporarily loses the capacity to import power from across the fault, but is self-sufficient at reduced load (3000-4000 MW, roughly 80% of pre-earthquake capacity), if natural gas supply is available.

Expert-Recommended Mitigation

- Very few low-voltage (below 115 kV) transformers are secured. Anchoring these transformers is the Number 1 mitigation measure.
- Replace more of the breakable ceramic insulators with polymer insulators.
- Consider conducting study where uncertainties exist regarding:

- to what extent liquefaction would affect underground transmission and distribution; and
- to what extent power demands will be reduced during some critical repair intervals.

vi. Fire Following Earthquake

Study by Charles Scawthorn, S. E., *Kyoto University, SPA Risk LLC*

"A Note of Fire Following Earthquake for the Southern San Andreas Fault *M7*.8 Earthquake Scenario"

Review Panel:
Donald Manning, retired chief, *Los Angeles City Fire Department*
Donald Parker, *California Seismic Safety Commission*
Gerry Malais, *Los Angeles City Fire Department*
Michael Reichle, *California Geological Survey*
Larry Collins, *Los Angeles County Fire Department, USAR Task Force 103*

Urban ignitions

Fire following earthquake can have devastating consequences, as tragically seen worldwide, notably after the 1906 San Francisco, 1923 Tokyo, and 1995 Kobe earthquakes. It is a significant problem in urban areas of southern California and doubles the fatalities ad economic losses in the ShakeOut Scenario earthquake. Fires occur following all earthquakes that damage human settlements, but are generally only a very significant problem in large metropolitan areas predominantly comprised of densely spaced, wood buildings. Modeling of fire following earthquake is an extremely nonlinear process, and a precise number of ignitions is neither possible nor pertinent. The important distinction is between situations that create a few to dozens of small fires versus many hundreds to thousands of fires that result in conflagration.

The earthquake in the ShakeOut Scenario combines all the requisite factors for conflagrations that, depending on circumstances, can be of uniquely catastrophic proportions: a large earthquake simultaneously causes numerous ignitions, degrades building fire-resistive features, drops pressure in water supply mains, saturates or damages communications, and impedes transportation through damage and debris on routes.

The key to modern fire protection is a well-drilled, rapid response by professional firefighters in the early stages of structural fires, arriving in time to suppress the fires while that is still relatively feasible. This is true for each single ignition. Most fire departments are not sized or equipped to cope with the number and geographic extent of ignitions following a major earthquake. In low-humidity conditions, an unfought ignition will grow into a room-sized fire within several minutes, and a fully involved, single- family structural fire within several more minutes. At this point, the fire is termed a large fire, meaning that more than one engine company is required to suppress it.

Figure 4-9. Photograph that shows examples of two neighborhoods at risk of fire following earthquake due to high-density, wood buildings.

In the ShakeOut Scenario, approximately 1,600 ignitions occur that require the response of a fire engine. Note that this is only about ten times as many ignitions as occurred due to the Northridge earthquake, although more than twenty times the area is involved here. The critical distinction, however, is not the precise number of ignitions but the order of magnitude—will there be dozens of small fires, or hundreds of fires that exceed fire department resources? In the ShakeOut Scenario, about 1,200 of the ignitions will be large fires, in which the first responding engine company is not able to adequately contain the fire. Fortunately, about one third of the large fires occur in areas of Riverside County, the San Fernando Valley, and Orange County where building density is relatively low, so that even though the fires are initially uncontrollable, their spread is limited due to large firebreaks. In these areas, large fires will grow into one or several conflagrations that destroy tens of city blocks. However, in the more densely built areas of the central Los Angeles basin (and parts of Orange County) there are large plains of relatively uniform, dense, low-rise buildings such as those in figure 4-9, and these provide a fuel bed where hundreds of large fires merge into dozens of conflagrations, destroying tens of city blocks. In turn, some of these conflagrations merge into one or several super conflagrations, destroying hundreds of city blocks.

The ShakeOut Scenario assumes typically dry, breezy November weather, but no Santa Ana wind conditions. Under the assumed conditions, a preliminary estimate is that the approximately 1,200 large fires result in an ultimate burnt area of 1-1.5% of the study area, or approximately 200 million square feet of residential and commercial building floor area. To put this number into context, this is equivalent to 133,000 single-family dwellings; however, commercial and other non-residential buildings will also burn.

Directly attributable to these fires following the earthquake is a large loss of hundreds to perhaps a thousand lives, shelter needs for one half to one million persons, building property loss of $40-$ 100 billion dollars, and content property loss of $25 billion. Because the economic loss is almost fully insured, it will result in severe distortions to the U.S. and global insurance industries. Other economic impacts include the loss of several billion dollars in local tax revenues.

Wildfire ignitions

Wildland fire following earthquake has not been explicitly evaluated for the ShakeOut Scenario. However there are numerous locations within the study area that are in wildlands or at the wildland-urban interface and are thus prone to wildland fire. Damage to electrical lines and petroleum product pipelines in Cajon Pass is particularly likely to cause a fire that spreads to brush-covered mountains. The need to fight urban fires will reduce the resources available to fight any wildland ignitions.

Fire fighting

When briefed on the report, fire chiefs and other firefighters emphasized that in preventing conflagrations, what is key is not the number of ignitions, but rather how and whether the ignitions are spread. Southern California firefighters have extensive experience in fighting conflagrations due to recurring wildland fires, and they were able to quickly identify numerous, varied options for mitigating the potential for conflagration following the ShakeOut earthquake. Most options focus on the crucial needs to increase water supply by identifying and accessing additional water sources; and to enable firefighters to reach fires quickly despite transportation system or other damages. The scale of this problem is so great and the impact potentially so large that the California Seismic Safety Commission is convening a special task force comprised of leaders of the firefighting community who will respond to the ShakeOut Scenario findings by discussing and recommending solutions in a report later this year.

E. Integration of Results, HAZUS, and Supplemental Studies

When the supplemental studies of physical damages were completed, we integrated or reconciled them with the HAZUS analysis. As summarized in Table 4-9 below, four of the studies—fire following earthquake, high-rise pre-Northridge welded- steel moment frame buildings, utility pipelines, and highways—found property losses (generally building repair costs and content losses) either omitted by, or significantly different from, loss estimates produced by the HAZUS analysis and we adjusted HAZUS results accordingly for use by ShakeOut modeling of emergency response, casualties, and economic consequences. Other property losses determined in the supplemental studies (ports, rail, airport, and surface streets) may be quite significant to the owners and stakeholders, but were very small in comparison with total property losses (generally less than $100 M) and were therefore not used to make HAZUS adjustments. In some of the latter cases, we noted time-element losses, such as direct business interruption or additional living expenses, for use in the Scenario's subsequent economic analysis.

Fire following earth quake—HAZUS was used to estimate fire following earthquake (FFE), but it was felt that an independent assessment of FFE would be worthwhile. A brief study was performed by a leading expert in fire following earthquake who estimated 1,600 ignitions, 1,200 large fires, 200 million square feet of burnt building, and approximately $40-$ 100 billion in building losses, plus $25 billion in contents loss and $7 billion in additional living expenses or business interruption costs. This figure has some slight overlap with shake-related damage, but the overlap is second order or less. Note that HAZUS appears to produce significantly lower loss estimates in this Scenario than its own documentation implies it should; the HAZUS developer agrees that there is a possible FFE implementation problem in HAZUS, so the HAZUS FFE estimates are not considered for comparison purposes here. **Therefore $65 billion is added to the HAZUS property loss estimate, and $7 billion in business interruption or additional living expenses is noted for use in subsequent economic study.**

Buildings—HAZUS estimates of damage to four classes of important buildings were compared with those of special studies by structural engineers. The classes are high- rise pre-Northridge welded-steel moment-frame buildings (denoted here by S 1H, using the HAZUS notation), non-ductile reinforced-concrete moment-frame buildings (C1L, C1M, and C1H), unreinforced masonry buildings, and woodframe construction.

High-rise Pre-Northridge welded-steel moment-frame. HAZUS has a reasonable map of the locations of high-rise pre-Northridge welded-steel moment-frame buildings (denoted here by S1H, using the HAZUS notation), but because of the treatment of long- period motion, and the high content in this earthquake of long-period motion, HAZUS will not produce a reasonable simulation of collapses to these buildings in this Scenario.

Table 4-9. Property Losses to be Added to HAZUS Loss Estimates in Billions of Dollars values are explained or supported in text following table

Subject	Property loss ($ billions)	Comment
Fire following earthquake	$65	Insured. Includes $40 B in buildings, $25 B in contents, and $7 B in business interruption and additional living expenses.
High-rise, Pre-Northridge welded-steel moment frame buildings	$2.2	Does not include business interruption. Largely uninsured.
Non-ductile reinforced con-crete moment-frame buildings	-	Study generally agrees with HAZUS estimate of damage; no addition
Unreinforced masonry buildings	-	Study generally agrees with HAZUS estimate of damage; no addition
Highways	$0.4	Repair cost: $0.4 B. Business interruption: $4.6 B,
Surface streets	-	Repair cost < $0.1B; no addition
Pipelines	$1.1	Water: $0.5 B, sewer: $0.4 B, gas: $0.1 B
Ports + airports	-	Repair cost < $0.1 B; no addition
Rail + light rail	-	Repair cost < $0.1 B; no addition
Telecommunications	-	Repair cost < $0.1 B; no addition
Hospitals	-	HAZUS estimate assumed
Total added property loss	$69	Additional $12 B business interruption and additional living expenses from FFE and highway damage

We therefore supplement the HAZUS-simulated damage to S 1H with the results of the supplemental study on these buildings. In the supplemental study, we posited five complete, pancake-style collapses of such buildings, involving 1.3 million square feet of commercial, hotel, and government occupancies, with 4,300 indoor occupants at the time of the earthquake, producing building and content losses of approximately $700 million. Approximately ten more of these buildings are estimated to be red-tagged, generally requiring demolition and replacements, and causing another approximately $1.4 billion in building and content losses. Approximately twenty are yellow-tagged. Equating yellow tagging with the HAZUS "moderate" damage state suggests a repair cost on the order of 10% of replacement cost of the buildings (but no content loss), or $140 million. **We therefore supplement the HAZUS estimate with $2.2 billion in building and content losses. In addition, casualties, direct and indirect business interruption losses from the collapses, red tags, and yellow tags will be added to HAZUS estimates in the ShakeOut Scenario's estimation of consequences.**

Non-ductile reinforced-concrete moment-frame. HAZUS has a reasonable map of the locations of non-ductile reinforced-concrete moment-frame construction, and produces a reasonable map of the locations of extensive and complete damage. It estimates that in the eight-county area, approximately 350 older (pre-1976) concrete buildings experience complete damage, mostly in Los Angeles, San Bernardino, and Riverside Counties. It can be inferred using HAZUS estimates of collapse rates among buildings with complete damage that approximately 35 of these experience some degree of collapse. The supplemental study posited 50 such collapses, which seems to generally agree with HAZUS. **We therefore do not supplement the HAZUS estimates of damage to non-ductile reinforced-concrete moment-frame buildings from the special study.**

Unreinforced masonry. HAZUS estimates approximately 300 unreinforced masonry buildings in the study area would experience complete damage; most of these are un-mitigated buildings in Riverside and San Bernardino Counties. Approximately 300 more are in the extensive damage state, suggesting approximately 500 unreinforced masonry buildings are damaged beyond economic repair. The supplemental study, by contrast, estimated 900, which agrees on an order-of-magnitude basis. **We therefore do not supplement the HAZUS estimates of damage to unreinforced masonry buildings from the special study.**

Woodframe. HAZUS estimates the spatial distribution, damage, and loss to woodframe buildings. A supplemental study proposed a relationship between era of construction and building code era, a HAZUS parameter. This relationship was incorporated into the HAZUS analysis, so the HAZUS damage and loss estimates for woodframe construction reflect the findings of this special study. ***We therefore do not further supplement the HAZUS estimates of damage to woodframe buildings.***

Highways. A detailed engineering and economic analysis using the REDARS software was performed in collaboration with Caltrans, producing estimated physical damage of $400 million, and consequent economic impacts of $4.6 billion, to be added to HAZUS results in the estimation of economic consequences in Chapter 7.

Water supply, wastewater, and gas pipeline damage. HAZUS estimates sewer pipeline damage of 224,000 leaks and 56,000 breaks, estimated here to cost $1,000 per leak repair and $3,600 per break repair, for a total of $430 million in repair costs to sewer pipe. It also estimates 280,000 leaks and 71,000 breaks to potable water supply pipeline. Applying the same per-break repair costs results in an estimated repair cost of $540M in repair costs to water supply. Gas pipelines are more robust, and are posited to cost on the order of $100 million to repair. The total pipeline repair cost estimate is therefore $1.1 billion.

Surface streets. Repair costs posited to be less than $0.1 billion.

Table 4-10. Expected Building Damage by Occupancy.
These are HAZUS Results; Our Supplemental Findings do not Have Impact on These Values

	None		Slight		Moderate		Extensive		Complete	
	Count	%	Count	%	Count	%	Count	%	Count	%
Agriculture	2,860	0.07		0.06	163	0.08	39	0.07	13	0.03
Commercial	136,030	3.44	21,299	3.54	11,249	5.46	2,629	4.86	799	1.77
Education	4,537	0.11	517	0.09	266	0.13	79	0.15	22	0.05
Government	2,138	0.05	275	0.05	206	0.1	71	0.13	30	0.07
Industrial	37,948	0.96	5,275	0.88	3,314	1.61	1,258	2.33	521	1.15
Other Residential	553,848	13.99	78,793	13.08	54,579	26.47	33,749	62.41	40,444	89.6
Religion	10,828	0.27	1,463	0.24	605	0.29	142	0.26	44	0.1
Single Family	3,209,557	81.1	494,398	82.08	135,803	65.86	16,112	29.79	3,267	7.24
Total	**3,957,746**		**602,020**		**206,185**		**54,079**		**45,140**	

Table 4-11. Expected Building Damage by Building Type (All Design Levels). These are HAZUS Results; Our Supplemental Findings do not Have Impact on These Values. RM is Reinforced Masonry, URM is Unreinforced Masonry, and MH is Manufactured Housing

	None		Slight		Moderate		Extensive		Complete	
	Count	%	Count	%	Count	%	Count	%	Count	%
Wood	3,723,116	94.07	555351	92.2	151,354	73.41	19,642	36.32	4,412	9.77
Steel	12,831	0.32	3500	0.58	2,608	1.26	592	1.09	123	0.27
Concrete	40,522	1.02	8521	1.41	5,773	2.8	1,518	2.81	493	1.09
Precast	12,615	0.32	1904	0.32	795	0.39	104	0.19	14	0.03
RM	59,819	1.51	5929	0.98	3,291	1.6	876	1.62	190	0.42
URM	10,843	0.27	1569	0.26	646	0.31	270	0.5	258	0.57
MH	97,999	2.48	25587	4.25	41,718	20.23	31,077	57.47	39,649	87.84
Total	**3,957,745**		**602,361**		**206,185**		**54,079**		**45,811**	

Rail. Repair costs of rail at fault crossing are required, however these are most likely trivial compared with business interruption costs associated with interruption of rail service for one to two weeks. To illustrate: it is posited that on the order of 10-20 rail breaks occur at fault crossings, each requiring repair or replacement on the order of 50- 500 m of track, at a cost of $1.5 million per km of rail, for a total on the order of $1- 10 million of repairs. **We therefore do not further supplement the HAZUS estimates of damage to rail.**

Light rail. Repair costs posited to be less than $0.1 B.
Other mass transit. Repair costs posited to be less than $0.1 B.
Airport. Repair costs posited to be less than $0.1 B.
Ports. Repair costs posited to be less than $0.1 B.
Telecommunications. Repair costs posited to be less than $0.1 B.
Hospitals. HAZUS' repair-cost estimate for hospitals is used.

Table 4-12. Debris in Thousands of Tons Per County. These Values Combine HAZUS and Supplemental Study Estimates

County	Brick, wood and others	Concrete and Steel	Total
Riverside	1,438	1,310	2,748
San Bernardino	1,904	2,072	3,976
Imperial	23	15	39
Ventura	7	5	12
San Diego	1	0	1
Kern	30	20	50
Orange	266	285	551
Los Angeles	1,086	2,363	3,449

Table 4-13. Expected Damage to Essential Facilities.
EOC is Emergency Operations Center

Classification	Total	At Least Moderate Damage > 50%	Complete Damage > 50%	with Functionality > 50% on Day 1
Schools	6,435	32	0	5,930
EOCs	22	1	0	16
Police Stations	444	16	0	386
Fire Stations	1,100	47	0	943

The above facilities were not studied in detail. The numbers are approximately correct but we do not have accurate data on the building types nor building conditions, which are key in estimating damage levels. It should be assumed that any buildings within MMI IX zones will have enough non-structural damage to impair functionality unless non-structural mitigation has been a priority activity.

F. Hazardous Materials

Study by Ronald T. Eguchi, and Shubharoop Ghosh, *ImageCat, Inc.*
"Impacts of a M 7.8 Southern San Andreas Fault Earthquake: A Hazardous Materials Release Scenario"

Overview of the Issues

It is generally acknowledged that a major earthquake in an industrialized, densely populated area of the United States could lead to the release of hazardous chemicals. A large post-earthquake release would present a threat not only to residents in the immediate vicinity of the source, but also to those of surrounding communities. Affected areas would then face a range of emergency management problems. For example, a major earthquake is likely to seriously impair community emergency response capability, making it difficult to effectively deal with secondary emergencies such as hazardous materials releases and fires. Tasks which are normally problematic, such as warning the public about a toxic release and evacuating people from areas that are hazardous, would be much more difficult following a major earthquake. Furthermore, communities are accustomed to responding to hazardous materials releases one at a time, while in an earthquake situation multiple accidents may occur simultaneously, greatly compounding resource problems.

Although there has never been a major incident involving hazardous materials in a U.S. earthquake, smaller releases have occurred in events that were moderate in size. An example is an accident at a chlorine repackaging facility in the 1987 Whittier Narrows Earthquake, in which nearly one ton of chlorine gas was released (FEMA, 1987). While awareness of the problem is growing, there has been little research to date on the seismic sources of hazardous materials releases, and seismic vulnerability models for chemical facilities are almost nonexistent.

The main challenge in approaching this problem from a community perspective is to develop a risk-assessment methodology that is sophisticated enough to provide the type of information needed for more effective hazard management, but is also cost-effective to apply on a regional basis. Conducting detailed seismic risk assessments and modeling potential failures in chemical facilities is very time consuming and expensive; few communities can afford to conduct such studies.

Adding to the complexity of the problem, highly hazardous materials number in the thousands and new products are constantly being developed. Before systematic analyses can be undertaken, it is necessary to determine which hazardous substances are likely to pose the biggest threat to the community in an earthquake. In this limited assessment, we have chosen to focus on two hazardous materials, chlorine and ammonia. These substances were selected because: 1) they are responsible for the majority of fatalities and casualties in U.S. hazardous materials incidents, 2) they are present in large quantities in our study area, and 3) they form clouds that can spread to adjacent areas and present a hazard beyond the plant gates.

Damage Estimates for the ShakeOut Scenario

Facilities store and use varying amounts of chemicals and are dispersed throughout the study region. In general, they are broken into three facility types based on chemical usage: chlorine storage facilities, ammonia storage facilities, and ammonia processing facilities. Chlorine storage amounts range from 4 to 1,000 tons, while ammonia storage varies from 2 to 206 tons. The special study addressed Los Angeles County alone. Extrapolating to the eight-county region on the basis of population and shaking intensity, the ShakeOut Scenario posits four major hazardous material releases: three chlorine gas releases in wastewater treatment plants in Los Angeles, Riverside, and San Bernardino Counties, and one ammonia release in a Los Angeles County refinery. The releases affect approximately 175,000 people in Los Angeles County, and 70,000 each in Riverside and San Bernardino Counties. This is the estimated number of people who experience at least one airborne hazardous material release exceeding the American Industrial Hygiene Association's (AIHA) Emergency Response Planning Guidelines (ERPGs) threshold criterion ERPG 3, defined as “the maximum airborne concentration below which it is believed that nearly all individuals could be exposed for up to one hour without experiencing or developing life-threatening health effects.”

5.Emergency Response and Communications by Dennis Mileti and James Goltz

A. Introduction

Emergency response can be defined as “the actions taken immediately before, during, and after a disaster occurs to save lives, minimize damage to property, and enhance the effectiveness of recovery” (Mileti, 1999). Given that our ShakeOut Scenario earthquake occurs suddenly and without warning, our discussion of response to a M7.8 southern San Andreas earthquake will be confined to the “during and immediately after” timeframe of this definition. The activities that comprise response can be classified in a number ways and from multiple vantage points. Our classification breaks 17 functions of response into 7 general classes of activities: crisis information (public information and responder communications), search and rescue, victim services (shelter, provision of food and water, and the management and distribution of donated goods and services), access management and law enforcement (control and security, and traffic control), the staffing and functioning of emergency operations centers, fire suppression, medical emergency response, and service restoration (repair of utilities, route recovery, and debris removal). Table 5-1 explains the acronyms and abbreviations used in this chapter.

Our approach to this section was to first review the social science research literature compiled over the past half century on emergency response. We then extracted the commonalities of emergency response that characterize the hundreds of disasters studied, and projected them onto this earthquake. In the matrix in Section 2, presented as Tables 5-2 through 5-18, we provide a generic look at how both organizations and victims of a large urban earthquake disaster would respond based on the accumulated knowledge contained in these studies. For some response functions we also provide narrative in which we incorporate specific scientific and engineering data, as well as loss estimates specific to the ShakeOut Scenario disaster. The information from the matrix and the narrative supplement one another and together provide a more comprehensive view of disaster response.

The disaster hypothesized in the ShakeOut Scenario would be one of the worst natural disasters in American history, striking an urban region populated by more than 20 million people with a building inventory valued at $2 trillion. In a real sense, this is one earthquake that creates multiple disasters, requires an unprecedented level of response, and affects the national and world economies. In the narratives in Section C, we convey how organizations and individuals respond immediately following this event—a disaster that for most emergency management organizations exceeds by far any emergency they have previously been called upon to address. For individuals who survive this disaster, it will be a period of challenge calling for values and norms that submerge individual self- interest in favor of collective survival.

While earthquake vulnerability has increased with population growth and an ever-expanding urban footprint, the tools and technologies available to respond to this disaster have changed significantly as well. Disaster response has benefited from an approach to emergency management that incorporates multi-agency mutual aid, a well-defined and integrated response structure, redundant and hardened communications, and computer- aided emergency information systems. Nevertheless, response to a major southern California earthquake will pose serious challenges to the best-tested emergency response capabilities and will expose gaps in planning and training. In this section, we will explore how current response capacity will meet, or fail to meet, the challenges presented by this earthquake.

B. Response Function Matrix Tables

NOTE that in the tables that follow, the term “Eastern Region” indicates the hardest-hit areas nearer the San Andreas Fault, such as the Coachella Valley and the Inland Empire, and the term “Western Region” refers to the parts of southern California farther from the Fault and without major structural damage.

Table 5-1. Acronyms and Abbreviations Used in Response Function Tables

Caltech	California Institute of Technology	ISO	Independent System Operator
Caltrans	California Department of Transportation	JIC	Joint Incident Command
CALWAS	California Alert and Warning System	LA	Los Angeles
CEPEC	California Earthquake Prediction Evaluation Council	LA Metro	bus and train provider
CERT	Community Emergency Response Team	NGO	Non-Governmental Organization
CGS	California Geological Survey	NIMS	National Emergency Management System
CHP	California Highway Patrol	OES	California State Office of Emergency Services
CISN	California Integrated Seismic Network	REOC	Regional Emergency Operations Center
CSWC	CA State Warning Center	S & R	Search & Rescue
DMAT	Disaster Medical Assistance Team	SCE	Southern California Edison Company
DMORT	Disaster Mortuary Team	SEMS	State Emergency Management System
DWP	Department of Water and Power	ShakeCast	software for delivery of ShakeMaps and notifications of shaking levels
EMSA	Emergency Medical Services Authority	Shakemap	Estimate of ground shaking based on seismic data and models
EOC	Emergenc y Operations Center	SMIP	Strong Motion Information Program
EQ	Earthquake	So CA	southern California
ESF	Emergenc Function y Response	SOC	State Operations Center
FEMA	Federal Emergency Management Agency	USAR	Urban Search and Rescue
FHA	Federal Highway Administration	USGS	United States Geological Survey
HAZUS	Loss estimation software developed by FEMA	WCATWC	West Coast and Alaska Tsunami Warning Center

Table 5-2. ShakeOut Scenario Emergency Response Function: Public Information

Organized and Spontaneous Response Summarized for Five Time Periods After ShakeOut Earthquake											
Impact (2-5 minutes)		30 Minutes		2 Hours		24 Hours		72 Hours		1 Week	
Organized	*Spontaneous*	Organized	*Spontaneous*	Organized	*Spontaneous*	Organized	*Spontaneous*	Organized	*Spontaneous*	Organized	*Spontaneous*
*Local media that can still function switch to ongoing event coverage.	**People seek information from electronic media.* **Informal discussions among those nearby.*	* Local and national media mobilized. *Focus is on localized incidents. *Science organizations report on catastrophic earthquake.	**Widespread information seeking.* **Frustration over lack of information.* **Public begins to realize scope of disaster.* **Focus shifts from self to community.* **Concern about others and massive number of attempts to contact others.*	*Electronic media fully converted to earthquake disaster coverage. *National news media sending correspondents to SoCA. All levels of government providing information for broadcast. *News crews have located some critical incidents and are reporting on them. *Focus mainly on large urban areas. *Reporting emphasizes major impacts. *Reporting is worldwide.	**Focus shifts from local to wider community based on media coverage of event.* **Vigorous discussion of media content.*	*News media have located and are reporting on major known structural failures. *FEMA, OES and Operational Areas have set up a JIC and have begun providing emergency period instructions. *Federal response focused on major urban area of LA. *Eastern <u>Region</u>: Local broadcasters providing emergency period information and instructions.	**Rumors are circulating.* **There is vigorous discussion of both local incidents and those in the wider community based on media (especially radio) reports.* **Wide variation in the quantity and quality of information available and communicated within communities and across the area of impact.* **In hardest hit communities, emerging concern that untransported dead will cause disease outbreak, despite official assurances to the contrary.*	*Local media now providing information about the events and information needed by victims to deal with the situations that they face. *National media have greater detail about specific damage incidents and casualties. *National media focus remains on reporting the disaster and its impacts; still emphasizes losses and damage in the greater Los Angeles area. *National and local media cover Governor visiting the disaster site.	**Victims' demand for information shifts from immediate needs to longer-term needs, e.g., from shelter to reconstruction of their homes.* **Some rumors have been laid to rest but other are rampant.* **Rumors of looting stem from victims taking essential, life-sustaining supplies from any place they can find them.*	*Local media give information about State and Federal programs for victims. *Information that is helpful to victims is still being reported to victims. *Local and national media continue to report on aftershocks. *The media provide updates on high-profile damage such as high-rise collapse. *Media focus on individuals and communities that "have a personal story to tell" like an auto mechanic who rescued a dozen people, a survivor who lost all of her family members, and so on. *National and local media present coverage of President and President-elect visiting the disaster site. *Numerous national and state experts from varied fields are interviewed about the event and human response to it. *A dramatic "one-week rescue" is reported across the nation and around the world. *OES and FEMA step up efforts to abate rumors, including establishment of rumor hot lines.	**Media images continue to fuel images of looting.* **<u>Eastern Region</u>. Fears of looting peak due to widely distributed damage and limited law enforcement to cover all sites.*

Mitigation Opportunity:

Provide information to residents regarding safe search & rescue techniques, reducing unnecessary injuries and deaths among rescuers and those being rescued.

Table 5-3. ShakeOut Scenario Emergency Response Function: Communication

Organized and Spontaneous Response Summarized for Five Time Periods After ShakeOut Earthquake											
Impact (2-5 minutes)		30 Minutes		2 Hours		24 Hours		72 Hours		1 Week	
Organized	*Spontaneous*	Organized	*Spontaneous*	Organized	*Spontaneous*	Organized	*Spontaneous*	Organized	*Spontaneous*	Organized	*Spontaneous*
*Face to face. *911 centers and phone systems saturated	**Face to face.* **Begin to turn on electronic media.*	*Telephone system and Internet not functioning. *Communication via amateur radio, satellite phone, hand-held radio devices.	**Massive attempts to use telephones.* **Limited to face to face communication.* **Small groups forming among neighbors, co-workers or others in close proximity.* **Most people outside the region of impact are unable to reach families.*	*Media reports feature a few, known critical incidents (such as S&R at a collapse site or a fire being fought at a high-rise building), leaving listeners/viewers with conception of total disaster. *All major governmental agencies at every level are communicating only through hardened channels (e.g., satellite telephones, radio).	**Radio reports being received and discussed by families and neighborhood residents.* **Content of communication among neighborhood residents is focused on information about disaster and own localized needs.* **People outside the region of impact still unable to reach families.*	*Within the impacted region, every operating system of communication will be in use and overwhelmed. *The only reliable means of communication is hardened communications channels.	**Ham radio operators within the impacted area have transitioned to disaster response and are assisting official responders with communications.* **People outside the region of impact are unable to reach families and friends in the region and NGOs are beginning to address these needs.*	*<u>Western Region</u>: Telephone service has been largely restored. *<u>Eastern Region</u>: Telephone service has been intermittently restored. *Official responders continue to use hardened communication channels. *Major NGOs launch national media campaigns to collect donations.	**Ham radio operators continue to assist official responders.* **The Internet is back up and running and available where electricity is available.* **Most people outside the region of impact are beginning to reach friends and family inside the region.*	*Response organizations are now able to take advantage of telephones and the Internet.	**There is increased, routine communication among people in the region and with people outside the impacted region.*

Mitigation Opportunity:

Additional hardened communications devices are warranted.

Table 5-4. ShakeOut Scenario Emergency Response Function: Search and Rescue

Organized and Spontaneous Response Summarized for Five Time Periods After ShakeOut Earthquake											
Impact (2-5 minutes)		30 Minutes		2 Hours		24 Hours		72 Hours		1 Week	
Organized	*Spontaneous*	Organized	*Spontaneous*	Organized	*Spontaneous*	Organized	*Spontaneous*	Organized	*Spontaneous*	Organized	*Spontaneous*
None.	**Begin to search for victims in immediate vicinity.*	*Local S&R teams mobilizing. *Notification made to national teams.	**Ad hoc S&R groups form from victim volunteers and extract people from debris of nearby, damaged buildings.* **They use hands and have little or no equipment.*	*Locally based S&R personnel converge and are deployed to local incidents. *Focus is on some of the more complex and difficult rescues. *National S&R teams begin to mobilize.	**Ad hoc S&R groups continue to extract people from debris.* **A vast majority of S&R is accomplished by these groups.* **Additional volunteers join in S&R and begin to use tools and available equipment, such as gloves, crowbars, and chainsaws.*	*Western Region: USAR teams arrive at El Toro, coordinate with REOC and are dispatched to impacted areas. *Eastern Region: Teams mobilizing; local fire services teams on site.	**95% of those rescued in this timeframe have been rescued by neighbors, family, and other victims on location.* **Enormous need for gloves and crowbars.* **Organization emerges among work groups, including leadership and a support network for each work group.*	*All USAR teams are now in the field. *Other teams from around the country and world are being mobilized, or have arrived at staging areas in the state and are being dispatched.	**The many victim volunteers who organized into spontaneous S & R groups continue to rescue others as well as recover bodies.* **They are finding live, trapped people at a declining rate.*	*All local, national, and international teams are fully deployed. *Most of their work now is to recover bodies.	**Most ad-hoc search and rescue groups have disbanded owing to fatigue and frustration over finding fewer living victims, and over having inadequate tools to perform their work.*

Table 5-5. ShakeOut Scenario Emergency Response Function: Sheltering

Organized and Spontaneous Response Summarized for Five Time Periods After ShakeOut Earthquake											
Impact (2-5 minutes)		**30 Minutes**		**2 Hours**		**24 Hours**		**72 Hours**		**1 Week**	
Organized	***Spontaneous***	**Organized**	***Spontaneous***	**Organized**	***Spontaneous***	**Organized**	***Spontaneous***	**Organized**	***Spontaneous***	**Organized**	***Spontaneous***
None	**Look at immediate physical setting and relocate to perceived safe area.*	None	**People doing localized damage assessment. Where damage is observed, consider remaining versus leaving.*	*Discussions taking place between OES, Red Cross and local government to identify numbers and locations of needed shelters plus set-up logistics.	**With aftershocks continuing, people have assessed damage, taking into account utility service and their sense of personal risk to decide whether to remain or evacuate.*	*A few Red Cross shelters are opened at schools and recreation centers. *Regional Care and Shelter Task Force established to develop shelter strategy. *Care and shelter locations are identified in the SoCA region to house evacuees. *Logistics managed by the State and FEMA.	**Unplanned shelters set up in parks, vacant lots, and city streets.* **Victims with the financial means to do so relocate to available hotels. Others locate to the homes of family and friends.* **Most people, particularly in Eastern Region, are camped outside on their property and reentering homes only strategically.*	*Most Red Cross shelters have been set up throughout accessible parts of the impacted region. *Supplies for the shelters are coming into staging areas around the region and are being distributed to all of those shelters. *Most shelters are at capacity.	**Unplanned shelters in parks, vacant lots, and city streets are in need of supplies but are in competition with organized shelters for those supplies.* **Unmet needs are widespread.* **Occupants of unplanned shelters are largely poor, minority people, and families with dependent children.* **There is demand from those with out-of-state, permanent residences to return home.*	*A regional network of shelters has been established. *All the shelters that will be established are in place. *Established shelters are receiving a steady flow of supplies.	**Unplanned shelters are becoming organized, and supplies are arriving one way or another.*

Mitigation Opportunity:

People need to be adequately equipped to remain on their properties for the length of time they are likely to be evacuated from dwellings.

Table 5-6. ShakeOut Scenario Emergency Response Function: Food and Water

Organized and Spontaneous Response Summarized for Five Time Periods After ShakeOut Earthquake											
Impact (2-5 minutes)		**30 Minutes**		**2 Hours**		**24 Hours**		**72 Hours**		**1 Week**	
Organized	***Spontaneous***	**Organized**	***Spontaneous***	**Organized**	***Spontaneous***	**Organized**	***Spontaneous***	**Organized**	***Spontaneous***	**Organized**	***Spontaneous***
None.	**Check water sources in situ.*	None.	**Checking stored supplies. If none stored, sources of water considered.*	*Red Cross and other NGOs have mobilized and begin to identify sources to provision shelters. *Regional water needs are being assessed.	**By now, residents in impact area have at least some food and water available.* **Disaster victims and organized responders are sharing food and water.* **Trained Red Cross volunteers begin to arrive and congregate. Because official shelter sites are not yet designated, they congregate at locations that seem likely to become shelters, such as certain schools.*	*Organized shelter operation has been established in areas where damage is not as severe. *Eastern Region: Establishing shelters in severely damaged areas is slow due to access, availability of food and water and personnel.	**Eastern Region: People begin to need water and will look to local government for supplies.* **Before supplies arrive, some people who go to grocery and convenience stores to collect needed supplies may be labeled as looters by local media and authorities.*	*A coordinated effort among local, State, and Federal government agencies to bring water and food into the region is underway. *Needed water has begun to arrive. *Most of it is on its way but not yet on location.	**Victims are running out of food and water and are making demands on local government.* **Victims still camped on their front lawns are still reliant on their own supplies, which are running out.*	*Consistent supplies of free food and water are coming through staging areas and are being distributed to organized shelters, spontaneous shelters, and people camped out on their front lawns. *There are pockets of rural residents where water and food supplies are not available.	**In areas where food and water are scarce, particularly in the Eastern Region, some people with supplies of food and water will be sharing with those who need it. Others will be selling supplies at inflated prices.*

Mitigation Opportunity:

Educate people that pool water is not a drinking water resource.

Table 5-7. ShakeOut Scenario Emergency Response Function: Donations Management

Organized and Spontaneous Response Summarized for Five Time Periods After ShakeOut Earthquake											
Impact (2-5 Minutes)		30 Minutes		2 Hours		24 Hours		72 Hours		1 Week	
Organized	***Spontaneous***	**Organized**	***Spontaneous***	**Organized**	***Spontaneous***	**Organized**	***Spontaneous***	**Organized**	***Spontaneous***	**Organized**	***Spontaneous***
None.	*None.*	None.	*None.*	*As media coverage of disaster continues, calls are received by media and communicated to EOCs offering donations of money and other resources. * Red Cross and other NGOs active in disaster response are gearing up to solicit and accept donations for disaster victims.	**Victims are beginning to think about what they will need to get through the next few days.*	*Donations of money, services, and material are coming in from all parts of the U.S. *A donation management system has been established and donations are being managed in a centralized manner.	**Local businesses have donated food, water, and services to victims.* **Others are working with local NGOs and government to identify priorities for distribution.*	*Donations of money, services, and material have intensified. *The donation management system is operating to buy what it needed, and distribute goods to those who need them. *Local and national NGOs are distributing resources as they are received.	**Businesses in the region continue to donate necessary resources to victims.* **Volunteers approach NGOs to assist with donation management and distribution.*	*Coordination between NGOs and local government is high. *NGOs are represented at EOCs. *The effective distribution of donations is being accomplished.	**Volunteers from across the nation are approaching national NGOs to volunteer their services.*

Table 5-8. ShakeOut Scenario Emergency Response Function: Medical Services

Organized and Spontaneous Response Summarized for Five Time Periods After ShakeOut Earthquake											
Impact (2-5 Minutes)		30 Minutes		2 Hours		24 Hours		72 Hours		1 Week	
Organized	*Spontaneous*	Organized	*Spontaneous*	Organized	*Spontaneous*	Organized	*Spontaneous*	Organized	*Spontaneous*	Organized	*Spontaneous*
None.	**Check for injuries to self and others.*	*In situ damage assessment of medical facilities. *Convergence of medical personnel.	**Medically trained personnel volunteer.* **People trained in first aid active in situ.*	*Western Region: Hospitals receive and treat local patients. *Eastern Region: Medical personnel have arrived at damaged medical facilities and are performing triage outside. Some medical personnel are administer-ing medical assistance in situ. Medical transport is difficult or impossible.	**Most people have provided medical assistance for themselves and those nearby and now provide aid to the wider community.* **Some have volunteered for organized medical response.* **Sense of frustration over availability of medical resources.* **Western Region: Seriously injured have been transported to hospitals* **Eastern Region: Serious injuries are untreated or are being triaged.*	*EMSA takes lead in managing available regional and State medical services. *DMATs arrive at El Toro, coordinate with EMSA, and are deployed in the region. *Medical personnel converge on local hospitals.	**Sudden influx of patients seeking medical help, especially in Eastern Region. Medical personnel experience role strain as concerns emerge about their own family members.* **Informal mechanisms emerge to obtain info on families, enabling medical units to continue functioning.* **On site first aid being administered.* **People w/ CERT and/or first aid training assuming leadership positions.*	*Many medical staffers have worked without sleep since they first began responding to the disaster. *Seriously damaged hospitals have been evacuated, and open-air trauma centers have been set up in adjacent areas, especially in the Eastern Region. *Very short supply of equipment such as kidney dialysis machines. *Non earthquake disaster patients are being med-evacuated outside the impacted area, to hospitals in Nevada, Arizona, and other parts of California. *Undamaged hospitals have an influx of earthquake victims with crush injuries, broken bones, and trauma. This increased patient load is not distributed evenly, and some undamaged hospitals are dramatically overloaded, while others receive few patients. *Hospitals, particularly in the Eastern Region, are running low or have run out of medicines and supplies. *Western Region: Medical supplies have arrived at staging areas and are being distributed to functioning hospitals. *Eastern Region: Supply distribution is going on, but is much more problematic to accomplish. A few cities had pre-positioned emergency medical supplies throughout the community, and now quickly exhaust those extra supplies.	**People with first aid training continue to treat minor injuries in their neigh-borhoods.* **Continued attempts to transport more seriously injured people to nearby hospitals and/or to outside sites where hospital operations have been moved due to damaged buildings.*	*More uniform distribution of trauma patients to available hospitals. *Some of the hospitals that were evacuated due to non-structural damages have been deemed safe, and these hospitals are being re-populated. *Fewer new trauma patients are being admitted. *A Navy hospital ship has arrived at San Pedro Harbor to assist with treating patients.	**People with first aid training continue to treat minor injuries in their neighborhoods caused when people are injured at home or during aftershocks.* **Continued attempts to transport more seriously injured people to nearby hospitals and/or to outside locations.* **Survivors dramatically increase consumption of alcoholic beverages. This is a stress abatement response seen after all major disasters.*

Table 5-9. ShakeOut Scenario Emergency Response Function: Mortality Management

Organized and Spontaneous Response Summarized for Five Time Periods After ShakeOut Earthquake											
Impact (2-5 Minutes)		30 Minutes		2 Hours		24 Hours		72 Hours		1 Week	
Organized	*Spontaneous*	Organized	*Spontaneous*	Organized	*Spontaneous*	Organized	*Spontaneous*	Organized	*Spontaneous*	Organized	*Spontaneous*
None.	*None.*	None.	**Assessment of degree of injury to self and to others nearby.* **Remorse expressed over deaths and injuries.*	*Hundreds of deaths are expected, based on early projections of the scope and severity of damage using ShakeMap and HAZUS; and based on information communicated to EOCs from S&R teams. *DMORTs mobilized. *Local mortuaries alerted. *Local EOCs still unaware of modeled estimates.	**Eastern Region: Local S&R teams, both spontaneous and organized, locate victims who have perished.* **Families grieve and demand proper care of dead loved ones.* **Responders report deaths to local EOC,s but little can be done about the dead.*	*DMORTs arrive at El Toro, coordinate with REOC, and are deployed in the region. *Local coroner teams receiving and handling bodies.	**Western Region: As a result of building collapses in downtown Los Angeles, there is significant demand from thousands of loved ones to learn the status of building occupants.* **EOCs learning identities of some of the dead from S & R efforts, both organized and spontaneous.* **Eastern Region: Because deaths are more dispersed geographically, issues arise regarding collection and transportation of the dead to storage facilities, and determination of those facilities.*	*DMORTs and locally organized coroner teams have recovered and identified a majority of fatalities. *Processing of bodies continues, with approximately 30% of fatalities yet to be recovered.	**Family members arrange for the burial of their dead loved ones.* **Related activities include picking up bodies, notifying living family members, and making funeral-home arrangements.*	*Funeral homes are overwhelmed with the demand for their services.	**Surviving family members are focused on the mechanics of burial arrangements, a convergence of family members from outside the area to commemorate the dead, and grieving.*

Mitigation Opportunities:

Need to address mass casualties at the local level.

Currently, there are not enough body bags in the region for a catastrophic earthquake.

Table 5-10. ShakeOut Scenario Emergency Response Function: Emergency Operations Centers

Organized and Spontaneous Response Summarized for Five Time Periods After ShakeOut Earthquake											
Impact (2-5 Minutes)		**30 Minutes**		**2 Hours**		**24 Hours**		**72 Hours**		**1 Week**	
Organized	*Spontaneous*	Organized	*Spontaneous*	Organized	*Spontaneous*	Organized	*Spontaneous*	Organized	*Spontaneous*	Organized	*Spontaneous*
*Perceived need to activate local, regional, and State EOCs. *CSWC receives CALWAS messages from SoCA counties reporting an earthquake.	*None.*	*Attempted convergence of official responders. *CSWC begins notification of key State agencies, executive staff, and the Governor's Office. *Activation of SOC, REOC, and local EOCs. *Fire and Law Mutual Aid initiated.	*None.*	*State, regional, and local EOCs are fully activated (though not necessarily fully staffed) and communicating vertically and horizontally. *Federal ESF activated and mobilizing. *Major NGOs also activated and mobilizing.	**Eastern Region: Some realization that organized response will not arrive soon. Individual households have begun to provide assistance to neighbors in their immediate vicinity. Neighbors exchange basic first aid, S&R, and media reports.*	*Staging areas have been established at El Toro, Los Alamitos, Vandenburg, and San Diego. *Joint Federal/State disaster field office established in San Diego. *State provides disaster management teams to severely impacted counties of San Bernardino and Riverside. *SOC Advance Planning initiates relief and recovery planning process based on HAZUS data for lifeline restoration priorities, debris disposal plans, interim and long-term housing strategy, economic stabilization strategy, and private sector integration.	**Eastern Region: In isolated neighborhoods, emergent groups have formed to administer medical aid, conduct S&R, monitor information available via media and act collectively in the absence of organized assistance.*	*SOC, REOC, and Operational Area EOCs fully staffed and functioning. *Local, regional, and State EOCs are fully activated and working on 12-hour-on, 12-hour off shift basis. *These EOCs are communicating with each other and with the staging areas, and managing the mutual aid system, which is fully activated. *EOCs are also in communication with the joint Federal/State disaster field office in San Diego. *EOCs are organized according to NIMS. *Local, State, and Federal declarations of disaster have been proclaimed.	**As organized response takes over, emergent neighborhood response groups begin to dissipate.*	*EOCs are operating like well-oiled machines.	**No emergent EOC-like organizations are operational at this time.*

Mitigation Opportunities

Additional non-structural hazard mitigation is needed in EOCs, including secured equipment.

Additional mobile EOCs are needed as backups.

Table 5-11. ShakeOut Scenario Emergency Response Function: Disaster Intelligence

Organized and Spontaneous Response Summarized for Five Time Periods After ShakeOut Earthquake											
Impact (2-5 Minutes)		30 Minutes		2 Hours		24 Hours		72 Hours		1 Week	
Organized	*Spontaneous*	Organized	*Spontaneous*	Organized	*Spontaneous*	Organized	*Spontaneous*	Organized	*Spontaneous*	Organized	*Spontaneous*
*CISN posts preliminary magnitude and location. *Initial ShakeMap is generated. *Local government (police and fire) begin recon from mobile units. *Caltech and USGS confirm magnitude and location of earthquake to CSWC. *WCATWC issues bulletin indicating potential for local tsunamis due to underwater landslides.	**Strong motion felt over wide area of SoCA.* **Situation defined as an earthquake.* **Concern for safety to self and others.* **Concern over potential damage.*	*OES schedules conference call with CEPEC. *Updated ShakeMap is downloaded to federal, state and local agencies for posting in EOCs. *CSWC completes notification of executive staff, State agencies, and FEMA. *Governor and staff briefed. *USGS issues automated aftershock forecast. *CGS activates Earthquake Information Clearinghouse. *HAZUS is run for regional casualty and loss estimates including distribution of debris.	**Interpersonal attempts to define severity of the situation.* **Attempts to contact others outside immediate environment.* **Attempts to obtain information from media.*	*ShakeMap revised based on additional data. *HAZUS results shared among agencies. *SEMS/NIMS has been implemented and the Disaster Intelligence function has been activated at all EOCs. *The regional scope of the disaster is now known and being reported to the media. *CA Governor has declared state of emergency and requested Presidential Declaration. *Some major incidents have been identified. *<u>Eastern Region</u>: Communications still spotty due to severe damage.	**People who have phone service are calling to report problems such as a downed utility pole with exposed wires, or injured people needing assistance.* **People in damaged communities begin to understand regional scope of disaster, though media reports tend to portray a greater disaster than exists.*	*Understanding the event is still a challenge ("fog of war"). *Aftershocks larger than magnitude 5.0 continue to occur, creating additional damage. *Presidential Disaster Declaration approved.	**Most major local incidents and problems have been reported.* **Many aftershocks have occurred. People are apprehensive and discuss the situation with family and neighbors.* **Media reports, especially from out-of-the-area media, still portray disaster as much greater than it is.*	*Most major incidents in the region have been identified and all responders needed to address those incidents are on the scenes.	**Victims want to know about many things, including insurance payouts, safety inspections, if they can collect on insurance even thought they didn't have earthquake insurance, and whether they are safe from marauding bands of looters.*	*All major problems are known. *Rare new information becomes available about remotely located infrastructural problems, such as remote, compromised dams.	**Victims have begun to focus and seek information related to their intermediate futures, e.g., where are we going to live a few weeks from now? How will I get my house repaired?*

Table 5-12. ShakeOut Scenario Emergency Response Function: Fire Suppression

Organized and Spontaneous Response Summarized for Five Time Periods After ShakeOut Earthquake											
Impact (2-5 Minutes)		**30 Minutes**		**2 Hours**		**24 Hours**		**72 Hours**		**1 Week**	
Organized	*Spontaneous*	**Organized**	*Spontaneous*	**Organized**	*Spontaneous*	**Organized**	*Spontaneous*	**Organized**	*Spontaneous*	**Organized**	*Spontaneous*
None.	**Escape fire.*	*Begin to locate fires. *Priorities being considered.	**Localized fires being fought with any available resources such as fire extinguishers.*	*Western Region: Local fire departments are responding to ignitions within jurisdictional boundaries. *Eastern Region: Local fire departments are working to suppress fires on a prioritized basis while requesting mutual aid to fight additional fires. *Problems in fire suppression due to low water pressure (water system damage) and transportation. (Fire departments cannot get equipment across fault rupture or disrupted roads.)	**Small fires that were discovered early have been extinguished.* **Larger fires are either being fought by local fire departments or are burning out of control.* **While some neighborhood residents are assisting organized fire fighters, local emergent response is currently focused on other response needs (such as transport of injured).*	*First wave of fire mutual aid equipment and personnel arrive at some accessible locations of critical incident. *Other areas that have been seriously damaged and have fires are inaccessible.	**Emergent groups continue to fight smaller fires ignited due to aftershocks and accidents.* **Larger fires are now mainly being fought by organized fire services.*	*Most fire mutual aid in place. *Most of the major fires in the impacted areas are either being fought or have been extinguished.	**Groups of residents are monitoring their immediate areas for fires. They suppress the small fires themselves but report the larger ones to local fire departments.*	*Most major fires in the region have been extinguished. *Fire departments and mutual aid responders are available if needed.	**Groups of residents are still monitoring their areas for the eruption of fires.*

Mitigation Opportunity

Every home should have a functioning fire extinguisher.

Table 5-13. ShakeOut Scenario Emergency Response Function: Law Enforcement

Organized and Spontaneous Response Summarized for Five Time Periods After ShakeOut Earthquake											
Impact (2-5 Minutes)		30 Minutes		2 Hours		24 Hours		72 Hours		1 Week	
Organized	*Spontaneous*	Organized	*Spontaneous*	Organized	*Spontaneous*	Organized	*Spontaneous*	Organized	*Spontaneous*	Organized	*Spontaneous*
None.	*None.*	*At-home shifts converging. *On-duty personnel transitioning to disaster response.	*None.*	*Local law enforcement actively working to control traffic flow within jurisdictional boundaries, setting up perimeters around critical incidents, controlling entry and working detours around blocked streets and roads. *CHP and Caltrans attempting to identify highway and bridge damage. *Law enforcement volunteers assisting at local level.	**Business owners in some damaged areas begin to arrive to protect stores and inventories.* **Eastern Region: Sporadic incidents of residents directing traffic at intersections.*	*Western Region: Traffic signals have been restored, cordons established and law enforcement personnel now deployed to respond to community concerns about media reports of looting. *Eastern Region: Law enforcement spread very thin due to dispersed damage but community concern about looting has prompted local officials to request law enforcement mutual aid and suggest that the National Guard be activated and deployed.	**Local residents begin to realize that police are not available to protect property, and consider ways to protect selves and property, believing that they may be vulnerable.*	*Local law enforcement returns to regular duties as National Guard mobilization handles specialized, quake-related law enforcement duties.	**Local residents begin to see a greater police presence in their neighborhoods.* **Looting fears begin to abate.*	*Most local police departments and the California Highway Patrol continue with normal law-enforcement duties.	**Except for some isolated areas in the Eastern Region, police presence in neighborhoods returns to near normal.*

Table 5-14. ShakeOut Scenario Emergency Response Function: Control and Security

Organized and Spontaneous Response Summarized for Five Time Periods After ShakeOut Earthquake											
Impact (2-5 Minutes)		30 Minutes		2 Hours		24 Hours		72 Hours		1 Week	
Organized	***Spontaneous***	**Organized**	***Spontaneous***	**Organized**	***Spontaneous***	**Organized**	***Spontaneous***	**Organized**	***Spontaneous***	**Organized**	***Spontaneous***
None.	*None.*	None.	*None.*	*Law enforcement mobilizing to provide perimeter security for seriously damaged areas. *Private security forces deploy to protect businesses and communities and to deter looting, which is anticipated as a looming problem.	**Home and business owners consider need for protection against looting or home invasion.*	*Police and security personnel establish cordons around sites of collapse and severe damage, allowing entry only by authorized personnel, such as S&R teams. *Store owners demand access to their property and are admitted in some cases but not others. *Detours established where roads, streets and freeways have been seriously damaged or require inspection.	**Based on media reports that people are looting stores, people take measures to protect themselves and their property.* **These concerns inhibit some from leaving their property for shelters or the homes of friends.*	*Police and security personnel continue to maintain cordons around sites of collapse. *Business owners are allowed to re-enter their businesses on a very limited basis. *Safety inspection of businesses and homes for red- and yellow-tagging is underway.	**Official shelters have regular law enforcement.* **People resiaing at unplanned shelters are providing their own security.* **There are scattered reports of abuse by volunteers who are providing security.*	*National Guard performs some security functions that were earlier performed by local law enforcement. *Business owners have greater access to their businesses. *Partial reopening of some businesses.	**Issues of security at unplanned shelters lessen as National Guard takes over control and security function.*

Mitigation Opportunity

Deemphasize "looting" by facilitating availability of essential supplies at the community level.

Table 5-15. ShakeOut Scenario Emergency Response Function: Traffic Management

Organized and Spontaneous Response Summarized for Five Time Periods After ShakeOut Earthquake											
Impact (2-5 Minutes)		**30 Minutes**		**2 Hours**		**24 Hours**		**72 Hours**		**1 Week**	
Organized	***Spontaneous***	**Organized**	***Spontaneous***	**Organized**	***Spontaneous***	**Organized**	***Spontaneous***	**Organized**	***Spontaneous***	**Organized**	***Spontaneous***
None.	*None.*	*Air traffic diverted where runway damage is suspected. *Caltrans using computerized systems to identify possible freeway damage. *Trains halted pending damage surveys.	**Attempted convergence of family members via personal vehicles.* **Parents attempt to fetch kids from schools and daycares.*	*Many vehicles stranded due to road and bridge damage, or nonfunctioning traffic lights. *On some undamaged roads and streets, police direct traffic where power outages have disabled traffic signals.	**<u>Eastern Region</u>: Commuters have abandoned stranded cars and are walking toward nearby facilities.* **Local residents may direct traffic in lieu of organized response.* **Some attempts to rescue stranded motorists.*	*Teams of inspectors are in the field assessing damage to bridges and roadbeds. *Rail lines and airports also being inspected. *Some airports evaluated and functioning. *Some recovery of traffic signals in less damaged areas. *<u>Eastern Region</u>: Many road closures and vehicles that are abandoned by stranded drivers.	**<u>Eastern Region</u>: Motorized vehicles have very limited mobility due to road and bridge damage; situation worsened by abandoned vehicles and debris. Local traffic limited to emergency vehicles.* **Reunification of families still in process, many have been reunited.* **<u>Western Region</u>: Major incidents of structural collapse restrict movement but severe damage is more isolated here and there is more movement of people.*	*Damage inspection continues and abandoned cars are being towed off major highways. *<u>Western Region</u>: Traffic signals and temporary signage are in place and functioning in most communities. *<u>Eastern Region</u>: The process is ongoing to restore functionality.	**Victim ability to move within the impacted area is improving.*	*<u>Eastern Region</u> achieves traffic signal restoration and temporary signage that the Western Region had achieved in 72 hours.	**People take advantage of restored traffic flow to acquire supplies, sightsee, check on and visit with others, and obtain needed items from vendors that have reopened.*

Table 5-16. ShakeOut Scenario Emergency Response Function: Essential Services

Organized and Spontaneous Response Summarized for Five Time Periods After ShakeOut Earthquake											
Impact (2-5 Minutes)		30 Minutes		2 Hours		24 Hours		72 Hours		1 Week	
Organized	*Spontaneous*	Organized	*Spontaneous*	Organized	*Spontaneous*	Organized	*Spontaneous*	Organized	*Spontaneous*	Organized	*Spontaneous*
*Activation of automatic, emergency-triggered functions. *Widespread power outages throughout region. *SCE, DWP, and ISO Operations Centers initiate power shedding to balance grid. *Progressive blackout of region initiated to prevent cascading failures.	**Check utilities.*	*Convergence of utility personnel and early damage assessment. *Use of ShakeCast and SMIP data to set inspection priorities. *All critical facility and infrastructure operators (including utilities, lifelines, airports, harbors, hospitals) initiate damage assessment.	**Location of utility damage, securing tools, and assessing damage at home or work.*	*Initial reports of damage at most essential facilities and to infrastructure are communicated internally and some have been reported to local and regional EOCs. *Regional damage reports passed on to SOC.	**People and organizations seeking information on status of essential services: Will power be restored soon? Can hospital accept injured? Can we stay in our house or place of business? Will emergency response personnel be able to reach us?*	*Western Region: Some essential services have been restored. *Eastern Region: For most of the region there is no electricity, natural gas, or water; utilities struggle to restore services.	**People and organizations still seeking information on status of essential services. Will power be restored soon? Can hospital accept injured? Can we stay in our house or place of business? Will emergency response personnel be able to reach us?*	*The restoration of essential services continues. *Eastern Region: There are still significant outages.	**People still without an essential service have compensated for the loss. For example, they are using lanterns for light and outdoor stoves for cooking and heat.* **People inquire about repairs with utility workers they may see in their neighborhoods.*	*Western Region now has most utilities restored. *Eastern Region still suffers major outages.	**People have run out of the alternative sources of power and fuel that they used to compensate for utility outages in the immediate aftermath.* **Utility restoration in some areas is accomplished.*

Table 5-17. ShakeOut Scenario Emergency Response Function: Route Recovery

Organized and Spontaneous Response Summarized for Five Time Periods After ShakeOut Earthquake											
Impact (2-5 Minutes)		30 Minutes		2 Hours		24 Hours		72 Hours		1 Week	
Organized	***Spontaneous***	**Organized**	***Spontaneous***	**Organized**	***Spontaneous***	**Organized**	***Spontaneous***	**Organized**	***Spontaneous***	**Organized**	***Spontaneous***
None.	*None.*	*Major route obstacles being reported. *Locating heavy equipment for route recovery.	**Obstacles encountered by persons converging to home or to reunite with family.* **Traffic jams ensue.* **Some people who are traveling on roads and highways get stranded due to obstacles, road damage, or bridge damage.*	*Most major route disruptions and obstacles have been identified, characterized and reported to EOCs. *CHP, Caltrans and FHA working together to develop strategies and priorities for closures and detours. *Agencies become aware of pockets of stranded motorists who must be assisted.	**All street and road obstacles have been encountered by commuters.* **Some stranded motorists have phoned for help (where cell phones are working); thus the problem has come to the attention of local EOCs and those working on route recovery.* **Some stranded motorists remain close to their vehicles and are seeking assistance.*	*Critical access routes are identified, and priorities are set for debris removal by a transportation task force. *Route innovation involves use of helicopters and nonstandard vehicles for emergency response. *Eastern Region: Not much route recovery. The Palm Springs airport has been opened for emergency response only.	**Stranded motorists have been rescued.* **Some local streets have been cleared and local vehicle access has improved.* **There has been route innovation in use of bicycles and four-wheel drive vehicles.*	*Access has been restored to major highways with high priority restoration designations. *All major airports are now functioning. *LA Metro is back in full operation.	*None.*	*Greater recovery of secondary street routes has occurred through the use of detours and repairs. *Some bridges have passed inspection and passage has reopened on them. *Eastern Region: Interstate 10 remains impassable.	*None.*

Table 5-18. ShakeOut Scenario Emergency Response Function: Debris Removal

Organized and Spontaneous Response Summarized for Five Time Periods After ShakeOut Earthquake											
Impact (2-5 Minutes)		30 Minutes		2 Hours		24 Hours		72 Hours		1 Week	
Organized	*Spontaneous*	Organized	*Spontaneous*	Organized	*Spontaneous*	Organized	*Spontaneous*	Organized	*Spontaneous*	Organized	*Spontaneous*
None.	**Removal of debris in situ for life safety.*	None.	**Only to extent necessary to extract victims.*	*Becomes apparent that this will become a major response challenge, given the level and scope of damage. *Areas where debris is going to be biggest problem identified through HAZUS. *Efforts mobilized at local and regional EOCs, to decide what equipment and personnel are needed, and to coordinate efforts among agencies.	**Some people begin to clear debris in and around households.* **Where S&R was unnecessary or is no longer salient, victims begin to clear damaged household items and salvage needed resources.*	*Priorities being established for debris removal. *Top priority accorded to critical facility access and highways, ports, and airports.	**Residents in the impacted areas have completed debris removal in and around their own houses and have assisted others.* **Some residents are assisting official responders in debris removal in their neighborhoods.*	*Priority debris removal is initiated. *Ongoing debris removal has facilitated some degree of route recovery and the opening of some regional airports and harbors. *<u>Western Region</u>: Initial plan emerging for debris management associated with building collapses.	**Most people who have inhabitable homes have cleaned them.*	*Debris removal at high-priority, critical facilities is well underway, allowing restoration of transportation and other vital functions.	*None.*

C. Narratives

Public Information Narrative

First 5 Minutes

As the ground shook and many of the people at work in media centers had taken shelter under desks and tables, they knew that the stories they planned to work on that day were now scrapped and that they'd be covering an earthquake. Broadcasters on the air explained that an earthquake was happening to their viewers and listeners (as if those they were talking to didn't know that for themselves), as they got under their own desks and tables.

When the shaking stopped, people in the media centers did what was natural for people when the ground stops shaking: they checked on each other's safety. Within minutes, however, reporters scurried around their offices and used everything at their disposal to find out about the earthquake's magnitude, the location of its epicenter, the area affected, and, especially on the earthquake's affects and consequences. Technicians others checked on their station's ability to broadcast. They checked on electricity, equipment, and damage.

Those who were on camera and microphones at television and radio stations when the shaking started in the west end came out from their shelters to discover that they were still on the air. They began to account their person surprise, thoughts, and experience to their audiences. But broadcasters on television and in radio stations in the east end found that they were silenced; their capacity to broadcast had been lost. All of them quickly came to understand that in addition to covering an earthquake story that day that they were earthquake victims themselves.

Every functioning radio and television station throughout the southland immediately suspended normal broadcasting and turned to full coverage to the earthquake.

The first few minutes after the ground stopped shaking weren't that different for everyone else. People who were at home, work, shopping, driving in their cars and everywhere else comforted and checked on each other. Everyone focused on what had just happened to themselves and to the people around them. And after a brief few seconds of solitude, they began expressing their innate human need to talk about the event with the people around them. As if any of them were actually able to foresee such things, they told each other that it was an earthquake, not to worry, and that everything was going to be alright.

But while the need to talk with the people in their close physical proximity to make immediate sense out of what had just happened was being serviced, almost everyone was also quickly drawn to the media to learn about what had just happened and where. Everyone near a television tried to find coverage, people in cars and trucks scurried through their radio stations to find a broadcast about the earthquake.

But there wasn't any electricity in the east end and most televisions had been damaged. Most strapped down televisions in the west end were in fine working order, but many had no electricity. People were quickly drawn to their portable and car radios. Within minutes of the shaking, almost everyone in the southland was searching out information from the people they happened to be with and from the electronic media to which they had access.

Half Hour

By the time a half hour had passed, different scientific organizations had reported on the details of the quake and those specific details were being broadcast by the functioning media. The earthquake was reported to be a 7.8 magnitude event, the epicenter was reported as Bombay Beach. Media of all sorts quickly sent reporters to Caltech to interview seismologists to get as many more details from scientists first hand, and some arrived there within the first half hour.

Local and national media began to mobilize to cover the earthquake. Field reporters from one end of the earthquake stricken area to the other were focused on finding out about and covering local damage and reporting on it.

In the west end, however, attention began to be dramatically drawn to multiple collapsed high rise buildings in the Los Angeles area. This was, clearly, a big story that demanded coverage. In fact, news of the collapsed high rises had spread among the people who worked for local media organizations like wildfire, and, within the first 30 minutes, most radio and television reporters in the west end were engaged in a trying commute from where they were to those collapsed high rise buildings. Media coverage of other local damage, and the start of fires in many locations also began to be broadcast, but the "big story" had already become the high rise building collapses.

Now that a half hour had passed since the shaking had stopped, almost everyone who felt the earthquake was engaged in seeking out information about it. Most were frustrated over not being able to get more information because of failed electricity and broken media equipment. But slowly, the public was beginning to realize the scope of the disaster: it hadn't just happened to them, it happened across southern California. People had begun to realize that the "Big One" had, finally, in fact happened. As a consequence of this emerging realization, people began to shift their identification away from themselves as individuals, and began to identify with the entire affected community of victims. There concerns shifted away from concerns about themselves to being concerned about what had happened to their community and the entire human collective that comprises southern California. In no time at all, media accounts of the scope of the disaster had shifted everyone's identity from what it had been before the quake to "Southern Califonians." Although this shift would take a little longer to reach everyone, this mental shift began as a result of media coverage in the first half hour after the main shock.

And by now, everyone in southern California was trying to get in touch with loved ones and friends to find out if they were OK. Everyone who had one and could turned on their cell phones and tried to user them to call others to check on them and to let them know about themselves. Every other communication device, e.g., blackberry's, computers, and everything else was turned on in a virtual tsunami of interpersonal communication attempts. Anxiety increased as most attempts to use modern communication devices to reach loved ones failed. The use of personal communication devises were thwarted because of system overload and damage to the structural systems that support communication. Almost everyone desperately continued to try to find ways to communicate with their loved ones and friends.

2 Hours

Almost all scheduled television and radio broadcasting in the country was suspended as stations and channels devoted all of their broadcasting time to covering the earthquake.

National news media in New York City, Atlanta, and in other cities had already dispatched correspondents and their support teams to the southland. CNN, Fox News, and MSNBC broadcasts were now fully devoted to the earthquake and people in Washington, D.C., the rest of the nation and the world were beginning what would end up being glued to their broadcasts for many days to come. Most national news correspondents had now booked flights to Phoenix, Las Vegas, Bakersfield, and a few other cities. They planned on flying in there, and then driving into southern California. Others were on the road in vans. And most of the other major news organizations around the world had begun to cover the story. The world was beginning to learn that another mega-disaster had just struck the United States.

Although affected local and county governments are providing the reporters with information about local impacts for broadcast, some news crews have successfully relocated to sites of major damage and have begun broadcasting from those locations. But the collapsed high rises in the Los Angeles area are clearly the prime focus for most news coverage. And news reporting clearly is targeting major impacts in major urban areas for reporting. This is beginning to give people in other states and nations with the impression that the earthquake's affects, which were large and horrific, are actually much worse than actually experienced.

The national media has already reported that several major high rises have collapsed in Los Angeles and that those buildings were fully occupied when they fell. And television commentators who report of the failure of those buildings speculate about the number of people who might have been in them at the time that they went down. Comparisons are quickly made to New York City's Twin Towers that collapsed on 9/11, but most broadcasters who draw this comparison are quick to point out that most of the people in the World Trade Center Towers had time to evacuate unlike what had just happened in Los Angeles. Interviews with structural engineers in the east and in San Francisco offer insights into why there would have been such major structural failures in an earthquake.

But, now, fires have also been filmed by television news crews and pictures of them broadcast around the world. One film crew, broadcasting from a hill top in Los Angeles with a view of the greater Los Angeles downtown area out to the ocean broadcasts an account of the earthquake with multiple pillars of smoke rising into the sky. The claim is made that Los Angeles has begun to burn. And an analogy is drawn to the 1906 Great San Francisco Earthquake and the conflagration that followed.

Some of the hardest hit areas are in the more sparsely populated areas of southern California, but these locations and the losses that they have suffered have not yet been reported by the major media. National media have already begun to refer to the event as the "Los Angeles Earthquake," and the eyes of the world have turned to "The City of the Angels."

Those who have access to electronic media, especially television, have begun to shift their attention away from local issues to wider concerns about the earthquake based on media coverage. Victims throughout southern California have begun to vigorously discuss reports heard over the media. In our modern world, sometimes "the story" about an event has a larger impact on what people think happened that the facts about what actually did happen, and this is no exception. The media, and not the earthquake, have now begun to define the earthquake for viewers and listeners.

By now, hundreds of millions of people from around the nation and world are attempting to call people that they know and love in southern California. Despite appeals broadcast over the media by emergency response officials for the public to not use telephones except for emergency purposes, all telephone communications that could operate are inoperable because of overloaded. No calls go through.

Yet in the middle of media's focus on "the story" of major collapses, the fires, and reporting that has focused on the earthquake's impacts, many comments by broadcasters on both television and radio are made that do provide victims with useful information about what to do, how to do it, and preparing for aftershocks.

24 Hours

By the end of the first day after the main shock, the information that is available to people (both victims in southern California and onlookers elsewhere in the state, nation, and world) is the most distorted it has been so far or will be as more time goes by.

By now every major media organization in the nation has crews that have, somehow, found their way into southern California to cover the story. They have largely located to the sites of the major structural failures and are reporting on them, for example, the high rise collapses and the freeway bridges that have collapsed pinning burning cars and people.

One film crew at the site of a collapsed high rise is filming as the earthquake early warning system siren wails to warn search and rescue teams in the collapsed structure that an aftershock will strike in 30 seconds. They film 8 of the 11 search and rescue workers scramble out of the structure just before the aftershock hits. When the shaking stops they interview a member of the search and rescue team and reach the conclusion on the air that 3 members of the team of fire fighters did not make it out of the structure before the aftershock shifted the rubble that was once a proud high rise building.

The Governor's Office of Emergency Services' has Local Operational Areas in the southland and they along with the Federal Emergency Management Agency have now set up a "Joint Information Center" or "JIC" (a place for the media to come to be informed about events and provided information). Representatives of both organizations have begun providing the media with emergency instructions to pass on to the public. Special attention is being devoted to providing information to the public about the location of temporary shelters. Although a few media reporters are located at the JIC, many more are in the field filming and documenting first hand accounts of the quake's impacts, and documenting the experiences of victims.

Federal response is now almost exclusively focused on the Los Angeles metropolitan area. Just as initial media coverage of the 1989 Loma Prieta Earthquake in the Bay Area focused initial federal response on the City and County of San Francisco, media coverage of the "big stories" in Los Angeles have focused federal response on Los Angeles. Local broadcasters that can operate in the east end of the affected area are busy providing emergency information and instructions to the public.

Wide variation exists in terms of both the quality and quantity of information that is being made public and communicated within and across communities all over southern California serves as a basis for mis-information and rumors. Rumors are rampant and some of them are being fed and perpetuated by the media. These include the typical rumors that follow earthquake events of this magnitude. For example, people are passing along concern between each other that the un-transported dead will cause disease (despite official information to the contrary), fear over the possibility of becoming victims of looters is widespread but looting is not (this fear is fed by a few media reports of looting based on isolated observations of behavior that observers labeled as looting that a more critical eye would determine to not be looting at all, and no one is telling anyone how to safely dig through rubble to rescue trapped victims, even though that's exactly what many surviving victims are doing. The media continues to report on cases of trapped people rescued from the rubble. The estimates the media report on the number of dead is inflated.

People all over southern California are now engaged in vigorous conversations with each other about local incidents and events in other places in the southland; these conversations have been fed by media reports, stories handed on between people, and, especially, by stories over the radio.

72 Hours

Television and radio across southern California are now frequently interrupting their coverage of damage and human impacts stories by vigorously providing victims with much needed systematic information about how to deal with the range of situations that they face. For example, there are many broadcasts that tell people about what to do with bodies, basic sanitation, aftershocks and the risks that they will pose to damaged structures and people in them, the purification of water, food spoilage, and information on many more topics fundamental to helping them face their situations.

National media continue to provide 24-hour coverage of the earthquake's aftermath, but have assembled and are reporting on much greater detail about the full range of damage caused by the earthquake across the entire affected area , but they continue to highlight specific incidents of damage, casualties and injuries. A great deal of coverage is also given to people in temporary shelters across Los Angeles. But the focus of national news organization's coverage remains focused on losses and damage in the greater Los Angeles area. Occasional dramatic stories of damage and human interest are being aired about isolated dramatic events in the City of San Bernardino and its surrounding area. Special coverage is aired about a dam above the city that has a high risk of breaking and the need to evacuate a large section of the city.

The biggest story covered is now about the many fires in southern California. Covering fires in southern Californian is nothing new, but, this time, the fire storms are not limited to hill sides. Conflagrations, the likes of which have never been seen are ravaging flatland communities. One neighborhood after another is ablaze and the extent of damage is beyond comprehension. News helicopters are capturing it all on tape and the firestorms are being broadcast around the world. Half of the front page of USA Today is a photo of blocks of burning houses in Los Angeles. And what started as a reporter's exaggeration on the first day (that Los Angeles was burning) has now, or so it seems, come to pass. This story, however, is not just about the flames and the homes and businesses that they're burning, it's also about the impeded evacuation of millions of people since the question underlying the story is how can so many evacuate with so many roadways blocked by rubble from the earthquake.

In the midst of all this, the national and local media alike converge to cover the Governor's visit to the disaster site and the subsequent press conference that he holds. The Governor has, of course, been joined by his Director of the Governor's Office of Emergency Services, and mayors from several major impacted cities.

And by now, many of the millions of victims of the earthquake and fire have begun to demand information about their longer-term needs, for example, their attention has begun to shift from sheltering to reconstruction of their homes.

Some rumors have been laid to rest but others are rampant. Not the least of which is that insurance companies will turn their backs on the victims and not pay for losses from either the earthquake or fires. Members of California's Congressional delegation have been interviewed by national media who have clearly warned insurance companies to honor their responsibilities. But most victims don't really understand if their fire losses will be paid or not since the entire event began because of an earthquake, and most people did not have earthquake insurance, and many who did had recently dropped it because they thought the premiums were too high. Rumors of looting stem from victims taking essential life-sustaining supplies from any place they can find them.

1 Week

Local media have begun to provide information about state and federal programs for victims to assist in their recovery. And the media continue to provide information to victims useful for their dealing with the daily problems that they face. Media updates are provided on the high-profile damage including the high-rise collapses and the consequences of the conflagrations.

The media have now really focused on providing coverage of individuals and communities that "have a personal story to tell". Three such stories are aired around the world: an auto mechanic who rescued a dozen people by working for eight straight hours after the initial shaking, a survivor who's entire family survived the earthquake but who then lost everyone in the fire, and a man who, against all odds, had survived but only now was extracted from rubble in downtown Riverside.

National and local media present coverage of the President and President-elect visiting the disaster site. They both promise to help rebuild southern California. Numerous national and state experts from varied fields are interviewed about the event and human responses to it. The Governor's Office of Emergency Services and FEMA continue to abate rumors, including the establishment of rumor hot lines. But media images continue to fuel the looting rumor. In the eastern region, fears of looting peek due to widely distributed damage and limited law enforcement to cover all sites.

Search and Rescue Narrative

Search and rescue refers to efforts by organized and spontaneously assembled groups to locate and extricate persons trapped in debris, administer medical aid if necessary and coordinate the transportation of seriously injured persons to hospitals. In the early stages of the disaster, perhaps during the first 3-6 hours after the earthquake, nearly all search and rescue efforts are performed by family members, neighbors and community residents in the immediate vicinity of buildings that have collapsed or partially collapsed. Most of the "live" rescues will be accomplished in the first few hours by these spontaneous search and rescue groups who work with limited resources and skills but have the advantage of physical proximity to collapsed structures. Organized search and rescue teams require time to mobilize and will be hampered by severe damage to transportation and communications lifelines as well as secondary effects of the earthquake including fire following and potential dam collapse.

Upon the cessation of shaking, people in the southern California region who have experienced severe shaking will survey their surroundings; if at work, they will check on the well-being of their co-workers, if at home the members of their household and if in a public place, others who happen to be in the same vicinity. This basic "emergent" norm, calling for assistance to those who have become victims of the earthquake is highly salient during the immediate aftermath of the earthquake and provides the motivation for those who have survived the earthquake with minimal injury to devote themselves to those who are in need of assistance. This assistance is first expressed as a desire to help those who are trapped in debris and possibly injured. Most of these early rescues carried out by spontaneously assembled groups will be relatively uncomplicated, requiring few if any tools with first aid administered by those who happen to be present. If transportation is available and routes to local hospitals are unencumbered, the injured will be transported to nearby hospitals.

During the first six hours or so, search and rescue by survivors of the earthquake will be particularly intense in the areas that have experienced the most significant ground motion including eastern Los Angeles County and the cities of San Bernardino, Riverside, Palm Springs, Lancaster and Palmdale. Initially, the focal points will be collapsed or partially collapsed multi-family residential buildings and businesses in the older sections of these cities but given the relatively unsystematic approach of spontaneously organized search and rescue groups, activity will be focused on the immediate area in which groups form rather than according to any plan. For these groups operating in the first six hours, the full extent of the area damaged is unknown and with limited access to information about broader impacts, activity is highly localized.

Organized search and rescue teams, both local and national, will begin mobilizing as soon as the shaking has ceased. Search and rescue teams attached to local fire services agencies within southern California, will be the first organized search and rescue responders to appear at the scenes of collapsed structures. A few hours will be required for these groups to assemble, gather needed equipment, obtain instructions from EOCs (which are also in the process of mobilization and disaster intelligence gathering), and determine whether routes to locations where rescues are needed are unencumbered by debris. In many cases, these teams will be required to deploy without knowing whether direct routes to known collapses are navigable or not. In some cases in which local search and rescue team members are at home when the earthquake occurs, their efforts to report to duty stations may be impeded by damage in the vicinity of their homes, uncertainty regarding possible bridge and street damage between home and fire station and, if their skills are known in the community, attempts by spontaneous rescuers to recruit these professionals to assist in immediately local searches and rescues.

Once locally organized search and rescue teams arrive at the scenes of collapsed structures, they are likely to encounter ongoing search and rescue efforts by spontaneously formed groups who provide informal briefings to the organized teams and continue to work in cooperation with the organized responders, assume subordinate roles in the effort or redeploy to other collapse scenes where organized teams are not present. In most cases, spontaneous teams will have rescued many people trapped in the debris of badly damaged or collapsed buildings before organized teams appear on the scene. These rescues will be the less problematic ones that require minimal skills and simple tools. The remaining rescues which fall to the organized teams will be more difficult, requiring both training and more specialized rescue equipment. These rescues are also likely to involve more serious injuries and both greater on-site medical expertise and the need for close coordination with emergency vehicles available to transport the injured to hospitals.

At the end of the first day, the overall extent of the disaster is known as well as the fact that there are hundreds of sites throughout the region that require rescues. At this point, spontaneous groups of responders continue to work (though fatigue is depleting their ranks), locally based search and rescue teams are fully deployed with EOCs seeking assistance through the fire mutual aid system, and national teams are being mobilized by FEMA. While technologies such as ShakeMap and HAZUS, as well as traditional reconnaissance measures that include windshield surveys and media monitoring, have contributed to understanding the disaster, the situation continues to evolve. Immediately after the earthquake, fires have ignited at hundreds of locations and more fires continue to be reported. Bridge closures, damage to freeways and surface streets and debris in roadways have frustrated the timely and efficient deployment of search and rescue teams to objectives. Difficulties communicating between field units and EOCs has also impaired the ability of teams to accomplish search and rescue activities and teams are frustrated that successful rescues are not accompanied by rapid transportation of the critically injured to hospitals due to road conditions and poor communications.

Fire is becoming a major impediment to rescue efforts. Within 24 hours of the earthquake several large fires are burning in the Cities of Riverside, San Bernardino, Santa Ana and East Los Angeles. Though all local fire units and equipment including units from adjacent areas mobilized through the fire mutual aid system are engaged in fighting these fires, they are yet uncontained and there is danger that several individual large fires will merge into a major conflagration. These fires have occurred in areas where search and rescue is needed or already in progress. Some search and rescue teams as well as spontaneous volunteers have been forced to abandon efforts to rescue people trapped in the debris due to rapidly approaching fires. Though teams have accelerated their efforts in these fire endangered areas and are working under considerable stress, many have been ordered to leave by fire units despite the fact that those trapped in the debris will die if fire reaches the site of rescues.

Other secondary hazards have also hampered search and rescue efforts including an evacuation below a San Bernardino County dam that has been compromised by damage from the earthquake, several sites of hazardous materials releases and a continuing series of large aftershocks. Approximately seventeen hours after the earthquake, dam inspectors discover transverse cracking and muddy water emerging at the toe of the Lake Gregory dam near Crestline north of the City of San Bernardino. After lengthy discussions between the dam inspectors, city officials and the San Bernardino County EOC, it is decided that the endangered population below the dam will be evacuated. A major dilemma for decision makers is the ongoing rescue effort at the site of a large collapsed residential facility where senior citizens, many of whom are disabled, are housed. There are estimated to be 60 or more persons needing rescue from the facility when the order to evacuate is issued. Rescuers, faced with the prospect of leaving the scene, decide to ignore the evacuation order and continue rescue operations.

On the day following the earthquake, the Governor's Office of Emergency Services requests that members of the California Integrated Seismic Network deploy portable instruments to the epicentral area of the earthquake near the Salton Sea and provide warnings to search and rescue teams that are working in collapsed and partially collapsed buildings. By the third day, this system is in place and operational. For rescue sites in Palm Springs and desert cities, the warnings precede the arrival of strong ground motion by 8-10 seconds but in San Bernardino and parts of Los Angeles, the warnings are up to 30 seconds. These aftershock warnings are communicated to rescue sites in real- time with sirens placed near rescues that sound when an aftershocks above magnitude 5.5 are detected. By the end of the search and rescue phase of this disaster, dozens of warnings have been broadcast to rescue workers, saving the lives of many.

Toward the end of the first week, thousands of people have been rescued but search and rescue continues at hundreds of sites. Federally deployed Urban Search and Rescue Teams have arrived from throughout the country and some international teams have reported to staging areas and have been deployed as well. Despite the large and dispersed nature of the rescue efforts, news media attention has focused on 5 collapsed high rise buildings, three of which are in downtown Los Angeles, one in Costa Mesa and one in the City of San Bernardino. These buildings are believed to have been fully occupied at the time of the earthquake and their complete collapse has resulted in a convergence of rescue teams, equipment and journalists. Live rescues however have been few and, after an initial period in which a few survivors are located, the effort turns to recovery of the dead.

Search and rescue operations will continue for nineteen days (even though people who are not rescued within 24 hours are unlikely to survive) and coincide with recovery of the dead. After approximately one week, most are recoveries rather than rescues. In all, 2700 official search and rescue team members will have been deployed and tens of thousands of volunteers, some with CERT training will have contributed to the rescue and recovery effort. Thirteen rescuers have lost their lives due to the collapse of a high rise building during a large aftershock. The Regional Emergency Operations Center at Los Alamitos reports that there have been approximately 45,000 live rescues in the eight counties impacted by the earthquake and the many large aftershocks. The extensive fires and large conflagration in central and eastern Los Angeles prevented the rescue of others who died before rescuers could free them from collapsed structures. Regional news media have documented the heroism of both volunteers and official search and rescue team members who defied an evacuation order following the discovery of severe damage to the Lake Gregory Dam and continued to rescue people trapped in the collapse of a senior citizens housing complex.

Victim Services Narrative

In this section, we consider shelter for evacuees, the provision of food and water to victims and the management of donations targeted for those impacted by the earthquake and extensive fires. Shelter is usually regarded as formally organized public shelters that are opened and operated by the Red Cross and state and local government agencies at community centers, schools, recreation facilities and other public buildings that are undamaged by the earthquake but geographically near areas of damage. But informal shelters are also likely to characterize a very large earthquake affecting an entire region. These shelters may take the form of tent camps in open spaces or shelter in situ by residents reluctant to leave their property despite damage and multiple utility outages.

Food is unlikely to be a major problem immediately following the earthquake as most people will have sufficient stored food to last for two-three days and water may be obtained from stored bottled water and drinkable water from water heaters. But as household supplies dwindle, there will be a need for water and food directed both to neighborhood residents and to organized shelters. These same supplies will be more problematic for informal shelters that may not be known to authorities. Donations of money and relief supplies will begin arriving soon after the earthquake and will require management at the state and regional levels and distribution networks at the local level.

Evacuation and Shelter Narrative

Decisions regarding evacuation will not follow immediately after the earthquake. Those whose homes have been damaged will be concerned first with their own safety and that of family members who are not present. Once united, families will assess the situation and base their decision to leave or remain in their homes on several factors—the level of damage, the type and number of utility outages, their emotional state and the actions taken, or perception of the actions taken, by neighbors. Clearly, not all post- earthquake evacuations will be discretionary as severe damage, hazardous materials releases, potential dam failure and fire, or the likelihood of fire, will prompt evacuations ordered by authorities.

Evacuations will take place over the first several days following the earthquake. In the eastern portions of Los Angeles County, the Cities of San Bernardino and Riverside and the desert cities of the Coachella Valley, large numbers of people whose homes have been seriously damaged and are without electricity, water and natural gas are likely to evacuate on the day of the earthquake and seek shelter with friends and relatives whose homes have not been damaged. Debris in the streets, non-functioning traffic signals and damage to bridges will limit the mobility of many who wish to leave the area and some will either remain close to their damaged homes or relocate to public shelters which will open during the late afternoon of the first day.

As emergency operations centers ramp up to respond to the disaster, responsibility for shelter and mass care fall to mass care coordinators at the local and state levels who coordinate with FEMA at the federal level (Federal Response Plan, Emergency Support Function 6). In addition to the identification of shelter sites and activation of shelters for evacuees, these responders are responsible for the identification and provision of needed services and supplies to shelter locations including: personnel to manage the shelters, food and water, security, shelter identification and routing (including signage), information regarding the placement of pets and service animals and notification to the public regarding the location of the nearest shelters. Mass care coordination is mobilized quickly in this disaster as over one hundred shelters and associated services will be needed on the day of the earthquake.

Early evacuees may set up shelters in parks or open spaces before official shelters are opened and some will remain in these locations rather than move again. By the end of the first day, a total of 120 shelters have been opened and are staffed by Red Cross volunteers and emergency services workers from local and state government. The shelters receive food and water shipments that are directed to shelter locations by local emergency operations centers which have mobilized across the region. Because of the regional scope of the disaster, shelters open with minimal supplies and for some, particularly those in the most impacted areas and those which have opened without official sanction, these shortages are acute and require innovation on the part of shelter managers and leaders.

Initially, shelters house only a few evacuees but as fires spread, more and more people arrive and by noon of the second day, most are full and it is clear that more shelters must be opened. Over the next 48 hours, an additional 383 shelters are opened as fires spread requiring the evacuations of thousands of people whose homes are in danger of fire or have already burned. An additional 30,000 people are in need of shelter due to evacuations ordered after the discovery of damage to a dam above the City of San Bernardino. Shortages of food, water and medicine cause some discontent at the shelters but overall, evacuees are compliant with shelter rules, cooperative and willing to volunteer when needed.

The buildings and open areas selected for shelters are varied and include neighborhood schools, churches, recreation centers, non-essential government buildings, parks, vacant lots and fairgrounds. The need for shelter in western San Bernardino and Riverside Counties is greater than other impacted areas and several tribal nations offer Indian casinos as shelters, and pending safety assessments of these facilities, several begin accepting evacuees on the day after the earthquake. Where buildings are not available, tents are erected in parks and at fairgrounds. Tents are supplied by the National Guard which has been mobilized and is assisting in the response effort in a variety of capacities including the provision of security at official shelter locations, delivering water and food, directing vehicle and foot traffic to nearby shelters and regulating building entry in damaged commercial areas.

Although discouraged by officials at local emergency operations centers, approximately 25 "unofficial" shelter sites are set up in vacant lots and community parks around the region. These shelter locations struggle to gain the attention of local officials for tents, security, sanitary and medical supplies and staff. Many of these locations receive evacuees that official shelters have rejected or have forced to leave for various reasons including unruliness, alcohol abuse or violation of shelter rules. Despite difficulties, leaders emerge at these shelters and achieve moderate success in maintaining order and grudging recognition at local emergency operations centers. Nevertheless, the provision of supplies and services to these locations is slow, sporadic and often inadequate.

There are thousands of people in the 8-county region impacted by the earthquake whose homes have been damaged and are without utilities but remain on the property surrounding their homes rather than seek shelter elsewhere. Many fear that their homes will be looted if they leave, have no relatives or friends nearby who might after them refuge and reject the idea of going to a public shelter. Although they do not require overnight accommodation at shelters, this population makes demands on mass care resources as supplies of water, food, medication and sanitary supplies are exhausted. By the third day after the earthquake, shelter operators establish "pick up points" adjacent to shelters and adjust resource requests to local EOCs to assure adequate supplies for this in situ evacuation population.

Many families who have evacuated or are considering evacuation must make decisions regarding their pets and animals. Some official shelters accept pets and others do not and emergency operations center staff work with local animal care and control officials to identify animal shelters, veterinary hospitals, boarding kennels, pet stores and fairground facilities to house animals that shelters are unwilling or incapable of accepting. Some evacuees resort to unofficial shelters or decide to remain on their property due to concerns about the care and safety of their pets. In response to the issues of pet care, EOCs launch public information bulletins asking evacuees to consider their pets in evacuation decisions, provide locations where pets can be housed and assign personnel to shelters to act as referral sources for animal care operations.

The percentage of victims needing public shelter is larger in this earthquake than other disasters due to several factors: extensive urban fires have destroyed thousands of residential units to which evacuees cannot return, many victims of the earthquake are low income people who have fewer options for alternate shelter, widespread utility outages have made it difficult to inhabit homes that may have suffered only minimal damage and the many large aftershocks (including some that are greater than M6) have magnified fears that considerable danger remains and fuels discomfort in remaining in structures that may be further damaged or collapse.

As a result of earthquake damage, widespread utility outages, fear and discomfort due to ongoing large aftershocks, extensive fires, and a potential dam collapse, more than 270,000 people have been displaced from their homes in the 8-county area affected by the M7.8 earthquake. In some cases, those who have left their homes will return within a week; however, extensive fire damage will necessitate intermediate and long-term housing for approximately 120,000 people. A total of 503 public shelters will be necessary to house 175,500 people who cannot find shelter elsewhere, mainly with relatives, friends and in hotels in the region.

Food and Water Narrative

Immediately after the earthquake, food and water are less problematic than they will become in the timeframe of 72 hours to a week after the earthquake. In approximately three days, many households in the impacted region have exhausted their supplies of food, and water, in many cases, is needed even sooner. For those forced to evacuate their homes, food and water are provided as part of shelter operations though the delivery of these resources is difficult logistically in areas of southern California where damage is particularly severe. These areas are eastern Los Angeles County, the cities of San Bernardino and Riverside and the desert cities of the Coachella Valley.

Drinking water is a very high priority and shortages of potable water are particularly acute since the earthquake has caused many breaks in water pipes that will require lengthy and extensive repairs and water will be needed to fight the many fires that are burning throughout the region. Although areas within southern California that have had minimal impacts from the earthquake are providing some drinking water to heavily damaged areas, it is not sufficient to meet the need and a major effort is mobilized to transport water by land, sea and air.

The provision of food and potable water is coordinated at the state level at the State Operations Center in Mather near Sacramento. Drinking water supply needs are communicated from the county EOCs to the Regional Emergency Operations Center at Los Alamitos and then communicated to the SOC. A mass care coordinator at the SOC works with other state and federal agencies to identify water supplies, particularly from bottled water companies throughout the western US, and arranges for transportation to staging areas where water is distributed to shelter locations and to water supply delivery points accessible by residents who have not evacuated their homes but are in need of drinking water. Surface transportation of food and water supplies is hampered by road and bridge closures in the worst hit areas

Donation Management Narrative

In the hours following the earthquake, as the scope of the disaster becomes known beyond the southern California region, people from around the nation seek information on how they can assist disaster victims. Within 24 hours, offers of international aid are communicated to the US State Department from around the world. Thus, as local, state and federal agencies mobilize to address the immediate issues of life-saving emergency services, organizations and agencies responsible for donation management also move quickly to establish communications and coordination. The magnitude of the disaster requires a donation management response that will engage multiple federal agencies, state and local government agencies, private corporations and national volunteer organizations.

One of the key organizations is the National Voluntary Organizations Active in Disaster (NVOAD) an umbrella organization of established and experienced voluntary organizations that provide disaster services and most importantly, foster cooperation, coordination and collaboration among disparate voluntary organizations. The organizations that make up NVOAD are the first to receive calls with offers of donations which vary greatly from items vitally needed to those that are of little if any value. Leaders of NVOAD quickly establish contact with the State Operations Center at the headquarters of the Governor's Office of Emergency Services at Mather Field near Sacramento. This initial contact is followed by NVOAD' s dispatch of liaison personnel to the State Operations Center and to the Regional Emergency Operations Center at Los Alamitos in eastern Los Angeles County. Local affiliates of NVOAD send representatives to the Operational Area Emergency Operations centers in the eight counties affected by the earthquake.

Donation Management is be carried out primarily by local and state emergency management officials and NVOAD working cooperatively with support from the Department of Homeland Security according to the Federal Response Plan. Within the first 24 hours, a State Donations Coordinator is appointed by the Director of OES and a Federal Donations Coordinator is appointed by the Director of the Department of Homeland Security and together they assemble a Donation Coordination Team made up of NVOAD organizational representatives and local government officials in the eight impacted counties. Using software provided by the Department of Homeland Security, the team develops a plan for managing donations, both domestic and foreign. Based at the State Operations Center at Mather, the team first works with the Joint Information Center and media affairs staff to craft public service announcements, press releases and other public information materials regarding donations and volunteers.

Consistent with the donation management plan, donors are encouraged to provide cash rather than goods though there will be specific calls over the next two weeks for specific types of personnel, equipment, medical supplies and other resources which will be handled by the Donation Management System. Volunteers will be asked to register with one of the NVOAD recognized organizations rather than reporting directly to disaster sites. By the end of the first week cash donations of $42 million have been received. In addition, the US State Department has informed the Federal Donation Coordinator that many nations have offered monetary assistance as well as equipment, personnel and other resources for response and recovery. These donations are handled by the US Customs Service in cooperation with the State Department's Office of Diplomatic Contingency Programs.

Fire Suppression Narrative

Although fire services have broader responsibilities than fire suppression, the salience of fire in the ShakeOut Scenario is such that it merits its own section independent of other fire functions including emergency medical response and search and rescue which are documented in other sections. The fire situations discussed in this section is largely derived from the work of Charles Scawthorn, "A Note on Fire Following Earthquake for the Southern

San Andreas Fault M7.8 Earthquake (SoSAFE), March 3, 2008 and various focus group discussions of post-earthquake fire and fire control.

In the immediate aftermath of the earthquake, overturned heat sources (lamps, candles, kitchen burners, etc.), abraded and shorted electrical wiring, spilled chemicals, sheared natural gas lines and other sources, cause hundreds of fires around the region. Some of these fires are suppressed by residents and people at work with fire extinguishers and available water sources: however, many more will require trained fire fighters with appropriate equipment to fight larger fires. Due to damage to the telecommunications system and saturation of available working phone lines, many of the calls to fire departments will be delayed or simply not get through. In some cases, fire stations will be geographically near the fire and residents may be able to reach the station by foot in time to summon fire fighters but delays and distance will mean that small fires will grow into large ones. In other cases, fire fighters will self-dispatch to an observed smoke column.

As fire fighters begin to respond to reports of fires, they encounter obstacles. These obstacles in some cases are non-structural damage to fire stations and fire fighting equipment. Once on the road, they are likely to encounter debris in the street, nonfunctioning traffic signals, bridges that have been damaged by the earthquake and are of questionable safety and residents who attempt to divert the engine and firefighters to other locations where fires have broken out. In addition, fire fighting efforts will be hampered, particularly in the areas of heavy shaking, by water pressure drops due to pipe breaks and tank failures. For many of the responding fire fighting teams, a single crew and engine will be insufficient to control or suppress the fire. If they are successful, they will move on to the next fire. Very early on, in the first hour following the earthquake, all fire fighters in the region are ordered to report for duty, off-duty personnel will double the number of firefighters within 3-6 hours and triple it within 12-24 hours.

Within 24 hours, all fire service resources including personnel and approximately 2000 fire engines in the 8-county region are completely committed and local mutual aid will be unavailable given the number and extent of fires. Many volunteers will assist in fire suppression and Community Emergency Response Team (CERT) members who have been trained in fire suppression will work within their communities to fight fires and assist professional fire fighters in suppressing the many fires in the region. Emergency Operations Centers at the local and regional levels will receive requests to activate the fire mutual aid system which will summon fire resources from around the state. Outside resources will include additional personnel, hoses, foam, hand tools and light equipment as well as heavy equipment including cranes, bulldozers, backhoes and additional fire engines. Although summoned through mutual aid during the first day, it will require several hours to mobilize these out-of-region resources and highway damage and ongoing bridge inspections (closures) will further delay the arrival of aid though many of these outside resources will arrive at staging areas within 12 hours.

Most fires reported in the first 24 hours are reported in single family and multifamily residential structures but there are also fires in laboratories, chemical plants and oil refineries. These latter fires will require strike teams with special equipment and some will burn for several days. With hundreds of fires burning, more ignitions being reported hourly, and the fire services fully deployed with mutual aid resources not yet available, several large uncontrolled fires are burning and will later merge into several major conflagrations and one super conflagration in central Los Angeles which destroys hundreds of city blocks. In addition, fires have broken out in the urban-wild land interface due to downed power lines and most of these fires are not being fought as large fires in urban areas have been prioritized.

As large fires spread, residents near the fires either self-evacuate their homes and places of business or are urged to do so by firefighters, if they are on the scene. Those who experience the earthquake at work, attempt first to contact their families, then to drive home. Some will be unable to do so due to debris in the streets and other transportation obstacles and many families will be separated for hours and even days. Those who self-evacuate ahead of spreading fires without a clear idea of safe evacuation locations are in danger of injury and death. With large scale fires, even response agencies are uncertain regarding safe evacuation sites. Hundreds of people who are trapped in collapsed buildings awaiting rescue will die in fires before teams can locate and extricate them.

Within 72 hours following the earthquake, several conflagrations have developed from the convergence of large fires in the Cities of San Bernardino, Riverside, Santa Ana and south central Los Angeles. The largest of these fires is located south of downtown Los Angeles and is rapidly becoming recognized by both responders and the media as the largest and most serious fire in the region. This fire has prompted the evacuation of 130,000 persons in south central Los Angeles requiring the opening of an additional 320 public shelters to house 95,000 evacuees from this fire who cannot find shelter elsewhere. Although the mutual aid system has summoned both fire fighters and equipment from the entire state as well as neighboring states, their deployment has been delayed by transportation and communication problems and their effectiveness has been adversely impacted by loss of water pressure particularly in San Bernardino.

The largest fires are spreading rapidly and have not been slowed or stopped by potential fire breaks such as open spaces and highways. It is clear to Incident Commanders that standard methods of urban fire suppression will not be effective in controlling these fires. As reports are received indicating that most of the largest fires have not been contained, are spreading and requiring additional evacuations, other fire fighting strategies are considered. A decision is made to launch an aerial attack on the largest fires utilizing both helicopters and fixed-wing aircraft to fight these fires. Air drops begin, with additional aircraft requested by the OES Regional Emergency Operations Center at Los Alamitos. Five large fires in central Los Angeles, San Bernardino, Riverside and Santa Ana are targeted for air drops.

After six days of coordinated air and ground-based fire fighting efforts, most of the large structure fires have been contained, though fires at chemical plants and refineries continue to burn as do wild land fires that spread while attention was focused on urban fires. As the largest urban fires are contained, strike teams are re-deployed to join other units fighting fires at several refineries and chemical plants. In addition, while major fires were being fought in residential areas, fires at the urban wild land interface were burning out of control and are now major fires. These fires have destroyed residential and commercial buildings and caused evacuations in northern Los Angeles, Orange and Riverside Counties. Fires have been responsible for a total of 230,000 evacuations in the 8-county region affected by the earthquake, a number that dwarfs the number of evacuees who left their homes due to damage, utility outages and other non- fire related factors. Fires will have accounted for 885 deaths and $90 billion in property losses.

6. CASUALTIES BY KIMBERLEY SHOAF

Contributor: Merritt Schreiber, UCLA

Introduction

Earthquakes have the capacity to be incredibly damaging both to the physical infrastructure of a community and to the population living there. Morbidity and mortality (illness/injuries and deaths) in earthquakes result from a number of forces: those directly associated with the ground motion; those associated with secondary hazards such as fires or dam failures; and those resulting from the loss of infrastructure that maintains health. This section will explore the health impacts (both physical and emotional) related to the ShakeOut Scenario event of a *M*7.8 earthquake on the southern portion of the San Andreas Fault.

Describing Injuries

Casualty estimation methodologies generally provide estimates of injuries in categories that are not useful for healthcare preparedness efforts. In order for healthcare planners to make use of casualty estimates, the results must be provided in a format that provides them with information on the types of resources that might be required. At a minimum this information should include the types of injuries (or the mechanism of the injury) and the level of care required. Figure 6-1 shows the different levels of care that ideally could be identified in casualty estimation (Seligson and Shoaf, 2005). Additionally, information for healthcare planners would describe if the injuries were predominately blunt force trauma, crushing trauma, burns, or piercing trauma. The source of these mechanisms would help planners to identify specific healthcare resources that would be needed such as burn beds or suture materials.

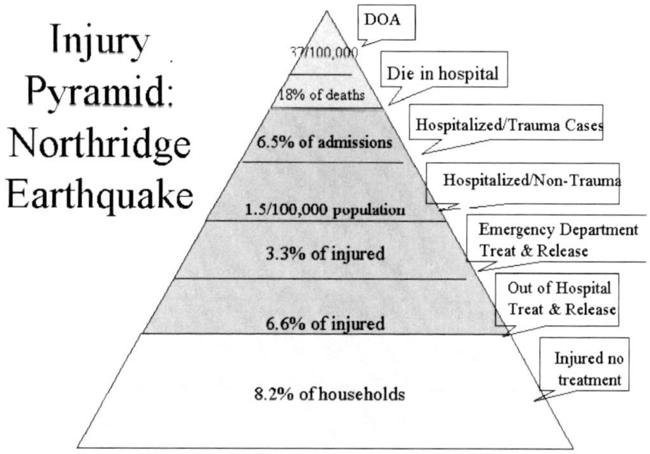

Figure 6-1. Injury Pyramid Example from the 1994 Northridge Earthquake

In this chapter, the estimated injuries will be reported per injury mechanism (such as fire, building damage, automobile crash) and in three categories: fatal, severe (requiring specialized care), and treat and release.

Physical Injury Results

The *M*7.8 ShakeOut Scenario earthquake would result in thousands of physical injuries both directly from the ground motion and the resulting damage as well as from secondary hazards. The numbers of injuries presented below were estimated utilizing a number of factors. Overall, the earthquake is expected to result in almost 50,000 injuries requiring treatment, 750 injuries requiring specialized trauma or burn care, and almost 1,800 deaths.

Injuries from Initial Ground Motion and Damage

Injuries from initial ground motion and the resulting damage were estimated utilizing a post-processing adjustment to HAZUS results. This adjustment is based on research done following the Northridge earthquake and data from other California earthquakes studied by the UCLA Center for Public Health and Disasters (Seligson and Shoaf, 2005). The HAZUS results provide the basis for the casualty estimates for all ground-motion-induced building damage, except the damage to steel-frame buildings.

As shown in Table 6-1, Los Angeles and Riverside Counties will experience similar and significant numbers of injuries. However, San Bernardino County will experience a number of injuries equal to those in the other two counties combined.

Ground motion and the resultant damage will result in approximately 260 deaths. This is 8 times the number of deaths experienced in the 1994 Northridge earthquake (Peek-Asa and others, 1998). Again, San Bernardino County is hardest hit, with more than 130 deaths resulting from building damage from ground shaking (excluding steel- frame highrises).

In all, nearly 50,000 people will be injured enough to need some sort of treatment as a result of ground motion and building damage. The majority of these injuries will require treatment at outpatient locations or emergency departments and patients will be released to return to their homes or other locations. As discovered in other earthquakes, the majority of these injuries are to the extremities, primarily the legs and feet and include lacerations, minor orthopedic injuries, and mild closed head injuries (Shoaf et al, 1998).

Approximately 1,000 people will need to be transported to the emergency department by the Emergency Medical Services (EMS) system. This excess of transports represents a significant increase over the normal baseline of EMS transports, especially for San Bernardino County (4.5 times their daily baseline load) and Riverside County (2.5 times their daily baseline load).

Perhaps the most significant results for injuries are the number of expected trauma injuries. Trauma injuries are those injuries which generally require specialized care and surgery. California has a trauma system in place with trauma hospitals designated in most counties. The eight county impact region has 31 designated trauma centers, each of which generally handles one or more traumas on a normal day. Trauma centers are designated as Level I through IV, with only Level I and II having on-duty specialists 24 hours a day, 365 days a year. Approximately 63 individuals will require trauma care for their injuries. This is approximately 3.5 times the number of trauma injuries that resulted from the Northridge earthquake (Peek-Asa and others, 1998) and approaches the capacity for trauma care in the region.

Injuries from Damage to Steel-Frame Buildings

For the ShakeOut Scenario, collapse-related damage estimates for high-rise steel frame buildings were developed outside of HAZUS; therefore, no HAZUS casualty estimates for these structures are available and an alternate means of estimating associated casualties was required. These alternate casualty estimates have been developed only for high-rise steel frame buildings that are assumed to collapse completely. It has been assumed that casualties in high-rise steel frame buildings in non- collapse damage states will be captured within the basic HAZUS damage and casualty assessment.

The ShakeOut Scenario damage for high-rise steel frame buildings postulates the collapse of five high-rise steel frame buildings. Unfortunately, there is no empirical casualty data available for earthquake-related steel-frame building collapse from which to calculate injury estimates. There has only been one documented complete collapse of a modern steel-frame building in an earthquake, the Pino Suarez building in the 1985 Mexico City earthquake. The damage mode for that building was reportedly a "side- sway" collapse and not a "pancake" collapse, as is being hypothesized in the ShakeOut event. Furthermore, there is no documented casualty data specific to that building that could be used to develop an empirical casualty model. It should be noted that the most recognizable steel-frame building collapse—the collapse of the World Trade Center buildings in the 2001 terrorist attacks—do not provide a reasonable parallel to high-rise steel-frame building performance in this earthquake event; the World Trade Center building size, design and configuration are significantly different than the buildings being considered in the ShakeOut Scenario, and the presence of the jet-fuel-induced fire had a significant impact on the building's performance, which would not be similar to expected earthquake performance. Therefore, the calculations of the casualty estimates for high- rise steel-frame buildings in the ShakeOut Scenario have been based on a casualty model developed to reflect pancake collapse of mid-rise concrete structures. For comparison, the casualty model utilized here predicts more fatalities than the default fatality model for high-rise steel frame buildings deployed by HAZUS.

Table 6-1. Injuries Resulting from Building Damage

County	Fatal	Trauma	ED Visits	Outpatient	EMS transports
Los Angeles	66	16	4,100	7,700	234
Imperial	0	0	0	0	1
Kern	0	0	0	100	1
Orange	1	0	700	1,500	22
Riverside	61	15	4,100	7,400	251
San Bernardino	132	32	7,400	13,400	469
San Diego	0	0	0	0	0
Ventura	0	0	0	0	0
8 County Totals	**260**	**63**	**16,300**	**30,100**	**978**

Table 6-2. Injuries Resulting from Collapse of Steel Buildings

County	Steel Buildings	Fatal	In Patient (Trauma)	Emergency Department	Outpatient
Los Angeles	Collapse # 1, 3, 4	242	65	107	315
Imperial		0	0	0	0
Kern		0	0	0	0
Orange	Collapse # 2	105	28	46	136
Riverside		0	0	0	0
San Bernardino	Collapse #5	92	25	41	119
San Diego		0	0	0	0
Ventura		0	0	0	0
8 County Totals		**439**	**117**	**194**	**570**

Table 6-3. Injury Rates Per 100 by Injury Pyramid Category, Level of Building Damage and Building Height, 1999 Kocaeli Turkey Earthquake (N=517)

Level of building damage[1]	Total collapse		Partial collapse		Total
Building height	5-10 stories	1-4 stories	5-10 stories	1-4 stories	
Death on arrival (DOA)[2]	12.7 (n=33)	0.0 (n=0)	2.0 (n=3)	0.0 (n=0)	7.0 (n=36)
Died in hospital	0.4 (n=1)	0.0 (n=0)	0.0 (n=0)	0.0 (n=0)	0.2 (n=1)
Hospitalized	3.5 (n=9)	1.7 (n=1)	0.0 (n=0)	0.0 (n=0)	1.9 (n=10)
Hospital care: treat and release[3]	5.8 (n=15)	3.4 (n=2)	0.0 (n=0)	0.0 (n=0)	3.3 (n=17)
Out-of-hospital care: treat and release[4]	17.0 (n=44)	8.6 (n=5)	7.4 (n=11)	3.9 (n=2)	12.0 (n=62)
Injured but no treatment sought	10.0 (n=26)	5.2 (n=3)	8.7 (n=13)	9.8 (n=5)	9.1 (n=47)
Not injured	50.6 (n=131)	81.0 (n=47)	81.9 (n=122)	86.3 (n=44)	66.5 (n=344)
Total N	**N=259**	**N=58**	**N=149**	**N=51**	**N=517**

[1]Partial collapse involved ceiling/roof collapse, wall collapse, floor collapse, and/or foundation destruction but did not result in destruction of the entire building.

[2]Those who sustained fatal injuries and did not seek treatment are included in this category.

[3]Those who sought treatment for their injury at a hospital but did not report the type of care received (i.e., outpatient care vs. admission) are included in this category.

[4]Those who a) sought treatment for their injury but did not indicate the source of care, b) sought treatment from multiple sources, and c) sought out-of-hospital care but did not report the type of care received (i.e., outpatient care vs. admission) are included in this category.

Table 6-3 provides the rates of injuries in collapsed concrete buildings based on data gathered following the 1999 Kocaeli earthquake in Turkey. In order to estimate injuries from collapse of steel-frame buildings of 10 to 15 stories, we utilized the rates for the total collapse category for 5- to 10-story concrete buildings (Shoaf and others, 2005; Seligson and others, 2006).

Altogether, these five buildings result in an additional 439 deaths, 117 injuries requiring trauma care, and an additional 800 individuals seeking care from emergency departments and other sources of medical care. Note that these additional 117 trauma injuries now brings the trauma total to 180 injuries requiring surgery and other specialized trauma care across the region, about 3 times the average daily census of trauma injuries.

Injuries from Fire Following Earthquake

In addition to damage directly from ground motion, secondary hazards such as landslides, hazardous materials releases, and fire following earthquake present additional risk for injuries. Injuries and deaths from residential fires have decreased dramatically in the United States in the last few decades. The majority of this reduction is the result of the increased utilization of smoke detectors and adequate fire suppression. In spite of this, people are still injured in residential fires. In 2006, fire departments responded to 412,500 home fires in the United States, which claimed the lives of 2,580 people and injured another 12,925. Increased risk for dying in a fire is attributed to young children, older adults, persons living in substandard housing, and persons living in rural areas (CDC Factsheet).

To calculate the deaths and injuries resulting from the fires, the numbers of single-family-dwelling equivalents was used as the base for this estimation of casualties because the fire locations are not specific enough to apportion other occupancy types. We make the assumption that the number of casualties in residences due to fire will, within the same order of magnitude, approximate the number in other occupancies. The populations exposed to the fire are assumed to be those who would be at home at 10:00 a.m. on a weekday morning. These are most likely mothers with young children and the elderly. Therefore the exposed populations was calculated by the percentage of households in each county represented by those populations, multiplied by 2 per household for mothers with young children and 1.5 per household for the elderly. Approximately 3% of residential fires in the United States result in an injury or death. In rural areas, the risk of injury or death is 2.7 times higher than the U.S. average, primarily due to fire department response times greater than 5 minutes. For each injury-causing fire, 51% result in mortality, 29% in significant injuries requiring specialized care (burn beds), and 39% in injuries treated in and released from an emergency department.

The results add 916 more deaths to the earthquake total and 564 more injuries to be treated in emergency departments. Most significant are the 414 injuries requiring specialized treatment in burn units and/or intensive care units for respiratory injuries. There are 92 beds available in the Southern California region for burn patients. An additional 33 beds are available in Northern California. Given that this event generates more than 3 times the number of burn beds in the State, most of these patients will need to be airlifted to available burn units outside of California.

Transportation Related Injuries

Transportation related injuries can add significant numbers of deaths and injuries in an earthquake. In a study of the 1989 Loma Prieta earthquake, Shoaf and others (1998) found that approximately 25% of the injuries reported in a population survey resulted from transportation-related incidents. In the Northridge earthquake (Peek-Asa and others, 1998) about 10% of the deaths were associated with transportation incidents. The transportation injuries and deaths include those resulting from motor vehicle crashes due to stoplights being out, ground motion reducing driver control, as well as infrastructure damage such as broken roadways or bridge failure.

Transportation related incidents in the earthquake increase the death toll by an additional 164 fatalities. The need for trauma care from transportation-related incidents is increased by an additional 163 cases, and more than 1,000 additional people will seek care in an emergency department from injuries resulting from transportation incidents.

Total Injury Counts

As a result of this earthquake, it is estimated that that there will be a total of 1,780 deaths. Approximately 750 people will require specialized treatment for either traumas or burns. Nearly 50,000 people will be injured to an extent that they will seek treatment for those injuries from an emergency department.

Table 6-4. Injuries Resulting from Fire Following Earthquake

County	Fatal	In Patient (Trauma/Burn/ICU)	Emergency Department
Los Angeles	647	292	398
Imperial	0	0	0
Kern	0	0	0
Orange	255	115	157
Riverside	8	4	5
San Bernardino	6	3	4
San Diego	0	0	0
Ventura	0	0	0
Eight-County Totals	**916**	**414**	**564**

Table 6-5. Injuries Resulting from Transportation Incidents

County	Transportation Fatalities	Transportation Traumas	Transportation ED Visits
Los Angeles	105	81	519
Imperial	0	2	6
Kern	1	17	37
Orange	1	5	138
Riverside	2	5	96
San Bernardino	53	45	102
San Diego	2	6	139
Ventura	0	2	38
Eight-County Totals	**164**	**163**	**1,076**

Table 6-6. Injury Totals Per County

County	Total Deaths/County	Total "Serious" Injuries/County	Total Non-Fatal Injuries/County
Los Angeles	1,060	453	13,593
Imperial	0	2	8
Kern	1	17	155
Orange	362	148	2,825
Riverside	71	24	11,624
San Bernardino	283	105	21,170
San Diego	2	6	146
Ventura	0	2	39
8 County Totals	**1,780**	**757**	**48,322**

Table 6-7. Mental Health Impacts

County	Mental Health Impacts - Distress	Mental Health Impacts - Disorders (depression, anxiety, PTSD)
Los Angeles	3,814,440	110,400
Imperial	59,660	24
Kern	188,860	73
Orange	1,144,560	35,834
Riverside	717,060	19,042
San Bernardino	733,020	68,743
San Diego	1,124,040	34
Ventura	218,880	12
8 County Totals	**8,000,520**	**234,162**

Mental Health Results

Just as disasters affect the physical health of the population, so they also impact the mental well-being of the population. Similar to physical injuries, emotional injuries also vary in their severity from minor impacts that are self-limiting to new incidence of moderate to severe disorders. The severity of the impact is related to a number of factors, including pre-existing mental health status and existing coping strategies. One of the major predictors of severity however, is the exposure to various disaster conditions. These conditions include: losing a family member; feeling your life is in danger; having a significant injury; or being evacuated from your home. Each of these exposures is cumulative and increases the likelihood of a diagnosable disorder. Self-Limiting "Emotional Injuries"

Studies following disasters, including both natural disasters and terrorism-related events, suggest that much of the mental health impact on the population is a mild, time- limited distress symptom (that is, insomnia) that is not reflective of a clinical disorder. Significant portions of the population experience this disorder. Research following the 1994 Northridge earthquake (Bourque and others, 2002) found that 38% of the population of Los Angeles

County suffered from emotional injuries that were described as a distress symptom that was not consistent with a clinical disorder. Extrapolating that 38% to this event would suggest that approximately 8 million residents of the eight- county region would experience such distress symptoms (Table 6-7).

Exposure to certain disaster conditions would increase the likelihood that individuals would develop a condition that would be diagnosable as a mental health disorder. Research indicates that the majority of these disorders would be classified as depression or anxiety disorders. Given the severity of the damage, the numbers of deaths, and the large portion of the population that would need to evacuate their homes for a period of time, it is estimated that more than 200,000 individuals would experience a new mental health disorder.

7. Regional Economic Consequences by Anne Wein and Adam Rose

Contributors:

- Rich Bernknopf, USGS
- Stephen Levy, Director, Center for Continuing Study of the California Economy
- Daniel Ponti, USGS
- Sungbin Cho, Consultant, Imagecat, Inc.
- Ben Sherrouse, Geographer, USGS
- David Hester, Geographer, USGS
- Judith Steele, Consultant, Judith Steele, Ph.D., Consultant Services
- Richard Champion, Mathematician, USGS
- David Strong, Physical Scientist, USGS
- Rick Gailing, Principal Engineer, Southern California Gas Company
- Mike Morel, Operations Management, Metropolitan Water District
- Ron Tognazzini, Consultant, Tognazzini and Associates
- Charles Scawthorn, Professor, Department of Urban Management, Kyoto University
- Hope Seligson, Associate, MMI Engineering
- Dan Wei, Post-doctoral student, USC
- John Vaughn, Programmer, Economic Development Department
- Michael Smith, Economist, Surface Transportation Board
- Keith Porter, Structural Engineer, University of Colorado
- Erdem Karaca, Associate, Swiss Reinsurance America Corporation
- Tim Richison, Chief Financial Officer, California Earthquake Authority
- Richard Holden, Regional Commissioner, U.S. Department of Labor, Bureau of
- Labor Statistics
- Lori Dinitz, Operations Research Analyst, USGS
- Pauline Cotelle, Student, Université Paul Valéry, Montpellier III

Table 6-8. Summary of Mental Health and Casualty Issues Per County PTSD is Post-Traumatic Stress Disorder

County	Mental Health Impacts - Distress	Mental Health Impacts - Disorders (depression, anxiety, PTSD)	Total Deaths	Total "Serious" Injuries	Total Non-Fatal Injuries	Number of Displaced Households /Revised HAZUS	Number of Persons in "Shelters" /Revised HAZUS	Displaced from Fires	Displaced from Dam Evacuations	Average Household Size	Total Displaced Households
Los Angeles	3,814,4 40	110,4 00	1,05 9	45 3	13 ,454	27,339	11,852	94,000	0	2. 9	121,3 39
Imperial	59,660	24	0	2	6	22	9	0	0	3. 2	22
Kern	188,860	74	2	13	140	30	10	0	0	2. 9	30
Orange	1,144,560	35,834	363	146	2,787	2,917	1,126	37,000	0	2.9	39,917
Riverside	717,060	19,043	73	22	11,598	18 ,893	8,48 5	1,00 0	0	2.6	19,893
San Bernardino	733,020	68,744	284	103	21,141	43,047	19,132	1,000	30,000	2.8	74,047
San Diego	1,124,040	35	2	4	10 7	0	0	0	0	2. 7	0
Ventura	218,880	12	0	1	29	3	1	0	0	2. 9	3
Eight-County Totals	8,000,520	234,165	1,782	745	48,322	92,251	40,615	133,000	30,000	2.9	255,251

Review/support

- Leonard Gaydos, USGS
- Laurie Johnson, Principal, Laurie Johnson Consulting

Goods Movement Panel Participants

- Ron Guss. CEO, Intermodal West, Inc., California Trucking Association
- Capt. Richard McKenna, Marine Exchange
- Arul Arulmoli, Principal, Earth Mechanics, Inc.
- Alan Thompson, Planning and Policy, Southern California Association of Governments
- Michael Armstrong, Government and Public Affairs, Southern California Association of Governments
- K. Murthy, Asst Chief, Construction, Metropolitan Transport Authority
- Lu Hersh, Technical Services Manager, Alameda Corridor Engineering Team
- Sue Lai, Senior Transportation Engineering, Port of Los Angeles
- Doug Thiessen, Managing Director Engineering, Port of Long Beach
- Peter Yin, Senior Structural Engineer, Port of LA
- Larry Cotrill, Master Planning, Port of Long Beacb
- Gill Hicks, President, Gill V. Hicks and Associates, Inc.

SoCalfirst Panel Participants

- Ren Powers, Business Continuity and Disaster Management, City National Bank
- Bill Lloyd, City National Bank
- Vicki Saylor, Business Continuity, Countrywide Bank
- Marlene Roberts, Acting senior advisor for Critical Infrastructure Protection, Federal Deposit Insurance Corporation,
- Mark Hoffman, Executive Director of Critical Infrastructure Development and Response, Infragard
- Julie Davenport, Wells Fargo and Company
- Barry Edelman, Enterprise Risk Management, Zenith Insurance Company
- Patrick Allen, Crisis Management Team, BNP Paribas
- Dawn Matthews, Director of Security, California National Bank
- Frank Paczona, LBS Financial Credit Union

Economic Conseqeuences Panel Participants

- Hasan Ikhrata , Director, Southern California Association of Governments (SCAG)
- Frank Wen, Senior Economist, Growth Forecasting, SCAG
- Ping Chang, Program Manager, Monitoring and Assessment, SCAG
- Liz Wojdak, Senior GIS Analyst, SCAG
- Bob Huddy, Program Manager, Transit/ITS, SCAG

- Alan Thompson, Senior Regional Planner, Security and emergency preparedness, SCAG
- Michael Armstrong, Program Manager, Aviation, SCAG
- Sungbin Cho, Transportation Consultant, ImageCat, Inc. (and SCAG)
- Naresh Amatya, Acting Manager,Transportation Planning/Programming, SCAG
- Ryan Kuo, Associate Regional Planner, Transportation, SCAG
- Hsi Hwa Hu, Transportation Modeler, SCAG
- Jerry Nickelsburg, Economist, Anderson Forecasting, UCLA
- Steve Levy, Director, Center for Continuing Study of the California Economy

A. Introduction

Within the ShakeOut Scenario, the objectives of the economic consequences effort have been to:

1. Demonstrate that scientific information can be transformed into economic consequences to help predict the effect of earthquake-related shocks on the regional economy;
2. Examine the resilience of the regional economy in the event of the earthquake, and
3. Elicit feedback about the value and uses of the information for planning and decision-making.

In the United States, the geographic scope of the economic impacts of most natural and manmade disasters is regional rather than local or national. This is not said to diminish the individual suffering or the national concern. The scope is regional because even local impacts ripple spatially to the boundaries of larger economic trading areas and because the vast size of the United States limits the impact of forces, whether man-made or natural. Thus, the appropriate geographic area for analysis is often the county or county group (though not always within the boundaries of a single state).

The interdependence of the economy extends disaster losses beyond the area of the initial stimulus. One view of interdependence is the production pyramid, which characterizes the economy as stacked building blocks. Primary commodities, such as minerals, agricultural crops, and forest products, are at the foundation of this economic edifice because they are at the starting point of the production process. Intertwined with all the layers are roads, utilities, and communication networks that provide the lifelines of logistic support for even the most basic economic activity. While all goods and services in the economy are interdependent, infrastructure may be the most critical component.

The last time a comparable earthquake occurred in southern California, the year was 1857—California had recently become a State and the Civil War had not yet begun. Needless to say, the current economy has evolved substantially since that time. Our study seeks to predict the economic consequences of the ShakeOut Scenario earthquake in the year 2008 by posing several key questions, including: What will the sources and magnitude of economic losses be? What infrastructure will be most critical to the determination of the regional economic losses? How significant is the event in relation to the large size of the regional economy as a whole? Our analysis of economic losses is based on the use of input-output (I-O) analysis, a widely used tool of regional economic impact analysis. This modeling approach is adept at tracing economic interdependencies that can cause total regional economic impacts to be several times greater than direct impacts.

In order to utilize results from ShakeOut geologists, geographers, and engineers, we first transformed their estimates of damage (Stock Loss) into estimates of reduced lifeline service capacity for geographical areas (counties, Instrumental Intensity zones) and/or industrial sector because infrastructure possesses features of resilience, such as flexibility and the ability to rebound, that are not captured by damage estimates alone. In addition, we disaggregated regional fire damaged building occupancy estimates to counties. We transformed highway-bridge damages into regional traffic delays and lost trips for industrial sectors. We determined the effect of infrastructure damage on port operations. The physical damages and lifeline service outages represent the earthquake- related sources of shock to the regional economy that result in four types of economic losses. Table 7-1 provides examples of direct and indirect, stock and flow economic losses.

We next provide perspectives on the baseline economy of the eight county study region in the ShakeOut Scenario, and derive the inputs (the economic shocks) to our regional economic analysis. The input-output analysis culminates with estimates of the economic losses associated with each shock, highlighting the key sources of total economic loss, and their relative significance to the region. This is followed by an exploration of the insurance coverage of the largest source of building losses, residential building losses. In the discussion and conclusion, we return to and reflect upon the three objectives of this chapter.

Many of the studies that were conducted for the ShakeOut Scenario are available as reports on-line. For details go to http://urbanearth.usgs.gov/scenario08.

B. The Southern California Economy

The Southern California economy is explored in this section through three types of analyses: sector analyses of economic activity; location quotients; and net worker and income flows between counties.

Comparison of Employment and Payroll Among Industrial Sectors

A commonly used view of the economy is through pie charts of the relative industrial sector activity as measured by employment and payroll (figures 7-1 and 7-2). For the purpose of this overview, the USGS used Economic Development Division (EDD) business, establishment, employee, and payroll data, for the 4th quarter, 2006. (The input- output analysis instead used a different dataset from IMPLAN, a software for regional economic analysis.) This is the most recent quarter data set consistent with the timing of the November 2008 Golden Guardian exercise, which will use the ShakeOut Scenario. These data were compiled by two-digit North American Industry Classification System NAICS code, at the zip code level. A trade-off between spatial detail and industrial sector detail resulted in suppression of 10% of the data (see Sherrouse and others 2008a for details).

As shown in figure 7-1, the top two sectors contributing to the region's payroll are Public Administration and Professional, Scientific and Technical Services. Comparing the percentages (the slices of the pie) between figure 7-1 and figure 7-2, Retail and Accommodation and Food Services are lower-wage sectors (because their percentages in the Employment chart is larger than in the Payroll chart) while higher-wage sectors include Information, Finance and Insurance, Professional, Scientific, and Technical Services, and Arts, Entertainment and Recreation (because these sectors have payroll percentages that are higher than their employment percentages).

Table 7-1. Comparison of Economic Losses for Stock vs. Flow, Direct vs. Indirect Losses

	STOCK (*measure*)	**FLOW (*measure*)**
DIRECT	Building or highway damage from earthquake shaking or fault rupture (*replacement cost*)	Business interruption due to relocating the business or power outage (*sales revenue, income, output*)
INDIRECT	Building damage from fire following earthquake (*replacement cost*)	Business interruption at the ports due to damage to the railways (*sales revenue, income, output*)

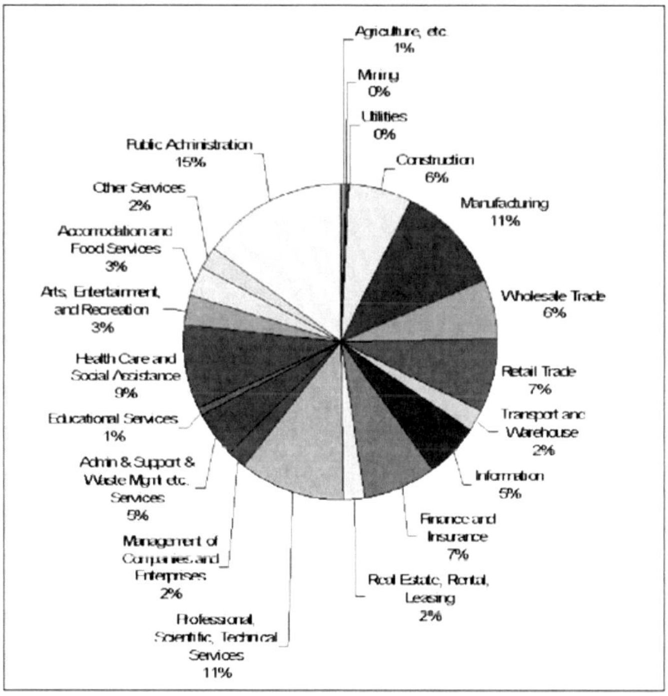

Figure 7-1. Industrial sector payroll for the eight county total payroll of $99 billion (10% suppressed data), 4th quarter 2006, Source California Economic Development Department.

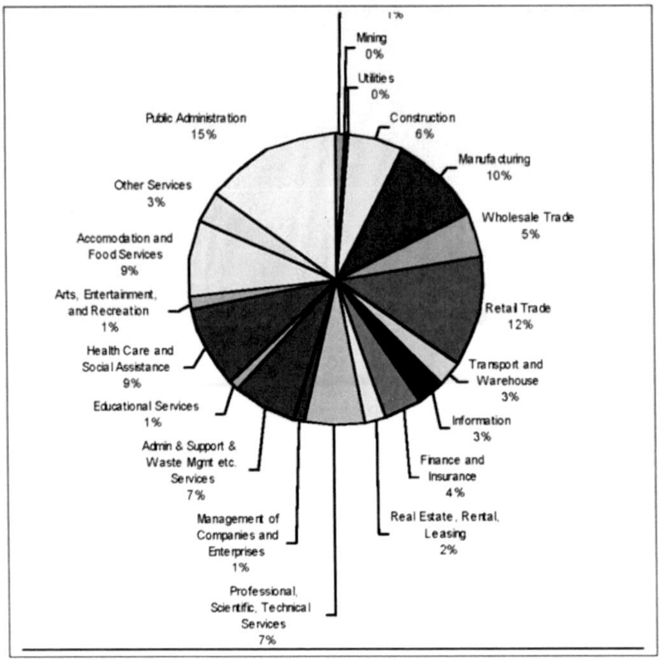

Figure 7-2. Industrial sector employment for the eight county total employment of 8 million (10% suppressed data), 4th quarter 2006, Source California Economic Development Department.

Comparisons with Other Economies Using Location Quotients

Is the regional economy of the eight counties in the ShakeOut Scenario study region typical or unusual? The Bureau of Labor Statistics Location Quotient (LQ) efficiently identifies distinguishing characteristics of an economy by comparing sectors of industrial employment in an area of interest relative to a base, which is another area such as a nation or state:

$$LQ_i = \frac{\text{\% of Local Employment in Industry } i}{\text{\% of Base Employment in Industry } i}$$

- When the ratio LQ = 1, the proportion of industry concentration is equal in the area of interest and the base.
- When LQ > 1, the proportion of industry concentration is greater in the area of interest than it is in the base.
- When LQ < 1, the proportion of industry concentration is lower in the area of interest than it is in the base.

Location quotients have been used to identify local economic strengths and competitive advantages, opportunities, and industry clusters. Shields (2003) offers the following interpretation of location quotient values:

- LQ > 1.0 indicates that the economy is self-sufficient, and may even be exporting the good or service of that particular industry. As a rule of thumb, LQ > 1.25 identifies exporting/stronger industries. Sometimes, however, the industry may require more workers than average to produce a level of output necessary to meet local needs. In that situation, the local industry or workforce is inefficient and the industry may be relatively weak rather than relatively strong in the industrial sector.

Table 7-2. Supersector Location Quotients for the ShakeOut Scenario Study Region Using the Nation and State as Bases

Industry	Calif LQ Base=U.S.	8 County LQ Base=U.S.	8 Cty LQ Base=CA
Natural Resources & Mining	1.95	1.14	0.59
Construction	1.05	1.02	0.97
Manufacturing	.91	0.97	1.06
Trade, Transportation, & Utilities	.94	0.96	1.02
Information	1.33	1.46	1.10
Financial Activities	.98	1.00	1.02
Professional & Business Services	1.09	1.09	1.00
Education & Health Services	.80	0.78	0.98
Leisure & Hospitality	1.00	1.01	1.01
Other Services	1.38	1.41	1.02
Unclassified	1.14	0.03	0.03

Table 7-3. Supersector Location Quotients by County

Industry	Imperial	Kern	LA	Orange	Riverside	San Bernardino	San Diego	Ventura
Natural Resources & Mining	17.79	15.74	0.21	0.28	1.82	0.45	0.65	5.52
Construction	0.72	1.34	0.65	1.16	2.31	1.26	1.25	1.10
Manufacturing	0.49	0.46	1.03	1.06	0.87	0.97	0.76	1.10
Trade, Transportation, & Utilities	1.17	0.89	0.97	0.86	1.04	1.30	0.87	0.89
Information	0.34	0.45	2.14	0.85	0.55	0.52	1.26	0.80
Financial Activities	0.47	0.56	0.96	1.40	0.63	0.72	1.06	1.11
Professional & Business Services	0.40	0.74	1.08	1.30	0.78	0.93	1.26	0.94
Education & Health Services	0.46	0.65	0.87	0.67	0.67	0.82	0.75	0.67
Leisure & Hospitality	0.69	0.80	0.93	1.07	1.20	0.89	1.22	0.96
Other Services	2.17	1.02	1.73	0.91	1.21	1.31	1.30	0.95

- LQ < 1.0 suggests that the region tends to import the good or service or it has a weaker presence in the local economy. Here, the rule of thumb is that LQ < 0.75 indicates an importing or weaker industry.

The location quotients in Table 7-2 suggest that the ShakeOut study region is like the U.S. economy, with LQs near 1, except for lower concentrations of Health Services and higher concentrations of Information, as well as the smaller sector, Other Services (figures 7-1 and 7-2). When compared with the nation, the region appears to be exporting the Information industry. When compared with the state, the study region's economy is like California's with the exception of the agricultural sector; the region has a relatively weak agricultural sector compared to the rest of the state.

As seen in Table 7-3, location quotients can also help to distinguish the economies from county to county. Note in particular:

- the predominance of agriculture in Kern, Imperial and Ventura counties;
- the county construction location quotients reflect higher growth rates in Riverside, Kern, and San Bernardino Counties, although not for Imperial County. (See population growth rates in figure 7.3.);
- the concentration of the transportation industry in San Bernardino County;
- the regional information sector concentration is primarily in Los Angeles County
- the financial industry is strong in Orange County, although recent declines in the financial industry may have weakened this concentration (Jerry Nickelsburg, economist, Anderson Forecast, UCLA, personal communication); and
- for Orange and San Diego Counties, the Professional and Business services sector is an exporting industry.

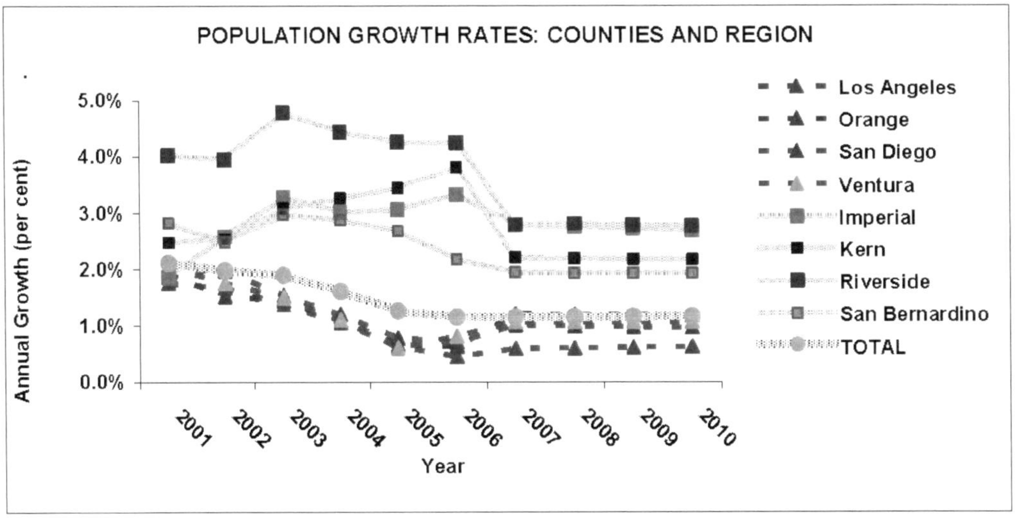

Figure 7.3. Population growth rates by county and region. Sources: U.S. Census Bureau and the State of California, Department of Finance.

The ports of Los Angeles and Long Beach are the largest and busiest on the West Coast; in fact, projected growth of operations at the port at San Pedro has been identified by the Southern California Area of Governments as one of the top three issues for longterm, regional planning. Despite this activity, at the level of analysis in Table 7-3, the Transportation sector displays only an average concentration in the region. However, when Champion and Wein (2008) conducted more detailed industrial sector classification, the activity becomes apparent in the large LQs of sectors related to port activity:

- Support activities for water transportation in Los Angeles County (LQ of 3.99)
- Warehousing and storage in Kern (1.49), Riverside (1.84) and San Bernardino (2.37) Counties;
- Truck transportation in Kern (1.26), Imperial (1.30), and San Bernardino (2.62) Counties;
- Local freight trucking and other special local trucking in Kern, Riverside, San Bernardino and Imperial Counties.
- Long distance freight trucking (2.62) and courier services (2.49) in San Bernardino County.
- Freight transportation arrangement in Los Angeles and Imperial Counties.

More details on the Location Quotient analysis of the MHDP eight county region is documented in Champion and Wein (2008).

Net Flows of Workers and Total Earnings

A third perspective on the regional economy comes from examination of census year 2000 net worker and total earnings flow by county of residence and industry by David Hester and Ben Sherrouse, Rocky Mountain Geographic Science Center, USGS. The net flows were calculated for only those workers who reside and work in different counties. The analysis reveals three types of net flow counties within the region:

- Counties that are primarily **net sources** of jobs and income:
  - Los Angeles (all sectors)
  - Imperial (except for agriculture and transportation)
- Counties that are primarily **net providers** of employees and influx of wages:
  - Riverside (all sectors)
  - San Bernardino (all sectors)
  - Ventura (all sectors)
  - Kern (except agriculture and federal government); and
- Counties that are **mixed** (that is, a net source of jobs and income for some sectors and a net provider of employees and influx of wages for other sectors):
  - Orange
  - San Diego.

In figs. 7-4(a-f), each type of county is illustrated. In these figures:

- Positive values for WORKERS indicate *net inflow:* there are more workers who commute into the county to work in the industry than workers who commute out of the county to work in the industry.
- Negative values for WORKERS indicate *net outflow*: there are more workers who commute out of the county to work in the industry than workers who commute into the county to work in the industry.
- Positive values for EARNINGS indicate *net inflow:* workers who commute out of the county to work in the industry have higher aggregate earnings than workers who commute into the county to work in the industry; and
- Negative values for EARNINGS indicate *net outflow:* workers who commute into the county to work in the industry have higher aggregate earnings than workers who commute out of the county to work in the industry.
- WORKERS WITH NO EARNINGS DATA are de-emphasized as they are of limited use in the context of this report.

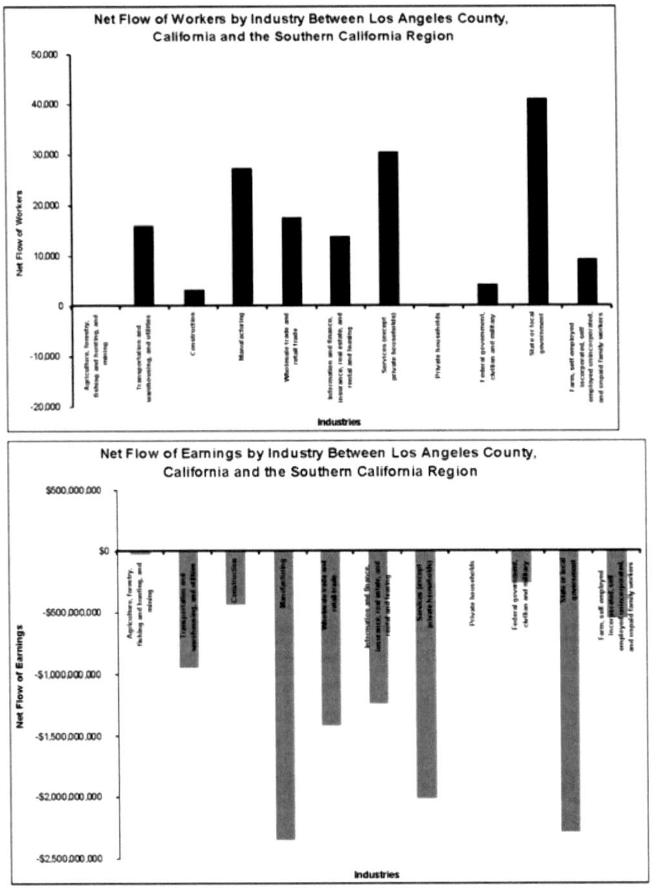

Figure 7-4. Continued on next page.

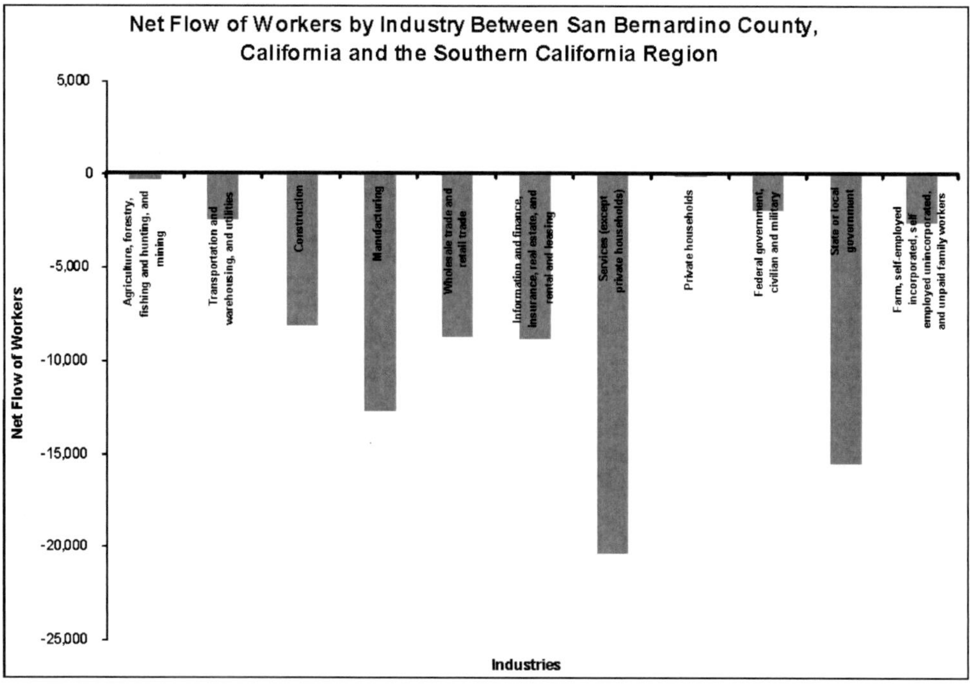

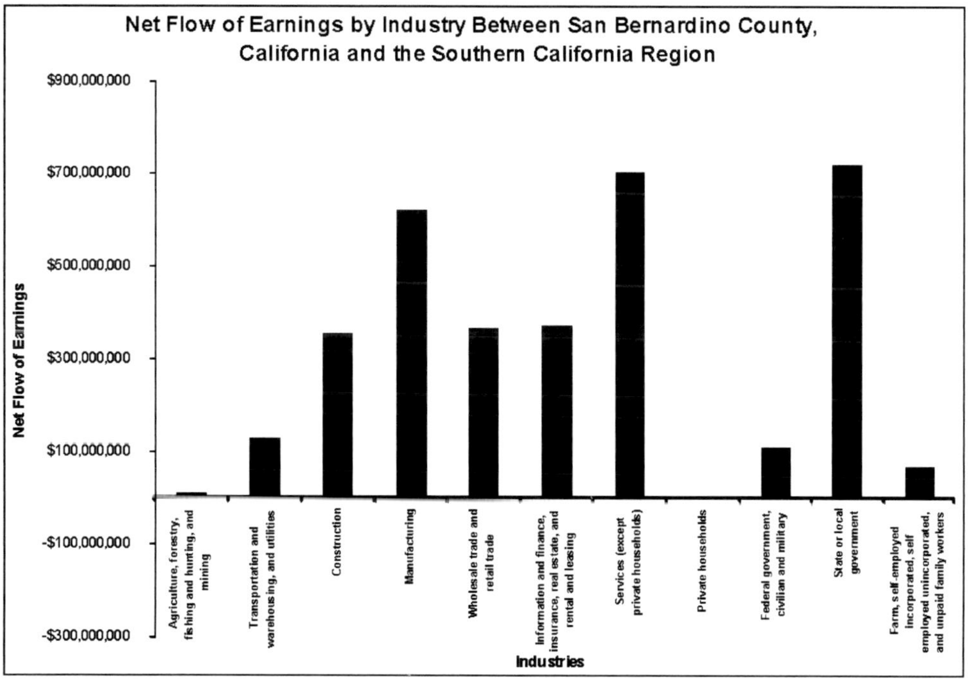

Figure 7-4. Continued on next page.

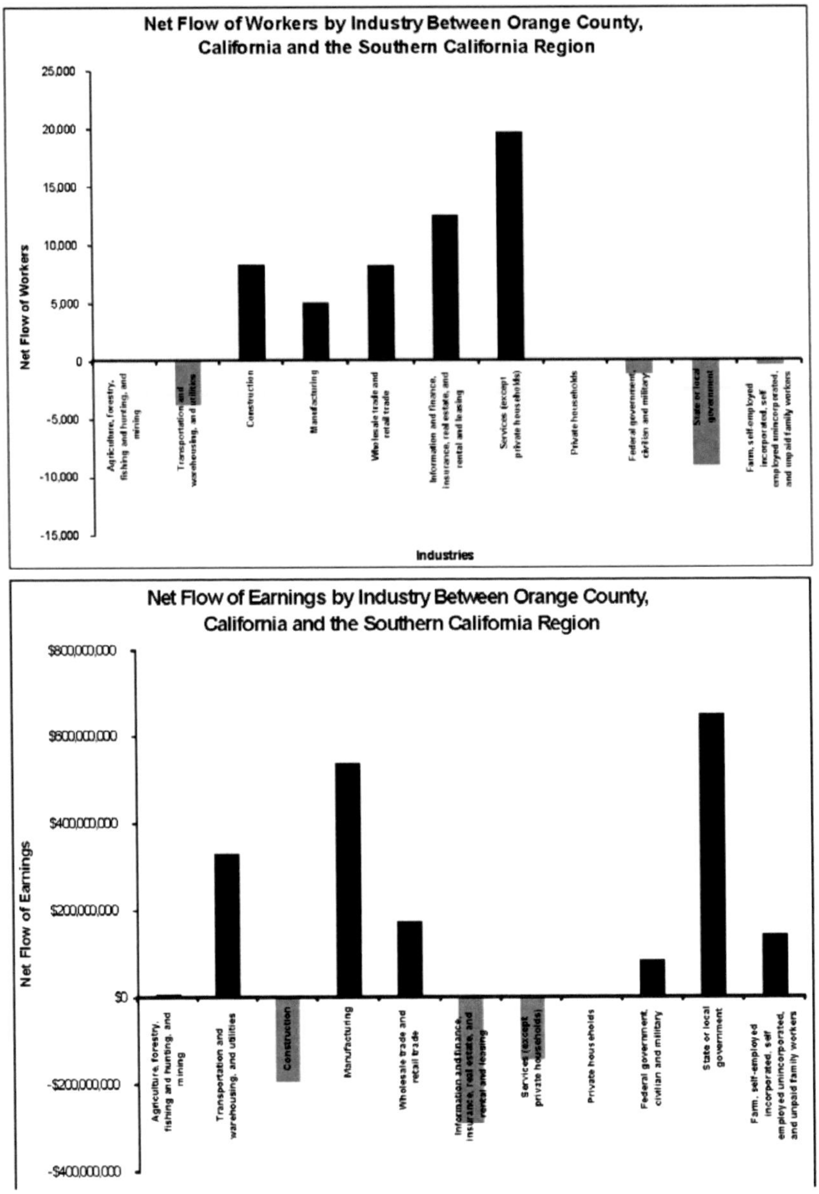

Figure 7-4. Net Flow of Workers and Earnings in counties that are examples of net sources, net providers, or mixed sources and providers

Los Angeles County is a net provider of jobs and income to San Bernardino, Riverside, Ventura, and Kern Counties.

A full table of results is available in Appendix I.

These analyses provide the backdrop to our analysis of economic losses and will also be used in pending work to evaluate regional recovery after the ShakeOut Scenario

C. Earthquake Shock Inputs for the Regional Economic Model

The shocks to the regional economy are the economic inputs that are severely affected by the ShakeOut Scenario. The evaluated shocks are captured in figure 7-5 and include:

- building damages (including high rise building collapses) from shaking,
- building damages from fire following earthquake,
- power service outage,
- water service outage,
- gas service outage,
- highway transportation (lost trips and delays), and
- disruption of port operations due to infrastructure damage.

These economic shock inputs for the regional economic model and serve as an interface between damage estimates and economic loss estimates. Given physical building and infrastructure damage estimates, further work was required:

- to determine the service outages for power, water and gas,
- to disaggregate damages to economic sectors for high rise buildings, and fire following earthquake,
- to transform highway-bridge damages into traffic impacts (delays and lost trips) by county and economic sector, and
- to estimate the effects of infrastructure damage on port operations.

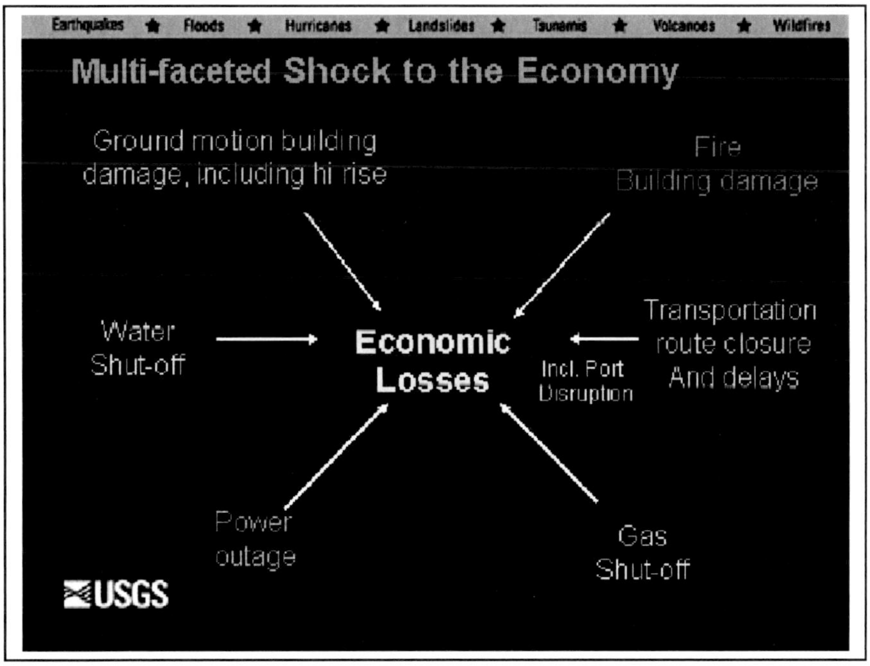

Figure 7-5. Multiple shocks to the regional economy

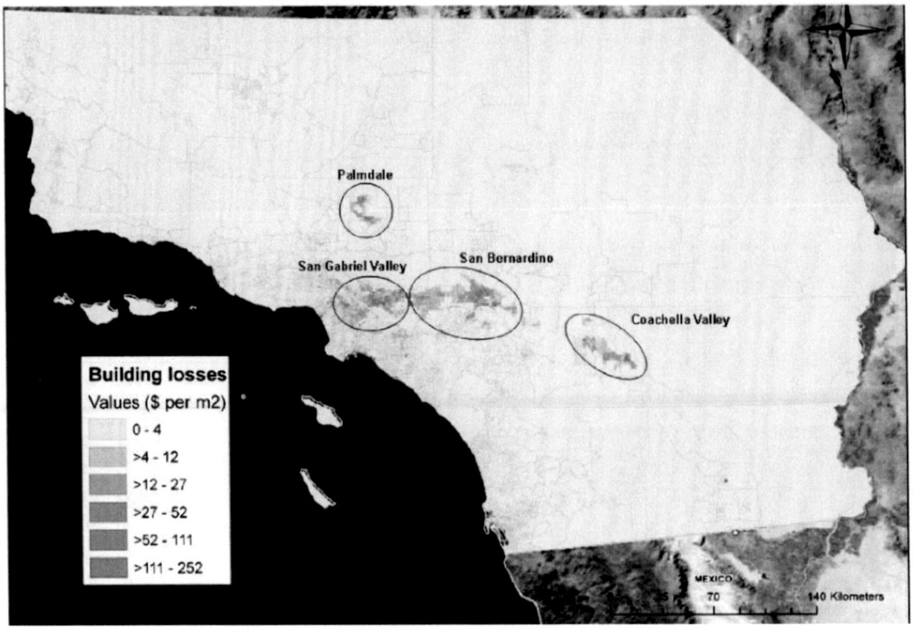

Figure 7-6. Building loss pattern in dollars per square meter

The methods used to derive the shock inputs to the regional economic model are described in this section in the order listed above with a couple of additions:

- a section that integrates lifeline service outages including telecommunications, and
- a commuter fault crossing analysis that supplements the highway-bridge system traffic impacts.

Building Damage Due to Shaking

HAZUS was used to estimate building damages and losses due to shaking from the ShakeOut Scenario for the eight county region. An assessment of high rise building damages was coordinated by Keith Porter and added to the HAZUS building loss estimates. Using HAZUS building loss data, provided by Hope Seligson, a regional building loss pattern is depicted by mapping building loss density (total building loss/area) for each census tract.

The map of southern California in figure 7-6. shows the interaction between the earthquake ground motions and the build environment. The four highly impacted areas are Coachella Valley, San Bernardino area, Palmdale area, and portions of the San Gabriel Valley.

HAZUS estimates $33 billion of building losses in the eight-county region. The source of the building losses are revealed in figure 7-7:

- residential building damage is the source of over half of the total building losses (single family homes are responsible for over half of the residential losses (29% of all building losses)) ;
- 31% of the total building losses involve commercial buildings, predominantly used for professional and technical services, retail and wholesale;

- 11% of the total building losses pertain to industrial buildings, light industry and construction offices, in particular.

There are three reasons for the larger losses from certain building occupancy classes (indicated by shaded areas in figure 7-7):

1. Relatively high square footage of the occupancy class: For example, in figure 7-8 single family residential square footage overwhelms the square footage of any other occupancy class. The building losses associated with the single family occupancy class are relatively large despite claiming the lowest complete and extensive damaged rate (0.5%) of all occupancy classes in the region;
2. Relatively poorly constructed buildings: For example, in figure 7-8, mobile homes and construction offices register the highest extensive and complete damage rate of 25-30% out of all of the occupancy class buildings in the region. It will be revealed later that these damage rates go above 90% in at the census tract scale.; and
3. More economic activity (and therefore, more built space) in the higher impact zone of the earthquake: For example, in figure 7-9, disregarding the utility sector (with a small number of employees), the most sensitive (% of employees in MMI 9+) sectors include manufacturing, wholesale, retail and construction. Notice that the information sector (a regional strength at a national level), is the least directly impacted by the earthquake. Using more spatially reliable Economic Development Division (EDD) employment data to explain HAZUS building losses assumes that HAZUS has generally placed the occupancy class building inventory in the right location. The only discrepancy appears to be the professional, scientific and technical service sector because employees appear to be less exposed than HAZUS building damage results suggest.

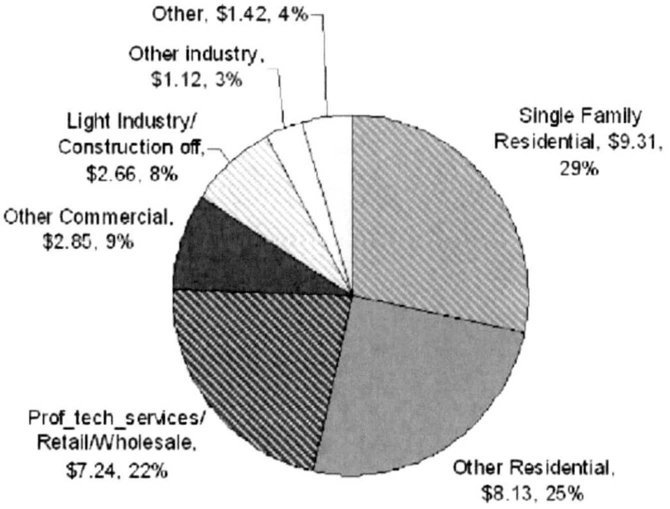

Figure 7-7. Building Losses by Occupancy Class

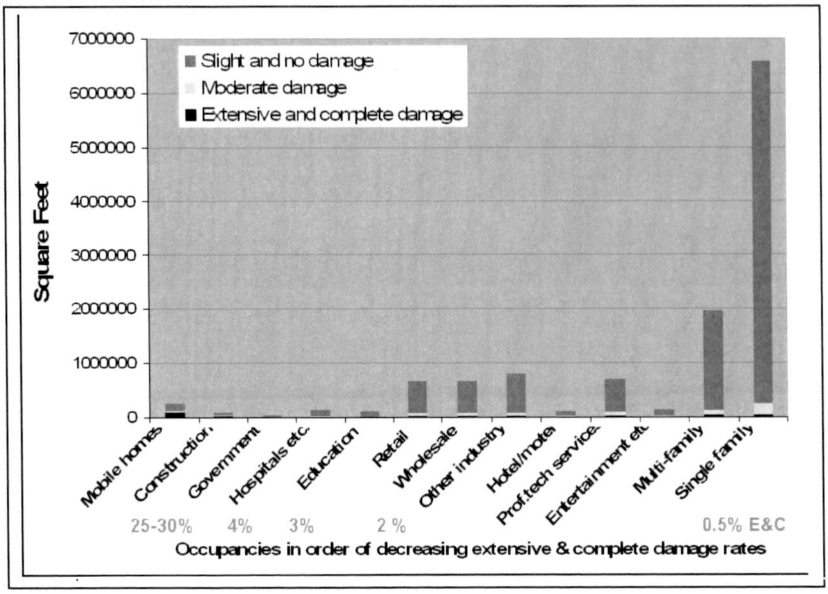

Figure 7-8. Damage by square footage of building occupancies and extensive and complete damage rates

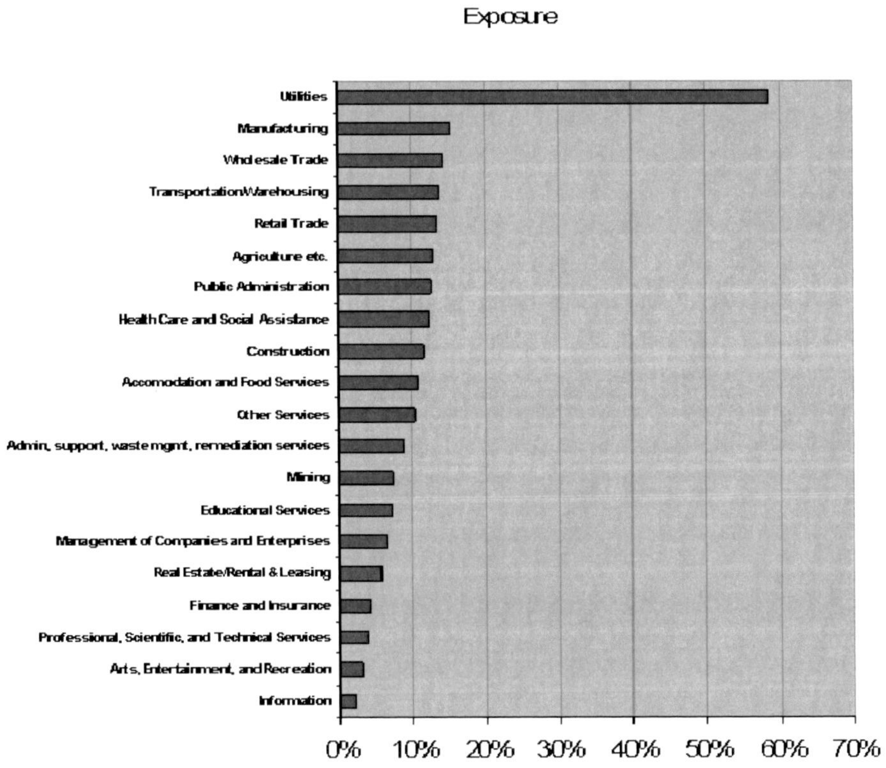

Figure 7-9. Industrial sector sensitivity: percentage of sector employees exposed to Instrumental Intensity IX and X. California Economic Development Department 2006 4th quarter data, (Data tables compiled by Ben Sherrouse & David Hester.)

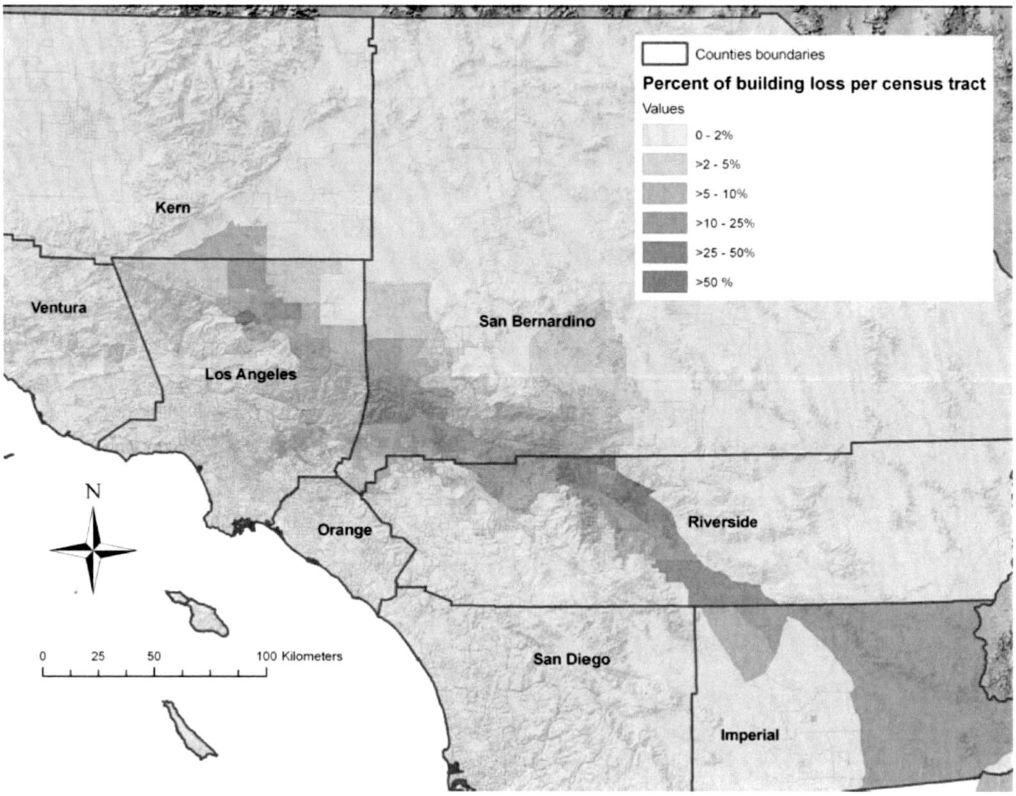

Figure 7-10. Proportional building loss by census tract.

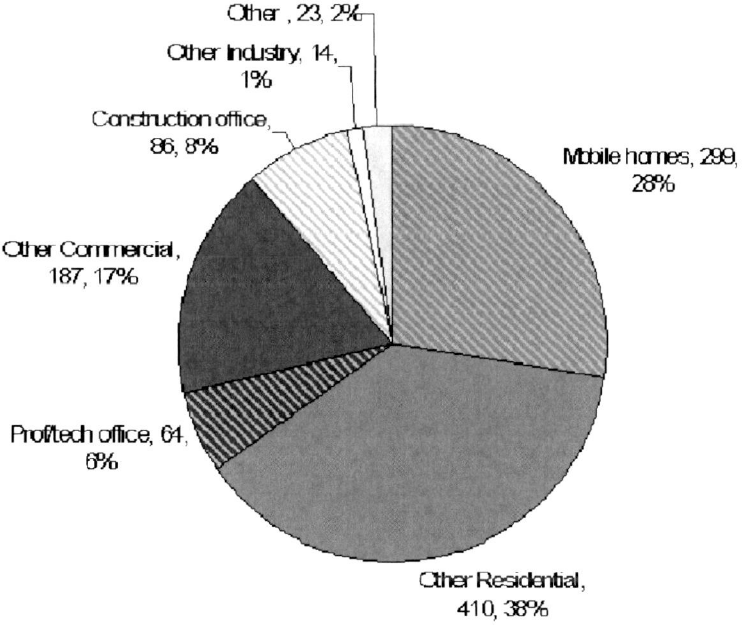

Figure 7-11. Building Losses in a high impact area

Table 7-4. Tall Building Damage Posited in ShakeOut Scenario

Number	Damage	HAZUS 0cc	Function	Sq. ft.	County
1	Collapse	COM4	Mixed office	300,000	Los Angeles
2	Collapse	COM4	Mixed office	240,000	Orange
3	Collapse	RES6	Warehouse	300,000	Los Angeles
4	Collapse	GOV1	Govt office	240,000	Los Angeles
5	Collapse	GOV1	Govt office	220,000	San Bernardino
6	Red tag	RES4	Hotel	200,000	Los Angeles
7	Red tag	RES4	Hotel	220,000	Los Angeles
8	Red tag	RES3	Apartments	360,000	Los Angeles
9	Red tag	RES3	Condos	400,000	Los Angeles
10	Red tag	COM4	Mixed office	260,000	Los Angeles
11	Red tag	RES6	Senior living	300,000	Los Angeles
12	Red tag	RES5	Jail	260,000	Riverside
13	Red tag	COM4	Mixed office	280,000	Riverside
14	Red tag	RES3	Apartments	320,000	Riverside
15	Red tag	RES4	Hotel	240,000	Riverside

Although the extensive and complete damage rates are low for the region, in high impact areas the relative building losses are high. figure 7-10 maps the percentage of building loss (building loss/building exposure value) for each census tract. Compared to figure 1, more census tracts are highlighted, particularly the larger less densely populated census tracts around the densely damaged areas depicted in figure 1. The explanation for the high damage rates in some of these regions is mobile homes and/or construction offices. figure 6, displays the building loss breakdown for a community in a high impact area.

High-rise buildings are susceptible to long period ground motions that are not accounted for in HAZUS damage calculations. Table 7-4 lists collapsed and red-tagged high-rise buildings posited for the ShakeOut Scenario.

High rise building losses amount to an additional $2.2 billion of building damage and approximately $0.7 billion of property damage. We assume that the replacement of a collapsed or red-tagged building takes 18 months to 3 years, but replacement for non-government buildings is not automatic because these buildings are mostly uninsured for earthquake.

Building Damage Due to Fire Following

Fires following an earthquake completely damage buildings that are not as severely damaged by shaking. For the ShakeOut Scenario, Scawthorn (2008) estimates approximately 1,200 large fires in the eight counties. About a third of these large fires occur in Imperial, Kern, Riverside and San Bernardino counties where building density is relatively low. Even though the fires in these counties initially are uncontrollable, their spread within the built environment would be limited to several city blocks. However, of concern, are fires in Orange County and the central Los Angeles basin, where a large plain of relatively uniform dense low-rise buildings provides a fuel bed such that dozens to hundreds of large fires are likely to merge into dozens of conflagrations. These fires could destroy tens of city blocks, and several of these large fires could merge into one or several super conflagrations that could destroy hundreds of city blocks (Scawthorn, 2008).

To create a fire following input for the regional economic model, it was necessary to allocate fire damage to economic sectors. We were able to allocate fire damage HAZUS occupancy classes by county using the number of fires in each county, total square footage burnt in the region and percent of the burnt square footage allocated to building uses, and square footage of the HAZUS occupancy class in each county. The equations used to allocate fire damage to occupancy class are illustrated below for the residential occupancy classes.

Fire Damage

Fire damage corresponds to the HAZUS complete damage state, and the time to replace a burnt building is assumed to be the same. The percentage of square footage completely damaged by shaking for each occupancy class is reported in column 6 of Table 7-5. Our fire damage allocation equations yield the additional percentage of square footage that is completely damaged by fire in column.7.

Fire following damages are significant in Los Angeles and Orange County to the extent that fire following damage dominates damage from shaking. The fire damage addition to complete damage in San Bernardino and Riverside is relatively small. Figure 7- 12 compares complete fire damage with extensive and complete damage from shaking. Fire damage is most prevalent in the residential, professional, scientific and technical services, and entertainment (includes restaurants) and other sectors.

Table 7-5. Allocation of Fire Damage to Counties and Occupancy Classes

County	Structure description (Scawthorne)	HAZUS occupancy class	Sq. ft. burned	Total Sq. ft. in occupancy class	% Sq. ft. completely damaged by shaking	Addt'l % Sq. ft damaged by fire
San Bernardino	One or Two Family Residential	RES 1, RES2, RES3A	244403	780029906	2.61%	0.03%
	Multi-Family Residential	RES3B-F	28795	79004674	4.36%	0.03%
	Public Roadway	IGNORED				
	Office & Power Prod/Dn	COM4	8695	29990024	4.21%	0.03%
	Primary / Secondary School (x10)	EDU1	16089	254787699	3.34%	0.01%
	Restaurant	COM8	3966	9526624	3.17%	0.04%
	Commercial	COM1-3,5-7,8,10	1770	108842457	3.48%	0.00%
	Other & Unknown	GOV1,2, IND1-6,REL1	22702	102529406	5.95%	0.02%
Riverside	One or Two Family Residential	RES 1, RES2, RES3A	235808	752597266	3.28%	0.03%
	Multi-Family Residential	RES3B-F	25447	69817242	2.63%	0.04%
	Public Roadway	IGNORED				
	Office & Power Prod/Dn	COM4	9788	33759501	1.77%	0.03%

Table 7-5. (Continued)

County	Structure description (Scawthorne)	HAZUS occupancy class	Sq. ft. burned	Total Sq. ft. in occupancy class	% Sq. ft. completely damaged by shaking	Addt'l % Sq. ft damaged by fire
	Primary / Secondary School (x10)	EDU1	13165	62484676	1.26%	0.02%
	Restaurant	COM8	4956	11903892	1.63%	0.04%
	Commercial	COM1-3,5-7,8,10	1450	89195658	1.91%	0.00%
	Other & Unknown	GOV1,2, IND1-6,REL1	12520	56544408	5.77%	0.02%
Orange	One or Two Family Residential	RES 1, RES2, RES3A	13636745	1176286477	0.04%	1.16%
	Multi-Family Residential	RES3B-F	3027392	224489074	0.00%	1.35%
	Public Roadway	IGNORED				
	Office & Power Prod/Dn	COM4	1344263	125311045	0.04%	1.07%
	Primary / Secondary School (x10)	EDU1	924687	118617046	0.00%	0.78%
	Restaurant	COM8	362611	23541349	0.00%	1.54%
	Commercial	COM1-3,5-7,8,10	111411	185204786	0.01%	0.06%
	Other & Unknown	GOV1,2, IND1-6,REL1	1151525	140561103	0.28%	0.82%
Los Angeles	One or Two Family Residential	RES 1, RES2, RES3A	75883043	2576445680	0.07%	2.94%
	Multi-Family Residential	RES3B-F	48918365	1427816286	0.21%	3.42%
	Public Roadway	IGNORED				
	Office & Power Prod/Dn	COM4	10637254	390308991	0.20%	2.72%
	Primary / Secondary School (x10)	EDU1	5046059	254787699	0.15%	1.98%
	Restaurant	COM8	1628467	41614333	0.15%	3.91%
	Commercial	COM1-3,5-7,8, 10	1885369	1233651241	0.30%	0.15%
	Other & Unknown	GOV1,2, IND1-6,REL1	12813254	615637476	0.33%	2.07%
TOTAL	All except roadway and vacant lots		178000000			
	Roadway and vacant lots		22000000			
TOTAL			200000000			

Assignment from structure description to HAZUS occupancy class is consistent with Adam Rose's sectoring schem HAZUS school square footage multiplied by 10. Orange County HAZUS school inventory 1/10th of actual (pers. comm., Hope Seligson) and HAZUS school square footage for 8 counties low compared to LAUSD square footage (personal communication, Robert Kamm, LAUSD)

Method to allocate fire building damage to occupancy classes: illustrated for residential occupancy

The building square footage completely destroyed is the sum of that completely destroyed by fire, and that completely destroyed by shaking, taking into account the possibility of "double counting" or "burning the rubble". The summation assumes that complete damage due to shaking and fire are independent such that these estimates are an upper bound of the complete damage state due to the combined effects of fire and shaking. (A lower bound would be the larger of the complete damage due either to fire, or shaking, but in this case, upper and lower bounds are quite close).

Estimation of Fire Following Earthquake (FFE) building damage for occupancy classes is based on the estimated area burnt in the region (200 million square feet of residential and commercial building floor area, equivalent to 133,000 single family equivalent dwellings (SFED)) and the percentages of burnt square footage assigned to building occupancies including 1- or 2- family residential and multi-family residential: Scawthorn (2008) assigns 71% of the burnt square footage to residential occupancy, 45% single/double homes and 26% multi-family. The allocation of the burnt square footage building occupancy *k*, $BSQFT_k$ and county *i* is weighted by the proportion of burnt SFEDs assigned to the county $PSFED_i$ (from Table 7-6) and the proportion of the residential occupancy class square footage in the county, $PSQFT_k$:

Burn allocation weight for occupancy class *k* in county *i* is calculated as $BAW_{ik} = PSFED_i \times PSQFT_k$

The burn allocation weight is normalized to derive the county burn allocation (CBA), the square footage of occupancy *k* burnt in county *i*:

$$CBA_{ik} = \frac{BAW_{ik}}{\sum_{j \in Counties} BAW_{jk}} \times BSQFT_k .$$

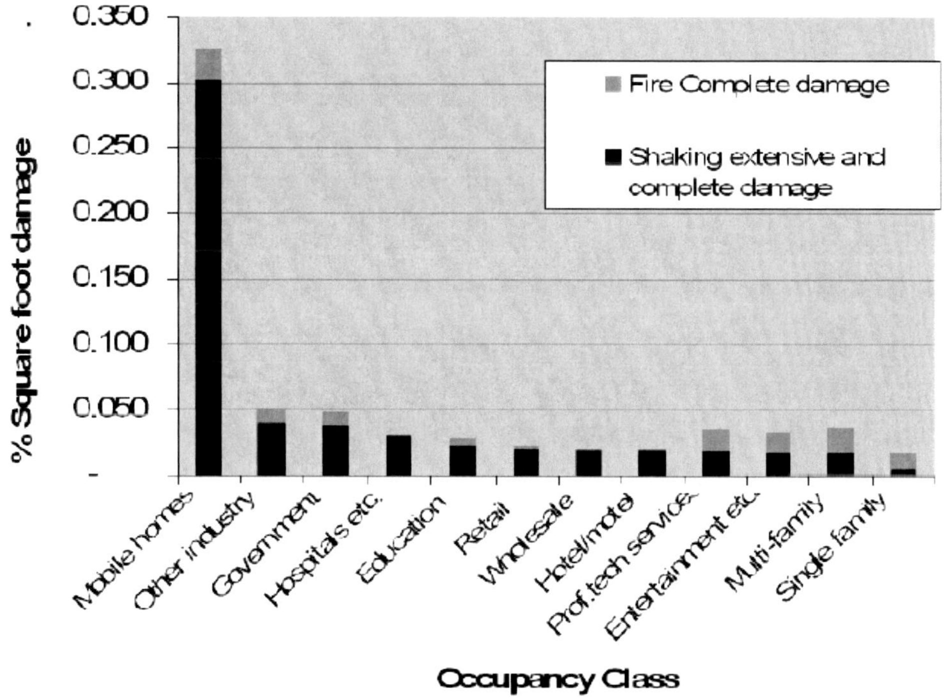

Figure 7-12. Comparison of fire building damage rates with (extensive and complete) shaking building damage rates

Table 7-6. Estimated Ignitions, Large Fires and Final Burnt Single Family Equivalent Dwellings. (from Table 4 in Scawthorn, 2008)

County	Estimated Number of ignitions	Est. Number of Large Fires	Est. Burnt SFED (thousands)
Imperial	131	45	Negligible
Kern	167	82	Negligible
Los Angeles	612	583	94
Orange	206	165	37
Riverside	239	157	1
San Bernardino	234	151	1
Ventura	18	0	Negligible
Total	**1,606**	**1,182**	**133**

The residential occupancy class square footage is included in the weighting scheme to avoid burdening an occupancy class in a county that has relatively less square footage compared to other counties, as is the case for multi-family residential in Orange County. Table 7-6 provides the residential SFED burnt in the counties. Again, the estimated number of large fires and estimated burnt SFED is of most significance to Los Angeles County (583 fires, 94,000 burnt SFED) and Orange County (165 fires, 37,000 SFED).

The calculation of additional fire following damage to residential occupancy classes assumes that shaking and fire damage occur independently, and that burnt buildings are completely damaged. Consequently, a larger number of structures that have sustained no or slight damage due to shaking will be completely damaged by fire following. The estimate of additional loss due to FFE is calculated using the formula:

$$\text{Additional \% completely damaged from FFE} = \\ (\text{\% burned and/or completely damaged from shaking}) - (\text{\% completely damaged from shaking})$$

where

$$\text{\% burned and/or completely damaged from shaking} = \\ 1 \quad (1 \quad \text{\% burned}) \times (1 \quad \text{\% completely damaged due to shaking})$$

The results for complete damage from FFE for four counties and two residential occupancy classes are tabulated in the last two columns of Table 7-7. We do not have the detail to make the calculation at the census tract level. The upper (tabulated in the 5th column of Table 7-2) and lower (maximum of sq ft burnt and sq ft shaken) bounds on the completely damage state are close because either there is a small amount of complete damage due to shaking or a small amount of fire damage.

Power Outage and Service Restoration

Power outage and service restoration estimates for the eight county region were derived from the power panel discussion and follow up with Ron Tognazzini. At the onset of the

ShakeOut Scenario, it is posited that power is immediately lost to all eight counties. The problem is a failure in the transmission and distribution of imported power. Generation facilities are expected to fare well.

Table7-7. The Percentage of Residential Occupancies Completely Damaged from Fire Following Earthquake and/or Shaking for Four Counties

County	Occupancy	% sq. ft. burnt	% sq. ft. completely damaged from shaking	% sq. ft. completely damaged from FFE and/or shaking	Additional % sq. ft. completely damaged from FFE	Additional sq.ft. completely damaged by fire
San Bernardino	1 or 2 family residential	0.03%	2.61%	2.64%	0.03%	238,035
	Multi-family residential	0.04%	4.36%	4.40%	0.03%	27,540
Riverside	1 or 2 family residential	0.03%	3.28%	3.3 1%	0.03%	228,070
	Multi-family residential	0.04%	2.63%	2.66%	0.04%	24,479
Orange	1 or 2 family residential	1.16%	0.04%	1.20%	1.16%	13,631,353
	Multi-family residential	1.35%	0.0%	1.35%	1.35%	3,027,324
Los Angeles	1 or 2 family residential	2.95%	0.07%	3.01%	2.94%	75,831,398
	Multi-family residential	3.43%	0.21%	3.63%	3.42%	48,813,580

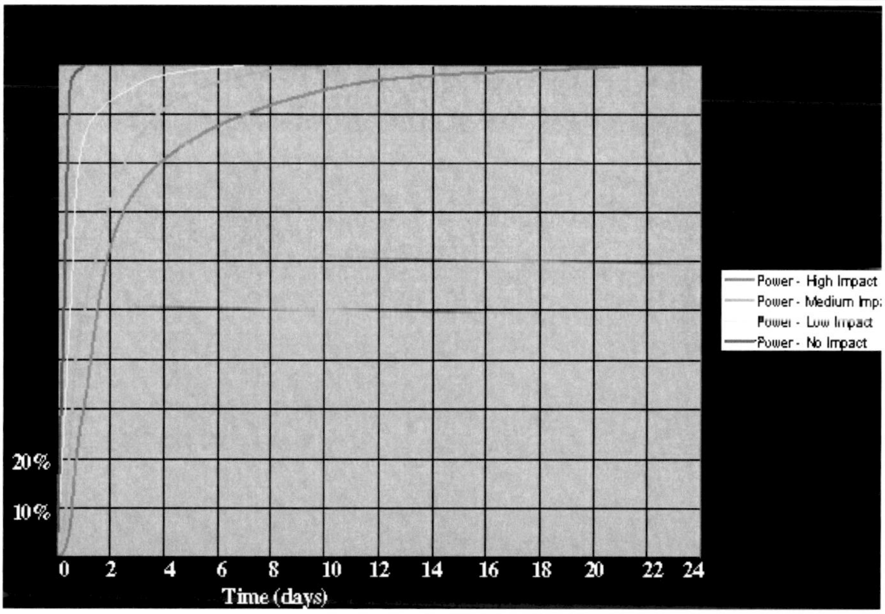

Figure 7-13. Power restoration curves for no, low, medium and high impact areas.

Table 7-8. Tabulation of Power Restoration at Points in Time for the Eight Counties

Electric power	Hours	Days : % demand restored	Weeks: % demand restored
Los Angeles, Riverside, San Bernardino County	Immediately lose power	1 day: 40% 3 days: 85%	3 weeks: 100%
Ventura, Orange, Imperial County	Immediately lose power	1 day: 50% 3 days: 85%	9 days: 100%
Kern, San Diego County	Immediately lose power	1 day: 85% 3 days: 95% 6 days: 100%	

In contrast to the analysis of water service (below), the nature of power system performance prevents them from being analyzed on an Instrumental Intensity scale because of the difference in their topologies, complexity of technologies, and vulnerabilities; the unique aspect of power outages during an earthquake is the effect on utilities that suffered no impact of physical damage from the earthquake. Figure 7-13 depicts power restoration curves for no, low, medium, and high impact areas. This graph is based on informed speculations about the performance of power systems and should be considered only for use in the ShakeOut Scenario and not specifically representative of the performance of any one utility or localized region serviced by a single utility.

For the ShakeOut Scenario, the counties were classified as follows:

- High impact: Los Angeles, Riverside, and San Bernardino Counties
- Medium impact: Ventura, Orange, and Imperial Counties
- Low Impact: Kern and San Diego Counties

Using the restoration curves, a tabulation of power outage and restoration times describes the restoration of power service in terms of hours, days and weeks.

It was noted that the power outage could be more far-reaching: If a cascading failure of the western (west of Rocky Mountains, British Columbia, Mexico) electric power grid results from the sudden loss in southern California, 11 isolated communities such as San Francisco and Portland may remain on line and other areas could restore power within three hours, except for a few isolated areas that have large industrial power loads without nearby generation, taking up to 12 hours to restore power.

Water Outage and Service Restoration

Water outage and service restoration estimates for the eight county region were derived from the water panel discussion and follow up with Ron Tognazzini and Mike Morel.

Despite aqueduct failures along the fault ruptures, Metropolitan Water District's (MWD's) water storage south side of the fault is adequate for 6 months for all its customers, assuming a 25% reduction in demand on average over the 6 months (expecting a 40% reduction in demand initially). MWD is a water wholesaler with 26 customers in about 5,200 square miles in six of the eight counties in the study region: Los Angeles, Orange, San Diego, Riverside, San Bernardino and Ventura counties. MWD's service area includes only the southwestern portion of San Bernardino County including areas such as: Chino, Ontario,

Upland, and Rancho Cucamonga. Most of these areas get some supply from Metropolitan as a supplement to local groundwater. The areas around San Bernardino proper are supplied by local agencies, either with groundwater, surface water, or by San Bernardino Valley Municipal Water District, which is a State water contractor like MWD. We have not obtained information about water infrastructure in Kern and Imperial Counties.

Six months of emergency storage is an outcome of Metropolitan's planning for emergency surface water storage and it is in addition to storage for operations and storage meant to help the region survive drought. In fact, one of the primary criteria in site selection for Diamond Valley Lake was its location in relation to the San Andreas Fault. Metropolitan's planning assumes that in a San Andreas event in Southern California, the aqueducts supplying the region would be out of service for up to 6 months. Agencies that have some ground water in Metropolitans service area would rely on Metropolitan to the extent that their systems (and Metropolitan's) were intact. Many of Metropolitan's customers have several connections with Metropolitan such that if one or more connections are out of service, Metropolitan can still supply water to other connections and customers can receive water from another pipeline. This is less true in the eastern area of the system. Even if the connection to MWD is intact or if damaged, repaired quickly, the system of the retail agency may have more extensive damage and it may take longer to restore the water service. Electricity for pumping is a considered to be a nonbinding constraint because pipeline repair takes longer.

In the past a 10-mile (from the fault) criterion has been used to delineate highly impacted areas for water distribution. Although some water response plans continue to be based on the magnitude of an earthquake and distance from epicenter, the water industry now has tools like ShakeMap, and to some extent ShakeCast, that provide spatial metrics based on estimates of ground shaking. Therefore, it is appropriate to describe the water service outage and restoration in terms of Instrumental Intensities.

Figure 7-14 demonstrates the concept of water customer outage and restoration for an area of MMI VIII and greater to illustrate four points:

1. Not all customers lose the water service, even in the highly damaged areas. Some water systems have replaced their seismically vulnerable piping and storage system with more seismic resistant materials to mitigate the impact of these events. Seismic mitigation measures are expensive and time consuming so not all water system customers will benefit from this concept. About half of the customers indicated in this graph never lose their water service, although they may be restricted by boil water notices and water conservation announcements to use their water carefully.
2. Water customers often have water available to them for a period of time after the earthquake, and may get a false sense of security that they will not be impacted by water shortages. In actuality, the water systems are bleeding out through leakage into the streets, but that takes a few hours to as much as a day for higher elevations andas much as 2 days for the lower elevations. The restoration curve in figure 7-14 assumes that it takes water systems a little more than a day to bottom out.
3. There are steps of water service restoration as entire blocks of customers are restored in urban areas, but in general the step increment is small, and when shown over a 6 month time scale restoration appears to be continuous.
4. The restoration curve is confined to an area of MMI VIII or greater with areas of potential for seismic ground deformations from landslides or liquefaction. Another

graph might be made for MMI VII or greater with the same ground deformation qualifiers. That graph would show fewer customers without water service and a faster restoration time.

Figure 7-14 is based on informed speculations about the performance of water systems and should be considered only for use in the ShakeOut Scenario and not specifically representative of the performance of any one utility or localized region serviced by a single utility.

A tabulation of water restoration at points in time for MMI VI and VII and MMI VIII+ are tabulated below. The information for MMI VI and VII were guesstimated in the absence of a definition for "spotty service".

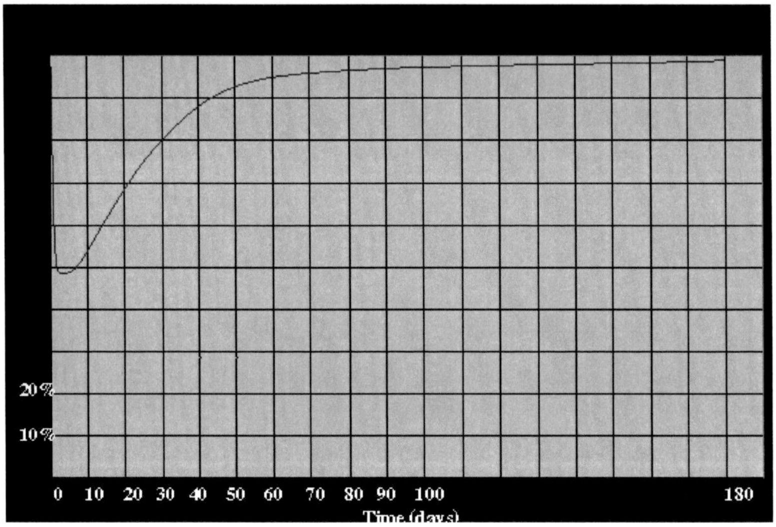

Figure 7-14. Water resilience of MMI VIII or greater. Y axis is percentage of customer demand restored.

Table 7-9. Tabulation of Pipelined Water Service Outages and Restoration at Points in Time for MMI Areas

Pipelined Water	Days	Weeks	Months
MMI VI and VII	80% customers have water service	2 weeks: 100% customers have water	
MMI VIII+ areas with landslide and liquefaction potential	50% of customers in MMI VIII+ areas are without water, but may have water for a short period of time after the earthquake	1 week: continuous restoration of the 50% of customers in MMI VIII+ begins 2 weeks: 60% of customers have water 3 weeks: 70% 4 weeks:78%	5 weeks: 85% of customers have water 6 weeks: 90%, 2 months: 95 % 6 months: the last 5% of customers in MMI VIII+ have water service

Finally, water service outage and restoration, by Instrumental Intensities, had to be related to the industrial sector and number of households in the MMI zones for each county before the information could be used as an input into the regional economic model. David Hester and Ben Sherrouse used the methods reported in Sherrouse and others (2008a) to measure the number of employees and payroll for 20 North American Industry Classification System (NAICS) sectors located in MMI zones. The Industrial sectors include 17 single 2-digit NAICS codes and Manufacturing, Retail Trade, and Transportation and Warehousing (that are actually aggregations of 2 to 3 of the 2-digit NAICS codes).

For example, Table 7-10 displays the 4th quarter 2006 payroll by industrial sector and MMI zones for Los Angeles county.

Table 7-10. Los Angeles County 4th Quarter 2006 Payroll by Industrial Sector within MMI Zones

SECTOR	MMI III-V		MMI VI and VII		MMI VIII-X	
	Quarterly payroll	Percent	Quarterly payroll	Percent	Quarterly payroll	Percent
Agriculture, For., Fish etc	$2,224,627	11.8%	$3,028,374	16.1%	$13,604,148	72.1%
Mining	$2,235,406	18.1%	$6,295,281	50.9%	$3,840,348	3 1.0%
Utilities	$0	0.0%	$8,784,553	5.2%	$159,699,719	94.8%
Construction	$347,498,252	17.3%	$917,492,844	45.7%	$742,024,063	37.0%
Manufacturing	$1,138,195,228	21.1%	$1,919,775,239	35.6%	$2,333,739,932	43.3%
Wholesale Trade	$552,382,609	17.0%	$1,374,811,021	42.3%	$1,319,595,506	40.6%
Retail Trade	$659,205,021	19.8%	$1,610,411,384	48.5%	$1,054,117,149	31.7%
Transportation & Warehouse	$556,674,251	38.5%	$466,534,781	32.2%	$424,520,109	29.3%
Information	$1,180,250,792	30.1%	$2,582,400,111	65.9%	$153,811,083	3.9%
Finance and Insurance	$784,962,936	22.2%	$2,357,021,556	66.7%	$391,049,803	11.1%
Real Estate, Rental/Leasing	$367,197,926	33.9%	$555,036,682	51.3%	$160,755,341	14.8%
Prof., Sci., and Tech. Services	$1,725,318,131	30.9%	$3,339,047,885	59.9%	$514,460,202	9.2%
Mgmt of Coys & Enterprises	$246,210,878	26.4%	$520,191,005	55.7%	$167,747,350	18.0%
Admin. and Support/Waste	$504,657,595	24.1%	$1,139,497,744	54.4%	$450,726,796	21.5%
Educational Services	$132,487,514	28.0%	$230,670,330	48%.7	$110,016,899	23.3%
Health Care & Social Assist.	$1,166,905,651	25.3%	$2,310,835,927	50.2%	$1,125,654,701	24.5%
Arts, Entertain. and Recreation	$969,002,938	37.3%	$1,548,093,158	59.5%	$84,167,500	3.2%
Accommodation and Food Svc	$469,335,415	31.9%	$685,347,837	46.5%	$317,857,659	21.6%
Other Services	$244,976,305	23.6%	$550,641,788	53.0%	$243,687,900	23.4%
Public Administration	$1,035,549,992	13.9%	$5,002,983,072	67.3%	$1,396,124,790	18.8%

Table 7-11. Residential Occupancy Class Square Footage in MMI Zones and Counties

Region	Single Family	% SF	Mobile Homes	% MH	Multi- Family	% M-F
< MM VI	2822050	42.9%	116849	45.8%	935233	38.4%
MMI VI-VII	1653578	25.2%	52001	20.4%	869252	35.6%
MMI VIII-X	2095369	3 1.9%	86337	33.8%	634065	26.0%
Imperial						
< MM VI	37064	90.4%	4954	70.7%	7207	93.5%
MMI VI-VII	3549	8.7%	1155	16.5%	463	6.0%
MMI VIII-X	402	1.0%	898	12.8%	35	0.5%
Kern						
< MM VI	191708	75.1%	18264	76.6%	27803	73.8%
MMI VI-VII	55995	21.9%	3681	15.4%	9111	24.2%
MMI VIII-X	7512	2.9%	1907	8.0%	767	2.0%
Los Angeles						
< MM VI	586556	24.8%	1152	15.3%	483893	29.6%
MMI VI-VII	889704	37.7%	1987	26.4%	726280	44.4%
MMI VIII-X	884722	37.5%	4400	58.4%	425569	26.0%
Orange						
< MM VI	408421	36.6%	7951	24.0%	71170	28.3%
MMI VI-VII	377719	33.8%	10106	30.5%	98014	38.9%
MMI VIII-X	329797	29.6%	15065	45.5%	82534	32.8%
Riverside						
< MM VI	150766	22.9%	19621	24.2%	12844	15.6%
MMI VI-VII	230420	35.0%	28605	35.2%	23754	28.9%
MMI VIII-X	277712	42.1%	33020	40.6%	45674	55.5%
San Bernardino						
< MM VI	50033	6.9%	6730	15.7%	5459	5.8%
MMI VI-VII	88580	12.3%	5926	13.8%	10220	10.9%
MMI VIII-X	583613	80.8%	30323	70.6%	78150	83.3%
San Diego						
< MM VI	1081113	100.0%	46878	99.9%	287566	100.0%
MMI VI-VII	215	0.0%	68	0.1%	18	0.0%
MMI VIII-X	0	0.0%	0	0.0%	0	0.0%
Ventura						
< MM VI	316390	94.3%	11299	90.4%	39292	93.5%
MMI VI-VII	7397	2.2%	473	3.8%	1392	3.3%
MMI VIII-X	11611	3.5%	723	5.8%	1337	3.2%

David Strong, USGS, recalculated HAZUS census tract residential square footage data into MMI zone data. In future, we would use household data rather than square footage data, if possible.

Natural Gas Outage and Restoration

Natural gas outage and restoration estimates were derived from conversations with Rick Gailing of Southern California Gas Company (SoCalGas). The posited earthquake pipeline damages involve natural gas transmission pipelines crossing the fault rupture, but there can be failures of transmission pipelines (and distribution pipelines) without significant loss of service. SoCalGas has proactively studied the ShakeOut Scenario, the rupture slips along the fault, in particular. Although multiple transmission pipeline breaks are a plausible outcome, the utility's response plan is to repair one of the transmission lines within three days to restore an external supply of natural gas in the region. Until transmission lines are repaired, underground storage fields in the region will be the source of natural gas supply to the region because gas field levels are high in November. There will be natural gas service outages resulting from distribution pipeline failures, but primarily from customer shut-offs. During the Northridge earthquake, 88% of natural gas service outages were customer shut-offs. As is typical in an earthquake, the public was erroneously informed by various media to shut off gas supply to residences, even when there was no evidence of gas leaks. Only one AM news radio station consistently advised against shutting off the gas unless there was some reason to suspect that leakage was occurring. Restoration of gas service was further delayed by people failing to notify the gas company that they had turned off their gas.

Unlike water, a relationship between MMI levels and gas service outages is unavailable. Rick Gailing recommended calculating a ShakeOut-to-Northridge earthquake impact ratio to compare the impacted customers from both earthquakes. We used the impact ratio and Northridge earthquake gas service outage and restoration information to formulate a gas outage and restoration story for the ShakeOut Scenario economic analysis. Initially, the impact ratio calculation did not factor in population changes for two reasons:

- New structures and their infrastructure are more resilient in the event of an earthquake.
- Since the Northridge earthquake, the utility has continued to upgrade its distribution pipelines, particularly in areas of high and very high liquefaction susceptibility.

However, population change would be relevant if the new housing construction in the Inland Empire, particularly Riverside County, requires earthquake shutoff valves at the gas service meter. We need to verify this regulation with both Riverside and San Bernardino Counties and determine if the requirement stands and how long it has been in place.

Given time constraints, the impact ratio was based on Northridge earthquake and ShakeOut MMI maps and Census 2000 population and household data. Using each set of population and household data, two estimates of the impact ratio were produced from:

- census tracts completely contained in MMI zones, and
- census tracts intersecting with the MMI zones.

A broad range of results indicated impact ratios ranging from 1 to 25; the ratios increased as the minimum MMI zone increased. (We observed that population was fairly equally distributed across the MMI zones in the Northridge earthquake, but higher proportions of population reside in the higher MMI zones of the ShakeOut Scenario.) We deployed the impact ratio of 5 associated with population captured by MMI8 and greater zones.

Northridge earthquake natural gas service outage and restoration statistics were assembled from a Technical Council of Lifeline Earthquake Engineering (1995) monograph and scaled up:

- 151,000 gas outages were reported in the wake of the Northridge earthquake. 88% of these were customer shut-off.
- For the ShakeOut Scenario it is assumed that (151,000 x impact ratio) customers lose gas service, and 88% are customer shut-offs. Steps of gas service restoration for all earthquakes are:
  1. Safety first
  2. Service Restoration
     i. leakage surveys
     ii. checking structure and appliance connections
     iii. restoring service where safe.
- The restoration of gas service, following the Northridge earthquake, was aided by personnel from neighboring utilities for 9 days:
  - 1 week: 84,000 customers (57%) restored
  - 9 days: approximately 103,000 (68%) restored. Further, we assumed that that 22,900 of self restored customers are restored within 9 days, yielding a total of 125,900 (83%) customers restored
  - 1 month: more than 141,900 (94%) customers restored
  - 9,100 (6%) customers not restorable due to structural property damage
- For the ShakeOut Scenario, the Northridge 9-day restoration period (when neighboring utilities are assisting) was scaled to 3 weeks for the initial restoration of 83% of customers, and the Northridge 1-month time period was scaled to 2 months regarding restoration of the remaining restorable customers.

Information at the SoCalGas web site and from Rick Gailing suggested that 95% of natural gas customers are households, 5% are small businesses and less than 1% are large customers. The large customers typically are served by more resilient systems. The latter assumption was challenged during the economic roundtable discussion with a couple of examples: (1) the City of Oxnard bus problems following a large landslide that broke a SoCalGas transmission pipeline and (2) SoCal Edison's dependence on natural gas for electric generation. From the utility's perspective, the City of Oxnard bus situation was a very unique incident in that the bus system is dependent upon natural gas having a particularly narrow Btu tolerance. Alternative supplies from the coastal area were available to serve the bus system; however, the natural gas heating value did not meet the City's stringent fuel specifications. Consequently, it is our understanding that the City of Oxnard brought in diesel buses to supplement their fleet until the gas supply outage was restored. There are no other known gas vulnerabilities in the region's bus systems. Regarding SoCal Edison's dependence on gas, we assume that the power restoration estimates factored in natural gas storage capabilities and back up potential. It is not to say that all delivery systems to large customers will be at full capacity, but there will also be reduced demand on natural gas and electric systems following a major earthquake event. As lifelines and transportation systems begin to recover, regional demand increases.

The big picture of gas service outages and restoration was disaggregated to the eight counties based on county and regional population data, and HAZUS residential and business count data within MM VIII zones and greater (for example, Tables 7-10 and 7-1 1).

The numbers in Table 7-12 are probably optimistic because:

- A gas shut-off requirement for new development in San Bernardino and Riverside county would increase the number of customers affected.
- It is likely that San Diego County would be impacted by this event. Gas supplies to the county would be limited due to the outage of the southern transmission supply system, requiring some of the larger customers to be curtailed in order to maintain supply to residential communities. However, gas consumption in the San Diego region would be down for a short time period following the event, compensating for the limited supplies. However, it would be more realistic to lower the percentage of San Diego County customers with gas service after the earthquake and the percentage restored after three weeks from 100% to 95% and 98%, respectively, in future runs of the economic model.

Lifeline Outage and Restoration Summary

Each lifeline was evaluated independently of each other. Figure 7-15 illustrates a comparison of lifeline service outages and restorations in a high-impact area. Telecommunications is posited to have the quickest restoration and the water restoration curve has a long restoration tail out to 6 months. Lifeline service is further complicated by interdependencies between lifelines. For example, telecommunication service depends on power and water services, and power, water, transportation and gas services depend on telecommunication service.

Table 7-12. Gas Outage and Restoration Input Data for the Regional Economic Model, 95% Residential Customers and 5% Small Business Customers

County	% restorable customers with gas service after the earthquake	% restorable customers with gas service after 3 weeks	% restorable customers with gas service after 2 months
Imperial	99%	100%	100%
Kern	98%	100%	100%
Los Angeles	83%	97%	100%
Orange	85%	99%	100%
Riverside	78%	96%	100%
San Bernardino	59%	92%	100%
San Diego	100%	100%	100%
Ventura	98%	100%	100%

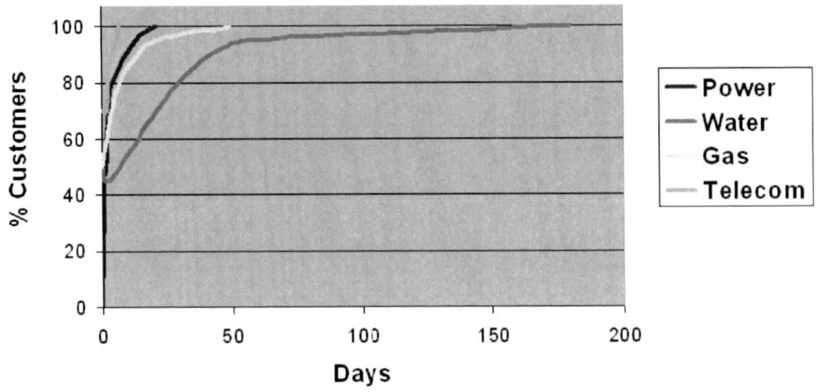

Figure 7-15. Posited Lifeline service outage and restoration for a community in a high- impact area.

Highway Transportation: Lost Trips and Delays

Sungbin Cho, ImageCat Inc., worked at the interface of the REDARS 2 Highway- bridge model and the regional economic model. He was able to disaggregate lost trips and delayed trips into four types of trips by county. It was only feasible to assign the lost trips (that is, not the delayed trips) to the industrial sector. In parallel, Ben Sherrouse and David Hester (2008b) used different data and methods to predict the number of commuters that will be unable to cross the fault rupture.

Inputs for the Regional Input-Output Model

Background

A research team performed transportation impact analysis for the Southern California transportation system network that extends 7 counties, and travel demand from 6 counties (Werner and others, 2008). Based on the deterministic analysis using a default recovery model in REDARS® 2.0 , they concluded the economic loss due to transportation disruption might be as much as $5 billion.

Objective

The objective of this traffic analysis is to provide a set of input data for an input- output model to perform a regional economic analysis of the transportation disruption from a M7.8 Earthquake along the southern San Andreas Fault.

Theoretical Background

The REDARS® 2.0 estimates the transportation impacts (or additional costs to the system) resulting from an earthquake-damaged network according to the following algorithm:

- For a normal network condition, the number of trips, d_1 between an origin and a destination with a travel time p_1 is determined by a demand curve (D), and a supply curve (S_1).
- A reduced supply of network capacity, due to an earthquake, shifts the supply curve, S_1 to S_2. This shift drives travel demand d_1 to d_2, while the travel time increases to p_2.

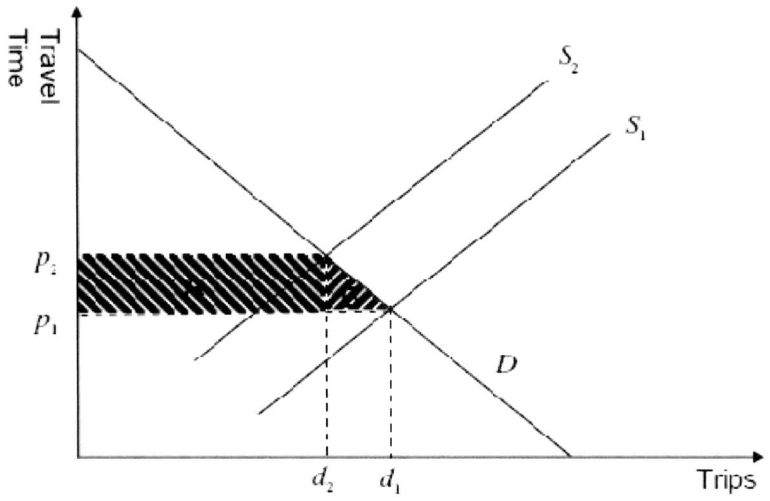

- The shaded area (A+B) represents surplus reduction to the system users from the earthquake. But area A and area B have somewhat distinctive characteristics.
- Area A is the additional travel delay costs incurred by the drivers who use the network system after the earthquake.
- Area B represents the loss from lost trips experienced by the drivers who used to use the system before but not after the earthquake.
- The conceptual supply and demand curves are identified, and explicitly implemented in a network model. The supply curve, S is the congestion function of traffic volume and network capacity. Travel cost between zone m and n is in following form:

$$p_{m,n} = p^0_{m,n} \cdot \left\{ \alpha + \beta \cdot \left(\frac{d_{m,n}}{c} \right)^{\gamma} \right\}$$

Where
p^0 = travel time between the zones at free flow speed,
d = traffic volume,
c = capacity of available routes, and
α, β,γ =coefficients (1.0, 0.15, and 4.0 respectively. From Bureau of Public Road)

- Note that D in the graph actually represents the inverse of demand curve, for x-axis is the number of trips, while y-axis is the congested travel time.
- The demand curve is a type of a gravity model—demand is proportional to the total trips originated from origin, O_m, and total trips terminated at the destination, D_n, and inversely proportional to travel cost or travel time.

$$d_{m,n} = D^{-1}(p_{m,n}) = \frac{O_m \cdot D_n \cdot a_m \cdot b_n}{\exp(\alpha + \beta \cdot p_{m,n})}$$

Where

$P_{m,n}$ = travel time between the zones (congested),
O_m = Trip originated from zone i,
D_n = Trip terminated to zone j,
a_m, b_n = calibrated zonal factors to keep the conservation constraints

$$O_i = \sum_j d^{i \to j}, \; D_j = \sum_i d^{i \to j}, \text{ and}$$

α, β = calibrated coefficients (> 0).

Tasks

To fulfill the objective, the following tasks are performed

- Modify REDARS transportation analysis model to report zone-to-zone travel time and demand change before and after earthquake.
- Refine the initial transportation analysis to better capture the economic meaning of trips. The original run was made with only one demand curve for each origin-and-destination pair. In this project, the trips are disaggregated into three passenger trip types and one freight trip type:
  - Passenger trips between home and work place (JHW)
  - Passenger trips between home and shopping destinations (JHS)
  - Passenger trips other than JHW and JHS (JHO), and
  - Freight (trucks)
- Calibrate demand curves for the four different trip types
- Aggregate the impacts (A and B) spatially (into zones),
- Convert the zonal impacts to industrial sectors.

Data Compilation and Process

Transportation analysis zones (TAZ) and the associated trip date (OD tables) are aggregated into a manageable number of data records for REDARS. The original transportation data from Southern California Association of Governments (SCAG) is highly disaggregated spatially into 4109 TAZ for the six counties of Los Angeles, Orange, Riverside, San Bernardino, Ventura, and Imperial. In travel demand data (OD), additional 81 TAZ are used for external zones to model boundary conditions. The OD tables and transportation network data are aggregated to 1,898 internal TAZ, and 40 external TAZ.

The SCAG 2003 transportation model socio-economic data (population, households, and number of jobs in the 4,109 internal TAZ for 13 industrial sectors) is aggregated into the 1,898 zones.

SCAG Production-Attraction (P-A) matrices are converted into OD trip matrices. The original form of the matrices from SCAG was in a P-A matrix, in which trips are counted only for the trip purposes, not for trips the drivers are actually making. For example, home-based-working trips in a P-A matrix represent the number of drivers who depart from home to the work place, and do not include return trips from work to home. To calculate trip reduction that is occurring along with travel time increase properly, SCAG P-A matrices are converted into OD matrices by considering the return trip ratio (SCAG, 2007). The baseline and post-earthquake P-A matrices are also used to allocate the impacts spatially.

Industrial activity data (output by industries, and household local purchase pattern by industries) for the Southern California six counties are based on the IMPLAN 2004 regional data set.

Spatial Distribution of Transportation Impacts

REDARS-estimated transportation impacts from the ShakeOut Scenario earthquake are aggregated spatially.

REDARS produces the following post-earthquake results in ten matrices:

- Baseline P-A (travel demand from Zone m to Zone n) for each of the four trip types, $d^0_{m,n}, \forall m \in N, n \in N$
- Baseline travel time, $p^0_{m,n}, \forall\, m \in N, n \in N$
- Post-earthquake P-A (travel demand from Zone m and Zone n) for each of the four trip types, $d^*_{m,n}, \forall\, m \in N, n \in N$
- Post-earthquake travel time, $p^*_{m,n}, \forall\, m \in N, n \in N$

Based on these ten matrices, the following method is applied to calculate the impacts from increased travel time and foregone trips. (For simplicity, indices for trip types, and days after earthquake are omitted):

1. Value of travel time increase in Zone n

$$A_n = k \cdot \sum_{m \in N} d^*_{m,n} \cdot \left(p^*_{m,n} - p^0_{m,n}\right),\ \forall\, p^*_{m,n} \geq p^0_{m,n}$$

2. Value of foregone trips in Zone *n*

$$B_n = k \cdot \sum_{m \subset N} \int_{p^0_{m,n}}^{p^*_{m,n}} D^{-1}(w)\, dw \ \ -A_n \ \ ,\ \forall\, p^*_{m,n} \geq p^0_{m,n},\ \text{and}\ d^*_{m,n} \leq d^0_{m,n},$$

where k is value of time ($13.45/hour for passenger trips, $71.03/hour for trucks)

Note that the losses are aggregated into destination zone n for the three passenger trip types, but the impacts from freight movement are accumulated into destination zones. See the Industrial Distribution Section for details.

Due to increased travel time, drivers in the system will spend additional driving time, a loss of $4.3 billion, until the system is fully recovered. Also, by not making trips, drivers will lose $0.7 billion.

Table 7-13. Loss from Travel Time Increase in Thousands of Dollars

County	Trip Types				Sum
	JHW	JHS	JHO	Freight	
LA	435,358	33,416	1,016,988	726,283	2,212,046
OR	93,675	1,894	112,023	91,402	298,994
RV	53,978	3,077	110,983	79,736	247,775
SB	371,841	9,519	564,411	315,805	1,261,576
VN	1,294	63	7,003	17,390	25,749
IMP	206	13	726	270	1,214
Outside	-	-	59,967	188,776	248,743
Total	956,352	47,982	1,872,101	1,419,662	**4,296,097**

Table 7-14. Loss from Foregone Trips in Thousands of Dollars

County	Trip Types				Sum
	JHW	JHS	JHO	Freight	
LA	35,108	1,236	114,507	51,530	202,381
OR	856	32	7,025	19,883	27,796
RV	30,163	33	30,261	2,929	63,386
SB	8,175	243	370,977	40,218	419,614
VN	0	1	562	1,627	2,191
IMP	1,056	0	192	667	1,914
Outside	-	-	8,555	13,150	21,705
Total	75,358	1,545	532,079	130,004	**738,987**

An Excel file —Impact by County.xls—contains spatially disaggregated transportation impacts calculated according to the method stated above. Worksheets in the excel file are:

- **3dImp act** : impacts in the first 3 days of earthquake happening (moderately damaged bridges are restored during this time period);
- **12dImpact** : impacts of 4 -12 days since the earthquake happened (extensively damaged bridges are restored during this time period);
- **49dImp act** : impacts of 13 - 49 days since the earthquake happened (closed roads due to surface rupture are opened during this time period);
- **140dImpact** : impacts of 50 - 140 days since the earthquake happened (unusable 3-span bridges are restored during this time period); and
- **221dImpact** : impacts of 141 - 221 days since the earthquake happened (all closed roadway is opened during this time period).

Columns in each worksheet are:

- ***jhwR*** : Impacts from reduced home-to-work trips in dollars;
- **jhwT** : Impacts from increased travel time of home-to-work trips in dollars;
- **jhwSum** : sum of jhwR and jhwT;
- **jhsR** : Impacts from reduced home-to-shop trips in dollars;
- **jhsT** : Impacts from increased travel time of home-to-shop trips in dollars;
- **jhsSum** : sum of jhsR and jhsT;
- **jooR** : Impacts from reduced other passenger trips in dollars;
- **jooT** : Impacts from increased travel time of other passenger trips in dollars;
- **jooSum** : sum of jooR and jooT;
- **frR** : Impacts from reduced trucks in dollars;
- **frT** : Impacts from increased travel time of trucks in dollars; and
- **frSum** : sum of frR and frT.

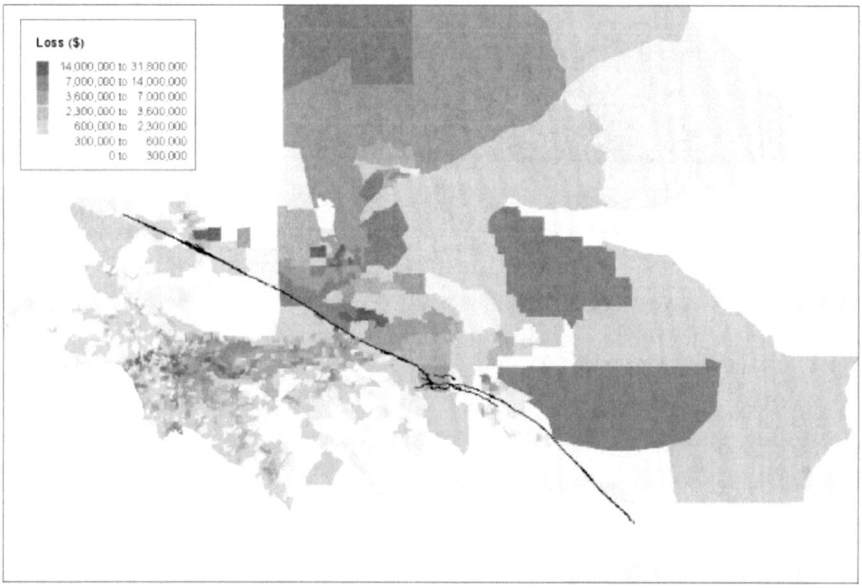

Figure 7-16. Map of southern California showing losses due to travel time increase. (Losses from passenger trips are accumulated to the destinations, while losses from freight movement are accumulated into the origins of trips.)

Industrial Distribution of Transportation Impacts

The losses (spatially distributed) due to foregone trips are allocated to industrial sectors, according to spatial distribution of industrial activities by sectors, along with other economic statistics:

- Trip types were considered in the allocation.
- The losses from home-to-work trip reduction were distributed to industry according to the spatial output distribution by industry.
- On the contrary, the losses from reduced home-to-other trips were distributed to industries based on household consumption pattern.

  These losses by industrial sectors will be entered into the regional input-output model, as the Final Demand Reduction to analyze the subsequent ripple effects to the regional economy.

  On the other hand, however, the losses due to increased travel time, which is calculated to $4.3 billion, could not be associated with specific industrial sectors. It does not mean that there would not be any economic ripple effects from these losses, but the input-output model simply cannot take the input:
- For any production activities that involve goods delivery (as intermediate goods movement, or as final demand shipment), increased travel time is considered to manifest as a price increase. Higher commodity price leads to less demand for the commodity in production, or consumption. Also, increased input price causes substitution of a less expensive alternative input.
- However, the original input-output model does not account for the effect of price, thus no substitution can be implemented, and therefore, cannot take such input directly.
- Even though it is possible to associate the impacts from travel time increase to industry by using the same method used to allocate the losses from trip reduction, the meaning of the allocated impact is not clear in the context of the input-output model.

  For a more comprehensive and thorough analysis of the transportation impact, advanced regional economic models, such as multi-regional computable general equilibrium can be used.

  Therefore, out of total $5.035 billion, only the intra-regional losses from trip reduction, $717 million will be distributed to regional industry:
- Distribution of Losses from Home-to-Work Trip Reduction (LJHW) are calculated according to the following assumptions:
  - Direct household income is not reduced by a worker's inability to get to work.
  - Employers take the loss, in the form of less demand for labor.
  - Spatially, the losses occur at the work place. Therefore the loss is accumulated at the destination of the working trip.
  - Loss is distributed to industries based on the composition of industrial output in the zone.
  - The loss from home-to-work trip reduction to industry i in zone k , county r, $LJHW_i^{k \in r}$ is calculated as follows

$$LJHW_i^{k\in r} = LJHW^{k\in r} \cdot \frac{x_i^r \cdot e_i^{k\in r}}{\sum_{j\in I} x_j^r \cdot e_j^{k\in r}}$$

$LJHW_i^{k\in r}$: Loss from home-to-work trip reduction, aggregated into zone k in county r
$e_i^{k\in r}$: Employment of industry i, in zone k that is part of a county r
X_i^r: Output of industry i, in county r. (From IMPLAN)

x_i^r: Output per employment of industry i, in county r, $x_i^r = \frac{X_i^r}{e_i^r}$
e_i^r: Employment of industry i, in county r. (From SCAG)

- Distribution of Losses from Home-to-Shop Trip Reduction (LJHS)
  - Assume the number of shopping trips is proportional to the shopping quantity, and this proportion is not changed even after the event.
  - Shopping trip reduction is translated into shopping reduction.
  - By reduction of shopping trips, households demand less grocery from retail sector.
  - Spatially, the loss resides at the shopping center.
  - Therefore, loss from Home-to-shop trip reduction is same to the value of forgone trips (B_n) with Home-to-Shop Demand curve.

$$LJHS_{Retail}^{k\in r} = LJHS^{k\in r} = B_{JHS}^{k\in r}$$

- Distribution of Losses from Home-to-other Trip Reduction (LJHO) is calculated according to the following assumptions:
  - Assume all home-based trips are associated to purchasing of goods and services.
  - Also assumed the proportion of purchasing goods and services per trip is not changed.
  - The losses will reside where the trips end, and accumulated into the destination zone.
  - Losses are distributed to industries according to household local purchase pattern.

$$LJHO_i^{k\in r} = LJHO^{k\in r} \cdot \frac{h_i^r \cdot e_i^{k\in r}}{\sum_{j\in I} h_j^r \cdot e_j^{k\in r}}$$

$LJHO^{k\in r}$: Loss from home-to-other trip reduction, aggregated into zone k in county r
$e_i^{k\in r}$: Employment of industry i, in zone k that is part of a county r
H_i^r: Household local purchase of industry i, in county r. (From IMPLAN)
e_i^r: Employment of industry i, in county r. (From SCAG)

h^r_i: Household local purchase per employment for the output from industry i, in county r, $h^r_i = \frac{H^r_i}{e^r_i}$

- Distribution of Losses from Freight Trip Reduction (LFREIGHT):
  - Assume all freight is for intermediate goods movement.
  - Interpret reduced intermediate goods movement as less demand for input.
  - Spatially, the losses reside where the delivery initiated. Therefore the losses are accumulated into origin zones.
  - Loss is distributed to industries proportional to output by only the industries that generate freight. Assume that such industries are Agriculture (AG), Construction (Const), Manufacturing (Manu), Wholesale (Whole) and Retail (Ret).
  - The calculation is as follows

$$LFREIGHT_i^{k\in r} = LFREIGHT^{k\in r} \cdot \frac{x_i^r \cdot e_i^{k\in r}}{\sum_{j\in F} x_j^r \cdot e_j^{k\in r}}$$

where F={AG, Const, Manu, Whole, Ret}

Commuting Across the Fault

Sherrouse and Hester (2008b) produced a spatial distribution of commuters crossing the southern San Andreas Fault (figure 7-17) and aggregated the commuter flows within and between counties (figure 7-18). They estimate that the percentage of county employees who live and work on different sides of the fault amounts to approximately 2% of Los Angeles County employees, 10% of Riverside County employees, and 16% of San Bernardino county employees.

Table 7-15. Losses from Trip Reduction by County and by Industry in Thousands of Dollars

Sector	Los Angeles	Orange	Riverside	San Bernardino	Ventura	Imperial	Total
AG	632	122	1,297	4,367	12	1,404	7,834
Const	8,100	3,296	3,396	9,728	275	25	24,819
Manu	47,007	9,583	7,198	33,590	996	33	98,408
WHOLE	14,189	3,419	1,322	21,223	182	18	40,353
RET	27,616	4,001	13,124	**79,691**	249	68	124,748
TRANS	10,800	2,424	2,410	24,029	31	176	39,870
INFOR	280	37	46	130	1	0	495
FIRE	9,458	592	3,007	22,404	42	6	35,509
PROF	38,365	1,948	10,149	**96,247**	241	4	146,954
EDUC	5,167	134	626	13,561	8	3	19,498
ARTENT	18,550	1,513	13,981	58,416	65	47	92,571
OTHSER	15,883	607	3,266	40,218	84	9	60,067
PUBADM	6,246	95	3,559	15,512	3	123	25,538
SUM	202,294	27,770	63,381	419,116	2,188	1,914	**716,664**

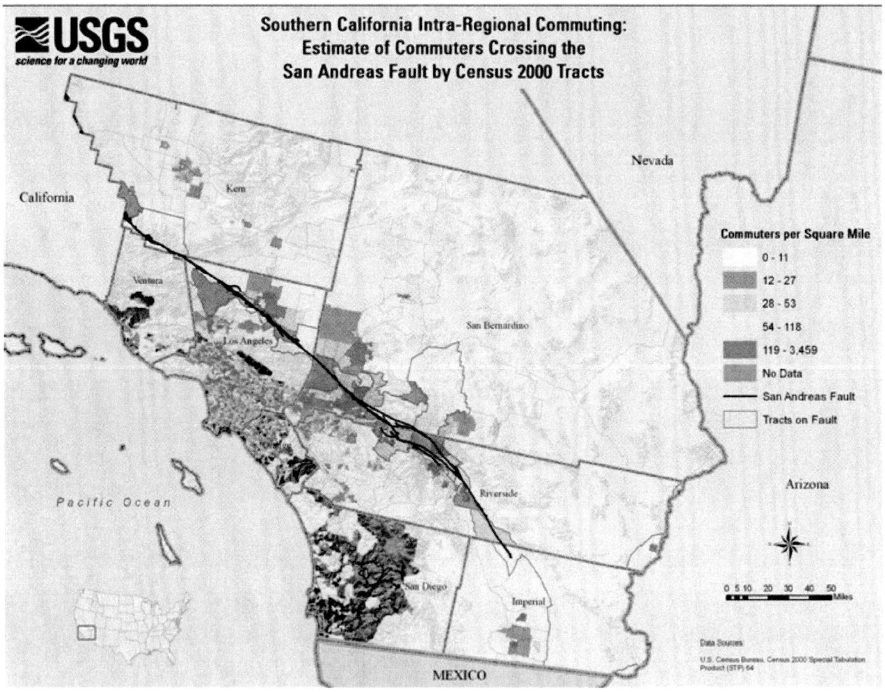

Figure 7-17. Commuters crossing the southern San Andreas Fault

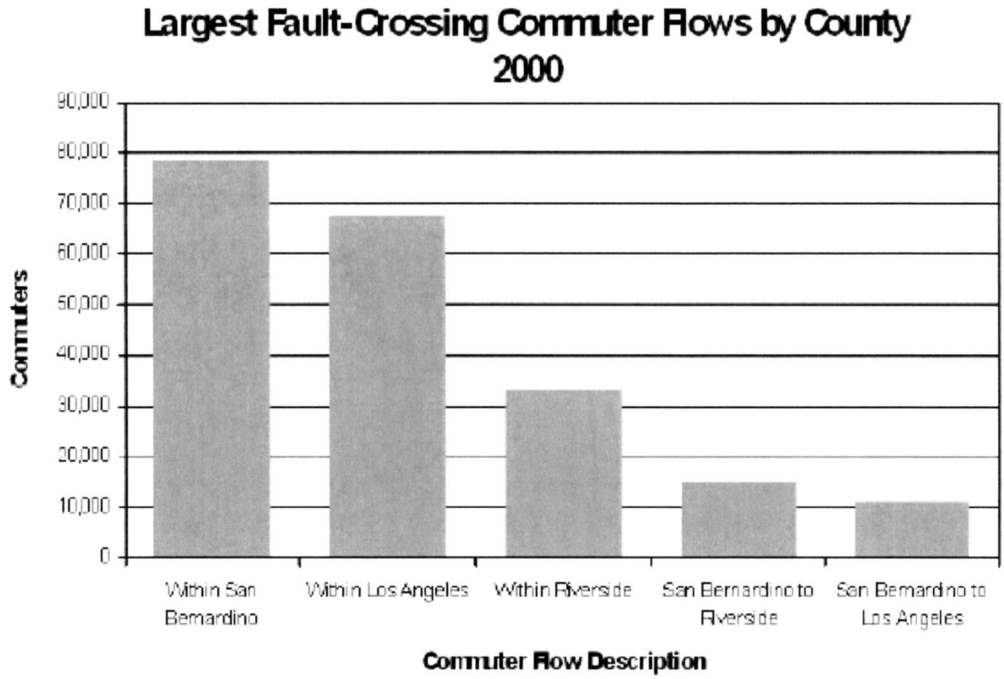

Figure 7-18. Number of commuters that cross the southern San Andreas Fault within and across counties

Comparison of Results

Sungbin Cho calculated the cost of lost home-work trips to county destinations, whereas Sherrouse and Hester (2008b) looked at the numbers of commuters (within and between counties) that would be unable to cross the ShakeOut Scenario fault rupture. While most of the commuters affected by the fault rupture are travelling within or to San Bernardino or Los Angeles Counties (figure 7-18), Cho attributes greater costs of lost home-work trips to Los Angeles and Riverside. Cho's analysis includes trips that are lost for other reasons (for example, congested highways) and factors in the value of time.

Disruption of Port Operations Due to Damage to the Rail System and Highway Network

A focus study of goods movement through the San Pedro Ports was conducted for the ShakeOut Scenario because of the regional, national and international significance of the San Pedro ports. The movement of goods through the Ports is one of the three priority concerns for the Southern California Association of Governments that covers six of the eight counties.[1]

The combined Ports of Los Angeles and Long Beach in San Pedro Bay constitute the world's fifth busiest container complex and the number one ranked harbor in the United States. As the leading gateway for trade between the United States and Pacific Rim nations, the combined Ports make a profound contribution to business and tax revenue, jobs and goods movement locally, regionally, nationally, and internationally.

International shipping lines are attracted to the world class facilities and infrastructure at both ports, which maximize the "one-stop shopping" concept of cargo transportation and delivery while offering modern, super-sized cargo terminals and an efficient train and truck intermodal network system. Even the combined capacity of all other west coast ports in North America cannot compete with the Los Angeles-Long Beach harbor complex (L. Cottrill and D. Thiessen, oral communication, 2/6/2008).

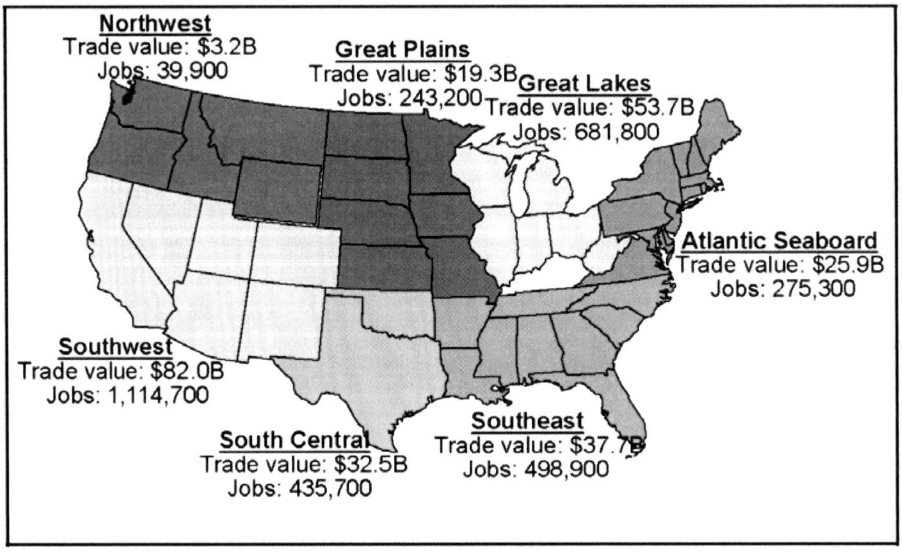

Source: BST Associates Trade Impact Report, 2007. B is billion.

Figure 7-19. National Economic Impact of San Pedro Ports

Table 7-16. Impact of International Trade through San Pedro Ports, 2005 Regional Summary

Rank	Region	States	($ millions) Trade Value	Jobs	($ millions) Income	($ millions) Taxes
1	Southwest	AZ, CA, CO, NV, NM, UT	$82,050	1,114,660	$39,240	$9,330
2	Great Lakes	IL, IN. KY, MI, OH, WV, WI	$53,640	681,860	$21,370	$5,630
3	Southeast	AL, AR, FL, GA, LA, MS, NC, SC, TN	$37,780	498,900	$14,840	$4,190
4	South Central	OK, TX	$32,580	435,710	$14,450	$3,940
5	Atlantic Seaboard	CT, DE, DC, ME, MD, MA, NH, NJ, NY, PA, RI, VT, VA	$25,940	275,230	$9,070	$2,690
6	Great Plains	IA, KS, MN, MO, NE, ND, SD	$19,260	243,220	$7,010	$2,070
7	Northwest	ID, MT, OR, WA, WY	$3,190	39,920	$1,130	$270
8	Alaska and Hawaii	AK, HI	$1,520	16,220	$450	$140
		Grand Total	$255,960	3,305,720	$107,560	$28,260

Source: BST Associates, Trade Impact Study Final Report, March 2007

Table 7-17. San Pedro Port Statistics

Port of Los Angeles	Port of Long Beach
7500 acres	3200 acres
43 miles of waterfront	10 piers
27 cargo terminals	80 berths
World Cruise Center	7.3 million TEUs
8.5 million TEUs	$140 billion cargo annually
190 million metric tons of cargo annually	87 million metric tons of cargo annually
$422.7 million operating revenue	$370.8 million operating revenue
Employment generated	Employment generated
Local 16,360	Local 30,000
Regional 1.1 million	Regional 316,000
National 3.3 million	National 1.4 million

Source: Ports of Los Angeles and Long Beach, 2007-2008

The capacity of the two Ports makes them both powerful and vulnerable at once. Given the national and international significance of Los Angeles-Long Beach harbors, the project team focused on identifying issues deriving from the high concentration of freight movement and the related logistics of their operations in the context of the ShakeOut Scenario. Meetings were held with key staff from both of the Ports, the Southern California Association of Governments, the Southern California Trucking Association, the Los Angeles County

Metropolitan Transportation Authority, and the Alameda Corridor Transportation Authority (ACTA) to obtain expert feedback on the anticipated impacts of the ShakeOut Scenario on goods movement in Southern California. The topics of discussion covered anticipated damage to transportation infrastructure and lifelines, the current goods movement system, disruption to the goods movement system (including interdependencies and resiliency of that system), supply chain effects of port disruption, effects on industries, businesses, labor, as well as on communities and regional, state and national economies. A port disruption input for the input-output analysis was derived and insights to increasing regional resilience were gained from a goods movement perspective.

Current Goods Movement System

The current goods movement system in the eight-county ShakeOut Scenario study region falls into three general categories:

1. Intra-regional (smaller trucks) - origin and destination within the region;
2. Inter-regional - originates within the region and moves out to domestic by truck, rail or air, and vice versa; and
3. International air and sea ports of entry and departure
   a. Arrives at port of entry and moves out to the region for consumption of the final product, or for further manufacturing and assembly, and vice versa;
   b. Arrives from international at the port of entry and moves out to domestic (directly by train or by truck and train via trans-load centers and warehouses), and vice versa;
   c. Passes through the region to domestic destinations, and vice versa (through the airports, for example).

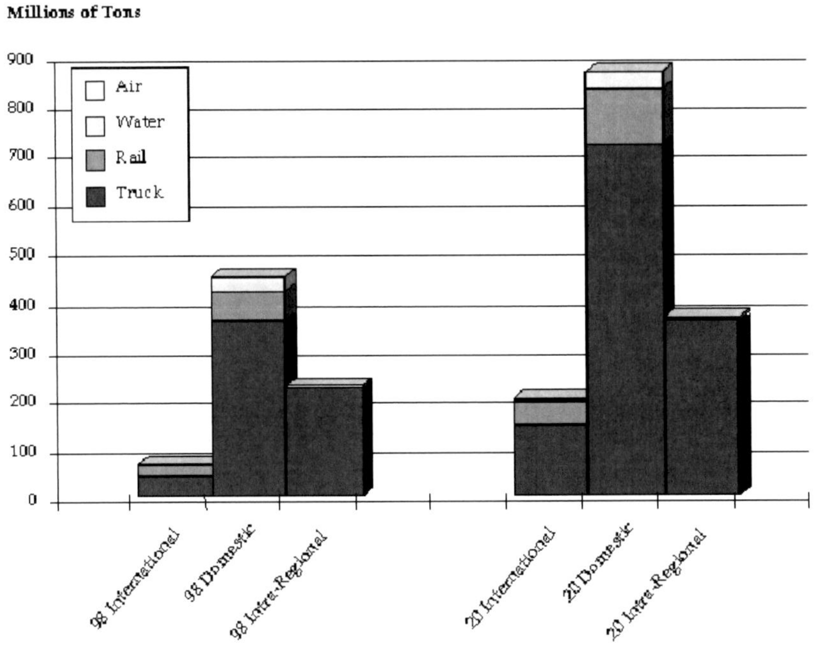

Figure 7-20. Freight in the Southern California Association of Governments region in millions of tons

The focus of the goods movement discussion involves 3a and 3b, the effects of infrastructure damage on the ports ability to operate and move goods to and from the ports, in particular.

Of the $269 billion (2005) waterborne trade through LA customs district, $256 billion trade value was through the San Pedros Ports, such that Port Hueneme handles less than 3% of the value through the LA customs district. Furthermore, 86% of the trade value is imports and figure 7-21 presents the modes of import cargo movement in 2004: 50% of port imports were delivered to the region by truck, and 50% of the imports were transported east of the Rocky Mountains for domestic distribution, but less than 1% of port imports are moved by long haul truck. Consequently, 49% of port imports were moved by train (16% directly, 11% transloaded (via truck to a warehouse before trucked to the train), 23% is trucked to the train). According, to a Surface Transportation Board report (STB, 2007) that was prepared for this project, the masked 2006 "revenue" associated with the 6.4 million cars and 150 million tons of cargo on the trains moving through Cajon Pass, Coachella Valley, and Palmdale is $9.3 billion.

Further information on current goods movement is available in the goods movement action plan (Wilbur Smith Associates, 2006).

Efficiencies in the Southern California freight transportation system have enabled companies to go offshore because costs are greatly reduced. It is cheap to move goods from China. Further increases in efficiency are expected to be incremental. Ships, rail, highways, warehouses, double-stacked container trains and intermodal cargo systems have become a finely tuned machine, bringing cargo into the region and beyond. Most of the goods arriving at the ports, destined for outside of the region, cross the San Andreas Fault rupture of the ShakeOut Scenario.

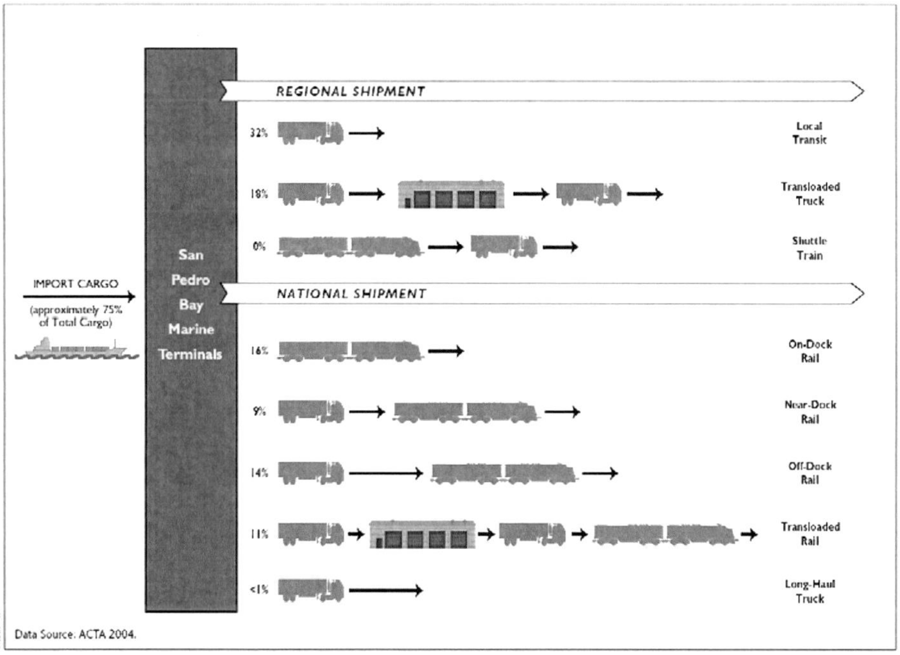

Figure 7-21. Destination and movement mode of import cargo arriving at the San Pedro ports

Summary of Damages to Transportation Infrastructure and Lifelines Affecting the Ports

Power service outages, described for Los Angeles County, suggest that service is restored to most customers, especially those out of the high impact areas, such as the ports, within three days. Telecommunications are operable except for congestion and delays. Gas and fuel are assumed to be available. No water distribution problems are expected in the port area.

According to engineers from both harbor departments, it is unlikely that a *M*7.8 event on the southern San Andreas Fault will cause significant damage to the physical infrastructure of the ports due to the long distance from the epicenter. Harbor structures tend to be "stiff," responding to 0.4 - 0.5 *g*, and mostly have short periods of motion, except for taller buildings and cranes. Although there could be some liquefaction, the impact would be lessened since most structures are built on pilings.[2] A possible area of vulnerability might occur in the power grid that serves the harbors. While the ports have some power generating capacity of their own, losses in the larger regional power system will have negative consequences since dockside cranes depend primarily on power from the grid.

Likewise, the airports are assumed to be functioning once power, telecommunications, and gas are restored. (LAX ground support operations depend on natural gas for environmental reasons.) Airports in the higher impact areas, such as Palmdale and San Bernardino, may be without power and internet for longer.

Although the Alameda corridor is out of the ShakeOut Scenario earthquake impact zone, the port activity is primarily affected by rail damage to the lines through Cajon Pass and along the Salton Sea (Byers, 2008), and highway damage within the Baldwin Park Region, all routes crossing the fault rupture (including the I-10 and I-15) and within the vicinity of Indio, Palm Springs, San Bernardino, and Palmdale (Werner and others 2007, REDARS *report*). Although highway-bridge repair estimates are considerably longer and more complicated, railroad damage in the Cajon Pass should be repaired within two weeks after the event (Byers, 2008; oral communication G. Hicks, 2008). Repairs to rail shipping lines will be a top priority. Union Pacific (UP) and Burlington Northern Santa Fe (BNSF) will cooperate to complete rail repair as effectively, as possible, and the trains will resume with some speed restrictions. Additional information acquired from Bill Byers (personal communication) regarding rail repairs include:

- Tracks in the area subject to liquefaction could be restored to an acceptable alignment within a week, based on extent of damage and repair time for the Southern Peru Copper Corp. railroad after the June 23, 2001 earthquake. However, it will probably be necessary to bring in ballast and surface the tracks. This can be accomplished in about 12 hours after the area is accessible from one end while work is still going on at other locations. Liquefaction may not increase total time to open the line and, at worst, should not increase the time by more than 12 hours.
- After passage of the first train at 10 mph, speed would be initially restricted to 25 mph through the affected area. Since the speed of westbound freight trains is permanently restricted to 30 mph through about half of the area and 35 mph through the rest and eastbound freight trains would have their speed limited to about 20 mph by tonnage on the heavy grade, the speed restriction would have little effect on running time. Based on removal of speed restrictions following the 1999 Hector

Mine earthquake where the initial 25 mph authorized speed was increased to 40 mph after five days, trains should be moving normally about a week after the line is opened.

Restoration of the highway-bridge systems is discussed in Werner and others (2008). Pavement damage is repaired within one to two months. Bridge damage continues to block road segments and reduce capacity in the five damage zones for several months.

Warehousing is a critical component of the goods movement system. While there is little storage in the harbor, there are millions of square feet of warehousing in the region, mostly outside of Los Angeles. It is primarily of tilt-up construction which is highly susceptible to earthquake damage. Although a good warehouse and distribution system attracts importers to a port in the long term (L. Cottrill, personal communication, 2008), the warehouse system is not expected to be a limitation in the short term because freight can be loaded on and off without warehouses. (R Guss, personal communication, 2008). Most of the trans-loading occurs between the ports and downtown. Furthermore, the HAZUS estimates of 0.05% of warehouses (0.04% of sq. footage) completely damaged, and 1.8% of warehouses (1.5% of sq. footage) extensively damaged fall within the same range of the vacancy rate of 2.7%, 7% of the warehouses moderately damaged, and 78% undamaged. Warehousing can operate 24/7.

The truck and rail system is near capacity under normal conditions. Rail piggybacks (double-decked trains) are already operating close to maximum efficiency. Over time, system resiliency in the face of disruptive events is decreasing as the goods movement continues to increase.

Disruption to and Resiliency of the Goods Movement System

We discussed the effects on goods movement within three time periods: immediate, short and longer term,

I. IMMEDIATE TERM

Immediately following the ShakeOut earthquake, major concerns and response will be:

1. **Prioritization of goods movement for food, water, and medical supplies.** The region is not self-sufficient in food. Much of the food for the region is trucked in from San Joaquin Valley and Imperial County. Food will be trucked and railed in from north and south. Ports will be open. Prioritization of containers off the ships will be complicated by mixed freight as cargo is dispersed on multiple shipping lines and in thousands of containers.

2. **Demand change for commodities and the distribution of goods** (rapid relocation of offices and distribution of emergency supplies). Relief supplies may come in through the ports, I-5, and from San Diego. Equipment will carry in generators and transformers. The current demand for goods movement will be reduced.

3. **Reconfiguration of a new and dynamic distribution system.** Panel participants asked: Who will be in charge? Who will negotiate priorities and make decisions?

II. SHORT TERM—Resiliency and disruption of port operations

There will be a logistics crisis during the first one to two weeks after the earthquake because rail systems will be down, the I-15 and I-10 truck routes will be blocked, and there will be congestion through other north-south routes, especially along the I-5.

While the rail system is under restoration, the alternative modes of transportation are likely to include ramping up the airports, ship diversion, and trucking. It is argued below that the capacity of all of these potential substitutes and the capability to sort cargo is very limited, resulting in a backlog of ships that will take months to clear.

Airports: As an alternative to port arrivals, there will be over-flights via air freighters to domestic destinations beyond Southern California. Consequently, the airports will ramp up freight traffic of high valued items after basic emergency supplies are delivered to the region. Ontario Airport is underutilized with one-third spare capacity. Small airports such as March, San Bernardino, and Chino will be available. Los Angeles International (LAX) is close to operating at capacity, but freight could be brought in at night. Airport capacity will be limited by availability of freight planes, pilots, and air traffic controllers and their ability to get to work. Access to Palmdale and Victorville airports will be hampered by highway damage. However, air cargo tonnage is insignificant compared with sea-borne freight.[3] Furthermore, most waterborne cargo is bulky and not well suited or require a luxury cost to go airborne (D. Thiessen, personal communication, 2008). Consequently, air freighting is a limited alternative. Hangers could be used as storage.

Ship Diversion: Some freight from the ports will be diverted. The 10 day 2002 labor lockout led to business being lost from130 of 2,000 ship arrivals over a 4- month period. Port operations were not irreversibly suspended, but turnaround times were extended, and the same is likely to occur in the event of the ShakeOut earthquake (D. McKenna, personal communication, 2008). Non-essential and non-perishable consumer products will sit, although some could move by barge. Non-local commodities are more likely to be diverted, if possible. Perishables will be diverted to the Port of Oakland in Northern California, if possible. Ship Diversions will be limited by:

- **Lack of capacity at alternate sea ports**—Seattle/Tacoma will quickly become overwhelmed, having little reserve. Oakland is even smaller. For example, the Total Los Angeles/Long Beach (LA/LB) Port container import count for 2007 is 7.01Million. The total Oakland import count 2007 is 799K, or 11.4%of LA/LB import count. The total Seattle- Tacoma import count for 2007 is 1.51M or 21.5% of Los Angeles/Long Beach. The total Portland import count for 2007 is 104.7K or .015%. There is also possible diversion to Vancouver and Prince Rupert in Canada, however the story is similar relative to capacity. In addition, these ports all have commitments to their regular customers and have traditionally prioritized handling their cargo before diverted cargo from the San Pedro Ports. These alternative ports would soon be overwhelmed by any significant cargo diversion. Furthermore, it would probably take several weeks for shipping Companies to arrange for significant diversion to the Gulf or East Coast because of the "string" system of ships in their pipeline. East Coast Ports are not ready to accept the cargo via an all water Suez route. Mexico could become an option in a few years, but that remains to be seen (D. McKenna, personal communication 2008).

- **Vessel size limitations** —Large ships (exceeding 5,000 TEU or more than 106 feet wide), which have become the norm in Pacific Rim trade, cannot navigate through the Panama Canal. There are not that many alternatives for large, deep draft ships. The Panama Canal can not handle the quantity and size of the ships (D. Thiessen, personal communication 2008). There will be some ability to barge goods from ships offshore to various destinations along the coast. Some freight could be barged north to Port Hueneme and the San Francisco Bay area, and south to San Diego. The potential to move from rail shipment to back loading to a short sea shipping concept is not likely to happen in a significant way any time soon (D. McKenna, personal communication 2008).

- **Limited inland rail from other ports**—Neither Oakland nor ports in Mexico have the rail capacity to distribute additional freight over land.

Diversion is not a solution. After a major disaster, most cargo will either pile up in the docks in Los Angeles/Long Beach (LA/LB) or wait offshore (more likely) at anchor. The nation has become very dependent on the San Pedros Ports and the goods movement industry has built major distribution warehousing in southern California. The San Pedro ports alone handle something like 65% of total west coast trade (D. Thiessen, personal communication 2008).

Trucking. Although trucking dominates the movement of goods throughout the SCAG region, most of the trucking is intra-regional. The large volume of truck traffic will have to be absorbed. The current San Bernardino and Riverside share of the trucking pie is even greater than indicated on the 2003 pie chart (R. Guss, personal communication 2008). There is a concentration of long haul trucking out of San Bernardino. Of the three major truck routes reported in the action plan (I-5 from LA/LB Harbors to Victorville (I15) and the I-15 to Mexico) only two of the 12 route bundles use the I-5 which carries about 25% of the trucks. The east-west I-10 truck route will have suffered severe damage.

- **All out-bound highway routes will be utilized.** Some limitations on restricted routes may be lifted (through Malibu, for example), although many alternative routes will not be *able to handle the size and weight of* the larger freighter haulers. Trucking will shift temporarily to the *south to access eastbound* routes, and freight could move on the US-101 to Oxnard. It is likely, that truck routes will operate on designated routes 24/7, or at night. There will be exemptions for hours of service.

- **Truck dispatching, which depends on the Internet, will be hampered.** If cell phones and Nextel systems are overwhelmed, hand-held communication will be used where only one person can talk at a time.

Trucking will be a limited solution to moving port imports out of the region while rail is inoperative.

Sorting cargo: All three alternatives—air, diversion and trucking—are further complicated by the challenge of sorting freight. There are estimates that only 50 to 60% of cargo is consumed within the region and the rest is "discretionary" cargo headed outside the Los Angeles Basin. It is unlikely that regional and out-of-region cargo arrives on separate ships. The location of goods on the ships may be unknown. For example, cameras and specific items sat out in the harbor during the lockout because their location was unknown. The sorting of the containers happens at the terminals. A system for separating local and national cargo for this earthquake scenario does not currently exist (D.McKenna, personal communication 2008).

> In an emergency, shipments can be broken up among the various ports and cargo prioritized. Standard lease penalties for failure to move freight would be waived. These actions will not eliminate all the ships waiting to get enter the harbor and there will be a back up of several weeks to gain a berth space. Proprietary terminals operated by international and competitive companies may not cooperate with collective priorities unless they receive some financial compensation. Berth specialization will also be an issue. Some shipping consortia (the Grand Alliance, for example) may be more flexible.

Clearing the backlog: While physical damage to port infrastructure may be minimal or non-existent, disruption of the larger transportation system, rail and highway, will be very disruptive and severely limit the movement of goods out of the harbor area to shipping destinations both near and far away for two weeks. The inability to move goods via truck and train through the region to distribution warehouses will back up goods movement at the ports, which have only about one week of storage. After the 2002 West Coast lockout, shipments took months to clear and 350 ships were out in the bay. Space and labor constraints limited the ability to work down the backlog. While the labor and terminal space availability is better today (in 2008), the volume of cargo has grown. So, we could expect similar delays if there was a major disruption to the inland transportation network (D. McKenna, personal communication 2008).

a) Supply Chain Effects of Port Disruption

Every day port operations are down, multiple days are required to recover. China trade shipments are huge, comprised of 8,000 to 10,000 TEUs, including consumer goods that are time sensitive and/or perishable.

A one-week closure would most likely result in an inventory adjustment. (The 2002 September/October 10-day west coast labor lockout disrputed over $6 billion in trade through the San Pedro Bay ports and required an inventory adjustment. While the lockout did not cause a major diversion of shipping, it required a month to absorb the backlog).

A one-month closure (not the case for the ShakeOut Scenario) will require more significant shipment diversion to other places (L. Cottrill, personal communication 2008). Ships that are most likely to divert are those carrying time-sensitive consumer goods. Some pharmaceuticals can go through airports. Penalties would be waived for failure to move freight. Downtime of 30 to 45 days would have severe impacts, but as noted above a 2 week significant disruption is more likely following this earthquake.

The ports are sensitive to the season and the shipment press begins in February and reaches its peak September through October as goods come in for the holiday shopping season. Low activity is December, January, and February. The November ShakeOut Scenario is on the tail end of the busy period.

Likely effects on the supply chain involve:,

- ***Crude oil supply to other State.*** Half of the crude oil comes in over the San Pedro Bay wharves—one million barrels per day. The product goes to Phoenix and Las Vegas. Considering the limited tank storage available, it would not take much to disrupt supply (L Cottrill, personal communication 2008).

- ***Industries and large companies.*** Just-in time industries will be vulnerable. Although, given the current delays of an already congested system, some industries are operating as just-in-case. Some industries will shut their doors. For example, Honda, General Motors, Toyota and Mitsubishi temporarily shut down auto plants due to the 2002 labor lockout because of parts shortages. Honda reported an $83 million loss in wages, shipping fees, and other costs. After the ports reopened, the automakers worked overtime to catch up. They also had to airfreight parts from Japan, or divert cargo through Canada and Mexico. Their suppliers also idled workers. Final assembly industries requiring many parts from many sources (for example, air conditioners) will be vulnerable. Orange County high-tech may be affected by supply chain and freight costs.

- ***Small businesses.*** Experience from the labor lockout suggests that small business may be within days of going out of business, or go out of business if ships are backed up for more than two weeks. Small businesses that are dependent on a shipment that cannot get unloaded are vulnerable. Businesses that have already spent holiday advertising budget may be affected if they cannot get their goods.

- ***Lower value industries*** such as the Agricultural sector may be affected by high costs of transporting produce by alternative and more expensive means. Perishable goods, such as fruit, fish, and meats may spoil if they cannot get to market. For example, during the labor 2002 lockout produce exporters were forced to dump stranded fruits and vegetables on the domestic market at lower prices. Dole Food Co. reported a loss of $250,000 to $500,000 as a result of perished foods.

- ***Shipping lines.*** Shipping lines will incur costs and lose business. For example, during the 2002 labor lockout, shipping lines incurred extra expense for crews, fuel, and capital costs spent on vessels idled by the lockout and not producing revenue. Container Shipping cited one estimate that put the losses for ocean carriers at $400 million to $600 million. Once the ports reopened, long shipping queues, major traffic congestion and gridlocked railways forced shipping lines to cancel Far East shipments and decline new bookings.

- ***Air freight.*** The airline industry will pick up some high-value cargo and those willing to pay higher freight rates in critical industries. The ability to fill the void will be limited by the overall capacity of the air freight industry in comparison to the ocean shipping and trucking businesses.

- ***Warehousing and logistics.*** The logistics industry as a whole will be affected (especially wholesaling and trucking).

b) Impacts on Local, Regional, State, and National Economies

(ii) Port-Related Employment Impacts

In 2006, direct international trade created 35,000 additional jobs, moving the annual average employment to 485,100 workers. Many of these jobs tend to be high- wage, and are found in a wide variety of activities, including vessel operation, services to vessels, cargo handling, surface transportation (rail and truck), air cargo, trade finance, freight forwarding, customs brokers, insurance, and government agencies (Econ. Trade Trends, 2007).

Will workers be able to get to their jobs in the face of damaged streets and highways? Will there be demographic impacts in the construction/repair labor force as happened following Hurricane Katrina with workers arriving from Mexico and Central America?

The following implications for labor were suggested by the goods movement panel:

- Labor is expected to be available at the ports. Not many port employees live inland and would be affected by expected highway delays. Port gangs will be temporarily out of work however, if cargo cannot move off the docks due to supply chain disruption.
- There will be a short term job loss in the warehouse districts. Distribution and warehouse labor needs will drop.
- Pilots and air traffic controllers will be in greater demand. More air traffic controllers and pilots will be needed.
- Wages will go up and may attract more skilled people to the logistics sector, which may ultimately improve the logistics workforce.
- Following the 1994 Northridge earthquake, the City of Los Angeles passed legislation requiring debris removal contracts be awarded to local vendors to keep windfall revenues within the community (Ad Hoc Committeee, 1995).
- People may move out of the stricken area for the short term, but native Californians are expected to stick around as they are more accustomed to earthquakes and aftershocks. Southern California is a transient society. Non-natives may migrate out, while foreign labor may migrate in, depending on the types of jobs available during recovery and reconstruction.

(iii) Local Communities

Communities in San Bernardino County will be subject to economic output and income losses due to industrial sector concentration in long-haul trucking.

Torrance has many logistics companies, economic activity is mostly related to LAX. Otherwise, Torrance is a bedroom and working community. Honda and Toyota are present. Other communities may also be affected, but time has not allowed further investigation.

(iv) Eight-county region

Airports and international trade contribute to 13% of SCAG's six-county economy, but overall, the event is not expected to have a major impact on this sector contribution. The economic output loss from disrupting goods movement through the ports in the eight-county region is evaluated by the regional input-output model in the section D of this chapter, using the following "reasonable" port disruption input for the first 7 weeks after the earthquake occurs:

0-3 days: The ports operate at 0% of capacity (*no power, general chaos*)

4 days-2 weeks: The ports operate at 10% of capacity (*no rail, limited alternatives*)

2-7 weeks: The ports operate at 50%-100% of capacity, ramping up from 50% to 100% (*rail starts back up, more highways open up, ports can start to clear the back log*)

85% of the difference between demand and supply during the 7 weeks is recaptured.

15% of the economic activity is lost to ship diversions, perished products, cancelled Far East shipments and declined bookings.

The input relies heavily on the experience of the 2002 labor lockout, especially the congestion period that followed because the ports are not completely shut down for 10 days. The input for the input-output analysis was also informed by the percentage and types of freight moving through the San Pedro Ports of entry and exit, and between the seaport and the region, in particular. Examples of the data for this input are in Tables 7- 18A and 7-18B, compiled by Ben Sherrouse and David Hester from the U.S. Department of Transportation (DOT), Federal Highway Administration (FHWA), Freight Analysis Framework (FAF) Version 2.2, produced in cooperation with the Bureau of Transportation Statistics (BTS) and contractors. It includes the Commodity Origin- Destination database which estimates commodity flows and related freight transportation activity among states, sub-state regions (particularly Metropolitan Statistical Areas (MSA)), and major international gateways. Sherrouse and Hester used the 2006 provisional database. FAF data should be handled with precaution, but there are plans to make FAF a more effective tool for analyzing freight transportation (U.S. DOT, 2004).

Table 7-18A. The Volume and Value of Commodity Flows by Truck That Originated in the Los Angeles-Long Beach-Riverside Metropolitan Statistical Area and Arrived at a U.S. Sea Port for International Shipping in 2006

Sea Port Commodity Flows Departing by Truck from Los Angeles-Long Beach-Riverside, CA in 2006		
Commodity	**Tons**	**Value ($ millions)**
Animal feed	154,091	$926.11
Base metals	483,009	$580.58
Cereal grains	2,510	$46.79
Chemical prods.	6,399,340	$2,784.29
Coal	180	$7.76
Coal-n.e.c.	157,055	$2,078.28
Crude petroleum	1,560	$5.19
Fertilizers	3,824	$27.40
Gasoline	12,317	$46.33
Live animals/fish	103,750	$84.87
Logs	4,789	$31.44
Machinery	9,656,430	$1,243.86
Metallic ores	117,370	$735.53
Mixed freight	1,147,795	$502.69
Natural sands	100	$18.10
Newsprint/paper	821,992	$827.80
Nonmetal min. prods.	11,455	$81.95
Nonmetallic minerals	1,901	$54.33
Other ag prods.	2,338,833	$3,087.71
Paper articles	421,033	$257.08
Waste/scrap	25,856	$167.11
Wood prods.	4,869	$9.63
TOTAL	21,870,057	$13,604.82

(v) State of California

The California map of San Pedro Port dependence from the BST report (2007, figure 7-22) is a good illustration of how the port-related economy looks in California, with a high-tech industry concentration in the Silicon Valley area. Oakland dominates in exports (especially, in agriculture), exporting more tonnage. Northern California generally is more dependent on air transport.

(vi) National

USGS analysis of the Bureau of Labor Statistics Location Quotient illustrates how the activity of the transportation sector is average compared to the rest of the Nation, and compared to California. But, the location quotient highlights the concentration of economic activity related to the ports and warehouse and distribution in the region. (Champion and Wein, 2007).

Table 7-18B. The Volume and Value of Commodity Flows That Were Shipped to a U.S. Sea Port from an International Origin and Arrived by Truck in the Los Angeles-Long Beach- Riverside Metropolitan Statistical Area in 2006

Sea Port Commodity Flows Arriving by Truck in Los Angeles-Long Beach-Riverside, CA in 2006		
Commodity	**Tons**	**Value ($ millions)**
Animal feed	17,686	$79.21
Base metals	3,111,217	$4,464.60
Cereal grains	16,600	$171.91
Chemical prods.	7,746,684	$3,059.06
Coal	20	$1.34
Coal-n.e.c.	101,993	$1,164.80
Crude petroleum	933,389	$3,105.58
Fertilizers	20,143	$105.60
Gasoline	415,398	$1,604.81
Live animals/fish	712,693	$590.17
Logs	32,065	$306.03
Machinery	131,811,859	$21,874.65
Metallic ores	1,656	$1.58
Mixed freight	4,150,519	$1,732.69
Natural sands	20,090	$1,988.63
Newsprint/paper	16,331	$15.95
Nonmetal min. prods.	535,599	$3,992.05
Nonmetallic minerals	1,391	$23.80
Other ag prods.	1,156,190	$2,879.50
Paper articles	1,120,657	$872.46
Waste/scrap	7,608	$36.65
Wood prods.	97,819	$244.49
TOTAL	152,027,605	$48,315.57

The national significance became evident during the 2002 labor lockout. Economic impacts from the lockout were widely felt across the United States:

"Major U.S. manufacturers that export through the west coast ports were significantly impacted, such as: General Electric in New York, West Virginia and Indiana; Caterpillar Tractor and Mitsubishi Motors in Illinois; Bose Corporation in Massachusetts; Cascade Agricultural Trading in Washington; and Nissan Motors and Kimberly Clark in Tennessee."

As a result of the shutdown, most people now better understand their own connection to the San Pedro Ports, the importance of moving freight, and the resulting ripple effect through the economy. A study by the Los Angeles County Economic Development Corporation estimated that the combined 10-day lockout and 23-day backlog disrupted trade valued at $6.28 billion just at the Ports of Long Beach/Los Angeles. In comparison to the 10-day West Coast lockout, which many economists estimated cost the U.S. economy $1 billion per day, the recent Hollywood writers' strike had mostly local impacts, and cost $2-3 billion for about 90 days, or a relative impact of 1/30th the scale of the Ports lockout. SCAG has been arguing that the region's transportation system qualifies for Federal funds because of the national value and benefits of the transportation corridors through these areas. Ocean shipping costs have dropped dramatically since 1989. This is a factor in "off-shoring" of manufacturing jobs along with wage rates. More intermediate goods are being produced in low-cost producer countries requiring less labor input in the United States. The low cost of shipping has resulted in a goods movement system such that raw materials, parts, assembled products, and customization of a final product may cross the ocean several times. Off-shore manufacturing has resulted in double and triple handling of materials and parts before final assembly; therefore, there are often no substitute manufacturers for the finished product within the United States. The import- export ratio at the Port of Los Angeles and Long Beach is three to one.

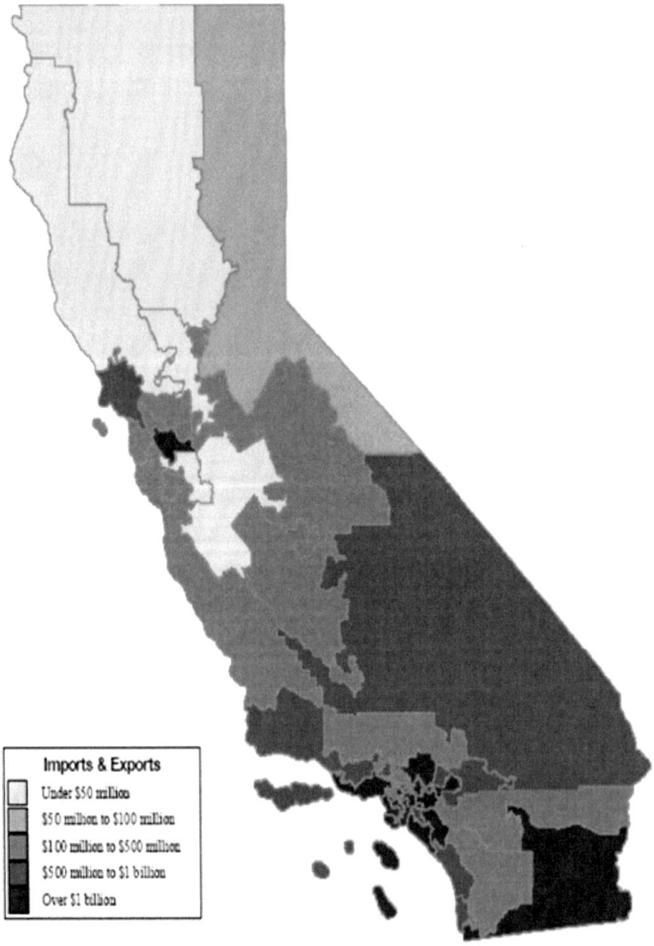

Source BST Associates (2007).

Figure 7-22. Value of all trade moving through the San Pedros ports by state senate district

(vii) International

The 23-day backlog following the 10-day, 2002 West Coast lockout illustrated how sensitive the global supply chain is to disruptions in the movement of cargo at just one node in the network (in this case, ports), which has the same effect as severe terminal, highway or railway capacity constraints. Transport delays impact the cost of doing business, the environment, and our nation's ability to compete internationally. (SoCal MTSAC, 2003)

III. LONGER TERM

The earthquake event will, most likely, cause temporary diversion of shipping rather than a permanent change. In the long term, port operators do not expect the San Pedro ports to lose large-scale activity like the Port of Kobe, Japan. Following the Kobe earthquake, the port activity dropped from the top 5 to the top 30 in the world because it is an island nation in an area with many alternative ports; China and Korea were able to pick up capacity. The continent of North America is different because the alternatives are limited by location, as well as highway and rail capability. Los Angeles/Long Beach is the fifth busiest port in the world, the busiest port in the United States, (followed by New York/New Jersey which is a distant second place) and has 6 times more activity than Oakland. Oakland and Seattle cannot readily bring substitute capacity on-line. Houston has been growing, but the ports are smaller. Vancouver, B.C., is small and, like other ports in the US and Canada, has environmental impact concerns about continued growth. Experience from the labor lockout demonstrated that the "swallows returned quickly" to Southern California as that is where the cargo wants to flow.

If the Suez Canal proved to be an alternative this could result in a permanent change to shipping more to the East Coast: an outcome that would be welcomed by some in the region; San Bernardino and Riverside are overly burdened already.

Warehouses in Southern California serve the state and the entire U.S. Each retailer has a scheme for distribution. Their ability to distribute depends on the number and location of gateways to which they have access. Practically, every retailer has a warehouse in the region. Ports capture market share, in part, by promoting local services. They attract customers by demonstrating the availability of warehousing services. Repair of earthquake damaged warehouses is critical activity for the ports in the longer term.

There is the possibility that reconstitution of the logistics sector could create new efficiencies for the movement of goods and the event may increase in wage rates, may attract better people and help to improve the efficiency of the logistics sector

Panel Recommendations and Insights

1. There will be management issues:
   a. Management of ships including redirection to other ports as well as offloading strategies for acceptance at Los Angeles.
   b. Prioritization of goods movement day and night.
   c. Need interagency coordination in advance:
      i. Pipelines are handled by LA City, but only for the City. Who handles for other jurisdictions?
      ii. Interjurisdictional traffic management.

iii. Food and water supply.
iv. Need better information on the availability of natural gas, given storage challenges, except for natural gas bladders. Buses and ground support at airports run on natural gas so there is no storage capacity to keep them moving following the earthquake.

3. Cooperation could be anticipated. Examples:
   a. An interagency task force comprised of Federal Department of Transportation (DOT), CalTrans, City of Los Angeles DOT, and the Governor's Office of Emergency Services was established to expedite repairs to the damaged freeway system following 1994 Northridge earthquake (Office of Chief Legislative Analyst, 1995). This task force should be reconstituted to include Los Angeles County as well as San Bernardino and Riverside cities and counties.
   b. A helpful group will be the Coast Guard and FMI, the area maritime security committee that is tied to police and OES. Captain Wheaten and Chris Hogan do port evacuations for terrorist threats. Industry is also participating.
   c. California Trucking Association should be part of the response and step up to move food and water, and provide emergency dispatch centers. Cal Trucking should also partner with the American Trucking Association in this effort.
   d. Because port unloading is managed by individual companies and is competitive, interorganizational agreements will be necessary for guidance during the recovery period. It is possible that State or Federal intervention may be necessary to ensure such agreements are in place quickly and appropriately.
   e. At many locations, railroads have maintenance roads on their rights-of-way. Use of these roads for emergency repairs could probably be arranged between a utility and the railroad, with some restrictions related to possible liability and interference with the railroad's repair, or other, operations. Before allowing utility use, the railroad would probably require a railroad protective insurance policy, as most general liability policies are not valid within 50 feet of an operational railroad track. It would be well for utilities that might want such access to explore the matter with the railroad in advance to determine requirements and limitations. Access on the railroad's roadbed would probably not be possible. (Bill Byers)
3. Although Federal response will come in many forms, including FEMA reimbursements, local agencies will need to be ready to be on their own and to be prepared with contingency plans.
4. Contingency planning including prioritization must be encouraged at all levels of government, as well as within business and residential communities.
5. Look for technological increases to resiliency: A developing capability in the port is marine power ships that can reverse power back into the grid and restore power in the port. This is a proven possibility. Manufacturing could use technology to manufacture more directly.
6. "Just in case" inventories: 3-day supplies of food and water will be inadequate. If everyone was prepared with 2 weeks' worth of emergency supplies it would greatly relieve the logistics of food and water delivery.
7. Voluntary evacuation from the area could reduce consumer demand on limited resources.

8. Publicize the ShakeOut Scenario findings; decision makers at all levels need to know what might happen.
9. Address the problem of money being poorly spent. Pre-think priorities and budget for them
10. Get a mountain bike with a basket, store enough water and food for more than three weeks, and consider raising a vegetable garden in your backyard.

D. Regional Economic Impacts of the ShakeOut Scenario Earthquake

The eight-county southern California region hosts a trillion dollar economy in terms of Gross Regional Product and more than $1.6 trillion in terms of Gross Output (roughly approximate to total sales revenue). We would expect that the ShakeOut Scenario earthquake will cause billions of dollars worth of property damage and business interruption (BI) losses. However, a key question is: how significant is this in relation to the large size of the regional economy as a whole?

This section summarizes the regional economic analysis of the impacts of the ShakeOut Scenario. The basic modeling approach is input-output (I-O) analysis, the most widely used tool to perform regional economic impact studies (Rose and Miernyk, 1989; 2004). Essentially, this is a detailed, comprehensive, double-entry bookkeeping record of all production activity. Almost every country in the world has constructed an input-output table, usually through an exhaustive census or at least an extensive survey. There is a rich literature on ways to use non-survey data-reduction, or "down-scaling," methods to generate tables for political jurisdictions at various sub-national levels. Our model is based on a solid set of data and is disaggregated to reveal details for 26 economic sectors of the southern California region. See tables 7-19A and 7-19B for the I-O table constructed for this study.

The sectors of the economy are labeled in the left-hand column of Table 7-19A and 7-19B as sellers of goods and services, and labeled at the top of the table, in the same order, as buyers. Looking across row 3, we see a tabulation of the dollar value of the services of Utilities sold to every other sector. Manufacturing and Real Estate and Rental are the two main business users of utilities, but note the largest single category of buyers is Households. The Utilities column shows the dollar value of various inputs needed to produce this valuable good.

Interdependence among the economic sectors is readily portrayed in this double- entry bookkeeping tabulation. Also, we can formally measure it by identifying and calculating backward and forward linkages between all sectors. If a sector affected by the electricity outage reduces its production by 25 percent, we say this is a direct business interruption (BI) effect. (Note also that this result can take place even if the factory property is unscathed by an earthquake, as long as its lifeline service is disrupted.) Because the factory then reduces its order for each input by 25 percent, the firms producing those inputs in turn will do the same, as will their suppliers, and so on, as the original perturbation ripples through the economy. The sum total of these ripples is some multiple of the original shock; hence, the origin of the term "multiplier" effect.

I-O models provide a great deal of basic information, insight, and computational ability. They also have significant limitations, such as linearity, absence of behavioral considerations, absence of markets and prices, and lack of formal constraints. Still, I-O models are useful in providing ball-park estimates of very short-run responses to property losses and infrastructures disruptions.

This analysis utilizes the results of several other researchers and professionals who have collaborated to perform loss estimates for the various aspects of the built environment, including ordinary buildings, high-rises, and various types of infrastructure services (electricity, water, gas, transportation including the ports). These loss estimates are described in the previous section regarding earthquake shock inputs for the regional economic model. Together, we have translated property damage estimates into BI. We then applied adjustments, in the form of business recapture factors, to the initial estimates to account for resilience. Resilience refers to the ability of businesses to mute the potential losses by various types of standard and adaptive behavior, such as conservation of scarce inputs, relocation, and recapturing lost production at a later date, as well as hastening recovery (Rose, 2007; Rose and others, 2007).

How significant is resilience in practice? Rose and Lim (2002), in a study of the aftermath of the 1994 Northridge earthquake, inferred direct business resilience of 77.1 percent to electricity disruptions. Various studies by Adam Rose using both I-O and CGE models in cases of utility service disruptions (water in the aftermath of a hypothetical earthquake in Portland, Oregon, and electricity disruption in a hypothetical New Madrid earthquake and the Northridge earthquake in Los Angeles) found direct business resilience to be between 85 % and 95 %. These same studies found market resilience to range between 50 % and 80 %. This means that omission of resilience factors can lead to an overestimate of BI losses by factors as high as five to ten. The recapture factors used in this study are time dependent and range from 0% (for loss of owner-occupied dwellings) to 98% (for manufacturing processes) of output losses within the first 3 months. The recapture factors decline for extended periods of output loss.

Table 7-20 displays the absolute and percent output loss estimates by industrial sector. In absolute terms, the sectors most affected are Manufacturing (especially Food & Chemicals), Professional, Scientific and Technical Services, Entertainment and Recreation, Wholesale and Retail Trade, and Owner-Occupied Dwellings. These sector losses can largely be explained by five key factors: the size of sector (in terms of output), the extent to which other sectors are dependent on it directly and indirectly (multiplier impacts on the sector), the extent the sector's buildings suffer fire damage, their reliance on water, and/or the recapture factor. The payroll pie chart, figure 7-2, indicates that the output from Professional, Scientific, and Technical services and Wholesale and Retail Trade is relatively high. Figure 7-12 of fire damage by sector reveals the bigger hit to Professional and Technical Services, Entertainment and Recreation (through restaurants) and Residential Building occupancies. The importance of water is particularly high for Food and Chemicals and for Entertainment and Recreation. In addition, the recapture factor is lower for Entertainment and Recreation and zero for Owner-Occupied Buildings.

The sectors most sensitive to the earthquake shocks in percentage terms are Hotels, Water Utilities, Owner-Occupied Dwellings, and Mining, Minerals Processing, and Metals Manufacturing. Sector sensitivity is strongly influenced by the sector's reliance on water and the recapture factor (lower for Hotels and zero for Owner-Occupied dwellings).

Table 7-19A. Southern California Eight-County Input-Output Table in millions of 2004 Dollars. Sectors Aggregated to Two-Digit NAICS codes (Continued in Next Table)

Sector	Description (with 2-digit NAICS code)	1	2	3	4	5	6	7	8	9	10	11	12
1	11 Ag. Forestry, Fish & Hunting	848	0	0	40	3,383	1	8	2	1	0	54	3
2	21 Mining	3	322	1,595	126	3,688	5	31	3	10	1	7	8
3	22 Utilities	68	89	27	189	3,225	456	128	946	387	210	2,044	437
4	23 Construction	22	1	411	56	510	174	103	309	440	225	1,286	293
5	31-33 Manufacturing	1,092	724	762	15,086	79,562	2,445	3,433	1,856	5,417	457	1,880	3,146
6	42 Wholesale Trade	240	106	139	2,177	17,095	2,047	778	400	1,406	109	290	559
7	48-49 Transportation & Warehousing	91	93	1,392	946	6,165	1,290	2,605	1,254	1,841	762	768	1,527
8	44-45 Retail trade	10	19	49	6,686	856	662	301	1,128	287	104	1,145	654
9	51 Information	26	36	118	863	3,164	1,258	424	1,310	15,742	1,187	1,327	3,677
10	52 Finance & insurance	89	89	207	935	3,328	1,158	891	1,480	1,278	17,358	2,810	1,858
11	53 Real estate & rental	243	282	91	780	2,374	1,703	785	3,514	2,116	2,195	5,980	4,052
12	54 Professional- scientific & tech svcs	106	195	550	3,441	0,927	3,007	1,012	3,503	6,730	3,394	3,899	0,073
13	55 Management of companies	4	237	12	79	7,035	1,605	207	3,954	473	558	221	363
14	56 Administrative & waste services	9	21	149	759	1,539	2,063	1,131	1,714	1,828	934	5,005	4,798
15	61 Educational services		2	71	9	211	89	28	58	199	35	40	132
16	62 Health & social services							8	0		0		12
17	71 Arts, entertainment & recreation	3	59	7	39	243	92	11	72	737	97	131	428
10	72 Accommodation & food services	3	6	129	400	1,226	300	351	446	451	644	696	1,464
19	81 Other services	56	11	36	679	3,238	687	369	947	1,211	309	827	819
20	92 Government & non NAICs	28	38	52	156	2,081	358	275	443	685	471	1,058	399
Total imports		2,174	1,880	6,841	10,806	71,026	5,482	4,441	8,630	17,905	11,341	8,205	10,071
Total Value added		7,088	5,887	14,776	47,372	80,762	54,245	25,704	55,224	64,305	63,878	78,974	66,967
Grand Total		11,000	10,021	24,381	91,271	380,823	78,396	42,936	84,452	120,801	104,413	116,368	100,607

Table 7-19B. Southern California 8-County Input-Output Table (continued) in Millions of 2004 Dollars. Sectors Aggregated to 2-digit NAICS Codes

Sector	Description (with 2-digit NAICS code)	13	14	15	16	17	18	19	20	Household	Other Final Demand	Grand Total
1	11 Ag, Forestry, Fish & Hunting		137	0	10	13	189	8	8	1,089	8,423	11,908
2	21 Mining	1	3	1	11	1	13	10	184	139	3,858	10,021
3	22 Utilities	358	271	50	412	279	858	447	372	18,347	2,982	24,381
4	23 Construction	248	71	204	234	174	236	350	1,988	0	83,959	91,271
5	31-33 Manufacturing	527	1,772	335	5,898	712	8,780	5,271	3,048	72,276	97,848	309,823
6	42 Wholesale Trade	85	483	90	970	142	1,489	870	557	23,335	25,158	79,358
7	48-49 Transportation & Warehousing	43	575	95	808	186	446	554	378	9,026	12,895	42,936
8	44-45 Retail trade	1	1,188	18	357	118	424	803	1,288	58,452	9,935	84,452
9	51 Information	1,297	875	285	1,128	419	638	931	300	12,241	72,782	120,001
10	52 Finance & Insurance	48	684	168	1,173	376	810	637	3,001	31,653	36,288	104,413
11	53 Real estate & rental	1,298	899	1,054	3,011	983	1,429	2,743	1,420	18,048	83,413	118,388
12	54 Professional,scientific & tech svcs	2,870	1,592	285	2,059	898	994	1,389	1,880	7,058	48,089	109,897
13	55 Management of companies		629	19	578	210	128	312	5	0	9,015	25,245
14	56 Administrative & waste services	31	1,844	299	2,345	524	477	1,294	1,017	2,418	14,570	44,701
15	61 Educational svcs		21	65	95	41	8	40	7	8,256	1,051	10,449
16	62 Health & social services		5	5	548	8	0	9		78,495	321	77,418
17	71 Arts, entertainment & recreation	10	82	28	60	1,493	172	128	10	8,559	13,323	25,779
18	72 Accommodation & food services	2	410	35	920	79	308	225	28	28,715	7,988	45,580
19	81 Other services	421	610	98	411	289	314	523	867	25,806	12,936	51,059
20	92 Government & non NAICs	171	170	34	247	130	313	248	292	72,966	92,770	173,379
Total Imports		2,180	3,772	1,063	7,164	2,002	6,779	6,800	6,198	100,887	62,704	(366,881)
Total Value-added		15,855	28,880	8,224	49,048	18,874	24,143	28,815	151,742	0	0	(889,025)
Grand Total		25,245	44,701	10,449	77,410	25,779	45,580	51,059	173,379	(754,350)	(1,030,249)	1,558,298

Table 7-20. Sectoral Output Losses from the Various Sources (with Recapture), in Millions of 2008 Dollars

Sector	Buildings	High-Rise	Fire	Power	Water	Gas	Transportation	Ports	Total	Total %
Agriculture	7	2	23	20	443	1	3	16	515	3.91
Construction	712	18	710	72	1,783	8	5	49	3,357	3.68
Food, Drugs & Chemicals	425	158	2,111	350	5,851	25	33	119	9,072	8.64
Mining & Metals/Minerals Processing & Mft.	56	24	407	58	1,349	18	5	36	1,954	10.78
High Technology	23	8	174	20	463	1	2	22	712	5.44
Other Heavy Industry	232	48	1,249	127	3,639	9	12	126	5,442	5.84
Other Light Industry	234	69	1,386	157	3,205	9	14	103	5,177	5.31
Air Transportation	15	16	189	35	226	1	4	3	488	7.48
Rail Transportation	6	6	41	12	109	0	1	2	178	9.83
Water Transportation	3	3	29	5	38	0	1	11	90	9.62
Highway & Light Rail Transportation	76	83	716	158	1,248	4	35	18	2,340	7.49
Electric Utilities	42	35	108	101	708	5	5	14	1,016	10.15
Gas Utilities	34	39	99	73	1,021	89	5	21	1,382	8.57
Water Utilities	1	1	3	1	41	0	0	0	47	15.97
Wholesale Trade	380	83	825	288	2,470	12	24	49	4,131	5.21
Retail Trade	431	127	914	364	2,401	21	47	40	4,344	5.14
Banks & Financial Institutions	89	37	279	101	652	6	7	11	1,182	4.63
Professional & Technical Services	1,085	720	5,647	1,050	6,268	73	82	120	15,045	4.67
Education Services	149	25	442	182	980	4	13	10	1,806	5.65
Health Services	1,349	429	905	509	3,215	17	30	43	6,498	9.16
Entertainment & Recreation	739	131	1 ,788	750	5,684	26	66	46	9,232	7.43
Hotels	249	368	63	50	456	2	4	3	1,196	16.37
Other Services	367	80	613	466	1,819	15	42	41	3,442	5.96
Gov't & Non-NAICS	193	430	1,177	232	1,506	11	15	33	3,597	3.90
Real Estate	618	95	808	1,254	2,885	202	43	24	5,928	6.08
Owner-occupied dwellings	533	121	1,733	913	4,567	253	17	37	8,173	12.27
Total	8,049	3,156	22,438	7,348	53,029	812	514	998	96,343	6.18

Three categories of direct economic impacts for the eight-county region as a whole are summarized in the first column of Table 7-21. These include property damage (direct and indirect stock loss) of more than $112 billion and BI (direct flow loss) of more than $54 billion and increased costs of doing business of more than $4 billion, for a total of $171.7 billion. These direct impacts are then fed into the I-O model to compute the indirect (or multiplier) flow effects that are added to the direct losses to obtain the figures in the second column of numbers in Table 7-21. Property damage and cost increases do not have multiplier effects, but BI multipliers increase the BI total to $96.2 billion, so that the total economic impacts are $213.3 billion.

To summarize:

- Property damage is only slightly greater than BI.
- Property damage from fire is 50% greater than property damage from shaking because fire is more devastating to building contents.
- BI from fire is greater than BI from shaking because fire repair takes longer than shaking repair .
- BI from water disruption is greater than 50% of total BI losses because of long duration of outage (4-6 mos. in heavily impacted areas).

Table 7-21. Total Regional Economic Impacts of Shake-Out Scenario Earthquake in Billions of 2008 Dollars

Indicator	Direct Impacts	Total Impacts
Building Damage	$32.7	$32.7
Related Content Damage	10.6	10.6
High Rise Building Damage	2.2	2.2
Related Content Damage	0.7	0.7
Fire Damage	40.0	40.0
Related Content Damage	25.0	25.0
Highway Damage	0.4	0.4
Pipeline (water, sewer, gas) Damage	1.1	1.1
Sub-total Property Damage	112.7	112.7
BI from Buildings	4.3	8.0
BI from High-Rise	1.7	3.2
BI from Fire	12.8	22.4
BI from Power	4.4	7.3
BI from Water	30.0	53.0
BI from Gas	0.6	0.8
BI from Transportation	0.3	0.5
BI from Ports	0.5	1.0
Sub-total BI (Flow)	54.6	96.2
Relocation Costs	0.1	0.1
Traffic Delay Costs	4.3	4.3
Sub-total Additional Costs	4.4	4.4
Total	$171.7	$213.3.

- Gas BI is low because it is not as extensively used in industrial and commercial sectors as are the other utilities; also, primarily residential.
- Port BI is low because of two assumptions: a fairly high recapture rate due to limited alternatives for ships to divert, and producers have sufficient inventories of goods to negate any adverse effects during the first 3 days.
- Total BI losses are 6% of annual gross output.

Even if we make a further adjustment of avoiding double-accounting of the several sources of business decline all at once, the total economic loss figure is still close to $200 billion. How significant is the figure in a relative sense? The $200 billion is more than three times the size of total losses estimated for the Northridge earthquake. Adjusting for ignored resilience factors (in the Northridge calculation) and 1994 dollars the total losses remain at about three times those of Northridge. The definition of a recession is two consecutive calendar quarters of negative economic growth. The earthquake will obviously cause this for the region. A higher standard is that the majority of U.S. recessions since World War II have been characterized by a 2% decline in economic output. The $96.2 billion gross output loss is a 6% loss in relation to Total Gross Output of the region and still well over 2.0% if we consider the longer period of impacts, a small proportion of which will linger past a single year. Also, the impacts fall far short of what might be characterized as a catastrophe—which we assume would be on the order of a decline of 10% or more. However, we should acknowledge some potentially significant matters of interpretation and omissions in the analysis. First, the percentage above represents output losses only and does not include the impact of the property damage, a factor over and above what is usually going on in a recession. A large portion of the property damage is likely to be recouped by insurance and outside aid. However, the diversion of capital from within the region from normal investment to repair and reconstruction would result in overall decreased output in subsequent years that we have not measured. Of course, some sectors, such as construction and its major suppliers will gain, while others will lose from this process. Also, in relation to "extended linkages" that may cause a non-linear downward spiral in the regional economy, systems linkages, or cascading failures, may make the region less inhabitable and less safe. Behavioral linkages, such as fear, may cause people to flee the region for an extended period. The ShakeOut Scenario could thus conceivably break the 10% threshold. It should also be kept in mind that a higher percentage of losses are likely in the most heavily impacted counties of San Bernardino, Riverside, and Los Angeles. Finally, even a 6% decline in employment represents employment for a couple of hundred thousand people in the eight-county region.

E. Insured Residential Losses by Richard Bernknopf, Anne Wein, Richard Champion, and Daniel Ponti

Introduction

The physical (structural and content) damages and business interruptions from rare events like large earthquakes are insurable. Insurance coverage is offered to residential and commercial property owners and tenants to protect themselves from wealth loss due to a damaging earthquake. Unless property owners take action before a disaster to protect themselves by mitigating or transferring some of the risk, they or the government will bear the brunt of earthquake losses. Losses resulting from earthquake- related shaking and ground deformation are covered by an earthquake insurance policy (Grossi and Kunreuther, 2005). This analysis of insurance coverage reports solely on residential earthquake insurance for the eight-county southern California region.

Consumers can transfer earthquake risk by purchasing earthquake insurance in advance. An individual's decision to purchase earthquake insurance requires a comparison of the benefits and costs of the hazard insurance with and without the potential loss (Kunreuther, 2004). A premium is paid to cover damage from a disaster for a prespecified period (that is, 1 year). The payoff to the insured is the loss minus the deductible that must be met before a claim can be paid.

Private insurers, reinsurers, and the California Earthquake Authority (CEA) are the key players in the California insurance market. Currently, the earthquake insurance market in southern California is split between CEA (74%) and non-CEA (26%) of the policies in the eight-county region (Richison, personal communication, 2007, Karaca, personal communication, 2008). The CEA, an agency run by the State of California, manages a fund that provides residential earthquake insurance coverage to homeowners and renters (Grossi and Kunreuther, 2005). Reinsurers are involved in the market because they provide protection to private insurers similar to the way that primary insurers provide coverage to residential and commercial property owners. In providing this backstop to their risk, the reinsurer charges a premium in return for providing an insurance company with funds to cover a stated portion of the losses it may sustain. Most insurers purchase reinsurance for covering earthquake losses to reduce their financial exposure.

Table 7-22. California Earthquake Authority Residential Earthquake Insurance Policies and Coverages

Residential occupancy class	Dwelling Coverage	Personal Property	Additional Living Expenses	Loss Assessment	Deductible	Increased Limits
Homeowner	Insured value stated on companion homeowner policy	$5,000	$1,500		10%-15%	$100,000 personal Property $15,000 additional living expenses and loss of use
Condominium	$25,000 building property			$25,000 or $50,000	10%-15%	$100,000 personal Property $15,000 additional living expenses and loss of use
Mobile home	Insured value stated on companion homeowner policy	$5,000	$1,500		10%-15%	$100,000 personal Property $15,000 additional living expenses and loss of use
Renter		$5,000	$1,500		10%-15%	$100,000 personal Property $15,000 additional living expenses and loss of use

Table 7-23. Residential Losses by Occupancy Class and Loss Category for the Region ($000,000)

Occupancy Class	Slight and Moderate Loss	Extensive and Complete Loss
Single family residence	$5,920.8	$3,392.4
Mobile home	$139.8	$1,867.9
Multifamily residence	$2,412.5	$2,944.6
Total	$8,473.0	$8,204.9

This summary contains a brief background on residential earthquake insurance in California, with a description, map, and tabular data about the distribution of residential property owners insured in the eight-county study region in and around Los Angeles. Residential damage estimates based on census tract HAZUS building and content loss output were used with zip code level insurance data to estimate the percent losses insured by county and for the region. Data limitations did not permit us to fully integrate CEA and non-CEA information for detailed occupancy categories (such as single family residences, multi-family dwellings and mobile homes). Some discontinuities in the data should be noted, such as (1) differences between HAZUS census tract and CEA / nonCEA zip code spatial units of analysis and (2) the dates of assembled damage data from the inventory of residential structures inflated to 2006 dollars in HAZUS and the latest data on insurance policies and exposures from CEA (2007) and non-CEA sources (2006).

Residential Earthquake Insurance in California

Recurring earthquake activity in California encouraged the state legislature to create the California Earthquake Authority to provide residential earthquake insurance and to share that responsibility with private insurers. In 1996, by act of the California Legislature, a reduced-coverage, residential catastrophic earthquake-insurance policy became available. The earthquake "mini-policy" protects a policyholder's dwelling—to provide a "roof over your head"—while excluding coverage for costly non-essential items, such as swimming pools, patios, and detached structures (some non-CEA policies provide for these non-essential items). The CEA policy is based on and authorized under the mini-policy law. Such policies are intended to help the policyholder avoid catastrophic loss while keeping premiums affordable for more consumers (CEA, 2008). Table 7-22 lists the current types of CEA policies and coverage by residential occupancy class.

In addition to the CEA, private insurers offer residential earthquake hazards policies that are outside the CEA policy. These private policies cover only homeowners and condominium owners.

Expected Residential Damage From the ShakeOut Scenario Earthquake

The residential building losses for the ShakeOut Scenario account for $17.44 billion or 53% of total building losses for the Scenario earthquake, because there is a very large number of residential units and square footage in the region. The distribution of the losses is across four states of loss as defined by HAZUS. Slight and moderate categories assume 10% or less loss while the extensive and complete categories assume greater than 10% loss. Table 7-22 lists the estimated loss for residential properties by occupancy class and loss category. It should be noted that for estimating insured losses only the extensive and complete loss column are included because they exceed the CEA and non-CEA deductibles.

Regional Residential Exposure, Policies, and Coverage

The insurance industry uses a metric known as the hazard insurance penetration rate to estimate how much insurance is in force at a point in time and how exposed a particular insurance provider could become in an event. The penetration rate is the proportion of single-family homes that have hazard insurance on an annual basis (Dixon and others, 2006). The penetration rate in this report is defined as the proportion of residential units that have earthquake insurance in the region. The proportion of earthquake policies is the sum of CEA (that is, homeowner, mobile home, and condominium, and renter (contents only), and non-CEA (that is, homeowner and condominium (structure and contents)) policies, a total of 723,486, divided by the total number of residential units summed over the number of units for all residential occupancy classes, a total of 5,030,485 in the HAZUS data set (Seligson, 2008). These residential occupancy classes include single family residences, mobile homes, and multifamily residences that may be owned or rented. There are more content policies than structure policies because CEA provides renter policies and non-CEA insurers provide separate contents policies. For the eight-county region, the residential unit penetration rate is estimated to be 14.4%.

For the ShakeOut Scenario, there were insured structural and content losses amounting to about 6% of the residential structural and content losses compared to the 14% of the residential units in the region that have earthquake insurance. Te CEA provides a renters' coverage also (see Table 7-23). The difference between percentages of the insured losses and total insured units is explained by:

- About equal losses split between the combined slight and moderate loss categories and the combined extensive and complete categories (in Table 7-22). Of total units, 16.5% suffer slight and moderate loss and fewer units (2% of total units) suffer extensive and complete loss. The remaining 81.5% of the units in the region suffer no loss. In the slight and moderate losses categories, the loss is the owner's responsibility because slight and moderate losses are assumed to be below deductibles such that half of the losses will be too small to file an insurance claim;
- The deductible; and
- Renter policies only cover contents.

Furthermore, the 6% insured loss is optimistic because it assumes that all those units suffering extensive and complete damage are insured.

An underlying feature of figure 7-24, the spatial distribution of insured losses, is that the percent insured loss is relatively greater in areas that are subject to the greatest damage. This is to be expected because locations in those zip codes, which, in most areas, are closer to the fault, have a greater proportion of extensive and complete damage to structures that, by definition, exceed the insurance deductible for both CEA and non-CEA policies. Furthermore, there is a 7.5% deductible for contents and specific coverage available for non-CEA policies. Both of these deductibles create a minimum threshold of damage that a dwelling must suffer to trigger financial mitigation by the insuring agent to the insured. These results raise several issues concerning the recovery of the region. Property owners and renters who suffer an uninsured loss will pursue compensation from the government in a variety of forms, most notably Individual Assistance as provided by the Stafford Act, and disaster loans provided by the Small Business Administration.

We have undertaken a "scenario based" modeling approach that employs a simulation of a large, damaging earthquake that provides the basis for assessment of the hazard, the physical damage to properties, and the cost of repairing and replacing structures to estimate the insured and uninsured loss. Insured loss is calculated from the structural damage estimates and insurance coverage terms, such as deductible, building contents, loss-of-use coverage, and coverage exclusions. Liquefaction and landslide damage have not been incorporated. It is also assumed that fire following earthquake damage is covered by a homeowners' policy (Scawthorn, 2008), although this assumption may no longer be applicable to all homeowners' policies.

The HAZUS and residential insurance data were intersected in a Geographic Information System (GIS). This intersection provides a "first cut" estimate of the level of insured losses for the ShakeOut Scenario. Table 7-24 lists the number of policies, estimated insured and uninsured residential building and content losses, and the percentage of insured losses by county and for the eight-county region. The level of property owner purchases of insurance policies, amounts of insured and uninsured losses, and percentage of insured losses vary widely across the counties in the region.

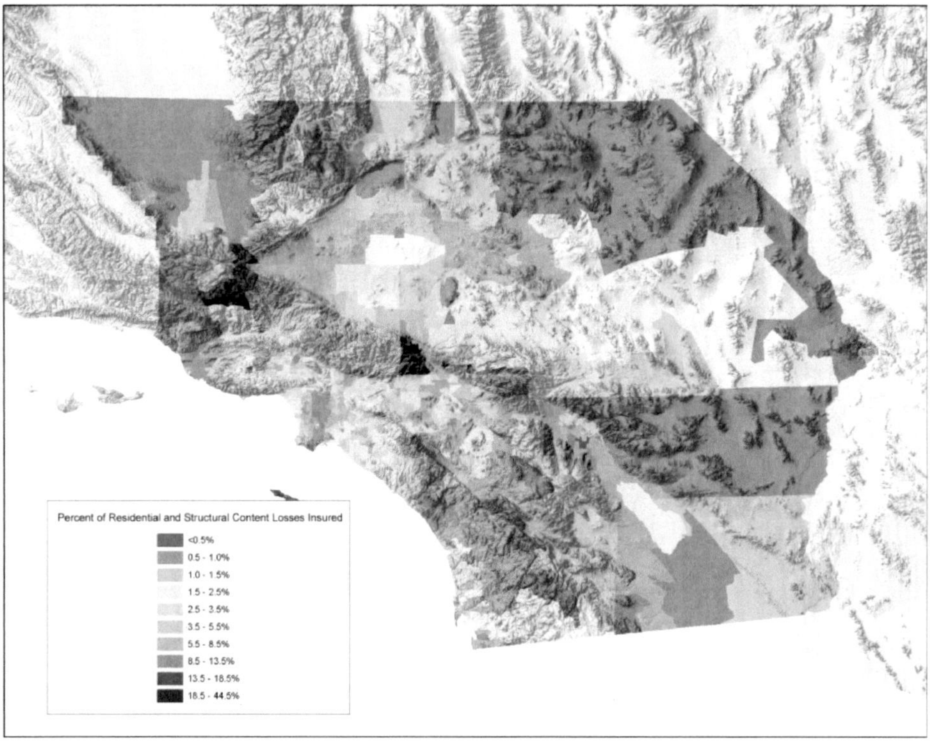

Figure 7-23. Percent of Insured Residential Structural and Content Losses by ZIP code

Table 7-24. Insured and Uninsured Residential (both structural and contents) Losses

County	# CEA & non-CEA policies	Insured Loss ($000,000)	Uninsured Loss ($000,000)	% Losses Insured
Imperial	378	$1.30	$165.75	0.78%
Kern	16,786	0.36	15.27	2.33
Los Angeles	317,191	345.13	6,313.97	5.18
Orange	124,214	21.00	877.34	2.34
Riverside	54,636	525.68	4,591.19	10.27
San Bernardino	46,760	387.77	7,660.46	4.82
San Diego	16,068	0.05	4.81	0.93
Ventura	37,453	0.66	20.65	3.07
Region	**723,486**	**$1,282**	**$19,649**	**6.12%**

F. Discussion

In light of research completed and numerous discussions held with ShakeOut colleagues and stakeholders, we reflect on the three objectives of the economic consequences portion of the ShakeOut Scenario.

Objective 1—Demonstrate the Transformation of Scientific Information into Economic Consequences

Estimation of regional economic losses from multiple sources of earthquake- related shocks is a novel and innovative contribution to planning. Collaboratively, we have expanded the horizons of modeling the economic losses from an earthquake event. The methodology has produced credible results and considerable community interest in both our results and our future efforts.

The process of transforming earthquake science outputs into seismic engineering inputs, then transforming engineering outputs to sociological and economic inputs, then transforming economic outputs to useful information for end users, has rarely been done across all these disciplines in a single project. (See Shinozuka, Rose and Eguchi, 1998.) Each step represents extensive and iterative collaboration with partners and experts. Some of the economic shock inputs were informed by models (REDARS® 2.0 for traffic delays and lost trips; HAZUS for ordinary building losses), while others involved educated guesses, derived from rules of thumb that remain subject to modification throughout the ongoing, deliberative process of refining them. For example, very recent feedback on the gas pipeline input suggests more of a population influence on gas shutoffs, and some gas outages in San Diego County that will increase the economic losses from gas service disruption.

Not all sources of economic shocks were pursued in this first stage of the project. We included the most significant, as indicated by the damage estimates (for example, fire) and/or political interest (for example, long-term planning at the ports). We omitted telecommunications because telecommunications services appear to be the least disrupted of all the lifeline services. We also omitted most considerations of labor constraints including morbidity and mortality effects. At meetings, some discussion about labor ensued in terms of getting employees to work. A commuter fault-crossing study estimated the percentage of employees in Los Angeles, Riverside and San Bernardino Counties that will not be able to cross the fault rupture until the pavement is repaired, but the performance of public transportation and recommencement of school (allowing parents to return to work) have not been considered. Other labor issues raised include increased demand for certain skills, such as airplane piloting and construction, and changes in the labor supply.

Objective 2—Examine the Resilience of the Regional Economy in the Event of the Earthquake

The economic analysis was conducted as a regional study because even local impacts can ripple spatially to the boundaries of larger regional economic trading areas, and because the vast size of the United States waters down the impacts at a national level. The regional analysis splits the economic results into the loss to buildings and structures (the destruction of part of the regional stock of assets) and the loss of economic activity (the flow impact), which is anticipated to occurs over a relatively short period of time (within 3 to 6 months) while the region regains more and more ability to function as normal.

In Section D of this chapter we refer to definitions of an economic recession to conclude that the economic losses from the ShakeOut Scenario earthquake have breached the economic recession threshold, but resilience assumptions prevent the earthquake from being an economic catastrophe. Furthermore, it was insightful to receive a reaction from Steve Levy (expert on the California economy) who framed the estimated regional losses in a variety of ways:

- To use a farming analogy, the $200 billion economic losses are like losing 6% of the harvest for a year.
- The $100 billion building and infrastructure stock losses are a fraction of the existing trillion dollar value regional economy, and a fraction of the hundreds of billions of planned infrastructure investment, and trillions if we include new homes and commercial buildings over the next 20 to 30 years.
- The region has recently incurred a different sort of shock, a $700 billion loss in home value, and may be in a mild recession.
- Hurricane Katrina was an economic catastrophe for New Orleans and for the State of Louisiana. The estimated 6% output loss for the southern California region is nowhere near the comparable magnitude of Katrina. Furthermore, the eight- county economy is, perhaps, 20 times greater than that of the New Orleans area and considerably more robust.
- During the great depression of the 1930s the unemployment rate was 25%.

The economic losses may not be large relative to the regional economy, but economists emphasize that the economic outcome is contingent upon assumptions of resilience, the ability of the economy to recover from a shock.

The research indicates that the extent of the economic and personal losses experienced in a major quake are heavily dependent on the speed at which core regional infrastructure systems can be restored and lost productivity can be recaptured. Losses, while never welcome, can be minimized if advance planning and response implementation are rapid and well thought out. The experience of Katrina, with all the delays in getting infrastructure rebuilt, provides a clear warning signal and direction for policymakers working to minimize the impact of an earthquake that is likely to hit Southern California sometime in the next 30 years.

Objective 3—Obtain Feedback on the Value and Uses of the ShakeOut Scenario

Panel participants responded to the ShakeOut Scenario by suggesting ways to increase resilience. Also, they identified further research that would further inform decision making. We held five panel discussions and forums with experts and professionals in these topic areas:

- Highway-bridges (HB)
- Goods movement through the ports (GM)
- SoCalfirst (critical infrastructure contingency planners) (SF)
- Community study in the Coachella Valley (CV)
- Economic Overview (EC)

Resilience Strategies

In addition to the value of the ShakeOut Scenario for an emergency response exercise, panel participants recognized the potential to use the ShakeOut Scenario to increase resilience by proactively:

- elevating issues to higher levels of decision-making (SF,GM);
- anticipating and preparing for disaster management challenges (HB,GM));
- anticipating beneficial and multi-agency cooperation, and expanding existing local emergency preparedness efforts out into the region (CV, GM, HB, EC);
- addressing emergency funding sources and priorities for use (HB, GM, CV, EC);
- promoting personal and institutional contingency planning and preparedness (GM,SF);
- stimulating flexible technological solutions (for example, technology to reverse power from ships to the grid)(GM); and
- informing the work of SCAG and other regional planning agencies about infrastructure planning and land-use/growth policies (EC).

Research Needs

Panel participants suggested that the ShakeOut Scenario provides a medium to communicate and prioritize future research funding. Stakeholders endorsed continued cooperation to ensure that the science based information is useful on receipt, and endorsed the following research topics:

- Setting critical infrastructure priorities (HB, SF, EC);
- Improving data and models for regional analyses (all);
- Estimation of economic consequences including tax base implications (SF, EC);
- Sensitivity analyses to address "what if" questions, and benefit-cost studies (HB,EC);
- Improved risk assessments to examine mitigation decisions (HB, SF, EC);
- Resilience, as it differs by sector and geographic location, as well as by scope, magnitude, and duration of a disaster (SF, EC);

- Recovery research (CV, EC); and
- National significance of the ShakeOut earthquake in terms of multiplier effects of disrupted port activity and cost to the government.

The next phase of this project will examine some of the determinants of recovery in the context of the Shakeout earthquake. For example, the availability of funding helps to speed recovery. As a prelude, a coarse estimation of insured losses suggests that, at most, 6% of ShakeOut residential building and content losses (the largest contributor to regional building shaking damage losses) are insured.

8.CONCLUSIONS

Our purpose in creating the ShakeOut Scenario is to identify actions that southern Californians could take now—*before* an earthquake—to avoid catastrophic impact *after* the inevitable earthquake occurs.. We modeled a magnitude 7.8 earthquake on the southern San Andreas Fault to analyze its impact on southern California regional economy and society to better understand how we can reduce future losses. The southern San Andreas Fault is the most likely source of a large earthquake in California and the M7.8 earthquake we modeled is large enough to cause significant damage to a large part of the urban environment.

We estimated the ground motions from this earthquake with physics-based computer simulations demonstrated that these results are consistent with the newest attenuation relations from the Next Generation Attenuation (NGA) relations. We validated our modeling results through comparison of multiple methods, use of distinct velocity models, and comparison with empirically based attenuation relations. In all, four teams were engaged to make independent models of the ground motions. Several features of the ShakeOut earthquake ground motions are consistent across all the models including:

- Very strong shaking (approaching 3 m/sec) near the fault;
- Strong shaking with medium to long durations (20-45 sec) in the basins near the fault including the Coachella, San Bernardino and Antelope Valleys;
- Damaging shaking (at least 0.5 m/sec) over large areas (~10,000 km2) of Los Angeles, San Bernardino and Riverside counties;
- Pockets of very strong shaking (≥1 .5 m/sec) with long durations (45-60 sec) in areas of the San Gabriel Valley and East Los Angeles.

The damage impacts of the ShakeOut Scenario earthquake were estimated using both HAZUS-MH and expert opinion through 13 special studies and 6 expert panels. The major losses for this earthquake fall into four categories: building damages, nonstructural damages, damage to lifelines and infrastructure, and fire losses. Within each category, the analysis found types of losses that are well understood—losses that have been seen in previous earthquakes and the vulnerabilities recognized, but not removed— and types of losses that had been less obvious, where the type of failure is only recently understood or the extent of the problem not yet fully recognized. The study also found numerous areas where mitigation conducted over the last few decades by state agencies, utilities and private owners, has greatly reduced the vulnerability.

The total damage to buildings is $35 billion. Most of the losses will be in building types that are no longer allowed by the building code. The damage to contents of the buildings is over $10 billion and much of these losses could be prevented by individual actions.

The lifelines that cross the fault will all break when the fault moves. This will disrupt the movement of water, petroleum products, telecommunications, and general transportation. Repair of the lifelines will be slowed because the lifelines all cross the fault at just a few passes in the mountains and therefore interact with each other. Away from the fault, however, California's investments in mitigation have paid off in increased robustness and resiliency of the region's lifelines. Significant vulnerabilities remain, however, especially in the water conveyance system where the worst hit areas may not have water in the taps for 6 months. This damage to the water system will also greatly increase the problems in fighting the fires that will follow the earthquake.

Southern California is unfortunately well situated to generate major fires following earthquakes. The number of ignitions that will create fires large enough to call the fire department is estimated at 1,600 ignitions of which 1,200 will be too large to be controlled by one fire company. The fire risk is increased by the damage to the water distribution system and by the traffic gridlock. We estimate $40 billion in damage to buildings and $25 billion in damage to building contents.

The magnitude 7.8 ShakeOut Scenario earthquake is modeled to cause about 1,800 deaths and $213 billion of economic losses. These numbers are as low as they are because of aggressive retrofitting programs that have increased the seismic resistance of buildings, highways and lifelines, and economic resiliency. These numbers are as large as they are because much more retrofitting could still be done.

The earthquake modeled here may never happen. Big earthquakes on the San Andreas Fault are inevitable, and by geologic standards extremely common, but probably will not be exactly like this one. The next very damaging earthquake could easily be on another fault. However, lessons learned from this Scenario event apply to many other events and could provide benefits in many possible future earthquake disasters.

The ShakeOut Scenario also found that previous efforts to reduce losses through mitigation before the event have been successful. There are dozens more actions and policies that could be undertaken at the individual and community levels to further reduce these losses. For instance, actions to improve the resiliency of southern California's water delivery system would reduce the loss from business interruption, as well as reduce the risk of catastrophic conflagrations. At an individual and business level, actions to secure non-structural items in buildings and retrofitting of existing structures will greatly reduce individual risk. Planning and preparedness can improve personal and business resiliency.

Over the next six months, the ShakeOut Scenario will be used to prepare for future earthquakes and for emergency response and preparedness exercises in the November 2008 Great Southern California ShakeOut. This process will encourage public discussion of these risks and possible solutions. The risks can be analyzed and described by scientists but the solutions will come from southern Californians themselves.

Many of the studies that were conducted for the ShakeOut Scenario are available as reports on-line. For details go to http://urbanearth.usgs.gov/scenario08.

REFERENCES

Aagaard, B. T., Brocher, T. M., Dolenc, D., Dreger, D., Graves, R. W., Harmsen, S., Hartzell, S., Larsen, S., McCandless, K., Nilsson, S., Petersson, N. A., Rodgers, A., Sjoegreen, B. & Zoback, M. L. (1906). in press, Ground-Motion modeling of the 1906 San Francisco Earthquake II: Ground-Motion estimates for the 1906 Earthquake and scenario events: *Bulletin of the Seismological Society of America*.

Aagaard, B., Hudnut, K. & Jones, L. (2007). SoSAFE kinematic rupture model: 10-page technical report.

Acharya, H. K. (1997). Influence of fault bends on ruptures: *Bulletin of the Seismological Society of America*, v. *87*, 1691-1696.

Ad Hoc Committee on Earthquake Recovery, (1995). *In the Wake of the Quake*, Report to the Mayor and City Council, Office of the Chief Legislative Analyst, Los Angeles City Council.

Agnew, D. C. & Sieh, K. E. (1978). A documentary study of the felt effects of the great California earthquake of 1857: *Bulletin of the Seismological Society of America*, v. *68*, 1717-1729.

Barberopoulou, A., Qamar, A., Pratt, T. L., Creager, K. C. & Steele, W. P. (2004). Local amplification of seismic waves from the Denali earthquake and damaging seiches in Lake Union, Seattle, Washington: *Geophysical Research Letters*, v. *31*, doi: 10.1029/2003GL018569.

Bardet, J. P., Tobita, T., Mase, N. & Hu, J. (2002). Regional modeling of liquefaction-induced ground deformation: *Earthquake Spectra*, v. *18*, 19-46.

Barrows, A. G. (1974). *A review of the geology and earthquake history of the Newport-Inglewood structural zone*, southern California: California Division of Mines and Geology Special Report 114, 115, scale 1, 12, 000.

Barrows, A. G., Kahle, J. E. & Beeby, D. J. (1985). *Earthquake hazards and tectonic history of the San Andreas Fault Zone*, Los Angeles County, California: California Division of Mines and Geology Open-File Report, 85-10 LA, 139.

Bodin, P., Bilham, R., Behr, J. Gomberg, J. & Hudnut, K. W. (1994). Slip triggered on southern California faults by the 1992 Joshua Tree, Landers, and Big Bear earthquakes: *Bulletin of the Seismological Society of America*, v. *84*, no. 3, 806-816.

Brune, J. N. (1970). Tectonic stress and spectra of seismic shearwaves from earthquakes: *Journal of Geophysical Research*, v. *75*, 4997-5009.

Brune, J. N. (2002). Precarious rock constraints on ground motion from historic and recent earthquakes in southern California: *Bulletin of the Seismological Society of America*, v. *92*, 2602-2611.

Bryant, W. A. compiler, (2005). Digital Database of Quaternary and Younger Faults from the Fault Activity Map of California, version 2.0: URL http://www.consrv.ca.gov/CGS/information/publications/QuaternaryFaults_ver2.h tm.

BST Associates, 2007, Trade Impact Study Final Report URL http://www.portofla.org/DOC/REPORT_ACTA_Trade_Impact_Study.pdf. [accessed April 4, 2008]

California Department of Conservation, (1997). Division of Mines and Geology, Probabilistic seismic hazard assessment for the State of California: Open-File Report 96-08, 60.

California Earthquake Authority, (2008). Insurance Policy Information: URL http://www.earthquakeauthority.com/index.aspx?id=13&pid=2. [accessed April 8, 2008]

California Geological Survey, (1997). Guidelines for evaluating and mitigating seismic hazards in California: California Geological Survey (California Division of Mines and Geology) Special Publication, 117, 74.

California Geological Survey, (2003a). *Seismic Hazard Zone Report for the Hi Vista 7.5-Minute Quadrangle*, Los Angeles County, California: California Geological Survey Seismic Hazard Zone Report, 087, 29.

California Geological Survey, (2003b). *Seismic Hazard Zone Report for the Lake Hughes 7.5-Minute Quadrangle*, Los Angeles County, California: California Geological Survey Seismic Hazard Zone Report, 103, 49.

California Geological Survey, (2003c), *Seismic Hazard Zone Report for the Littlerock 7.5-Minute Quadrangle*, Los Angeles County, California: California Geological Survey Seismic Hazard Zone Report, 099, 45.

California Geological Survey, (2003d). *Seismic Hazard Zone Report for the Palmdale 7.5-Minute Quadrangle*, Los Angeles County, California: California Geological Survey Seismic Hazard Zone Report, 105, 51.

California Geological Survey, (2003e). *Seismic Hazard Zone Report for the Ritter Ridge 7.5-Minute Quadrangle*, Los Angeles County, California: California Geological Survey Seismic Hazard Zone Report, 083, 53.

California Geological Survey, (2003f). *Seismic Hazard Zone Report for the Sleepy Valley 7.5-Minute Quadrangle*, Los Angeles County, California: California Geological Survey Seismic Hazard Zone Report, 106, 47.

California Geological Survey, (2003g). *Seismic Hazard Zone Report for the Whitaker Peak 7.5-Minute Quadrangle*, Los Angeles County, California: California Geological Survey Seismic Hazard Zone Report, 077, 47.

California Geological Survey, (2004). *Seismic Hazard Zone Report for the Lovejoy Buttes 7.5-Minute Quadrangle*, Los Angeles County, California: California Geological Survey Seismic Hazard Zone Report, 086, 43.

California Geological Survey, (2005a). *Seismic Hazard Zone Report for the Alpine Buttes 7.5-Minute Quadrangle*, Los Angeles County, California: California Geological Survey Seismic Hazard Zone Report, 092, 29.

California Geological Survey, (2005b). *Seismic Hazard Zone Report for the Del Sur 7.5-Minute Quadrangle*, Los Angeles County, California: California Geological Survey Seismic Hazard Zone Report, 101, 49.

California Geological Survey, (2005c). *Seismic Hazard Zone Report for the Little Buttes 7.5-Minute Quadrangle*, Los Angeles County, California: California Geological Survey Seismic Hazard Zone Report, 089, 17.

California Geological Survey, (2005d). *Seismic Hazard Zone Report for the Lancaster East 7.5-Minute Quadrangle*, Los Angeles County, California: California Geological Survey Seismic Hazard Zone Report, 094, 33.

California Geological Survey, (2005e). *Seismic Hazard Zone Report for the Lancaster West 7.5-Minute Quadrangle*, Los Angeles County, California: California Geological Survey Seismic Hazard Zone Report, 095, 45.

California Geological Survey, (2005f). *Seismic Hazard Zone Report for the Rosamond 7.5-Minute Quadrangle*, Los Angeles County, California: California Geological Survey Seismic Hazard Zone Report, 093, 33.

Campbell, K. W. & Bozorgnia, Y. in press, NGA ground motion model for the geometric mean horizontal component of PGA, PGV, PGD, and 5%-damped linear elastic response spectra for periods ranging from 0.01 to 10 s: *Earthquake Spectra*, v. *24*.

Campbell, K. W. (1981). A Ground Motion Model for the Central United States Based on Near-Surface Acceleration Data: Proceedings of Earthquakes and Earthquake Engineering: The Eastern United States, September 14-16, Knoxville, Tennessee.

Champion, R. & Wein, A. (2008). *Characterizing a regional economy:Bureau of Labor Statistics Location Quotients for industrial sectors in Southern California*, to be submitted for review.

Chiou, B. S. J. & Youngs, R. R. (2006). Chiou and Youngs PEER-NGA empirical ground motion model for the average horizontal component of peak acceleration and pseudo-spectral acceleration for spectral periods of 0.01 to 10 seconds: PEER Report.

Clahan, K. B. (2003). *Geologic map of the Oxnard 7.5-minute quadrangle*, Ventura County, California, A digital database: California Geological Survey, Preliminary Geologic Maps, scale 1:24,000.

Clark, M. M. (1984). *Map showing recently active breaks along the San Andreas Fault and associated faults between Salton Sea and Whitewater River-Mission Creek*, California: U.S. Geological Survey Miscellaneous Investigations Series Map, I1483, scale 1:24,000.

Cruden, David, M. & Varnes, David, J. (1996). Landslide types and processes in A. Turner, Keith; Schuster, & L. Robert (eds)., Landslides; investigation and mitigation: Special Report - Transportation Research Board, *National Research Council, Report*, *247*, 36-75.

Dean, R. G. & Dalrymple, R. A. (1991). *Water wave mechanics for engineers and Scientists: World Scientific*, Singapore, 353.

Department of Conservation - California Geological Survey (DOC-CGS), (2003a). Official seismic hazard zone map, Palmdale 7.5-minute quadrangle, Los Angeles County, California: CGS release date August, 14, 2003; scale 1:24,000.

Department of Conservation - California Geological Survey (DOC-CGS), (2003b). Official seismic hazard zone map, Ritter Ridge 7.5-minute quadrangle, Los Angeles County, California: CGS release date August, 14, 2003, scale 1:24,000.

Dixon, L., Clancey, N. & Overton, A. (2006). *The National Flood Program's Market Penetration Rate:Estimates and Policy Implications: Rand Corporation*, Santa Monica, CA, 116 p.

Eberhart-Phillips, D., Haeussler, P., Freymueller, J., Frankel, A., Rubin, C., Craw, P., Ratchkovski, N., Anderson, G., Carver, G., Crone, A., Dawson, T., Fletcher, H., Hansen, R., Harp, E., Harris, R., Hill, D., Hreinsdóttir, S., Jibson, R., Jones, L., Kayen, R., Keefer, D., Larsen, C., Moran, S., Personius, S., Plafker, G., Sherrod, B., Sieh, K., Sitar, N. & Wallace, W. (2003). The 2002 Denali Fault Earthquake, Alaska:A Large Magnitude, Slip-Partitioned Event: Science, 300(5622), 1113. [DOI: 10.1 126/science.1082703].

Economic Information & Research Department, International Trade Trends & Impacts, (2007). The Southern California Region 2006 Results & 2007 Outlooks, May 2007, Los Angeles Economic Development Corporation.

A. F., Espinosa, ed., 1976, The Guatemalan earthquake of February 4, 1976: A preliminary report: U.S. Geological Survey Professional Paper, 1002, 90.

Evernden, J. F. & Thomson, J. M. (1985). Predicting Seismic Intensities, Evaluating Earthquake Hazards in the Los Angeles Region - An Earth Science Perspective: J.I. Ziony, ed., U.S. Geological Survey Professional Paper, 1360, 151-202.

Federal Emergency Management Agency (FEMA), (2003). *Multi-hazard loss estimation methodology earthquake model*, HAZUS-MH MR3 Technical Manual: developed by Department of Homeland Security Emergency Preparedness and Response Directorate, FEMA Mitigation Division, Washington D.C., URL http://www.fema.gov/plan/prevent/hazus/hz_manuals.shtm, 699.

FEMA, (1992a). *Inventory of lifelines in the Cajon Pass, California: Earthquake Hazard Reduction Series*, 60, 97.

FEMA, (1992b). *Collocation impacts on the vulnerability of lifelines during earthquakes with applications to the Cajon Pass, California: Earthquake Hazard Reduction Series*, 61, 110.

FEMA, (1987). The Los Angeles - Whittier Narrows Earthquake of October 1, 1987: Federal/State Hazard Mitigation Survey Team Report: Prepared by the Federal Emergency Management Agency, Region IX; California Governor's Office of Emergency Services, Southern California Earthquake Preparedness Project and Planning Division.

Fisher, M. A., Normark, W. R., Greene, H. G., Lee, H. J. & Sliter, R. W. (2005). Geology and tsunamigenic potential of submarine landslides in Santa Barbara Channel: *southern California, Marine Geology*, v. *224*, 1-22.

Gardner, J. V. & Dartnell, P. (2002). Multibeam mapping of the Los Angeles, California Margin: U.S. Geological Survey Open-File Report 02-162, URL http://geopubs.wr.usgs.gov/open-file/of02-162/.

Goff, J., Dudley, W. C., deMaintenon, M. J., Cain, G. & Coney, J. P. (2006). The largest local tsunami in 20th century Hawaii: *Marine Geology*, v. *226*, 65-79.

Graves, R., Aagaard, B. & Hudnut, K. (2007). Broadband ground motion simulations for ShakeOut: Southern California Earthquake Center Annual Meeting, *Proceedings and Abstracts*, v.*XVII,* 1 15-116.

Graves, R., Aagaard, B., Hudnut, K., Star, L. & Stewart, J. (2008). *Broadband ground motion simulations for a Mw 7.8 Southern San Andreas Earthquake*: in preparation.

P. Grossi, & H. Kunreuther, editors, (2005). *Catastrophe modeling: a new approach to managing risk:* Springer, New York, 245.

Hansen, W. R. (1965). Effects of the earthquake of March 27, 1964 at Anchorage, Alaska: U. S. Geological Survey Professional Paper, 542-A, 68.

Hart, E. W., Bryant, W. A., Wills, C. J. & Treiman, J. A. (1990). The search for fault rupture and significance of ridgetop fissures, Santa Cruz Mountains, California: in The Loma Prieta (Santa Cruz Mountains), California Earthquake of 17 October, 1989, S. R. McNutt, & R. H. Sydnor, (Editors): *California Division of Mines and Geology Special Publication*, *104*, 83-94.

Heaton, T. H. (1985). *A model for a seismic computerized alert network: Science*, v.*228*, 987-990.

Holzer, T. L. & Youd, T. L. (2007). Liquefaction, ground oscillation and soil deformation at the Wildlife Array, California: *Bulletin of the Seismological Society of America*, v. *97*, 96 1-976.

T. L. Holzer, Ed., (1998). The Loma Prieta, California, earthquake of October 17, 1989 – Liquefaction: U. S. Geological Survey Professional Paper, 1551-B, 314.

Holzer, T. L., Bennett, M. J., Noce, T. E., Padovani, A. C. & Tinsley, J. C. III, (2006). Liquefaction Hazard Mapping with LPI in the greater Oakland, California, area: *Earthquake Spectra*, v. *22*, no. 3, 693-708.

Holzer, T. L., Bennett, M. J., Ponti, D. J. & Tinsley, J. C. III, (1999). Liquefaction and soil failure during 1994 Northridge earthquake: *Journal of Geotechnical and Environmental Engineering*, v. *125*, No. 6, 438-452.

Holzer, T. L., Noce, T. E., Bennett, M. J., Tinsley, J. C. III, & Rosenberg, L. I. (2005). Liquefaction at Oceano, California, during the 2003 San Simeon Earthquake: *Bulletin of the Seismological Society of America*, v. *95*, n. 6, 1-16.

Hudnut, K., Jones, L. & Aagaard, B. (2007). The southern California ShakeOut scenario, part 1: Earth science specification of a big one, *Seismological Research Letters*, v. *78*, 264.

Hudnut, K. W., L. Seeber, & Pacheco, J. (1989). Cross-fault triggering in the November 1987 Superstition Hills earthquake sequence, southern California: *Geophysical Research Letters*, v. *16*, n. 2, 199-202.

Jibson, R. W. (2002). A public health issue related to collateral seismic hazards:The valley fever outbreak triggered by the 1994 Northridge, California earthquake: *Surveys in Geophysics*, v. *23*, n. 6, 511-528.

Jibson, R. W., Harp, E. L., Schneider, E., Hajjeh, R. A. & Spiegel, R. A. (1998). An outbreak of coccidioidomycosis (valley fever) caused by landslides triggered by the Northridge, California, earthquake: *Geological Society of America, Reviews in Engineering Geology*, v. *12*, 53-61.

Johnson, C. E., Rojahn, Christopher, & Sharp, R. V. (1982). The Imperial Valley, California, earthquake of October 15, 1979: U. S. Geological Survey Professional Paper, 1254, 451.

Jones, L., Bernknopf, R. & Hudnut, K. (2007). The southern California ShakeOut scenario, part 2:Reducing losses from a really big one through improved tenacity: *Seismological Research Letters*, v. *78*, 281.

Jones, L., Sieh, K., Agnew, D., Allen, C., Bilham, R., Ghilarducci, M., Hager, B., Hauksson, E., Hudnut, K., Jackson, D. & Sylvester, A. (1991). Short-term Earthquake Hazard Assessment for the Southern San Andreas Fault, Southern California: USGS Open-File Report #91-32.

Joyner, W. B. & Boore, D. M. (1988). Measurement, characterization, and prediction of strong ground motion: Proceedings of Earthquake Engineering & Soil Dynamics II, Geotechnical Division of the American Society of Civil Engineers, 43-102.

Keefer, D. K. & Wilson, R. C. (1989). Predicting earthquake-induced landslides, with emphasis on arid and semi-arid environments in P. M. Sadler, & D. M. Morton, eds., Landslides in a semi-arid environment, with emphasis on the inland valleys of southern California: *Publications of the Inland Geological Society*, v. *2*, 118-149.

Keefer, D. K. (1984). Landslides caused by earthquakes: *Bulletin of the Seismological Society of America*, v. *95*, n. 4, 406-421.

Kennedy, M. P. (2003). Geologic map of the Vail Lake 7.5-minute quadrangle, *San Diego and Riverside counties, California:A digital database: California Geological Survey*, Preliminary Geologic Maps, scale 1:24,000.

Kennedy, M. P. (2006). Geologic map of the Boucher Hill 7.5-minute quadrangle, San Diego County, California:A digital database: California Geological Survey, Preliminary Geologic Maps , scale 1:24,000.

Kennedy, M. P. & Tan, S. S. (2005a). Geologic map of the San Diego 30' x 60' quadrangle, California:A digital database: California Geological Survey, Preliminary Geologic Maps, scale 1:100,000.

Kennedy, M. P. & Tan, S. S. (2005b). Geologic map of the Oceanside 30' x 60' quadrangle, California:A digital database: California Geological Survey, Preliminary Geologic Maps, scale 1:100,000.

Krishnan, Swaminathan, Ji, C., Komatitsch, D. & Tromp, J. (2006). Case Studies of Damage to Tall Steel Moment-Frame Buildings in Southern California during Large San Andreas Earthquakes: *Bulletin of the Seismological Society of America*, v. *96*, n. 4A, 1523-1537, doi: 10.1785/0120050145.

Kunreuther, H. (2004). Mitgating disaster losses through insurance, in Essays in the Theory of Risk-Bearing: in H., Kunreuther, & A. Rose, eds., *The economics of natural hazards*, vol. *II*, Edward Elgar, Cheltenham, UK, 489.

Lee, H., Ryan, R. E., Haeussler, H., Kayen, R. E., Hampton, M. A., Locat, J., Suleimani, E. & Alexander, C. R. (2007). Reassessment of seismically induced tsunamigenic submarine slope failures in Port Valdez, Alaska, USA: V., Lycousis, D. Sakellarion, & J. Locat, (Eds.), Submarine Mass Movements and Their Consequences, 357-365.

Lee, H. J., Greene, H. G., Edwards, B. D., Fisher, M. & Normark, W. submitted, Submarine landslides of the southern California borderland: Earth Science in the Urban Ocean: the Southern California Continental Borderland, H. J. Lee, & W. Normark, Eds., Geological Society of America Special Paper.

Lee, H. J., Normark, W. R., Fisher, M. A., Greene, H. G., Edwards, B. D. & Locat, J. (2004). Timing and extent of submarine landslides in Southern California: Proceedings, Offshore Technology Conference, Houston TX, OTC Paper 16744. 11.

Liao, S. S., Veneziano, D. & Whitman, R. V. (1988). Regression models for evaluating liquefaction probability: *Journal of Geotechnical Engineering*, v. *114*, n. 4, 389- 411.

Locat, J., Lee, H. J., Locat, P. & Imran, J. (2004). Numerical analysis of the mobility of the Palos Verdes debris avalanche, *California, and its implication for the generation of tsunamis: Marine Geology*, v. *203,* 269-280.

Lynch, D. K. & Hudnut, K. W. (2008). The Wister mud pot lineament: southeastward extension or abandoned strand of the San Andreas fault?: accepted *Bulletin of the Seismological Society of America.*

Makdisi, F. I. & Seed, H. B. (1978). Simplified procedure for estimating dam and embankment earthquake-induced deformations: *Journal of the Geotechnical Engineering Division, American Society of Civil Engineers*, v. *104*, n. GT7, 849-867.

Matti, J. C. & others, (2007). unpublished mapping of the greater San Gorgonio Pass region.

Mattison, M. E. & Barrows, A. G. (2003). *Liquefaction evaluation report for the Ritter Ridge 7.5-minute quadrangl*e, Los Angeles County, California: in CGS Seismic Hazard Zone Report, 083, 3-20.

McCulloh, D. S. & Bonilla, M. G. (1970). Effects of the earthquake of March 27, 1964 on the Alaska Railroad: U.S. Geological Survey Professional Paper 543-A, 41 p. McGarr, A., 1965, Excitation of seiches in channels by seismic waves: *Journal of Geophys. Res.,* v. *70*, 847-854.

Meltzner, A. J. & Wald, D. J. (1999). Foreshocks and aftershocks of the great 1857 California earthquake: *Bulletin of the Seismological Society of America*, v. *89*, n. 4, 1109-1120.

Miller, D. M. & Bedford, D. R. (2000). *Geologic map database of the El Mirage Lake area, San Bernardino and Los Angeles Counties*, California: U.S. Geological Survey Open-File Report, OF-00-222, scale 1:24,000.

Miller, F. K. & Cossette, P. M. (2004). *Preliminary geologic map of the Big Bear City 7.5' quadrangle, San Bernardino County, California*: U.S. Geological Survey Open- File Report OF-2004-1 193, scale 1:24,000.

Miller, F. K., Matti, J. C., Brown, H. J., Powell, R. E., Cossette, P. M. & Morton, G. (1998). Digital geologic map of the Fawnskin 7.5' quadrangle, San Bernardino County, California: U.S. Geological Survey Open-File Report, OF-98-579, scale 1:24,000.

Molnar, P. (1988). *Continental tectonics in the aftermath of plate tectonics: Nature*, v. *335*, 131-137.

Morton, D. M. & Kennedy, M. P. (2005). *Preliminary geologic map of the Sage 7.5' quadrangle, Riverside County*, California: U.S. Geological Survey Open-File Report, OF-2005-1285, scale 1:24,000.

Morton, D. M. & Matti, J. C. (2005). *Preliminary geologic map of the Hemet 7.5' quadrangle, Riverside County*, California: U.S. Geological Survey Open-File Report, OF-2004-1455, scale 1:24,000.

Morton, D. M. & Miller, F. K. (2006). Geologic map of the San Bernardino and Santa Ana 30' x 60' quadrangles, California: U.S. Geological Survey Open-File Report, OF-2006-1217, scale 1:100,000.

Morton, D. M. & Miller, F. K. (2003). *Preliminary geologic map of the San Bernardino 30'x60' quadrangle*, California: U.S. Geological Survey Open-File Report, 03- 293, URL http://pubs.usgs.gov/of/2003/of03-293/.

National Research Council, 1985, *Liquefaction of soils during earthquakes: Committee on Earthquake Engineering*, Commission on Engineering and Technical Systems, National Academy Press, Washington, DC.

Office of the Chief Legislative Analyst, (1995). In the Wake of the Quake: Report to the Mayor and City Council from the Ad Hoc Committee on Earthquake Recovery, Los Angeles City Council.

Olsen, K. B., Day, S. M., Minster, J. B., Cui, Y., Chouasia, A., Moore, R., Maechling, P. & Jordan, T. (2008). TeraShake2: simulation of Mw7.7 earthquakes on the southern San Andreas fault with spontaneous rupture description: Bulletin of the Seismological Society of America, in press.

Olsen, K. B., Day, S. M., Minster, J. B., Cui, Y., Chouasia, A., Faerman, M., Moore, R., Maechling, P. & Jordan, T. (2006). Strong shaking in Los Angeles expected from southern San Andreas earthquake: *Geophysical Research Letters*, v. *33*, L07305, doi: 10.1 029/2005GL025472.

Perez, F. & Bryant, W. A. (2007). unpublished mapping of the Mescal Creek and Mt. San Antonio 7.5-minute quadrangles.

Peterson, M. D., Bryant, W. A., Cramer, C. H., Cao, T., Reichle, M. S., Frankel, A. D., Lienkaemper, J. J., McCrory, P. A. & Schwartz, D. P. (1996). Probabilistic seismic hazard assessment for the state of California: U. S. Geological Survey Open-File Report, 96-0706, 33.

Plafker, G. (1967). *Surface faults on Montague Island associated with the 1964 Alaska earthquake*: U.S. Geological Survey Professional Paper, 1045, 42.

Plafker, G., Ericksen, G. E. & Fernandez, C. (1971). Geological aspects of the May 31, 1970, Peru earthquake: *Bulletin of the Seismological Society of America*, v. *61*, n. 3, 543-578.

Ponti, D. J. & Wells, R. E. (1991). Off-fault ground ruptures in the Santa Cruz Mountains, California: Ridge-top spreading versus tectonic extension during the 1989 Loma Prieta earthquake: *Bulletin of the Seismological Society of America*, v. *81*, n. 5, 1480-1510.

Powell, R. E. & Matti, J. C. (2000). *Geologic map and digital database of the Cougar Buttes 7.5′ quadrangle*, San Bernardino County, California: U.S. Geological Survey Open-File Report, OF-2000-175, scale 1:24,000.

Pridmore, C. R. (2003). Liquefaction evaluation report for the Palmdale 7.5-minute quadrangle, Los Angeles County, California: *CGS Seismic Hazard Zone Report, 105*, 3-20.

Proctor, R. J. (1968). Geology of the Desert Hot Springs-Upper Coachella Valley area, California: California Division of Mines and Geology Special Report, 94, 50, 1 plate, map scale 1:62,500.

Purvance, M., Graves, R., Brune, J., Aagaard, B. & Hudnut, K. (2007). Comparison between precariously balanced rocks and the ShakeOut simulation:Ground motion constraints and implications for electrical substation damage: 2007 SCEC Annual Meeting Abstracts.

Real, C. R., Pridmore, C. L. & Loyd, R. C. (2008). Multi-hazard demonstration project, preliminary liquefaction deformation analysis at lifeline crossings for the M7.8 southern San Andreas earthquake scenario: U.S. Geological Survey Open-File Report, 2008-1150G.

Rose, A. (2004). Economic Principles, Issues, and Research Priorities of Natural Hazard Loss Estimation: in Y. Okuyama, & S.Cheng, (eds), Modeling the Spatial Economic Impacts of Natural Hazards, Heidelberg: Springer.

Rose, A. (2007. Economic Resilience to Disasters: Multidisciplinary Origins and Contextual Dimensions, Environmental Hazards: *Human and Social Dimensions*, v. *7*, 1-16.

Rose, A. & Lim, D. (2002). Business Interruption Losses from Natural Hazards: Conceptual and Methodology Issues in the Case of the Northridge Earthquake, Environmental Hazards: *Human and Social Dimensions*, v. *4*, 1-14.

Rose, A. & Miernyk, W. (1989). Input-Output Analysis:The First Fifty Years: *Economic Systems Research*, v. *1*, 229-721.

Rose, A., Oladosu, G. & Liao, S. (2007). Business Interruption Impacts of a Terrorist Attack on the Electric Power System of Los Angeles:*Customer Resilience to a Total Blackout: Risk Analysis*, v. *27*, 513-531.

Roth, R. A. & Kavazanjian, E. Jr., (1984). Liquefactions susceptibility mapping for San Francisco, *California: Bull. of the Association of Engineering Geologists*, v. *21*, n. 4, 459-478.

Rust, D. (2005). Palaeoseismology in steep terrain: The Big Bend of the San Andreas fault, *Transverse Ranges, California: Tectonophysics*, v. *408*, 193-212.

Rymer, M. J., Tinsley, J. C., III, Treiman, J. A., Arrowsmith, J. Ramon, Clahan, K. B., Rosinski, A. M., Bryant, W. M., Snyder, A., Fuis, G. S., Toke, N. A. & Bawden, G. A. (2006). Surface slip associated with the 2004 Parkfield, California, earthquake: *Bulletin of the Seismological Society of America*, v. *96*, n. 4B, S 11-S27.

Rymer, M. J., Boatwright, J., Seekins, L. C., Yule, J. D. & Liu, J. (2002). Triggered surface slips in the Salton Trough associated with the 1999 Hector Mine, California, earthquake: *Bulletin of the Seismological Society of America*, v. *92*, n. 4, 1300-1317.

Sadigh, K., Egan, J .A. & Youngs, R. R. (1986). Specification of ground motion for seismic design of long period structures: *Earthquake Notes*, v. *57*, n. 1, 13.

Saucedo, G. J., Greene, H. G., Kennedy, M. P. & Bezore, S. P. (2003). Geologic map of the Long Beach 30' x 60' quadrangle, California:A digital database: California Geological Survey, Preliminary Geologic Maps, scale 1:100,000.

Southern California Association of Governments, SCAG Regional Goods Movement Knowledge Base, (Accessed, April 4, 2008), URL http://www.scag.ca.gov/goodsmove/knowledgebase/knowledgebase.asp.

Scawthorn, C. (2008). A Note on Fire Following Earthquake for the Southern San Andreas Fault M 7.8 Earthquake (SoSAFE) Scenario, 27.

Seed, H. B. & Alonso, J. L. (1973). *Effects of soil-saturated interaction in the Caracas earthquake of 1967: Earthquake engineering research at Berkeley*: Berkeley, Calif., University of California, College of Engineering Earthquake Engineering Research Center, Report 72-23, 47-50.

Seed, H. B. & Idriss, I. M. (1967). Analysis of soil liquefaction:Niigata earthquake: Proceedings of the American Society of Civil Engineers, *Journal of the Soil Mechanics and Foundations Division*, v. *97*, n. SM9, 83-108.

Seed, H. B. & Idriss, I. M. (1982). Ground motions and soil liquefaction during earthquakes: Earthquake Engineering Research Institute, Oakland, California, Monograph Series, 13.

Seed, H. B., Idriss, I. M. & Arango, I. (1983). Evaluation of liquefaction using field performance data: *American Society of Civil Engineers Journal of Geotechnical Engineering*, v. *109*, n. 3, 458-482.

Seed, H. B., Lee, K. L., Idriss, I. M. & Makdisi, F I. (1975). The slides in the San Fernando Dams during the earthquake of February 9, 1971: Proceedings of the American Society of Civil Engineers, *Journal of the Geotechnical Engineering Division*, v. *101*, n. GT7, 651-688.

Seed, H. B., Tokimatsu, K., Harder, L. F. & Chung, R. M. (1985). Influence of SPT Procedures in Soil Liquefaction Resistance Evaluations: *Journal of Geotechnical Engineering, American Society of Civil Engineers*, v. *111*, n. 12, 1425-1445.

Seligson, H. A., Eguchi, R. T., Tierney, K. J. & Richmond, K. (1996. Chemical Hazards, Mitigation and Preparedness in Areas of High Seismic Risk: A Methodology for Estimating the Risk of Post-Earthquake Hazardous Materials Release, Technical Report NCEER-96-00 13.

Sherrouse, Benson C. & Hester, David, J. in press, Potential Effects of a Scenario Earthquake on the Economy of Southern California: Baseline County-Level Migration Characteristics and Trends 1995-2000 and 2001-2010, U.S. Geological Survey Open-File Report 2008-XXXX, U.S. Geological Survey, Reston, Virginia.

Sherrouse, Benson C. & Hester, David J. in press, Potential Effects of a Scenario Earthquake on the Economy of Southern California: Intra-Regional Commuter, Worker and Earnings Flow Analysis, U.S. Geological Survey Open-File Report 2008-XXXX, U.S. Geological Survey, Reston, Virginia.

Sherrouse, Benson C., Hester, David J. & Wein, Anne, M. in press, *Potential Effects of a Scenario Earthquake on the Economy of Southern California:Labor Market Exposure and Sensitivity Analysis to a Magnitude*, 7.8 Earthquake: U.S. Geological Survey Open-File Report 2008-XXXX, U.S. Geological Survey, Reston, Virginia.

Sherrouse, Benson C., Hester, David, J. & Wein, Anne, M. in press, *Potential Effects of a Scenario Earthquake on the Economy of Southern California: Small Business Exposure and Sensitivity Analysis to a Magnitude*, 7.8 Earthquake: U.S. Geological Survey Open-File Report 2008-XXXX, U.S. Geological Survey, Reston, Virginia.

Shields, M. (2003). *Using Employment Data to better understand your local economy, Tool 3*. Use Location Quotients to Identify Local Strengths, Opportunities, and Industry Clusters, © The Pennsylvania State University 2003, Produced by Information and Communication Technologies in the College of Agricultural Sciences, Accessed: Mar 2, 2008, URL http://cecd.aers.psu.edu/pubs/Tool%203.pdf.

Shimazaki, K. & Ans Nakata, T. (1980). Time-predictable recurrence model for large earthquakes: *Geophysical Research Letters*, v. *7,* 279-282.

Shinozuka, A. Rose, & R. Eguchi, (eds.), (1998). *Engineering and Socioeconomic Impacts of Earthquakes: An Analysis of Electricity Lifeline Disruptions in the New Madrid Area*, Buffalo, NY: Multidisciplinary Center for Earthquake Engineering Research.

Sieh, K. (1978). Central California foreshocks of the great 1857 earthquake: *Bulletin of the Seismological Society of America*, v. *68*, n. 6, 173 1-1749.

Sieh, K. (1986). Slip rate across the San Andreas Fault and prehistoric earthquakes at Indio, California: Eos, Transactions, American Geophysical Union v. 67, p. 1200. K. Sieh, & P. Williams, (1990). Behavior of the southernmost San Andreas Fault during the past 300 years: *Journal of Geophysical Research*, v. *95*, 6629-6645.

Sieh, K., Stuiver, M. & Brillinger, D. (1989). A more precise chronology of earthquakes produced by the San Andreas Fault in southern California: *Journal of Geophysical Research*, v. *94*, 603-623.

Smith, D. P. (1979). San Gorgonio Pass Fault: California Division of Mines and Geology Fault Evaluation Report, FER-87, 7.

SoCal MTSAC, (2003). California Marine Transportation System Infrastructure Needs: California MTS Infrastructure Needs Report, Accessed April 4, 2008, URL http://www.mxsocal.org/Articles/70/1/Southern-California-MTS-AdvisoryCouncil.aspx.

Spotila, J. A., Niemi, N., Brady, R., House, M. & Buscher, J. (2007). Long-term continental deformation associated with transpressive plate motion: *The San Andreas Fault: Geology,* v. *35*, n. 11, 967-970.

Star, L. M. & Stewart, J. P. (2008). SHAKEOUT: Factors for Correcting Ground Motions at Large Distance from Empirical Models to be Compatible with Simulated Motions, 8.

Steinbrugge, K.V., Bennett, J. H., Lagoria, H. J., Davis, J. F., Borchardt, G. & Toppozada, T. R. (1987). Earthquake planning scenario for a magnitude 7.5 earthquake on the Hayward Fault in the San Francisco Bay Area: California Division of Mines and Geology Special Publication, 78, 245.

Stuart, W. D. (1986). Forecast model for large and great earthquakes in southern California, *Journal of Geophysical Research*, v. *91*, 1377 1-13786.

Surface Transportation Board, (2007). Quantifying the Amount of Rail Traffic Moving through Areas Possibly Affected by the San Andreas Fault: for the U.S. Geological Survey.

Taborda, R., Lopez, J., Urbanic, J., Bielak, J., O'Hallaron, D. & Ramirez-Guzman, L. (2007). ShakeOut and its effects in Los Angeles and Oxnard areas: abstract, Southern California

Earthquake Center, 2007 Annual Meeting Proceedings, University of Southern California, Los Angeles, California, pg. 163.

Tan, S. S. & Clahan, K. B. (2003). Geologic map of the Point Mugu 7.5-minute quadrangle, Ventura County, California: A digital database: California Geological Survey, Preliminary Geologic Maps , scale 1:24,000.

Tan, S. S. & Clahan, K. B. (2004). Geologic map of the White Ledge Peak 7.5-minute quadrangle, Santa Barbara and Ventura counties, California: A digital database: California Geological Survey, Preliminary Geologic Maps, scale 1:24,000.

Tan, S. S. & Irvine, P. J. (2005a). Geologic map of the Ojai 7.5-minute quadrangle, Ventura County, California: A digital database: California Geological Survey, Preliminary Geologic Maps, scale 1:24,000.

Tan, S. S. & Irvine, P. J. (2005b). Geologic map of the Santa Paula Peak 7.5-minute quadrangle, Ventura County, California: A digital database: California Geological Survey, Preliminary Geologic Maps, scale 1:24,000.

Tan, S. S. & Jones, T. A. (2006). Geologic map of the Matilija 7.5-minute quadrangle, Ventura County, California: A digital database: California Geological Survey, Preliminary Geologic Maps, scale 1:24,000.

Tan, S. S. & Kennedy, M. P. (2003). Geologic map of the Aguanga 7.5-minute quadrangle, San Diego and Riverside counties, California: A digital database: California Geological Survey, Preliminary Geologic Maps , scale 1:24,000.

Tan, S. S., Clahan, K. B. & Hitchcock, C. S. (2004a). Geologic map of the Camarillo 7.5-minute quadrangle, Ventura County, California: A digital database: California Geological Survey, Preliminary Geologic Maps , scale 1:24,000.

Tan, S. S., Clahan, K. B. & Irvine, P. J. (2004b). Geologic map of the Santa Paula 7.5- minute quadrangle, Ventura County, California: A digital database: California Geological Survey, Preliminary Geologic Maps, scale 1:24,000.

Tan, S. S., Clahan, K. B. & Rosinski, A. M. (2004c). Geologic map of the Saticoy 7.5- minute quadrangle, Ventura County, California: A digital database: California Geological Survey, Preliminary Geologic Maps, scale 1:24,000.

Tan, S. S., Jones, T. A. & Clahan, K. B. (2003a). Geologic map of the Pitas Point 7.5- minute quadrangle, Ventura County, California: A digital database: California Geological Survey, Preliminary Geologic Maps, scale 1:24,000.

Tan, S. S., Jones, T. A. & Clahan, K. B. (2003b). Geologic map of the Ventura 7.5-minute quadrangle, Ventura County, California: A digital database: California Geological Survey, Preliminary Geologic Maps, scale 1:24,000.

Tappin, D. R., Watts, P. & Matsumoto, T. (2003). Architecture and failure mechanism of the offshore slump responsible for the 1998 Papua New Guinea tsunami: J. Locat, & Mienert, (Eds.), Submarine Mass Movements and their Consequences, Kluwer Academic Publishers, The Netherlands, 383-389.

Technical council of Lifeline Earthquake Engineering, (1995). ed. J. Anshel, Shiff, Northridge Earthquake: Lifeline performance and post-earthquake response: Monograph, n. 8, ASCE.

ten Brink, U., Lee, H. J., Geist, E. L. & Twichell, D. Submitted, Assessment of tsunami hazard to the U.S. East Coast using relationships between submarine landslides and earthquakes: Marine Geology.

Tinsley, J. C. & Dupre, W. R. (1992). Liquefaction hazard mapping, depositional facies, and lateral spreading ground failure in the Monterey Bay area, central California: Hamada and O'Rourke, eds., 1992 Proceedings of Japan-U.S. *Conference on earthquake resistant design of lifeline facilities and countermeasures for soil liquefaction, 4th, Honolulu, Hawaii, May 27-29, 1992*, v. *1*, 71-86.

Tinsley, J. C., III, Egan, J. A., Kayen, R. E., Bennett, M. J., Kropp, A. & Holzer, T. L. (1998). Appendix: Maps and descriptions of liquefaction and associated effects: T. L. Holzer, ed., The Loma Prieta, California, earthquake of October 17, 1989: *U. S. Geological Survey Professional Paper*, *1551-B*, B287-B314, including 3 plates.

Tinsley, J. C., Youd, T. L., Perkins, D. M. & Chen, A. T. F. (1985). Evaluating liquefaction potential: Ziony, J. I., Editor, Evaluating earthquake hazards in the Los Angeles region: *U. S. Geological Survey Professional Paper, 1360*, 263- 315.

Tinsley, J. C. & Fumal, T. E. (1985). Mapping Quaternary sedimentary deposits for areal variations in shaking response: J. I. Ziony, Ed., Evaluating earthquake hazards in the Los Angeles region - An earth science perspective: *U.S. Geological Survey Professional Paper*, *1360*, 101-25.

Todd, V. R., Alvarez, R. M. & Techni Graphic Systems, Inc., (2004). Preliminary geologic map of the El Cajon 30' x 60' quadrangle, southern California: U.S. Geological Survey Open-File Report, OF-2004-1361, scale 1:100,000.

Todd, V. R., Busch, L. L., Foster, B. D., Hernandez, J. L. & Tan, S. S. (2006). Geologic map of the Ramona 7.5-minute quadrangle, San Diego County, California: A digital database: California Geological Survey, Preliminary Geologic Maps , scale 1:24,000.

Tokimatsu, A. M. & Seed, H. B. (1987). Evaluation of settlements in sands due to earthquake shaking: *Journal of the Geotechnical Division, American Society of Civil Engineers*, v. *113*, n. 8, 681-878.

Treiman, J. A. (2008). Multi-hazards demonstration project: fault rupture impacts at areas of lifeline concentration: USGS-OFR 2008-1 150E.

Treiman, J. A. (1994). The San Gorgonio Pass, Banning and related faults, Riverside County, California: California Division of Mines and Geology, Fault Evaluation Report FER-235, 47, 12 plates, 1:24,000.

Trifunac, M. D. (1976). A Note on the Range of Peak Amplitudes of Recorded Accelerations, Velocities, and Displacements with Respect to the Modified Mercalli Intensity Scale: *Earthquake Notes*, v. *47*, n. 1.

U.S. Department of Transportation, 2004, Federal Highway Administration, Freight Analysis Framework: Issues and Plans, Accessed April 4, 2008, URL http://www.ops.fhwa.dot.gov/freight/freight_analysis/faf/fafplan_sep04.pdf.

U.S. Department of Transportation, 2006, Federal Highway Administration, Southern California Regional Freight Study, Accessed April 4, 2008, URL http://www.ops.fhwa.dot.gov/freight/freight_analysis/reg_ind_studies/so_cal_stud y.htm

Varnes, D. J., Radbruch-Hall, D. H. & Savage W. Z. (1989). *Topographic and structural conditions in areas of gravitational spreading of ridges in the Western United States: U.S. Geol.* l Survey Prof. Paper 1496, 28.

Varnes, David, J. (1978). Slope movement types and processes, in R. L. Schuster, & R. J. Krizek, eds., Landslides – analysis and control: National Academy of Sciences *Transportation Research Board Special Report*, *176*, 11-33.

Wald, D. J., Quitoriano, V., Heaton, T. H. & Kanamori, H. (1999). Relationship between Peak Ground Acceleration, Peak Ground Velocity, and Modified Mercalli Intensity for Earthquakes in California: *Earthquake Spectra*, v. *15*, n. 3, 557- 564.

Weber, F H., Jr. & Kiessling, E. W. (1976). General features of seismic hazards of Ventura County, California, in Seismic hazards study of Ventura County, California: *California Division of Mines and Geology Open-File Report*, *76-5* LA, 8-218.

Weldon, R. J. (1986). The late Cenozoic geology of Cajon Pass: implications for tectonics and sedimentation along the San Andreas fault: unpublished PhD thesis, California Institute of Technology, Pasadena, CA.

Weldon, Ray J., Fumal, Thomas, E., Biasi, Glenn, P. & Scharer, Katherine, M. (2005). Past and Future Earthquakes on the San Andreas Fault: *Science*, v. *308*. n. 5724, 966-967, DOI: 10.1126/science. 1111707.

Werner, S. D., Cho, S. & Eguchi, R. T. (2008). Analysis of Risks to Southern Calfiornia Highway System due to M7.8 Earthquake along Southern San Andreas Fault. West Coast Port Lock Out, 2002, Accessed April 4, 2008, URL http://havenworks.com/labor

Williams, P. L., McGill, S. F., Sieh, K. E., Allen, C. R. & Louie, J. N. (1988). Triggered slip along the San Andreas Fault after the 8 July 1986 North Palm Springs earthquake: *Bulletin of the Seismological Society of America,* v. *78*, n. 3, 1112- 1122.

Wieczorek, G. F., Wilson, R. C. & Harp, E. L. (1985). Map of slope stability during earthquakes in San Mateo County, California: U.S. Geological Survey Miscellaneous Investigations, Map I-1257-E, scale 1:62,500.

Wilbur Smith Associates, (2006). Multi-County Goods Movement Action Plan, Draft Technical Memorandum 3, Existing Conditions and Constraints: Accessed April, 4, 2008, URL http://www.metro.net/images %20 1 00606.pdf.

Wills, C. J., Weldon, R. J. & Bryant, S. (2008). Appendix A. of WGCEP Report (version dated March 1, 2007): in review.

Wills, C. J. & Clahan, K. B. (2006). Developing a map of geologically defined site- condition categories for California: Bulletin of the Seismological Society of America, v. 96, n. 4A, 1483-1501.

Wilson, B. W. (1972). Seiches: *Advances in Hydroscience*, v. *8*, 1-94.

Wilson, M. E. & Wood, S. H. (1980). Tectonic tilt rates derived from lake-level measurements, Salton Sea, California: *Science*, v. *207*, 183-186.

Wilson, R. C. & Keefer D. K. (1985). Predicting Areal Limits of Earthquake Induced Landsliding, Evaluating Earthquake Hazards in the Los Angeles Region: *U.S. Geological Survey Professional Paper, Ziony, J. I., Editor, 3*, 17-493.

Wilson, R. I., Perez, F. G. & Barrows, A. G. (2003). Earthquake-induced landslide zones in the Ritter Ridge 7.5-Minute Quadrangle, Los Angeles County, California: in CGS *Seismic Hazard Zone Report*, *083*, 21-41.

Wilson, R. I., McCrink, T. P., Treiman, J. A. & Silva, M. A. (2008). Multi-hazard demonstration project, seismically-induced landslide hazard analysis and deformation at lifeline crossings for the M7.8 southern San Andreas earthquake scenario: USGS-OFR 2008-1150F

Youd, T. L. (1973). *Liquefaction, flow, and associated ground failure: U.S. Geological Survey Circular*, 688, 12.

Youd, T. L. (1978a). Major cause of earthquake damage is ground failure: *Civil Engineering*, v. *48*, n. 4, 47-5 1.

Youd, T. L. (1978b). Mapping liquefaction-induced ground failure potential: Proceedings of the American Society of Civil Engineers, *Journal of the Geotechnical Engineering Division*, v. *104*, n. GT4, 433-446.

Youd, T. L. & Perkins, D. M. (1978c). Mapping liquefaction-induced ground failure potential: Proceedings of the American Society of Civil Engineers, *Journal of the Geotechnical Engineering Division,* v. *104*, n. GT4, 433-446.

Youd, T. L. & Wieczorek, G. F. (1982). Liquefaction and secondary ground failure, in The Imperial Valley, California, earthquake of October 15, 1979: *U. S. Geological Survey Professional Paper*, *1254*, 223-246.

Youd, T. L. & Wieczorek, G. F. (1984). Liquefaction during the 1981 and previous earthquakes near Westmorland, California: *U. S. Geological Survey Professional Paper, 941-A*, A68-A74.

Yule, D. & Sieh, K. (2003). Complexities of the San Andreas fault near San Gorgonio Pass – implications for large earthquakes: *Journal of Geophysical Research,* v.*108*, B11, 23.

End Notes

[1] The other two areas are Orange County (urban density) and the Inland Empire (growth), Frank Wen, SCAG.

[2] Although the scenario is not positing damage to this area, there is a concern about local and near-field earthquake events which are well documented in the literature. Port facilities would be most vulnerable to a large event on the nearby Palos Verdes or the Newport- Inglewood faults. In the case of the 1994 M6.7 Northridge earthquake, for example, minor liquefaction caused damage to unreinforced masonry structures in one area in San Pedro. In the 1933 M6.4 Long Beach quake, unreinforced masonry buildings were badly damaged in Long Beach and numerous liquefaction incidents were identified (although engineers were not able to explain the phenomenon.

Although the scenario is not assuming damages to the port pipelines or storage areas, storage tanks in the harbor are vulnerable and could fail, requiring about a week to repair and bring back into operation. This is an important consideration because of limited storage tank capacity and interruptions to the system could delay processing and distribution operations. The harbor infrastructure also includes a large and complex system of pipelines to convey ocean-transported oil and petroleum products on shore to local refineries. Severe ground motion could cause failures to the pipelines which would be disruptive to the petroleum import system as well as cause environmental hazards. Larry Cottrill and Doug Thiessen, February 6, 2008.

[3] The American Association of Port Authorities is a resource for statistics for US waterborne trade. The US Customs Bureau may have break down the relative size of the different modes of cargo and Ports of entry.

In: The Big Shake Implications of a Major Earthquake in … ISBN: 978-1-61728-512-7
Editor: Christopher D. Monahan

Chapter 3

FORECASTING CALIFORNIA'S EARTHQUAKES: WHAT CAN WE EXPECT IN THE NEXT 30 YEARS?

U.S. Geological Survey

In a new comprehensive study, scientists have determined that the chance of having one or more magnitude 6.7 or larger earthquakes in the California area over the next 30 years is greater than 99%. Such quakes can be deadly, as shown by the 1989 magnitude 6.9 Loma Prieta and the 1994 magnitude 6.7 Northridge earthquakes. The likelihood of at least one even more powerful quake of magnitude 7.5 or greater in the next 30 years is 46%—such a quake is most likely to occur in the southern half of the State. Building codes, earthquake insurance, and emergency planning will be affected by these new results, which highlight the urgency to prepare now for the powerful quakes that are inevitable in California's future.

WHAT IS AN EARTHQUAKE RUPTURE FORECAST?

Californians know that their State is subject to frequent—and sometimes very destructive—earthquakes. Accurate forecasts of the likelihood of quakes can help people prepare for these inevitable events. Because scientists cannot yet make precise predictions of the date, time, and place of future quakes, forecasts are in the form of the probabilities that quakes of certain sizes will occur during specified periods of time.

In our daily lives, we are used to making decisions based on probabilities—from weather forecasts (such as a 30% chance of rain) to the annual chance of being killed by lightning (about 0.0003%). Similarly, earthquake probabilities derived by scientists can help us plan and prepare for future quakes.

Earthquake forecasts for California have been developed in the past by multidisciplinary groups of scientists and engineers, each known as a "Working Group on California Earthquake Probabilities" (WGCEP 1988, 1990, 1995, 2003). However, those forecasts were limited to particular regions of California. Because of this, WGCEP 2007 was commissioned to develop an updated, statewide forecast, the latest result of which is the Uniform California

Earthquake Rupture Forecast, Version 2, or "UCERF" (U.S. Geological Survey (USGS) Open-File Report 2007-1437, http://pubs.usgs. gov/of/2007/1437/). Organizations sponsoring WGCEP 2007 include the USGS, California Geological Survey, and the Southern California Earthquake Center. The comprehensive new forecast builds on previous studies and also incorporates abundant new data and improved scientific understanding of earthquakes.

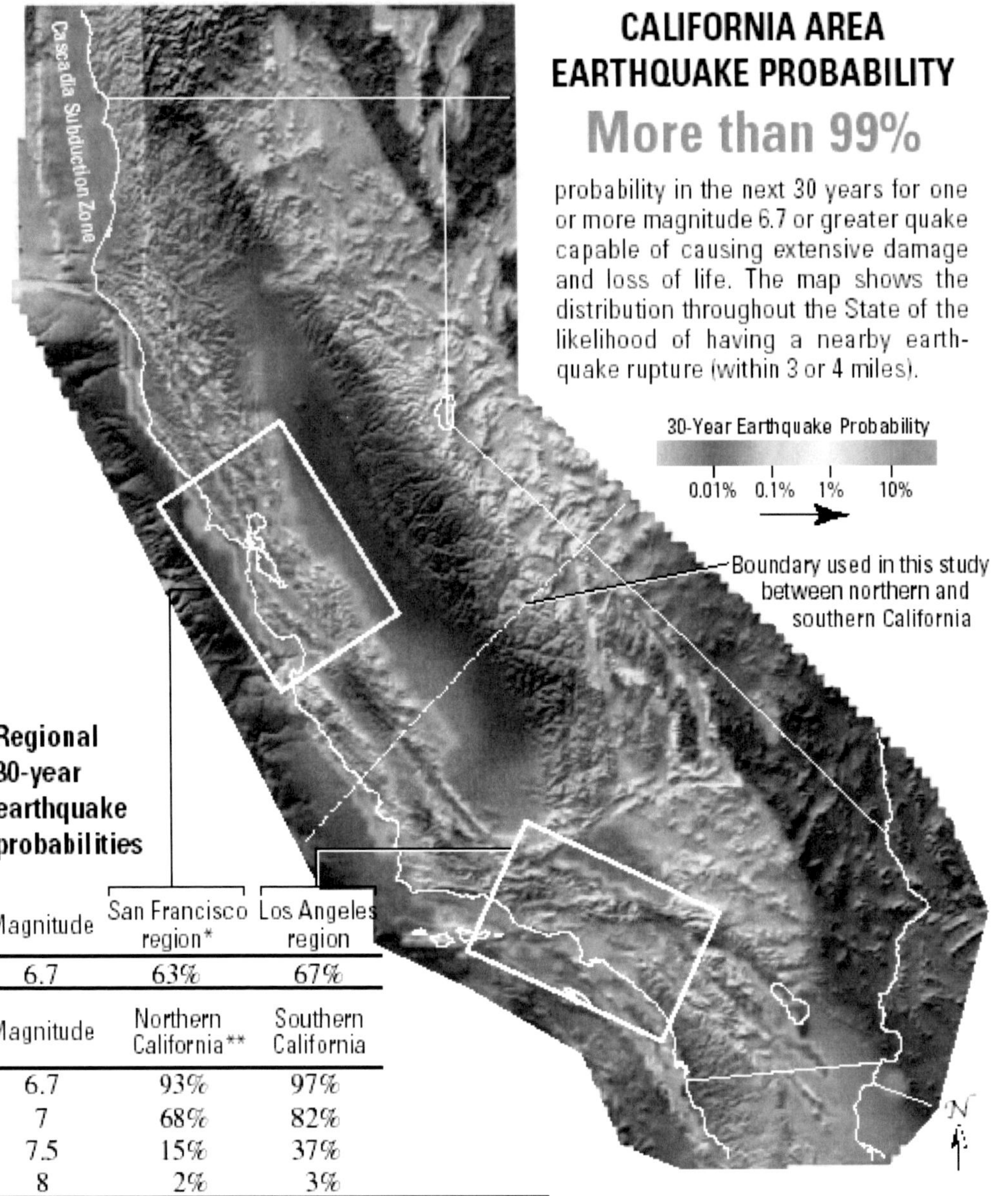

Magnitude	San Francisco region*	Los Angeles region
6.7	63%	67%

Magnitude	Northern California**	Southern California
6.7	93%	97%
7	68%	82%
7.5	15%	37%
8	2%	3%

*Probabilities from UCERF for the San Francisco region are nearly identical to the previous results from WGCEP 2003.

**These probabilities do not include the Cascadia Subduction Zone

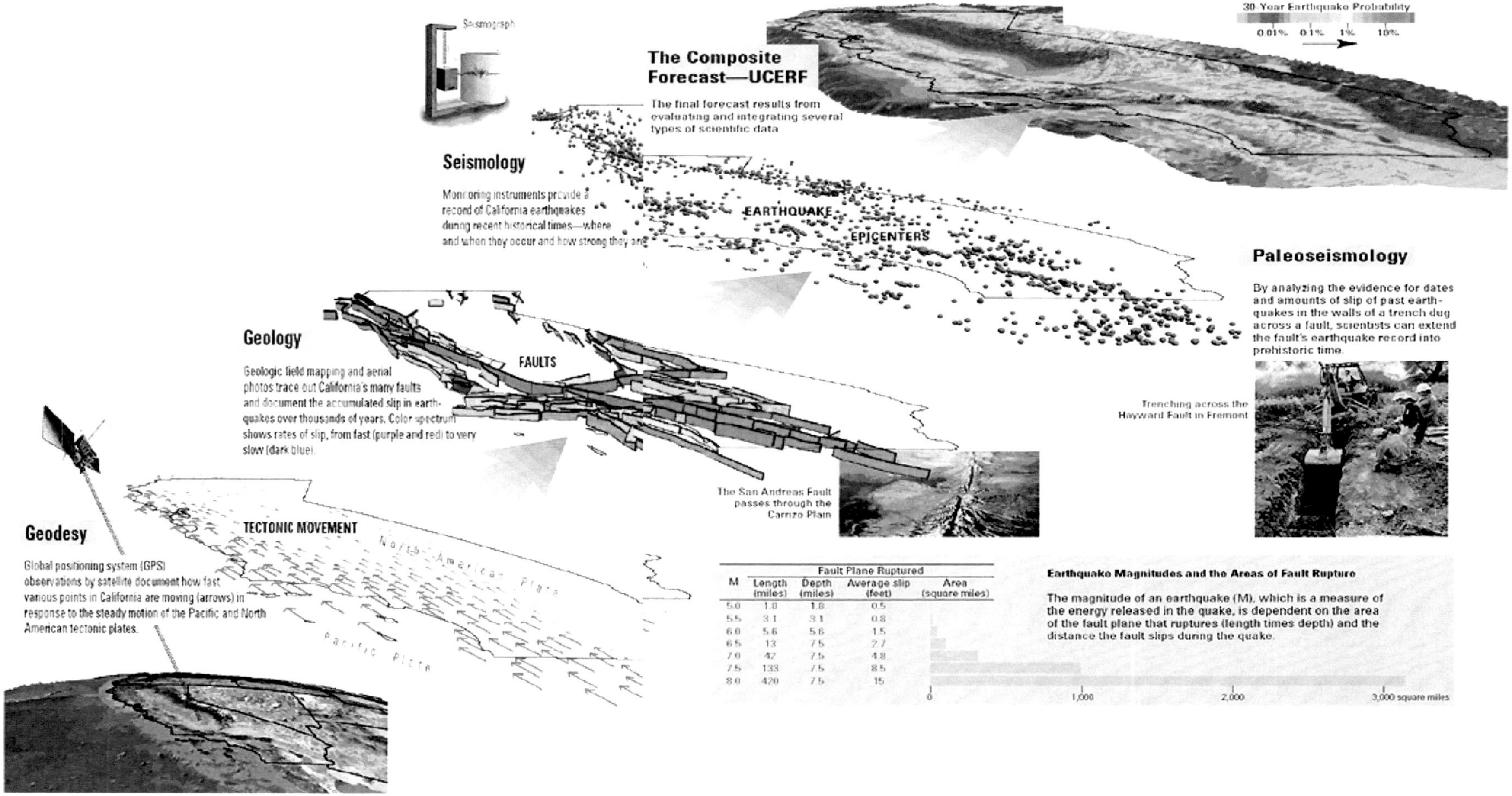

Earthquake Magnitudes and the Areas of Fault Rupture

The magnitude of an earthquake (M), which is a measure of the energy released in the quake, is dependent on the area of the fault plane that ruptures (length times depth) and the distance the fault slips during the quake.

M	Fault Plane Ruptured: Length (miles)	Depth (miles)	Average slip (feet)	Area (square miles)
5.0	1.8	1.8	0.5	
5.5	3.1	3.1	0.8	
6.0	5.6	5.6	1.5	
6.5	13	7.5	2.7	
7.0	42	7.5	4.8	
7.5	133	7.5	8.5	
8.0	420	7.5	15	

0 1,000 2,000 3,000 square miles

When an earthquake occurs, two things happen—a fault ruptures (a crack in the Earth's crust gives way and slips under tectonic pressure) and seismic waves, caused by this sudden fault motion, radiate out like ripples from a pebble tossed into a pond. The shaking that occurs as seismic waves pass by causes most quake damage. The strength of the waves depends partly on the quake's magnitude, which is a function of the size of the fault that moves and the amount of slip.

The UCERF study's goal was to determine probabilities for different parts of California of earthquake ruptures of various magnitudes, but not to estimate the likelihood of shaking ("seismic hazard") that will be caused by these quakes. This distinction is important, because even areas in the State with a low probability of fault rupture can experience shaking and damage from distant, powerful quakes.

How Did Scientists Make This Forecast?

California sits on the boundary between two of the Earth's major tectonic plates—the Pacific and North American Plates—which move inexorably past each other at a rate of about 2 inches per year. Much of this motion is accommodated from time to time by sudden slip on faults, producing earthquakes. Although the San Andreas Fault is the main locus of slip, hundreds, if not thousands, of other faults splay out from the plate boundary, spreading the threat of large earthquake ruptures through most of the State.

The new Uniform California Earthquake Rupture Forecast (UCERF) combines information from **geodesy** (precise data on the slow relative movement of the Earth's tectonic plates), **geology** (mapped locations of faults and documented offsets on them), **seismology** (occurrence patterns of past earthquakes), and **paleoseismology** (data from trenches across faults documenting the dates and offsets of past earthquakes on them). The first three kinds of data are shown here as layers in the diagram. All four kinds of data are combined mathematically to produce the final probability values for future ruptures in the California area, in regions of the State, and on individual faults.

Building on several previous studies and decades of data collection, UCERF was developed by a multidisciplinary group of scientists and engineers, known as the 2007 Working Group on California Earthquake Probabilities. Advice and comment was sought regularly from the broader community of earthquake scientists and engineers through open meetings and workshops. Where experts disagreed on aspects of the forecast, alternative options were accounted for in calculations to reflect these uncertainties. The final forecast is a sophisticated integration of scientific fact and expert opinion.

How Likely is a Damaging Quake in the Next 30 Years?

California straddles the boundary between two of the Earth's tectonic plates—as a result, it is broken by numerous earthquake faults. Taking into account the earthquake histories and relative rates of motion on these many faults, the UCERF study concludes that there is a probability of more than 99% that in the next 30 years Californians will experience one or more magnitude 6.7 or greater quakes, potentially capable of causing extensive damage and loss of life. For powerful quakes of magnitude 7.5 or greater, there is a 46% chance of one or more in the next 30 years—such a quake is twice as likely to occur (37%) in the southern half of the State than in the northern half (15%).

Smaller magnitude earthquakes are more frequent than larger quakes. According to the new forecast, about 3 magnitude 5 or greater quakes will occur in the California region per year, and a magnitude 6 or greater quake about every 1.5 years. These numbers do not include aftershocks that follow larger quakes—including them would roughly double the expected number of magnitude 5 or greater quakes.

STATEWIDE EARTHQUAKE PROBABILITIES

The numbers represent current best estimates. As earthquake science progresses, these probabilities will change. Actual repeat times vary considerably and only rarely will be exactly as listed in the table.

Magnitude	30-year probability of one or more events greater than or equal to the magnitude	Average repeat Time (years)
6.7	>99%	5
7	94%	11
7.5	46%	48
8	4%	650

*Not including Cascadia Subduction Zone

For the entire California region, the fault with the highest probability of generating at least one magnitude 6.7 or larger earthquake is the southern San Andreas (59% in the next 30 years). For northern California, the most likely source of such a quake is the Hayward-Rodgers Creek Fault (31% in next 30 years)—see USGS Fact Sheet 2008-3019. Quake probabilities for many parts of the State are similar to those in previous studies, but the new probabilities for the Elsinore and San Jacinto Faults in southern California are about half those previously determined. For the far northwestern part of the State, a major source of quakes is the offshore 750-mile-long "Cascadia Subduction Zone," which extends south about 150 miles into California. For the next 30 years there is a 10% probability of a magnitude 8 to 9 quake somewhere along the zone—such quakes occur about every 500 years.

The UCERF forecast was evaluated by an independent scientific review panel, as well as by both the California and National Earthquake Prediction Evaluation Councils, making it one of the most extensively reviewed earthquake forecasts ever produced. Uncertainties remain because the new quake probabilities are the result of evaluating and accommodating several earthquake theories. As scientific understanding of quakes improves, the probabilities will change.

The results of the UCERF study are a reminder that all Californians live in earthquake country and should therefore be prepared (see Putting Down Roots in Earthquake Country at http://www.earthquakecountry.info/roots/). The USGS has already used the UCERF to estimate California's seismic hazard, which in turn will be used to update building codes. Other subsequent studies will add information on the vulnerability of manmade structures to estimate expected losses ("seismic risk"). In these ways, UCERF will help to increase public safety and community resilience to earthquake hazards.

WHICH FAULTS CURRENTLY HAVE ELEVATED PROBABILITIES?

The new California earthquake forecast incorporates the concept that earthquake probabilities change with time. For example, a large quake may be less likely in the near future on a fault that has recently had one—but more likely on a fault on which the tectonic stresses have had much time to build back up. The forecast also incorporates adjustments for areas that have recently shown a change in the level of earthquake activity. Faults that have elevated current probabilities include the southern San Andreas and Hayward-Rodgers Creek Faults, though major quakes on these faults may still be decades away.

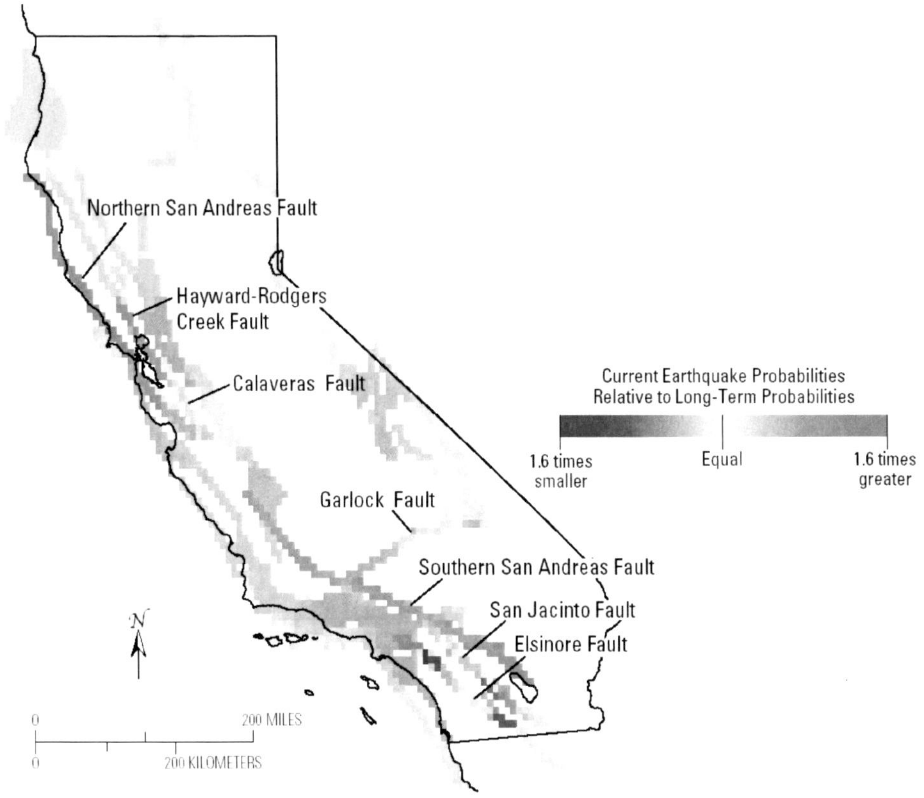

Earthquakes cannot be prevented, but the damage they do can be greatly reduced through prudent planning and preparedness. The ongoing work of USGS, California Geological Survey, Southern California Earthquake Center, and other scientists in evaluating quake probabilities is part of the National Earthquake Hazard Reduction Program's efforts to safeguard lives and property from the future quakes that are certain to strike in California and elsewhere in our Nation.

INDIVIDUAL FAULT PROBABILITIES

The UCERF report assigns individual probabilities to specific known major faults. Below are 30-year probabilities for seven of the faults for which scientists have the most data. Many other faults also have significant probabilities—in fact, the next big quake in California is just as likely to occur on one of the other faults in the State.

Fault	Probability of one or more magnitude 6.7 or greater quake
Southern San Andreas	59%
Hayward-Rodgers Creek	31%
San Jacinto	31%
Northern San Andreas	21%
Elsinore	11%
Calaveras	7%
Garlock	6%

In: The Big Shake Implications of a Major Earthquake in... ISBN: 978-1-61728-512-7
Editor: Christopher D. Monahan

Chapter 4

THE SHAKEOUT EARTHQUAKE SCENARIO: A STORY THAT SOUTHERN CALIFORNIANS ARE WRITING

U.S. Geological Survey

By Suzanne Perry, Dale Cox, Lucile Jones, Richard Bernknopf, James Goltz, Kenneth Hudnut, Dennis Mileti, Daniel Ponti, Keith Porter, Michael Reichle, Hope Seligson, Kimberley Shoaf, Jerry Treiman, and Anne Wein

INTRODUCTION

The question is not *if* but *when* southern California will be hit by a major earthquake—one so damaging that it will permanently change lives and livelihoods in the region. How severe the changes will be depends on the actions that individuals, schools, businesses, organizations, communities, and governments take to get ready. To help prepare for this event, scientists of the U.S. Geological Survey (USGS) have changed the way that earthquake scenarios are done, uniting a multidisciplinary team that spans an unprecedented number of specialties. The team includes the California Geological Survey, Southern California Earthquake Center, and nearly 200 other partners in government, academia, emergency response, and industry, working to understand the long-term impacts of an enormous earthquake on the complicated social and economic interactions that sustain southern California society. This project, the ShakeOut Scenario, has applied the best current scientific understanding to identify what can be done now to avoid an earthquake catastrophe. More information on the science behind this project will be available in *The ShakeOut Scenario* (USGS Open-File Report 2008-1150; http://pubs.usgs.gov/of/2008/1150/).

The "what if?" earthquake modeled in the ShakeOut Scenario is a magnitude 7.8 on the southern San Andreas Fault. Geologists selected the details of this hypothetical earthquake by considering the amount of stored strain on that part of the fault with the greatest risk of imminent rupture. From this, seismologists and computer scientists modeled the ground shaking that would occur in this earthquake. Engineers and other professionals used the shaking to produce a realistic picture of this earthquake's damage to buildings, roads, pipelines, and other infrastructure. From these damages, social scientists projected casualties, emergency response, and the impact of the scenario earthquake on southern California's economy and society. The earthquake, its damages, and resulting losses are one realistic outcome, deliberately not a worst-case scenario, rather one worth preparing for and mitigating against.

Decades of improving the life-safety requirements in building codes have greatly reduced the risk of death in earthquakes, yet southern California's economic and social systems are still vulnerable to large-scale disruptions. Because of this, the ShakeOut Scenario earthquake would dramatically alter the nature of the southern California community. Fortunately, steps can be taken now that can change that outcome and repay any costs many times over. The ShakeOut Scenario is the first public product of the USGS Multi- Hazards Demonstration Project, created to show how hazards science can increase a community's resiliency to natural disasters through improved planning, mitigation, and response.

THE SHAKEOUT SCENARIO IS NOT A PREDICTION

On November 13, 2008, the ShakeOut Scenario earthquake and projected damages will be used as a basis for public drills and emergency response exercises, and so this hypothetical earthquake has been arbitrarily assigned a date, a time, a plausible collection of aftershocks, and even local weather conditions. Because it is a scenario for planning, it provides detailed numbers, including casualties, collapsed buildings, and business losses. Yet, the ShakeOut earthquake is not a prediction. Certainly, a large earthquake involving this part of the San Andreas Fault is highly probable. In fact, scientists have determined that this is the most likely source of a large earthquake in all of California (see *Forecasting California's Earthquakes—What Can We Expect in the Next 30 Years?*, U.S. Geological Survey Fact Sheet 2008-3027; http://pubs.usgs.gov/fs/2008/3027/).

When the next San Andreas Fault earthquake does happen, some things are inevitable—the fault rupture will break any road, track, or pipe that crosses it, and intense shaking will damage or destroy buildings that weren't constructed to withstand it. The next earthquake will be different in details from the ShakeOut earthquake, and its total damages and losses will differ, because each earthquake produces its own patterns of shaking and damage. However, the widespread, regional effects will be similar, and so will the long-term social and economic impacts. Getting prepared for the ShakeOut earthquake will help southern Californians withstand other earthquakes of comparable size. If we take no additional actions for preparedness and mitigation, and the ShakeOut earthquake does occur, it will cause some 2,000 deaths, 50,000 injuries, $200 billion in damage, and severe, long-lasting disruption. These numbers can climb with each damaging aftershock.

LIFELINES CROSS THE FAULT

The Cajon Pass is one of five major "lifeline corridors" that are the veins and arteries through which economic life circulates in southern California. These lifeline corridors allow commuters, goods, telephone and Internet lines, electricity, water, gas, and fuels to move through the mountains that surround southern California. Because most of the corridors must cross the southern San Andreas Fault, future earthquakes are certain to sever the lifelines in one or more of these corridors and cause significant economic disruption by interrupting the movement of goods from the ports of Los Angeles and Long Beach. (USGS photo by Lucile Jones.)

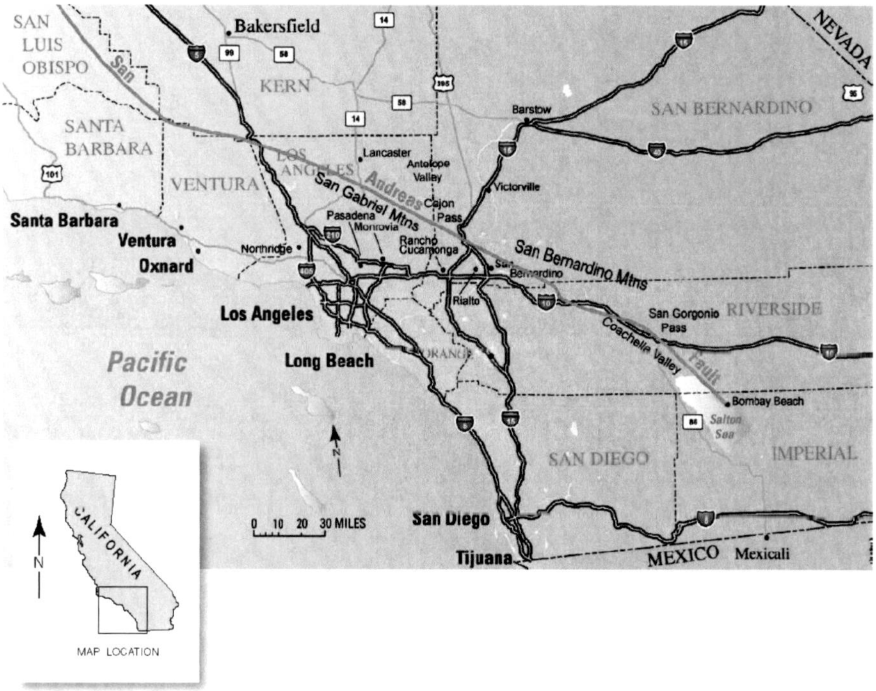

The San Andreas Fault slices through southern California and produces earthquakes that shape and reshape the region. Unlike many other faults, the southern San Andreas Fault produces no small earthquakes. Its next earthquake will disrupt the complicated economic and social systems that define southern California and will affect everyone, including those living and working in communities relatively undamaged by the initial violent shaking.

The Earthquake Advances

These computer-generated snapshots show ground motions for the ShakeOut Scenario earthquake, in meters per second (m/s; 1 m/s is about 3.3 feet per second). Yellow indicates higher amplitudes of ground motion. The snapshots show three points in time—30 (left), 60 (center), and 120 (right) seconds after the southern San Andreas Fault (dashed white line) first begins rupturing at Bombay Beach, on the eastern shore of the Salton Sea. Note that some areas remain orange colored for much longer than others, indicating extended, intense shaking in some of the sedimentary basins of southern California. (Simulation by Rob Graves of URS Corporation for the Southern California Earthquake Center on high-performance computers at the University of Southern California; images courtesy of Geoff Ely, University of California San Diego/San Diego Supercomputer Center.)

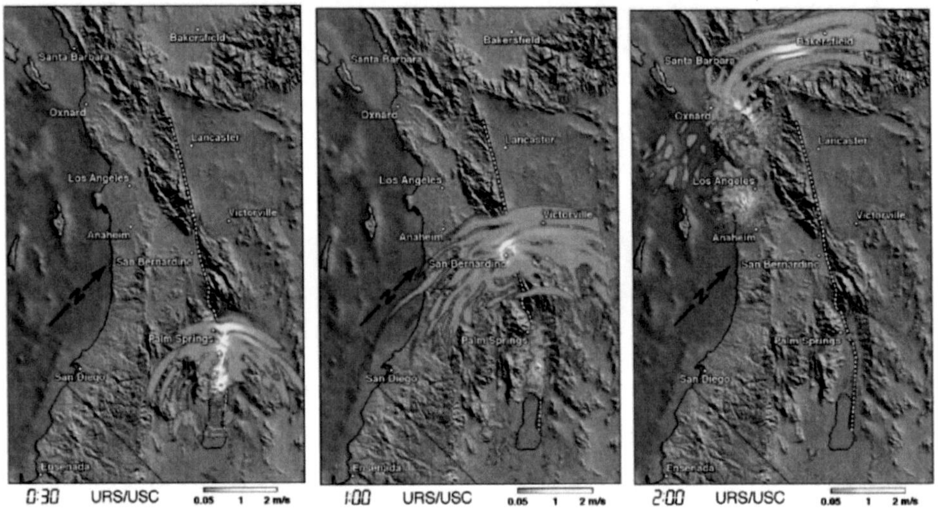

The ShakeOut Scenario Narrative

Much like a movie script, the following fictional narrative will guide those participating in the ShakeOut Earthquake Scenario public drills and emergency response exercises on November 13, 2008. More effectively than any statistics, this narrative describes what this magnitude 7.8 earthquake would be like in southern California if no additional actions are taken for mitigation or preparedness.

November 13, 2008

Thursday 9:50 a.m. (... 10 minutes before the quake begins...)

By mid-morning on this workday, 200,000 commuters have made their way from Kern, Riverside, and San Bernardino Counties into the Los Angeles area. These drivers trade a lengthy commute for the lower cost of housing in fast-growing communities like Victorville and Lancaster, on the far side of the San Andreas Fault. Others cross the fault in the opposite direction, to employers in high desert communities. The commuters have joined 7.5 million other southern Californians in workplaces constructed of steel, concrete, brick, or wood. Of the many millions of homes and workplaces, only a fraction are covered by earthquake insurance.

A steady flow of trains crosses the San Andreas Fault at multiple locations, moving goods between cargo ships at the Ports of Los Angeles and Long Beach and the rest of the country. Trucks are also on the move nonstop, carrying goods through narrow passes cut in the San Gabriel and San Bernardino Mountains. These "lifeline corridors" are the veins and arteries that sustain economic life in southern California. Sharing these passes with cars, buses, trucks, and trains are pipelines carrying natural gas and fuels; water conveyance tunnels, pipes, and aqueducts; electrical transmission towers and lines; and the telecommunications cables that connect people by phone and Internet—connecting banks and clients, suppliers and providers, buyers and sellers, friends and families, headquarters and field offices. Like the commuters, they all cross the San Andreas Fault.

Thursday 9:52 a.m. (... 8 minutes before the quake begins ...)

The San Andreas Fault slices through California, marking the boundary between the Pacific and North American tectonic plates. Along this boundary, the plates try to slide past each other, but near the surface they are locked by friction and deform instead of moving, storing up strain energy. Eventually and suddenly, the friction will give way and the plates will slip, creating a powerful earthquake. This earthquake will rupture the Earth's surface and release stored energy in seismic waves that travel out in all directions, shaking the ground as they go. On the southern San Andreas Fault, an earthquake rupture and energy release occur on average every 150 years—but the last time was more than 300 years ago.

The next earthquake is about to begin. It will not be the biggest earthquake that has ever occurred on the southern San Andreas Fault, but it will be big enough to change southern California markedly for untold years to come. It will wreak economic havoc on many who are unharmed by the initial shaking and damage.

Thursday 9:55 a.m. (...5 minutes before the quake begins...)

It is sunny with a light breeze. Fortunately, today there will be no Santa Ana winds. Throughout the region, cars flow to the pulse of traffic lights. Rush hour traffic has cleared, and the workday is well underway in offices, warehouses, factories, and stores. Many of the older buildings and even a few newer ones are constructed in ways that make them vulnerable to earthquake shaking. Schools are full of students, as well as furniture and equipment that will topple in a earthquake, and heavy objects that will become airborne.

Thursday 10:00 a.m. (... the quake begins...)

The San Andreas Fault suddenly awakens at Bombay Beach, northeast of the Salton Sea, and the rupture shoots northwest along the fault at 2 miles per second, sending seismic energy waves out in all directions. In an instant, the ground on the two sides of the fault is offset nearly 44 feet, changing the political and geographic boundary between Imperial and Riverside Counties.

Thursday 10:00:30 a.m.

As the earthquake's rupture front travels up the fault, it sends out seismic waves that shake the ground, shifting emergency generators, overturning computers, cracking airport runways, and igniting fires. By now, the thick sediments of the Coachella Valley are resonating, with the earthquake waves bouncing between the rock walls of the valley's edges. Strong shaking will continue here for nearly a minute.

The life-safety provisions of California's building codes have been improved over the years, and the many fairly new homes in the Coachella Valley suffer only minor damage. Yet every item inside these homes, if not secured, is heading to the floor. Shattered TVs and other home electronics create treacherous carpets of glass and cords.

Many older buildings suffer structural damage. Many older concrete buildings quickly collapse, trapping occupants. The rupture front continues its advance to San Gorgonio Pass and dismantles the ten miles of Interstate 10 freeway that straddles the San Andreas Fault. The eastern part of Riverside County is now cut off from the western part.

Thursday 10:01:00 a.m.

Most people in Los Angeles and Ventura Counties are not yet aware of what is happening as the earthquake pounds the Coachella Valley and heads their way. By now the first waves have crashed through the Cajon Pass, severing the I-15 freeway, bending rail lines, and derailing a train. Roads, previously throughgoing across the fault, now end abruptly and pick up again 15 feet to the right.The strong shaking also sends landslides across the rails and roads. Pipelines snap and electrical transmission lines fail. Spraying fuel ignites, causing an explosion.

Strong shaking begins to reverberate in the sediment- filled basins of the Inland Empire. Old warehouse districts and historic downtowns are crumbling, and many of their old, unretrofitted buildings have trapped or killed the people inside. Many older concrete buildings have collapsed, and many older woodframe buildings have shifted off their foundations, breaking gas and water lines in the process.

The Coachella Valley is still shaking.

Thursday 10:01:30 a.m.

Over geologic time, the motion of tectonic plates has pushed the mountains of southern California up, while fire, rain, and rivers have brought the mountains down, piece by piece, filling basins with sediments and creating low, flat areas. Like many cities, Los Angeles was built atop sediments. Some of the seismic waves now reach these sediments and find easy territory in which to move back and forth, shaking vigorously long after the waves fade elsewhere. Strong shaking will continue in Los Angeles for 55 seconds, to the shock of residents who remember the strong shaking during the 1994 Northridge earthquake, which lasted only 7 seconds.

The seismic waves that reverberate in the sedimentary basins are big, long waves. Many buildings ride them like boats in choppy seas, but some are not so resistant. The prolonged, strong shaking heavily damages and sometimes collapses hundreds of old brick buildings, hundreds of older commercial and industrial concrete buildings, many woodframe buildings, and even a few, fairly new high-rise steel buildings. The building damage causes tens of thousands of injuries and hundreds of deaths, and strands many thousands of people without homes or jobs.

Buried in the sediments are the water and sewer pipes that maintain the cities. Many of these pipes crack when the earthquake waves deform the ground.

In the newer houses, the primary damage is to contents. Kitchen floors have disappeared under heaps of cooking oil, syrup, flour, and smashed dishes, but there is no water in the taps to start cleaning.

Power is out so stoplights are dark and electric trains are suddenly immobilized. Buses, cars, and trucks become gridlocked, and many drivers will experience this earthquake as taillights for hours, to the horizon. Eventually, many will open their car doors and not look back as they begin the long walk home, perhaps envied only by people sitting in the dark, waiting to be rescued from stalled and stifling elevators.

Thursday 10:02 a.m. (...2 minutes after the quake began...)

At last, the fault has stopped rupturing, but seismic waves continue to advance into Bakersfield, Oxnard, and Santa Barbara—here the shaking is just beginning.

Across southern California, the power is out. Emergency generators that have been secured against earthquake shaking are still functional and now kick on. The shaking has finally stopped in the Coachella Valley—but the aftershocks are just beginning.

Throughout southern California in the next few months there will be tens of thousands of earthquake aftershocks large enough to feel. There will be dozens large enough to cause additional damage and to imperil victims and rescuers. Some of the aftershock damage will be to people's psyches. Big earthquakes are traumatic, and each new bout of shaking increases stress, especially in children who are cut off from their families.

In the areas of strong shaking, many mobile homes have collapsed off supports, snapping water, gas and sewer lines, and blocking rescue routes. Mobile homes installed snugly in shallow pits, or braced for earthquakes, are still intact.

The State highway system has fared well. A $6 billion investment in seismic retrofitting has paid off, and the only highway deaths have been in crashes caused by intense earthquake shaking. However, the long duration of shaking has taken its toll on bridges and overpasses within local jurisdictions, where the retrofitting process is not completed, or not yet begun.

No hospitals have seen complete collapses, but many hospital buildings are nonfunctional. Some hospital structures survived the shaking but must close due to nonstructural damage such as water pipes that break and flood.

Thursday 10:05 a.m. (... 5 minutes after the quake began ...)

The U.S. Geological Survey posts preliminary information about the earthquake. Learning that the magnitude is 7.8, the world turns its attention to southern California. Locally, news helicopters take to the air to begin spot coverage of the devastation. With power out, residents turn to their radios or talk to those they meet in the streets, searching for any information.

Across the region, phone systems, including cellular and 911, are unusable, overwhelmed by the vast number of attempted calls.

Thursday 10:30 a.m. (... 30 minutes after the quake began...)

Emergency operations centers are activating, and police, fire, and medical personnel shift into emergency response mode, focusing on localized incidents with any means available. They react quickly, according to their training and earthquake plans established in advance.

All over the region, a foreseeable tragedy unfolds. Buildings that engineers knew were going to perform badly, have performed badly. These are older buildings, constructed with little earthquake resistance. The experts have names for them—non-ductile reinforced concrete, tilt-up concrete, unreinforced masonry, soft-stories—and hundreds of these buildings have now followed their reputations into the dust. Thousands of other structures are still standing, but so gravely damaged that they can never be used again.

While the earth still shakes in places far from the earthquake's origin, people in the earliest hit areas are beginning to confront damaged buildings and to help those who are trapped or hurt. Lacking gloves, crowbars, and training, some people claw through debris with bare hands. Ultimately, 95 percent of those who are rescued will be rescued by other victims, as has been seen in earthquake disasters worldwide.

Air traffic is being diverted from southern California.

As people start to assess their situations, millions of them discover they are cut off from their families, with no way to learn the fate of their loved ones or homes. This realization also hits first responders as they move out to help; they understand that the disaster may seem to be over but is just beginning.

Fires are starting in countless ways. Power lines arc... gas appliance lines snap... chemicals spill and mix... a lamp hits a sofa, unnoticed with the power out and the earth shaking, then the power returns and the sofa starts to smolder... Most of the fires start small, but not all are discovered right away. In any case, the phones don't have dial tones. Even if they did, in a disaster this big and widespread, there are not enough emergency personnel to immediately respond to every call for help. Worse, response is slowed by roads that are impassable due to damage, building debris, or abandoned cars. Worse still, in many places the water system is damaged, leaving inadequate water pressure for fire fighting.

Once started, a small fire needs only minutes to engulf a home or workplace. Around southern Calif ornia on this day, there will be 1,600 fires large enough to warrant a 911 call. The stronger the shaking, the greater the number of fires ignited. In areas with densely packed, woodframe buildings, some of these ignitions will combine, spreading into conflagrations that burn dozens of blocks.

As soon as the shaking stops, experts race to inspect dams around the region. A few are found to be leaking at the toe—a sign of potential failure. Emergency responders are spread even thinner when they must begin the evacuation of downstream areas. No dam failures will occur in this particular earthquake.

Thursday 10:33 a.m. (... 33 minutes after the quake began...)

A magnitude 7.0 aftershock begins near the Salton Sea and ruptures to the south. Luckily, this is a relatively unpopulated area. Shak ing and its effects are felt throughout Imperial and San Diego Counties, as well as in Mexicali, Mexico. Damage to a dam in San Diego County requires an evacuation. Teams of firefighters from San Diego County had been getting ready to come north to help with the initial earthquake's aftermath, but are now diverted to respond to the strong aftershock affecting their own county.

Thursday 12:00 noon (...2 hours after the quake began...)

Smaller fires are merging into larger fires in parts of the region where shaking was high and wood buildings are in close proximity. World and national news coverage is focused on urban Los Angeles, especially on a few collapsed buildings. This media focus makes the damage seem even worse than it is, and also more localized. It will be several days before a clear picture emerges of damage around the region.

Fire departments in Arizona and the San Francisco Bay area start to mobilize, but mutual aid is hindered because so many roads into the affected region are impassable. By now, some hospitals are beginning to receive and treat the injured, but with routes and communications disrupted, ambulances struggle to reach victims and get them to hospitals.

November 14, 2008

Friday 03:17 a.m. (... 17+ hours after the quake began...)

A magnitude 7.2 aftershock begins near San Bernardino and ruptures west along the base of the San Gabriel Mountains. This earthquake is considerably larger than 1994's magnitude 6.7 Northridge earthquake, which killed 33 people and cost more than $40 billion. The rupture stops 18 miles east of Pasadena, near Monrovia. The location and size of this earthquake are devastating to the already-weakened infrastructure and overextended emergency response resources. The aftershock triggers damaging aftershocks of its own.

Friday 07:00 a.m. (... 21 hours after the quake began...)

By now, a Presidential Disaster Declaration has been issued, and Federal resources have been committed. The Federal Emergency Management Agency (FEMA), the California Governor's Office of Emergency Services (OES), and Operational Areas for emergency management have set up a Joint Incident Command Center. Communications remain difficult. Ham radio operators begin to assist official responders.

Since first hearing about the earthquake, people outside southern California have been trying to reach family and friends here. Very few have succeeded.

Friday 09:02 a.m. (...23+ hours after the quake began...)

A magnitude 5.6 aftershock rattles residential areas in Rancho Cucamonga. This is a bit bigger than the 1990 Upland earthquake that caused more than $10 million in damage.

Friday 10:00 a.m. (...24 hours after the quake began...)

Utility companies are working around the clock to restore services, yet most people in the areas of heaviest shaking lack electricity, natural gas, and water. Utility workers, like transportation crews, medical staff, and emergency responders, push themselves to do their crucial jobs despite concerns about their own families.

Donations of money, services, and material are arriving from all parts of the United States, and a few Red Cross shelters have opened at public schools and undamaged recreation centers, where food, water, and personnel are available. Most people, particularly in heavily damaged areas, are camped outside.

By now, most stranded motorists have been rescued, and some families are at last reuniting. To get around, emergency responders are using helicopters and any other means of transportation that works, while residents are using bicycles and four-wheel-drive vehicles.

November 15, 2008

Saturday 11:32 p.m. (... 2+ days after the quake began...)

A magnitude 5.7 aftershock occurs with an epicenter in Rialto. This is as large as the 1991 Sierra Madre earthquake that killed one person and caused $40 million in damage.

Fear of looting far exceeds the reality, yet by now many fearful rumors are circulating. Despite official assurances to the contrary, concern grows that if dead bodies are not recovered and transported away, they will cause disease outbreaks. Some are convinced that earthquake scientists are hiding knowledge that an even bigger earthquake is imminent.

November 16, 2008

(... 3 days after the quake began...)

It is getting easier for people outside southern California to make contact with friends and family here. Urban search and rescue teams continue to make rescues, but at a declining rate. Firefighters have extinguished most of the major fires, except where some conflagrations have merged into super-conflagrations—monster fires that consume everything for hundreds of blocks.

Many medical staff members have worked without sleep since they first began responding to the disaster. Seriously damaged hospitals have been evacuated, and open-air trauma centers have been set up in adjacent areas. There is a very short supply of medical equipment such as kidney dialysis machines. Some patients are being med-evacuated outside the region, to hospitals in Nevada, Arizona, and other parts of California. Undamaged hospitals have an influx of earthquake victims with crush injuries, broken bones, and trauma. This increased patient load is not distributed evenly, and some undamaged hospitals are dramatically overloaded, while others receive few patients.

By now, Red Cross shelters have been set up throughout the accessible parts of the affected areas. The donation and distribution of money, services, and material have intensified, yet unmet needs are widespread. A coordinated effort among local, State, and Federal government agencies is starting to bring water and food into the region. Our "just in time" economy does not stockpile goods in warehouses.

The National Guard has been mobilized to handle specialized, earthquake-related law enforcement duties, allowing local law enforcement to return to regular duties. The fear of looting, intensified by media emphasis, begins to abate. Police and security personnel maintain cordons around sites of building collapse, but tenants and owners are allowed to reenter certain other damaged buildings on a very limited basis.

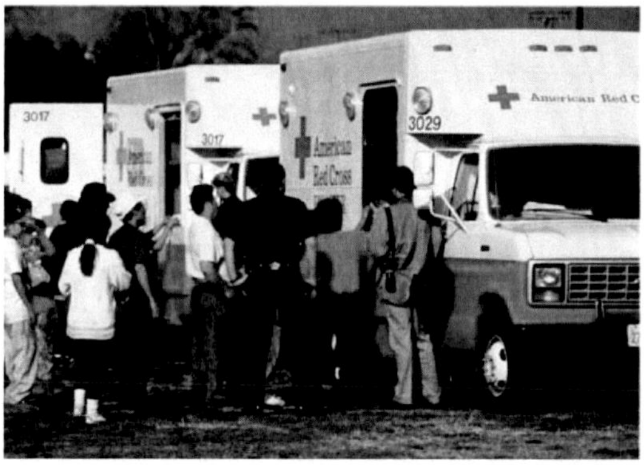

Mid-December, 2008

(...About a month later...)

By now, most gas and electric services have been restored, even in the heavily damaged areas closest to the San Andreas Fault. Landfills store millions of tons of fresh debris. Most of the major roads have reopened, but they are lined with heaps of debris. Small bridges that went down or remain damaged are keeping many local roads closed. Some freeways also remain closed for repairs, where bridge retrofitting prevented collapse but not all damage. Many commuters who work far from home are unable (and some are unwilling) to be back at work regularly. Tens of thousands of people are still without permanent shelter or jobs—or both—because their buildings sustained so much damage.

In the areas where shaking was not severe, most of the water service has been restored, although even in those areas a few communities continue to repair their water systems. Most people with running tap water have been warned to boil their water, because of confirmed or possible contamination due to cracks in the water conveyance system, particularly in areas where the sewer system also incurred damage. It will take a year or more for all the water systems to be certified as safe.

In scattered areas throughout southern California, people still cannot flush toilets because of damage to sewer pipes. Plenty of water is stored in the ground and in reservoirs, but near the fault it is not reaching some homes, schools, and businesses that need it. By now, many of these communities have determined that it will be cheaper and faster to replace rather than hunt for and repair every break in their water systems. So many communities have placed orders for pipes and connectors that the orders are heavily backlogged and there is debate about whose orders should receive priority.

A magnitude 6 aftershock with an epicenter in Cajon Pass offsets restored rail lines and disrupts the movement of goods from the ports once again.

In places without serious structural damage and with water service, businesses have reopened and are trying to get back to normal. Most had not secured their computers, their stock, or their files, or had these resources destroyed when unsecured water pipes broke. The few businesses with earthquake insurance now have the capital to move forward. Other businesses that survived the earthquake with little damage are nonetheless struggling because suppliers have not reopened or customers and employees are unable to reach them easily. Manufacturing plants that have resumed production are facing delays and additional costs to transport goods out of the region; some of their customers are considering other options.

Some residents have left the region to stay with family or friends, due to damage to schools and homes, or loss of jobs. At this point, most still expect to return.

Almost all the public K-12 schools and community colleges have reopened. Only a few of their buildings suffered structural damage, because they have special construction and inspection standards, regulated by the Field Act. In these schools, most of the nonstructural damage has been cleared, although undamaged furniture, equipment, and supplies are at a premium in many places. Class sizes are smaller where families have relocated to less damaged areas or are keeping kids at home. Not all teachers are back on the job.

Universities and private schools are not protected by the Field Act. Many have buildings that have not been retrofitted and thus have incurred significant damage. Some university students have now transferred outside the region.

May, 2009

(...6 months later...)

Water is back in faucets, sinks, and air conditioning units across the region, but it is too late for many businesses, especially smaller businesses that lacked the resources to wait out the bad times.

Businesses forced to close have a domino effect, and as the chances diminish for regaining jobs or finding new ones, more and more people are struggling to rebuild their lives.

By now, some faith communities have been forced to disband, due to severely damaged buildings and dwindling congregations.

Competition for building materials and construction crews is fierce. In communities that had been proactive in disaster planning, rebuilding is well underway. Other communities are still grappling with the complicated procedures and paperwork needed to launch post-disaster reconstruction.

Scattered around the region, there are multifamily residences that remain uninhabitable, with tenants still living outdoors.

Financial institutions face a growing number of loans in default, as businesses fail and individuals give up on recouping their losses.

November, 2010

(... 2 years later...)

Has this earthquake caused a disaster or a catastrophe?

Are things looking up? Is life getting back to normal for most people? Have most residents returned to their homes? Have most lifestyles and businesses resumed?

Or

Are things still going badly? Has recovery faltered? Have the systems that support life in southern California been too severely damaged? How many people have found their lives unbearably difficult since the earthquake struck? How many people have lost what matters most to them?

?

Southern Californians have the opportunity to work together to write this ending. In fact, by intention or not, they are writing this ending every day, through the decisions they make—or don't make—to get ready for earthquakes.

DISTRIBUTION OF SHAKING INTENSITIES IN THE SHAKEOUT SCENARIO EARTHQUAKE

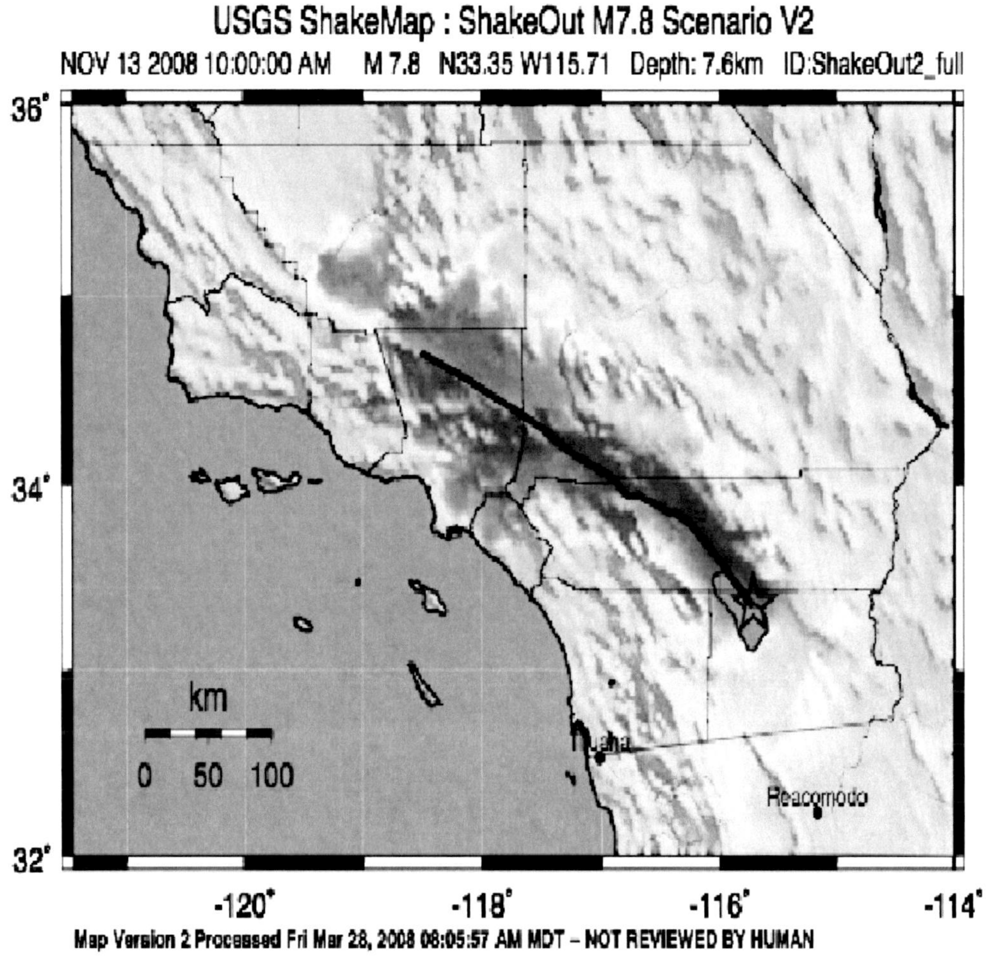

PERCEIVED SHAKING	Not felt	Weak	Light	Moderate	Strong	Very strong	Severe	Violent	Extreme
POTENTIAL DAMAGE	none	none	none	Very light	Light	Moderate	Moderate/Heavy	Heavy	Very Heavy
PEAK ACC.(%g)	<.17	.17-1.4	1.4-3.9	3.9-9.2	9.2-18	18-34	34-65	65-124	>124
PEAK VEL.(cm/s)	<0.1	0.1-1.1	1.1-3.4	3.4-8.1	8.1-16	16-31	31-60	60-116	>116
INSTRUMENTAL INTENSITY	I	II-III	IV	V	VI	VII	VIII	IX	X+

Although an earthquake has only one magnitude and epicenter, different locations will experience different shaking levels. The strongest shaking occurs very near the fault and dies off as seismic waves travel away. Away from the fault, natural basins filled with sediments trap some waves and create pockets of stronger shaking with longer durations. Not every basin traps waves in every earthquake. Consequently, in the ShakeOut earthquake, the San Fernando Valley experiences relatively low levels of shaking compared to the Coachella, San Bernardino, Antelope, and San Gabriel Valleys and the East Los Angeles area. The fault rupture modeled for the ShakeOut Scenario extends along 200 miles of the southern San Andreas Fault, so it would take this section of fault about 100 seconds to rupture. Ground shaking continues as the waves move out, and from the first fault motion near the Salton Sea to the last strong shaking in Ventura, the hypothetical magnitude 7.8 ShakeOut Scenario earthquake would shake southern California for more than 3 minutes. At most sites, strong shaking would last about 15 seconds, but in the basins strong shaking would continue for as much as a minute as the earthquake waves reverberate.

How Do We Write the Ending?

A major earthquake in southern California is inevitable, and it will be a regional disaster. Thousands of square miles will suffer heavy shaking, tens of thousands of people will be injured, hundreds of thousands of buildings will be damaged, and millions of lives and businesses will be disrupted. Whether or not the disaster becomes a catastrophe—an event that permanently changes the nature of life in southern California— will depend on the choices that every person makes between now and the day the earthquake occurs.

By examining the consequences of one hypothetical earthquake and the dynamic interactions among elements of our physical infrastructure and economic and social systems, the ShakeOut Scenario is helping to identify potential points of failure and places where relatively small efforts or investments *before* the next earthquake could yield tremendous benefit *after* the earthquake. Perhaps the biggest lesson of the ShakeOut Scenario is that our individual and collective decisions intertwine. The future vitality of southern California depends on the sum of personal decisions regarding earthquake preparedness and resiliency. The sum includes everyone—tenants and building owners, students and retirees, corporations and corner businesses, individuals and groups, neighborhoods, and governments. Some efforts can be made immediately, such as storing water and protecting families; some efforts will take time, such as increasing volunteer preparedness and collaborative planning; and some efforts will require money, such as retrofitting problem buildings or strengthening infrastructure.

These and many more steps can be taken, starting now, to protect all southern Californians from the earthquake disasters that are certain to come.

The USGS developed the ShakeOut Scenario at the request of community leaders and emergency responders who said that this was the kind of science they needed to improve southern California's earthquake resiliency. Through this and upcoming projects, the USGS and its partners are continuing to evolve the way that hazard science is done, to help protect all of us from natural disasters.

DISASTER OR CATASTROPHE?

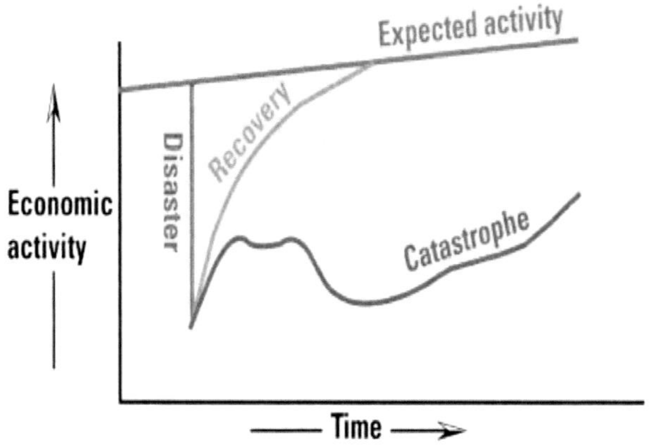

Under normal conditions (blue line), the economic activity in a region will gradually grow with time. When a disaster strikes (red line), assets are lost and many businesses shut down. As power and water service are restored, some businesses reopen, and an influx of insurance payouts and government assistance is used to hire contractors. This can lead to a rapid regeneration of economic activity and a return to economic health within a couple of years (orange line). To an economist, a disaster becomes a catastrophe (purple line) when the regional economy suffers a breakdown in resiliency and sinks into a depression that could last decades. To a sociologist, a disaster becomes a catastrophe when social, economic, and political systems suffer severe disruptions.

Additional Information

The ShakeOut Scenario (U.S. Geological Survey Open-File Report 2008-1150; http://pubs.usgs.gov/of/2008/1150/)

Forecasting California's Earthquakes—What Can We Expect in the Next 30 Years? (U.S. Geological Survey Fact Sheet 2008-3027; http://pubs.usgs.gov/fs/2008/3027/)

http://urbanearth.usgs.gov/ http://www.daretoprepare.org/

Prepare
Now to Ensure a Better
Outcome!!!

All Californians live in earthquake country and should therefore be prepared. Information in English and Spanish on earthquakes in southern California and how to prepare for them can be found in *Putting Down Roots in Earthquake Country* (*Echando Raíces en Tierra de Terremotos*), available at http://www. earthquakecountry.info/roots/.

Photographs in The ShakeOut Scenario Narrative (in order of appearance)— *Trucks on freeway*: courtesy Port of Long Beach; *Los Angeles cityscape*: copyright by and courtesy of Erik Arnesen; *Back to school after Hurricane Katrina*: courtesy FEMA; *Loma Prieta quake damage Salinas*: USGS by J.C. Tinsley; *Crushed bikes Loma Prieta quake*: USGS by C.E. Meyer; *Kitchen damage Northridge quake*: USGS; *Broken gas meter, Watsonville, Loma Prieta quake*: USGS by H.G. Wilshire; *Damaged apartment, Northridge quake*: courtesy FEMA; *2007 Peru quake victims*: copyright and by Michael Mullady; *LA Structure fire*: courtesy Los Angeles County Fire Department by Capt. Larry Collins; *LA firefighter being interviewed*: copyright and by Troy Case, courtesy of Los Angeles County Fire Department; *Nighttime fire*: courtesy Los Angeles County Fire Department by Capt. Larry Collins; *Helicopter, Northridge quake*: courtesy FEMA; *Northridge quake victims with tent*: courtesy FEMA; *Firefighter with chainsaw*: courtesy Los Angeles County Fire Department by Capt. Larry Collins; *Red Cross trucks*: courtesy FEMA; *Loma Pri eta quake damage*: USGS by C.E. Meyer; *Red-tag sign Loma Prieta quake*: USGS by John Nakata; *Bank- owned sign*: USGS by Dale Cox.

In: The Big Shake Implications of a Major Earthquake in… ISBN: 978-1-61728-512-7
Editor: Christopher D. Monahan

Chapter 5

HISTORICAL SEISMICITY IN THE SOUTH SAN FRANCISCO BAY REGION

William H. Bakun

SUMMARY

Locations (intensity centers) and moment magnitude **M** for 24 earthquakes (1858-1911) in the southern San Francisco Bay area are estimated from Modified Mercalli intensity (MMI) assignments. The uncertainties in location and **M** are generally large because there are few MMI assignments available. Preferred locations are selected to be consistent with a simple model (Oppenheimer *et al.*, 2008; Bakun *et al.*, 2008) for seismic activity on the central and southern Calaveras fault. Significant seismic activity can be explained by repetitive failure of the same fault areas in nearly identical earthquakes.

Significant earthquake activity occurred on both east- and west-Bay faults in the ten or so years before the 1868 east- Bay earthquake and before the1906 and 1989 west-Bay earthquakes.

INTRODUCTION

The 31 October 2007 **M**5.4 Alum Rock earthquake provided strong support for Oppenheimer *et al.*'s (1990) model for seismicity on the central and southern Calaveras fault. Nearly identical waveforms for the 2007 event and the 5 September 1955 M_L 5.5 earthquake (Oppenheimer *et al.*, 2008) showed that both events occurred on Oppenheimer *et al.*'s (1990) Zone V on the central Calaveras fault. Oppenheimer *et al.* (1990) had shown that waveforms for the 26 October 1943 M_L 4.9 and the 13 June 1988 **M** 5.0 earthquakes were also nearly identical and that both of these events were located on a patch of the central Calaveras fault which they labeled Zone IV. These recurring **M** 5-6 earthquakes suggest a simple model for seismic activity on the central and southern Calaveras fault. Significant seismic activity can be explained by repetitive failure of the same fault areas in nearly identical earthquakes. This report presents analyses of seismic intensity data necessary to address the question "Is the pre-instrumental seismicity in the region consistent with this simple model of seismicity?" This question is addressed directly by Oppenheimer *et al.* (2008b)

I use Bakun's (1999) list of significant San Francisco Bay region earthquakes, his Table 2 (updated to include events since 1999), as the starting point of this study. The locations and **M** given here are consistent with Bakun's (1999) analysis. The uncertainties in location are often large and Bakun (1999) sometimes chose not to list a preferred location or an intensity center for poorly located events. This report provides details of location and **M** and their uncertainties that were not given by Bakun (1999). In particular, the feasibility of location of historical earthquakes on Oppenheimer *et al.*'s (1990) Calaveras fault Zones I-VI is addressed.

My preferred locations in Table 1 are the epicenters for post-1911 earthquakes and my best guess for the epicenters of the pre-instrumental events. The intensity centers provided by the analysis are used for events with sufficient MMI assignments to reasonably constrain the intensity center locations. The intensity centers, however, are not good estimates of location for many of the events studied here. As an example, consider the earthquake that occurred on 30 March 1883 shown in Figure 13. The maximum intensities were assigned at a cluster of communities straddling the San Andreas Fault east of Monterey Bay. The earthquake generally was not felt at communities in San Luis Obispo County to the south, closer to the intensity center (36.1°N, 122.8°W). M_I at the intensity center is 7.1. A magnitude 7 earthquake in 1883 near the intensity center would have been felt all along the central California coast. For this earthquake, the intensity center is a computational artifact, and it should be ignored. MMI assignments at distant sites are necessary to constrain the location. A few MMI III assignments, if available, would constrain the intensity center to the region of the maximum MMI. Lacking those MMI assignments, the algorithm provides an implausible intensity center. My preferred location, on the San Andreas Fault, near the cluster on maximum MMI, honors the available data and ignores the implausible intensity center. I do not list implausible intensity centers in this report lest the casual reader mistakenly adopt these artifacts as useful information.

My preferred locations for some earthquakes are selected to be consistent with Oppenheimer *et al.*' s (2008b) model of repetitive failure of the same Calaveras fault areas in nearly identical earthquakes. Specifically, my preferred locations for earthquakes on 26 November 1858 (#1) and 2 January 1891 (#17) are in Oppenheimer's (1990) Zone V. Similarly, my preferred location for the earthquake on 20 June 1897 (#19) is in Oppenheimer's (1990) Zone II. Other locations are possible for these three events. Oppenheimer *et al.*'s (2008b) model can be extended through the pre-instruimental period with these preferred locations. My preferred locations for the other pre- instrumental events are elsewhere, *i.e.*, not on the Calaveras fault.

Table 1. Southern Bay Area Earthquakes

#	Date	Preferred	Location	M @ Preferred Location	M (95% Conf. Range)§	# of MMI
a) Historical Earthquakes						
1	26 Nov 1858	Zone V*	37.4°N, 121.8°W	6.2	5.7-6.6	7
2	4 July 1861	N. Calaveras fault	37.7°N, 121.9°W	5.9	5.5-6.2	13
3	26 Feb 1864	San Andreas fault	37.0°N, 121.8°W	5.6	5.0 -6.1	5
4	5 March 1864	Mission fault	37.5°N, 121.9°W	5.3	4.8-5.7	8
5	21 May 1864	San Andreas fault	37.2°N, 122.0°W	5.6	5.0-6.1	4
6	24 May 1865	San Andreas fault	37.0°N, 121.7°W	5.9	5.3-6.4	4
7	8 October 1865	Berrocal fault	37.2°N, 121.9°W	6.5	6.2-6.7	27
8	26 March 1866	San Andreas fault	36.9°N, 121.7°W	5.6	5.1-6.0	6
9	21 October 1868	Hayward fault	37.7°N, 122.1°W	6.8	6.5-7.0	70
10	17 Feb 1870	San Andreas fault	37.2°N, 122.0°W	5.7	5.3-6.1	10
11	2 April 1870	Hayward fault	37.8°N, 122.2°W	5.6	5.2-6.0	9
12	6 March 1882‡	San Andreas fault	36.8°N, 121.6°W	5.9	5.5-6.3	9
13	30 March 1883	San Andreas fault	36.9°N, 121.7°W	5.9	5.5-6.3	9
14	17 Sept 1888	San Andreas fault	37.0°N, 121.7°W	5.0	4.4-5,5	4
15	31 July 1889	Hayward fault	37.7°N, 122.2°W	5.6	5.2-5.9	17
16	24 April 1890	San Andreas fault	37.0°N, 121.8°W	6.3	6.0-6.5	29
17	2 Jan 1891	Zone V*	37.4°N, 121.8°W	5.7	5.3-6.0	16
18	13 Nov 1892	San Andreas fault	36.9°N, 121.7°W	5.8	5.4-6.2	11
19	20 June 1897	Zone II*	37.1°N, 121.5°W	6.3	5.7-6.6	39
20	6 July 1899	San Andreas fault	36.9°N, 121.7°W	5.7	5.3-6.1	11
21	30 April 1899	San Andreas fault	37.0°N, 121.7°W	5.8	5.4-6.1	15
22	11 June 1903	East Bay fault	37.3°N, 121.8°W	6.1	5.9-6.6	26
23	3 August 1903	East Bay fault	37.4°N, 121.9°W	6.1	6.0-6.7	22
b) Instrumental Earthquakes						
24	1 July 1911	Zone III*	37.3°N, 121.7°W	6.2	5.9-6.5	31
25	26 Oct 1943	Zone IV*	37.39°N, 121.77°W	M_L4.9	4.8-5.2	
26	9 March 1949	Zone I*	37.02°N, 121.48°W	M_L5.2	5.0-5.4	
27	5 Sept 1955	Zone V*	37.43°N, 121.78°W	M_L5.5	5.3-5.7	
28	28 Nov 1974	Busch fault	36.91°N, 121.48°W	5.1	4.9-5.3	
29	6 Aug 1979	Zone II*	37.10°N, 121.50°W	5.9	5.7-6.1	
30	24 April 1984	Zone III*	37.32°N, 121.70°W	6.2	6.0-6.2	
31	31 March 1986	Mt. Lewis	37.49°N, 121.69°W	5.7	5.5-5.9	
32	13 June 1988	Zone IV*	37.39°N, 121.77°W	5.0	4.8-5.2	
33	27 June 1988	Lake Elsman	37.13°N, 121.88°W	5.0	4.8-5.2	
34	18 October 1989	Loma Prieta	37.04°N, 121.88°W	6.8	6.6-7.0	
35	31 Oct 2007	Zone V*	37.43°N, 121.78°W	5.4	5.2-5.6	

* Calaveras fault

‡ There is a transcription error in Bakun (1999) where the date of this event is given as 28 June 1882.

§ Calculated for earthquakes #1-24 from the number of MMI assignments using Bakun and Wentworth (1999).

The uncertainty for instrumental **M** is discussed by Bakun (1999) on page 776.

INSTRUMENTAL EARTHQUAKES (1911 – 2008)

Earthquakes in the southern San Francisco Bay region that were recorded on seismographs (Table 1, Part b) are taken from Oppenheimer *et al.* (1990) and Oppenheimer *et al.* (2008). The 9 March 1949 ML 5.2 event and the 6 August 1979 **M** 5.9 Coyote Lake earthquake occurred on Zones I and II, respectively. The 1 July 1911 **M** 6.2 and the 24 April 1984 **M** 6.2 Morgan Hill earthquakes occurred on Zone III. The 26 October 1943 M_L4.9 and the 13 June 1988 **M** 5.0 earthquakes occurred on Zone IV. The 5 September 1955 M_L5.5 and the 31 October 2007 **M** 5.4 earthquakes occurred on Zone V. No earthquakes are known to have occurred in Zone VI. Historical (pre-1911) earthquakes in the region are listed in Table 1, Part a.

Intensity Data

Almost all of the MMI assignments used here are listed in Bakun (1998). J. Boatwright and H. Bundock (personal communication, 2008) have compiled more-extensive sets of MMI assignments for the 4 July 1861 northern Calaveras fault earthquake and for the 21 October 1868 Hayward fault earthquake. Their MMI assignments are used in the analysis of the 1861 event. Toppozada *et al. 's* (1981) MMI assignments and Boatwright and Bundock' s (personal communication, 2008) MMI assignments are used for independent analyses of the 1868 earthquake.

Analysis of Intensity Data

Following Bakun (1999), I estimate **M** from individual intensity observations for a trial epicenter using

$$M_I = \text{mean}\ (M_i), \tag{1}$$

where

$$M_i = \{(MMI_i - \delta MMI_i) + 3.29 + 0.0206\Delta_{h,i}\}/1.68, \tag{2}$$

MMI_i, MMI_i and $\Delta_{h,i}$ are the MMI value, the site correction and the hypocentral distance, respectively, at site i. A depth of 10 km is assumed. The δMMI are listed in Bakun (1998). δMMI_i is zero if a site correction is not available.

I find the misfit for each trial epicenter from

$$\text{rms}\ [M_I] = [\text{rms}\ (M_I - M_i) - \text{rms}_0(M_I - M_i)], \tag{3}$$

where $\text{rms}\ (M_I - M_i) = \{\ _i[W_i(M_I - M_i)]^2 / \ _iW_i^2\}^{1/2}$, $\text{rms}_0(M_I - M_i)$ is the minimum rms $(M_I - M_i)$ over the grid of trial epicenters, and W_i is Bakun and Wentworth's (1997) distance weighting function:

$$W_i = \begin{cases} 0.1 + \cos[(\Delta_i/150)(\ /2)] & \text{for } \Delta_i < 150 \text{ km} \\ 0.1 & \text{for } \Delta_i \geq 150 \text{ km.} \end{cases} \quad (4)$$

The intensity center is the trial source location for which rms [M_I] is minimum (Bakun, 1999).

The rms [M_I] contours provide a basis for identifying regions of likely epicenter location. The rms [M_I] contours bound the epicentral region. Bakun and Wentworth (1997) associated rms contour values with confidence levels that the epicenter was within the contour, as tabulated in the corrected Table 5b of Bakun and Wentworth (1999). M_I at trial locations within the appropriate confidence-level contours is the best estimate of **M** for these source locations.

Analysis of MMI for 24 Historical Earthquakes (1858-1911)

26 November 1858

MMI can be assigned at seven of the 16 localities for which felt reports (Toppozada *et al.*, 1981) are available. The maximum MMI was assigned at San Jose where brick walls were cracked and chimneys thrown down. Zones III-VI on the Calaveras fault are within the 67% confidence contour for location (Figure 1). MI is 6.2 for zones IV and V. The preferred location is in zone V at the epicenter of the 31 Oct 2007 event.

4 July 1861

MMI have been assigned at 13 sites (J. Boatwright and H. Bundock, personal communication, 2008). The maximum intensity is VIII-IX in Amador Valley. MMI VII-VIII intensities were assigned at residences in San Ramon Valley near the location of reported fissures along the Calaveras fault near Dublin (Toppozada *et al.*, 1981) and the intensity center (Figure 2). The intensity center, the preferred location, is located near the trace of the northern Calaveras fault. M_I is 5.9.

26 February 1864

MMI can be assigned at five of the 11 localities for which felt reports (Toppozada *et al.*, 1981) are available. The maximum assigned MMI VI was assigned at Monterey, San Francisco, San Jose, and Watsonville. Zones I-VI on the Calaveras fault are within the 67% confidence contour for location (Figure 3). Although a location on the Calaveras fault is possible, the maximum MMI at Monterey, San Jose and Watsonville are more consistent with a west bay fault. A location near the San Andreas Fault is preferred. M_I is 5.6.

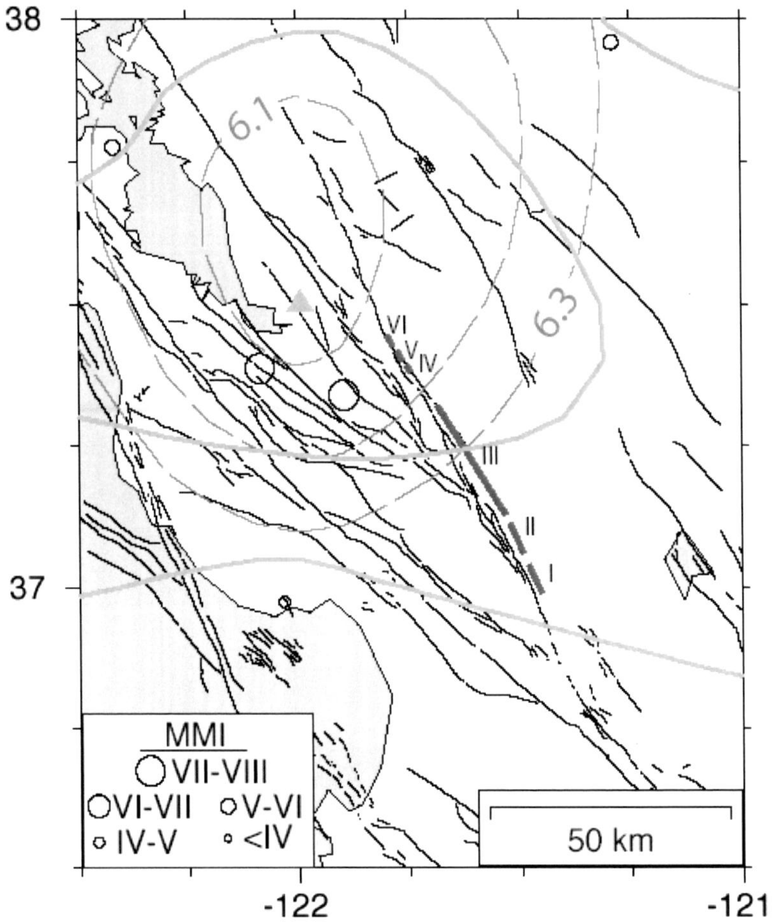

Figure 1. 26. November 1858 earthquake.
(MMI assignment-Site Correction) at 7 sites (some sites may be off the map) are shown as *circles* relative to Jenning's (1994) active faults (*black lines*). The contours of M_I (*dashed red lines*) are the best estimates of **M** from the MMI data for assumed epicenters on that contour. The rms [M_I] contours corresponding to the 67% (innermost contour) and 95% confidence contour (outermost contour) for location (Bakun and Wentworth, 1999) are shown as *solid green lines*. The intensity center (Bakun, 1999) is a *green triangle*. Locations of Oppenheimer *et al.*'s (1990) zones I-VI on the central and southern Calaveras fault are shown as *violet lines*.

5 March 1864

MMI can be assigned at eight of the 19 localities for which felt reports (Toppozada *et al.*, 1981) are available. The maximum MMI VI was assigned at San Francisco, San Jose, and Stockton. Zones I-VI on the Calaveras fault are within the 67% confidence contour for location (Figure 4). M_I is 5.3 for zones IV-VI. A location near the Mission siesmic trend is preferred.

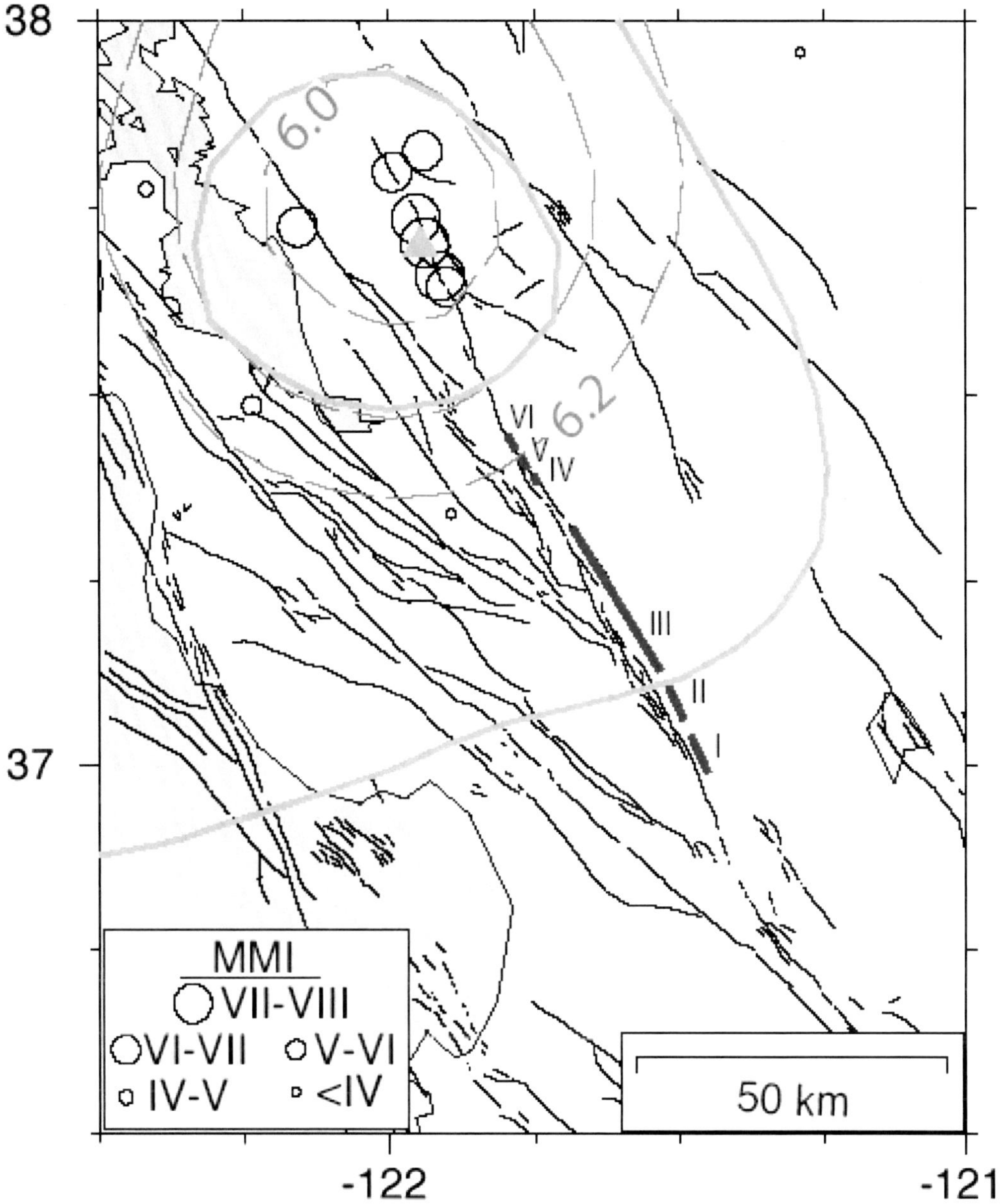

Figure 2. 4. July 1861 earthquake.
(MMI assignment-Site Correction) at 13 sites (some sites may be off the map) are shown as *circles* relative to Jenning's (1994) active faults (*black lines*). The contours of MI (*dashed red lines*) are the best estimates of **M** from the MMI data for assumed epicenters on that contour. The rms [M_I] contours corresponding to the 67% (innermost contour) and 95% confidence contour (outermost contour) for location (Bakun and Wentworth, 1999) are shown as *solid green lines*. The intensity center (Bakun, 1999) is a *green triangle*. Locations of Oppenheimer *et al.*'s (1990) zones I-VI on the central and southern Calaveras fault are shown as *violet* lines.

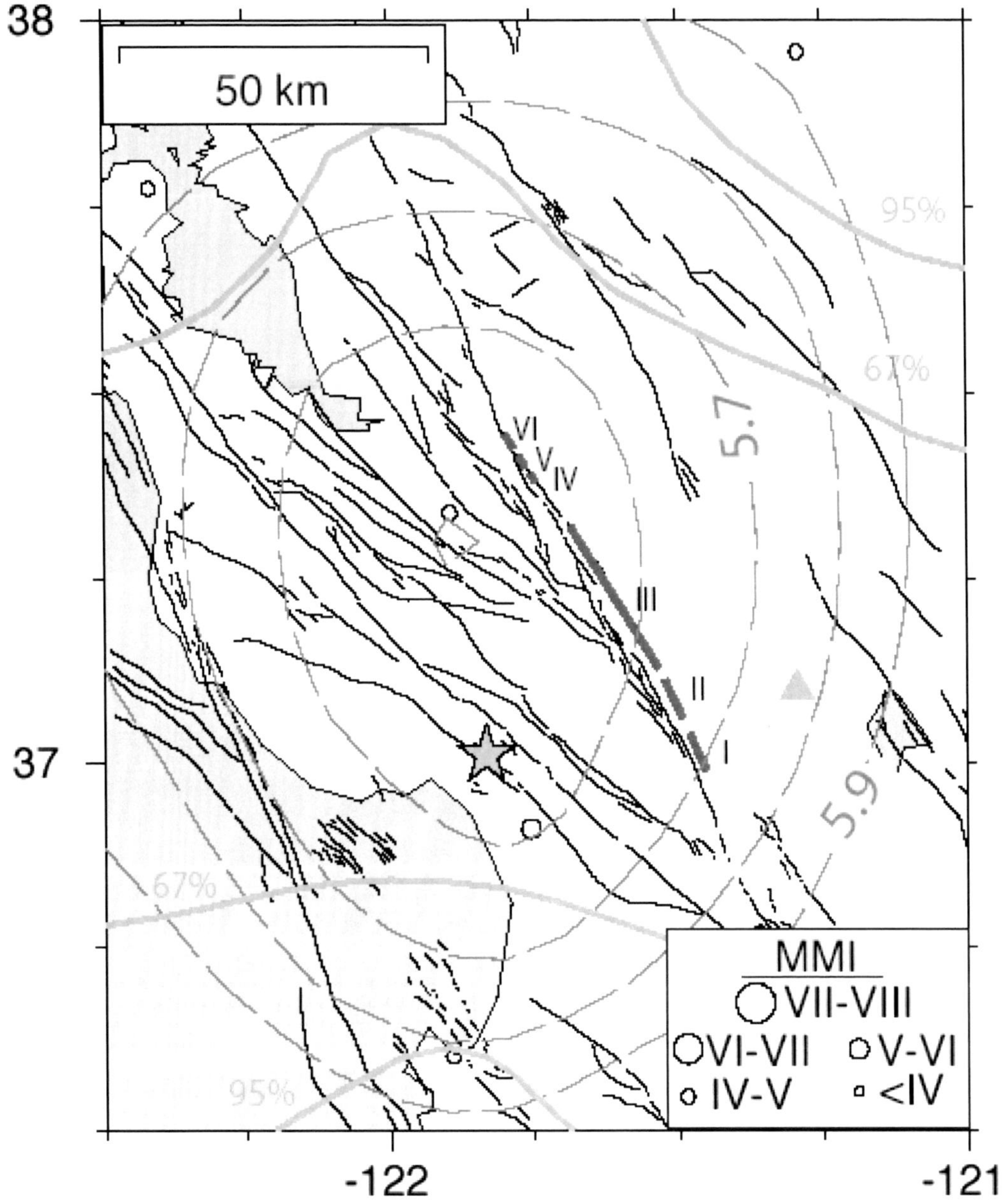

Figure 3. 26. February 1864 earthquake.
(MMI assignment-Site Correction) at 5 sites (some sites may be off the map) are shown as *circles* relative to Jenning's (1994) active faults (*black lines)*. The contours of M_I (*dashed red lines*) are the best estimates of **M** from the MMI data for assumed epicenters on that contour. The rms [M_I] contours corresponding to the 67% (innermost contour) and 95% confidence contour (outermost contour) for location (Bakun and Wentworth, 1999) are shown as *solid green lines*. The intensity center (Bakun, 1999) is a *green triangle*. The preferred location is the green star. Locations of Oppenheimer *et al.*'s (1990) zones I-VI on the central and southern Calaveras fault are shown as *violet lines*.

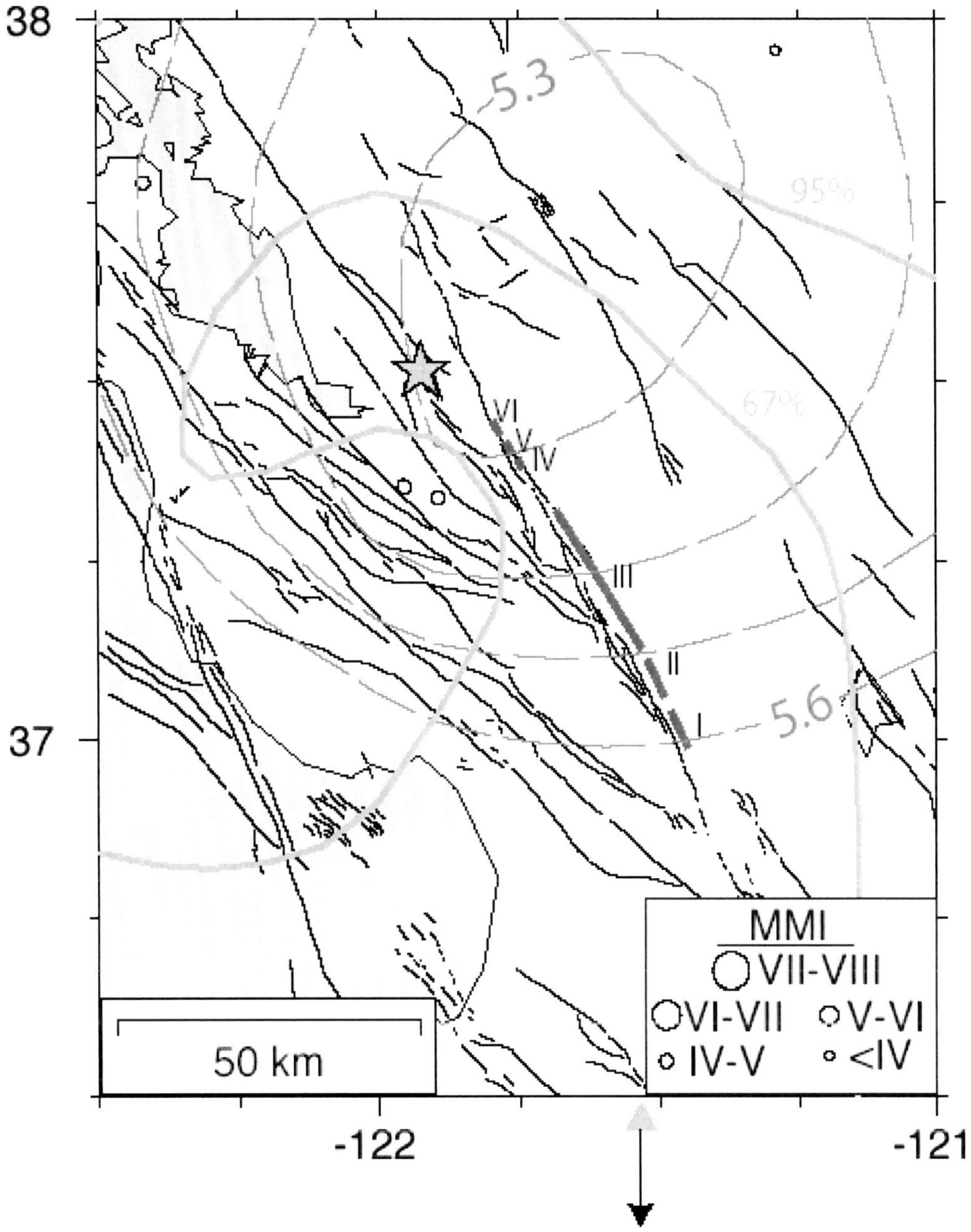

Figure 4. 5. March 1864 earthquake.
(MMI assignment-Site Correction) at 8 sites (some sites may be off the map) are shown as *circles* relative to Jenning's (1994) active faults (black lines). The contours of M_I (*dashed red lines*) are the best estimates of **M** from the MMI data for assumed epicenters on that contour. The rms [M_I] contours corresponding to the 67% (innermost contour) and 95% confidence contour (outermost contour) for location (Bakun and Wentworth, 1999) are shown as *solid green lines.* The intensity center (Bakun, 1999) is a *green triangle and id located off the map.* The preferred location is the green star. Locations of Oppenheimer et al.'s (1990) zones I-VI on the central and southern Calaveras fault are shown as *violet lines.*

21 May 1864

MMI can be assigned at four of the 10 localities for which felt reports (Toppozada *et al.*, 1981) are available. The maximum MMI VI was assigned at San Francisco and San Jose. Zones I-VI on the Calaveras fault are within the 67% confidence contour for location but a location near the San Andreas Fault, at the intensity center, is preferred (Figure 5). M_I is 5.6.

24 May 1865

MMI can be assigned at four of the 7 localities for which felt reports (Toppozada *et al.*, 1981) are available. The maximum MMI VI was assigned at San Juan Bautista. Zones I-III on the Calaveras fault are within the 67% confidence contour for location, the maximum MMI at San Juan Bautista is more consistent with a west bay fault (Figure 6). M_I is 5.9.

8 October 1865

MMI can be assigned at 27 localities for which felt reports (Toppozada *et al.*, 1981) are available. Bakun's (1999) analysis is adopted. The intensity center, the preferred location, is near the Berrocal fault (Figure 7). M_I is 6.5.

26 March 1866

MMI can be assigned at six of the 12 localities for which felt reports (Toppozada *et al.*, 1981) are available. The maximum MMI VII was assigned at Gilroy. Zones I-II on the Calaveras fault are within the 67% confidence contour for location, but a location closer to the San Andreas Fault is preferred (Figure 8). M_I is 5.6.

21 October 1868

MMI were assigned at 70 localities by Toppozada *et al.* (1981). Bakun's (1999) analysis of that data yields an intensity center on the Hayward fault and M_I is 6.8 (Figure 9a). J. Boatwright and H. Bundock (personal communication, 2008) have reexamined the accounts for the 1868 earthquake and have assigned MMI at 133 sites. The intensity center based on these MMI assignments is also on the trace of the Hayward fault, but farther south near the intersection of the Hayward fault and Mission seismic trend (Figure 9b). M_I is 6.5. The preferred location is at the intensity center in Figure 9a.

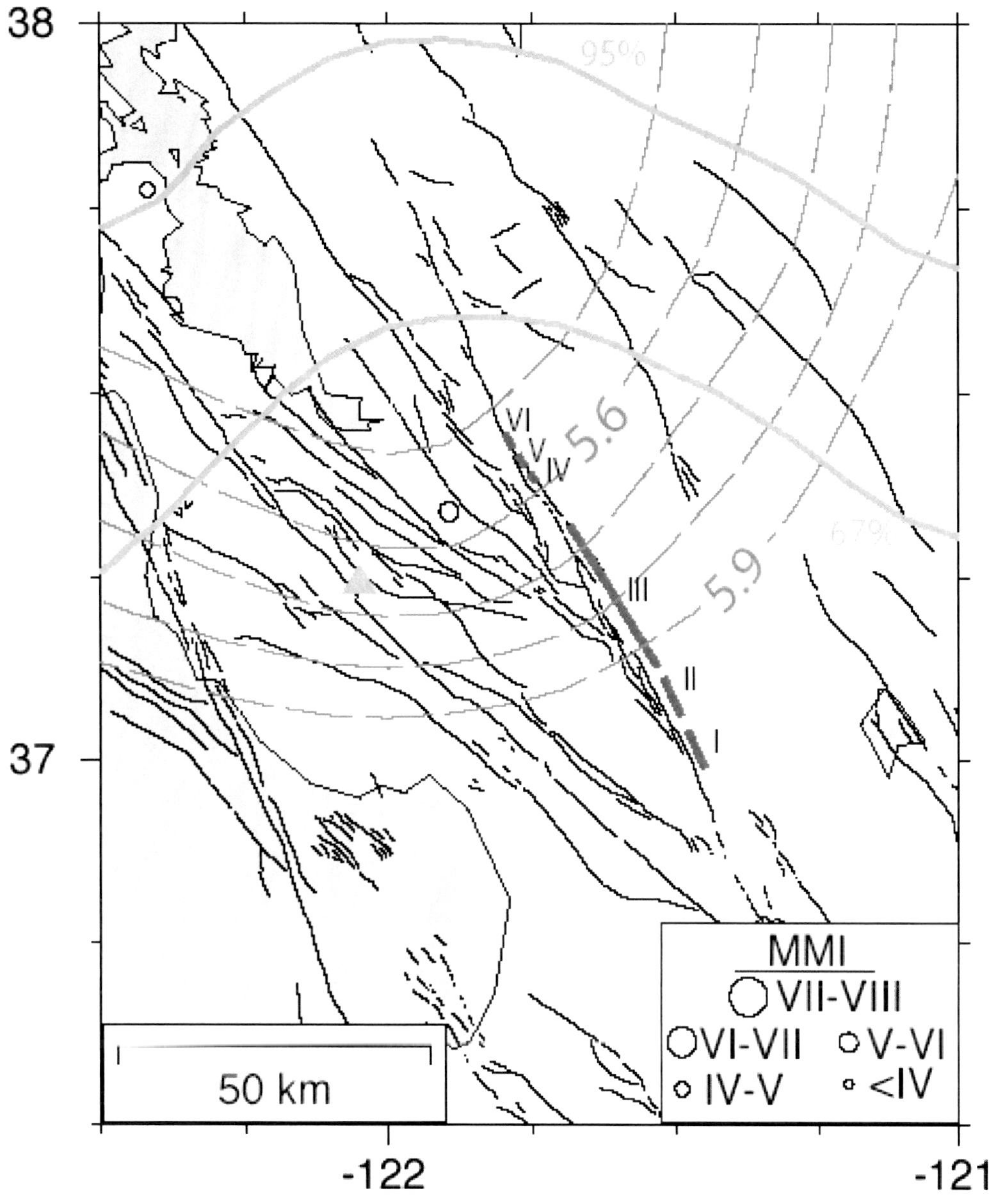

Figure 5. 21. May 1864 earthquake.

(MMI assignment-Site Correction) at 4 sites (some sites may be off the map) are shown as *circles* relative to Jenning's (1994) active faults (*black lines*). The contours of M_I (*dashed red lines*) are the best estimates of **M** from the MMI data for assumed epicenters on that contour. The rms [M_I] contours corresponding to the 67% (innermost contour) and 95% confidence contour (outermost contour) for location (Bakun and Wentworth, 1999) are shown as *solid green lines*. The intensity center (Bakun, 1999) is a *green triangle*. Locations of Oppenheimer *et al.*'s (1990) zones I-VI on the central and southern Calaveras fault are shown as *violet lines*.

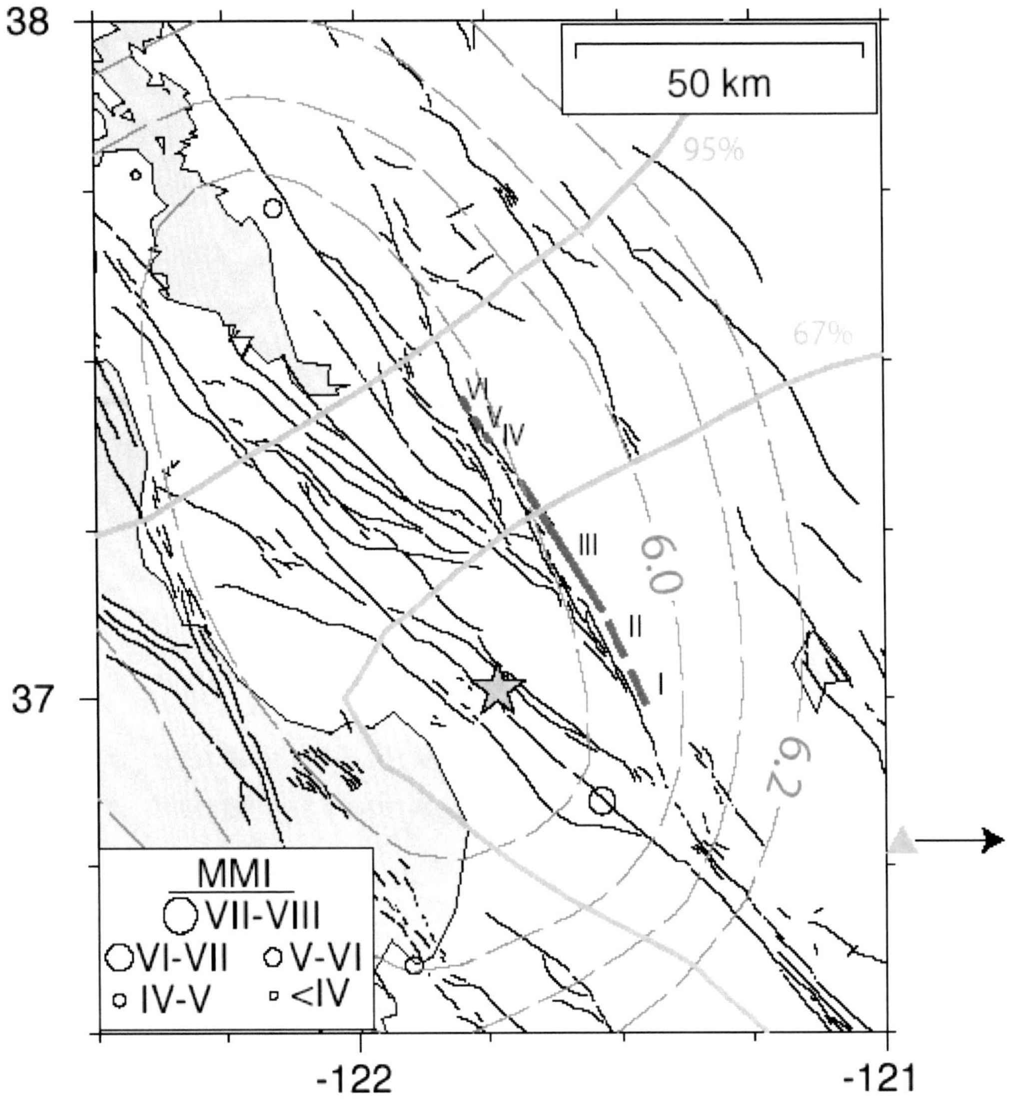

Figure 6. 24. May 1865 earthquake.
(MMI assignment-Site Correction) at 4 sites (some sites may be off the map) are shown as *circles* relative to Jenning's (1994) active faults (*black lines*). The contours of M_I (*dashed red lines*) are the best estimates of **M** from the MMI data for assumed epicenters on that contour. The rms [M_I] contours corresponding to the 67% (innermost contour) and 95% confidence contour (outermost contour) for location (Bakun and Wentworth, 1999) are shown as *solid green lines*. The intensity center (Bakun, 1999) is a *green triangle and is located off the map*. The preferred location is the green star. Locations of Oppenheimer *et al.*'s (1990) zones I-VI on the central and southern Calaveras fault are shown as *violet lines*.

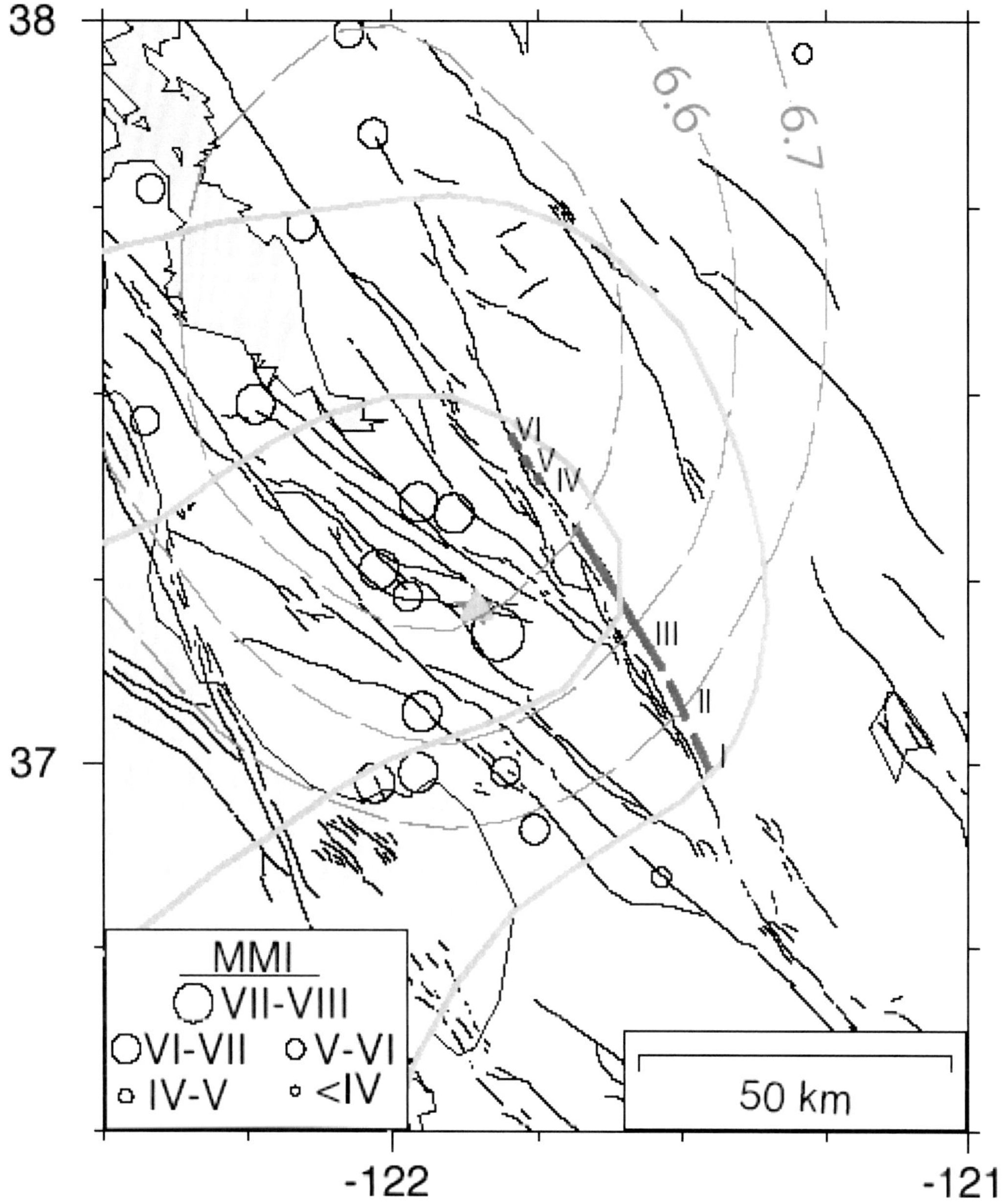

Figure 7. 8. October 1865 earthquake.
(MMI assignment-Site Correction) at 27 sites (some sites may be off the map) are shown as *circles* relative to Jenning's (1994) active faults (*black lines).* The contours of M_I (*dashed red lines*) are the best estimates of **M** from the MMI data for assumed epicenters on that contour. The rms [M_I] contours corresponding to the 67% (innermost contour) and 95% confidence contour (outermost contour) for location (Bakun and Wentworth, 1999) are shown as *solid green lines*. The intensity center (Bakun, 1999) is a *green triangle*. Locations of Oppenheimer *et al.*'s (1990) zones I-VI on the central and southern Calaveras fault are shown as *violet lines*.

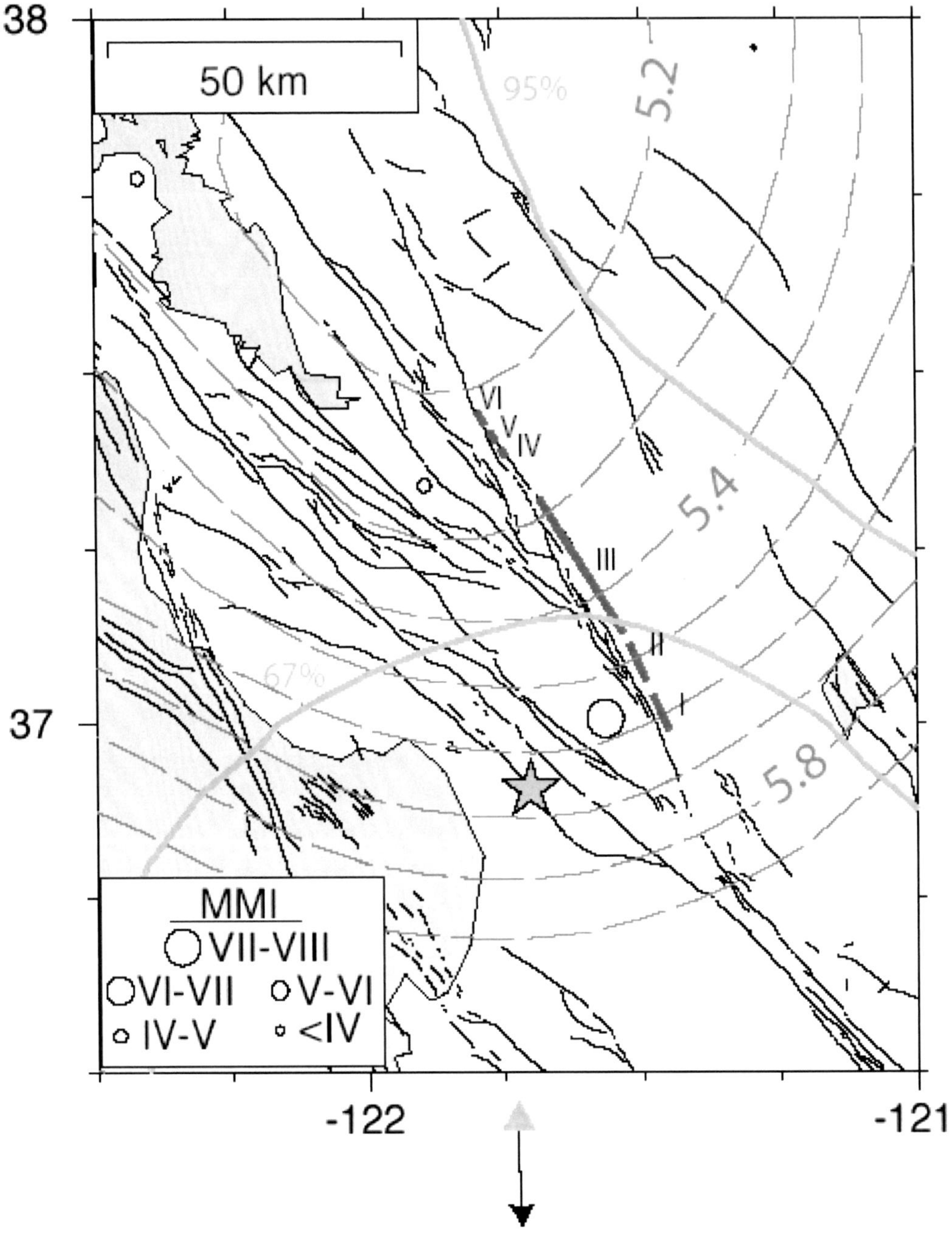

Figure 8. 26. March 1866 earthquake.
(MMI assignment-Site Correction) at 6 sites (some sites may be off the map) are shown as *circles* relative to Jenning's (1994) active faults (*black lines*). The contours of M_I (*dashed red lines*) are the best estimates of M from the MMI data for assumed epicenters on that contour. The rms [M_I] contours corresponding to the 67% (innermost contour) and 95% confidence contour (outermost contour) for location (Bakun and Wentworth, 1999) are shown as *solid green lines*. The intensity center (Bakun, 1999) is a *green triangle and is located off the map*. The preferred location is the green star. Locations of Oppenheimer *et al.*'s (1990) zones I-VI on the central and southern Calaveras fault are shown as *violet lines*.

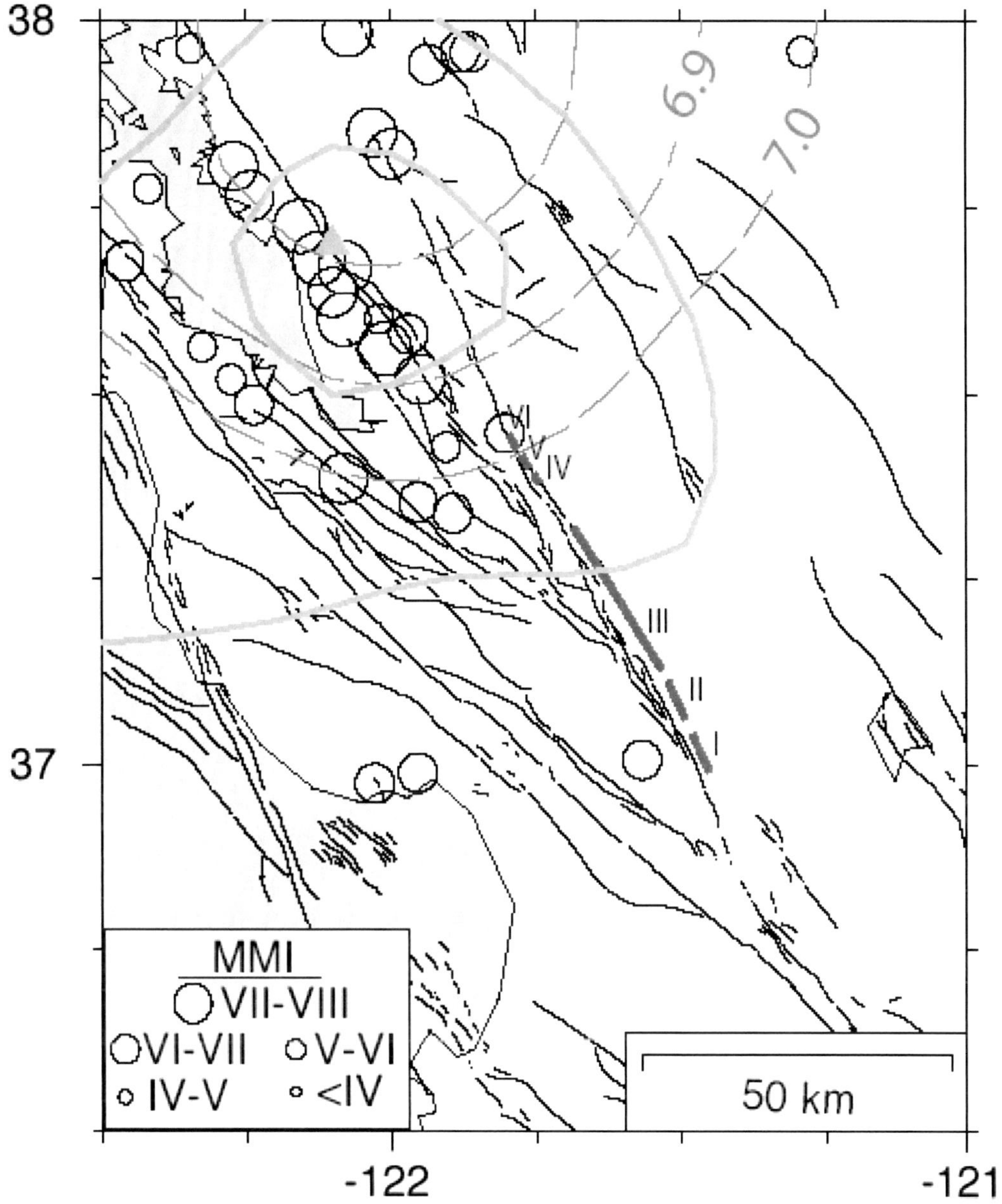

Figure 9a. 21. October 1868 earthquake.
(MMI assignment-Site Correction) at 70 sites (some sites may be off the map) are shown as *circles* relative to Jenning's (1994) active faults (*black lines*). The contours of M_I (*dashed red lines*) are the best estimates of **M** from the MMI data for assumed epicenters on that contour. The rms [M_I] contours corresponding to the 67% (innermost contour) and 95% confidence contour (outermost contour) for location (Bakun and Wentworth, 1999) are shown as *solid green lines*. The intensity center (Bakun, 1999) is a *green triangle*. Locations of Oppenheimer *et al.*'s (1990) zones I-VI on the central and southern Calaveras fault are shown as *violet lines*.

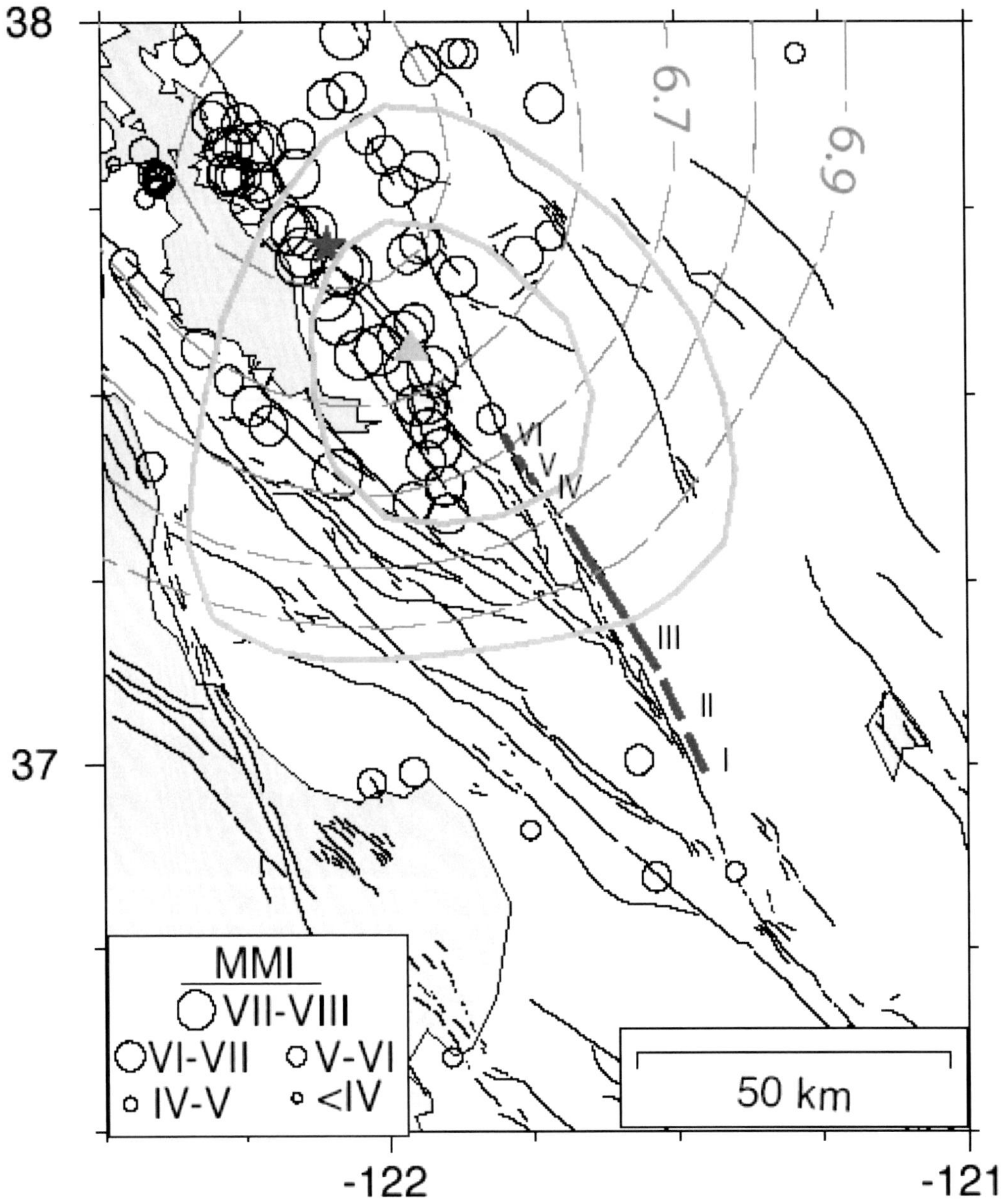

Figure 9b. 21. October 1868 earthquake.
(MMI assignment-Site Correction) at 133 sites (some sites may be off the map) assigned by J. Boatwright (personal communication, 2008) are shown as *circles* relative to Jenning's (1994) active faults (*black lines*). The contours of M_I (*dashed red lines*) are the best estimates of **M** from the MMI data for assumed epicenters on that contour. The rms [M_I] contours corresponding to the 67% (innermost contour) and 95% confidence contour (outermost contour) for location (Bakun and Wentworth, 1999) are shown as *solid green lines*. The intensity center (Bakun, 1999), a *green triangle, is located on the Hayward fault about 20-25 km south of Bakun's (1999) intensity center (violet star)* . Locations of Oppenheimer *et al.*'s (1990) zones I-VI on the southern Calaveras fault are shown as *violet lines*.

17 February 1870

MMI can be assigned at ten of the 21 localities for which felt reports (Toppozada *et al.*, 1981) are available. The maximum MMI VII was assigned at Los Gatos and Santa Cruz. Zones III-VI on the Calaveras fault are within the 67% confidence contour for location, but a location near the 8 October 1865 epicenter is perhaps more likely (Figure 10). The preferred location is at the intensity center. M_I is 5.7.

2 April 1870

MMI can be assigned at nine of the 14 localities for which felt reports (Toppozada *et al.*, 1981) are available. The maximum MMI VI was assigned at Oakland, San Leandro, San Francisco and Petaluma. The intensity center is on the 1868 Hayward fault rupture (Figure 11). The preferred location is at the intensity center. M_I is 5.6.

6 March 1882

MMI can be assigned at nine of the 18 localities for which felt reports (Toppozada *et al.*, 1981) are available. The maximum MMI VI was assigned at Salinas, Hollister, and Santa Cruz. The preferred location is on the San Andreas Fault (Figure 12). M_I is 5.0.

30 March 1883

MMI can be assigned at nine of the 25 localities for which felt reports (Toppozada *et al.*, 1981) are available. The maximum MMI VII was assigned at Old Gilroy and Sargents where chimneys were thrown down. The preferred location is on the San Andreas Fault (Figure 13). M_I is 5.9.

17 September 1888

MMI can be assigned at four sites with MM V at Gilroy, San Jose, and Santa Cruz, suggesting a west bay location. The preferred location is on the San Andreas Fault (Figure 14). M_I is 5.9.

31 July 1889

MMI can be assigned at seventeen of the 41 localities for which felt reports (Toppozada *et al.*, 1981) are available. The maximum MMI VII was assigned at San Leandro and West Oakland where chimneys were thrown down. The intensity center is near the 1868 Hayward fault rupture (Figure 15). The preferred location is at the intensity center. M_I is 5.6.

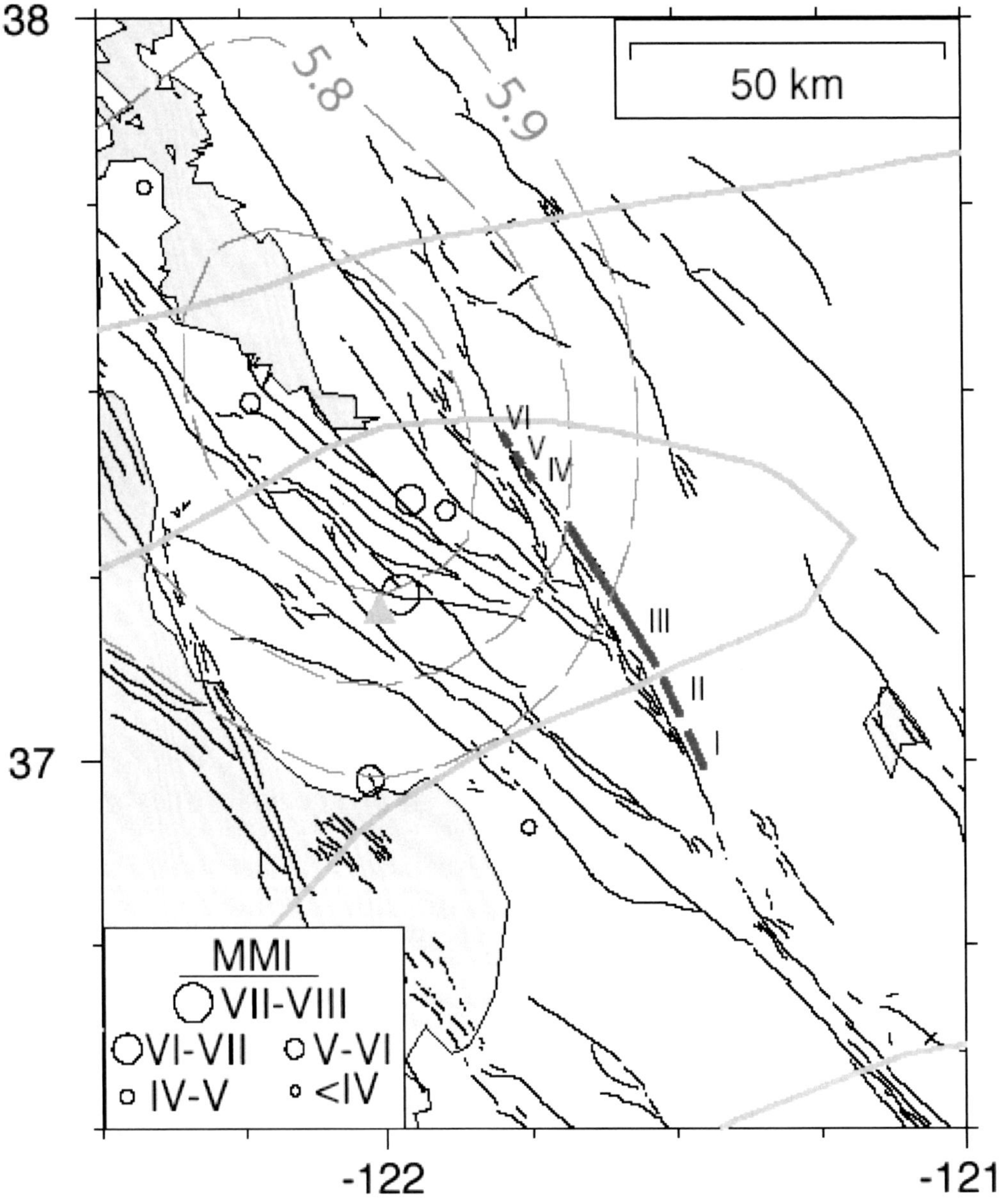

Figure 10. 17. February 1870 earthquake.
(MMI assignment-Site Correction) at 10 sites (some sites may be off the map) are shown as *circles* relative to Jenning's (1994) active faults (*black lines*). The contours of M_I (*dashed red lines*) are the best estimates of **M** from the MMI data for assumed epicenters on that contour. The rms [M_I] contours corresponding to the 67% (innermost contour) and 95% confidence contour (outermost contour) for location (Bakun and Wentworth, 1999) are shown as *solid green lines*. The intensity center (Bakun, 1999) is a *green triangle*. Locations of Oppenheimer *et al.*'s (1990) zones I-VI on the central and southern Calaveras fault are shown as *violet lines*.

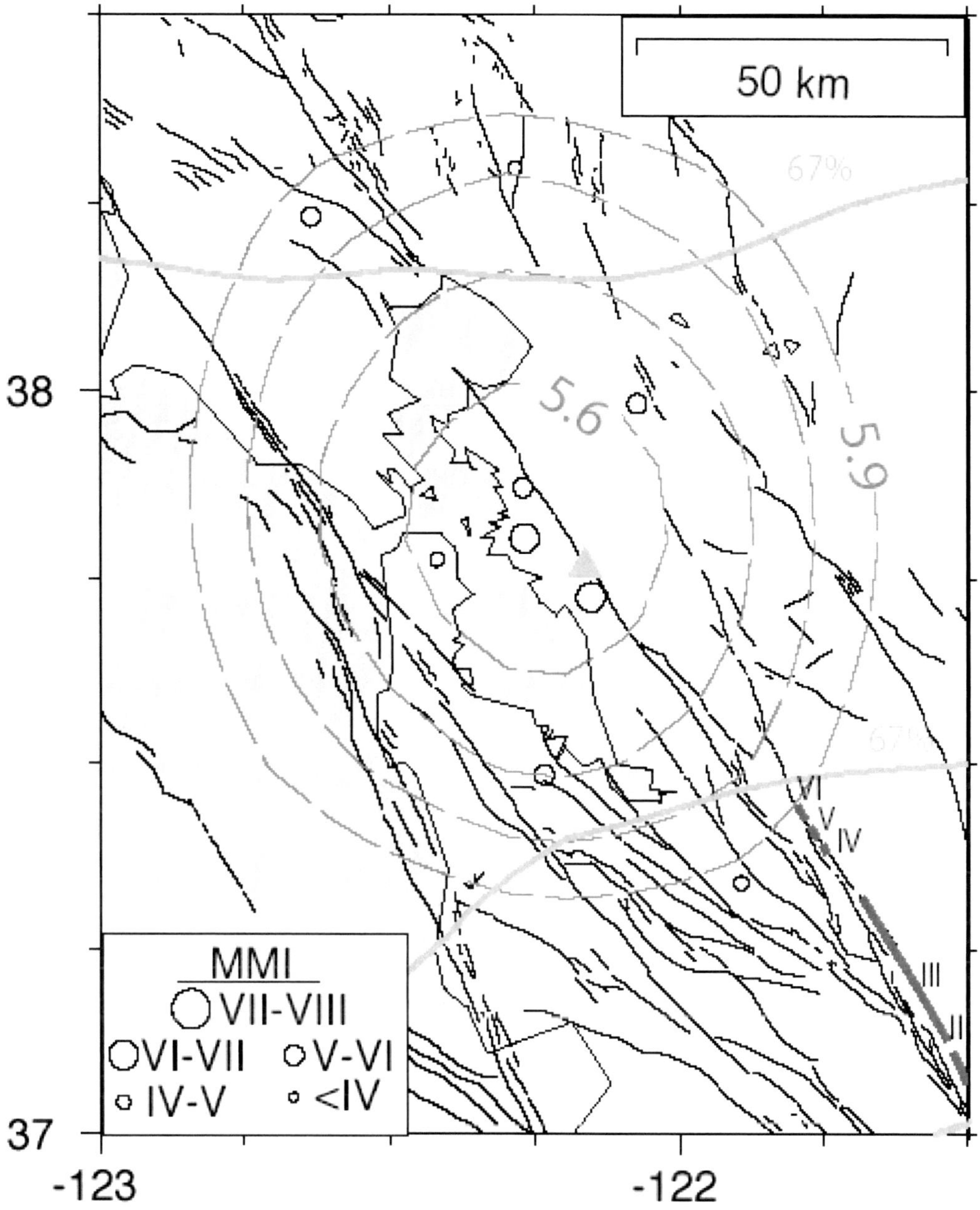

Figure 11. 2. April 1870 earthquake.
(MMI assignment-Site Correction) at 9 sites (some sites may be off the map) are shown as *circles* relative to Jenning's (1994) active faults (*black lines*). The contours of M_I (*dashed red lines*) are the best estimates of **M** from the MMI data for assumed epicenters on that contour. The rms [M_I] contours corresponding to the 67% (innermost contour) and 95% confidence contour (outermost contour) for location (Bakun and Wentworth, 1999) are shown as *solid green lines*. The intensity center (Bakun, 1999) is a *green triangle*. Locations of Oppenheimer *et al.*'s (1990) zones I-VI on the central and southern Calaveras fault are shown as *violet lines*.

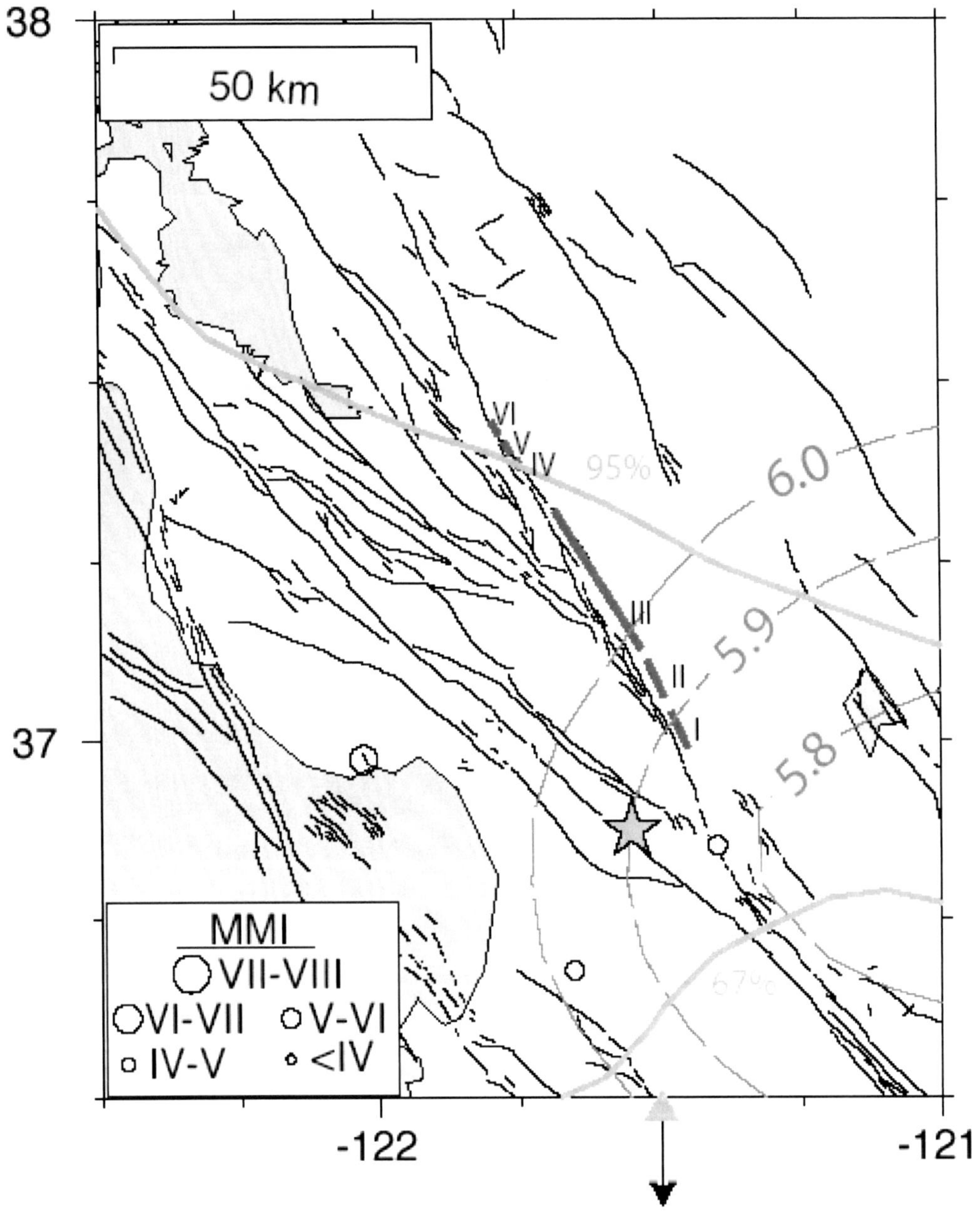

Figure 12. 6. March 1882 earthquake.
(MMI assignment-Site Correction) at 9 sites (some sites may be off the map) are shown as *circles* relative to Jenning's (1994) active faults (*black lines).* The contours of M_I (*dashed red lines*) are the best estimates of **M** from the MMI data for assumed epicenters on that contour. The rms [M_I] contours corresponding to the 67% (innermost contour) and 95% confidence contour (outermost contour) for location (Bakun and Wentworth, 1999) are shown as *solid green lines*. The intensity center (Bakun, 1999) is a *green triangle*. The preferred location is the green star. Locations of Oppenheimer *et al.*'s (1990) zones I-VI on the central and southern Calaveras fault are shown as *violet lines*.

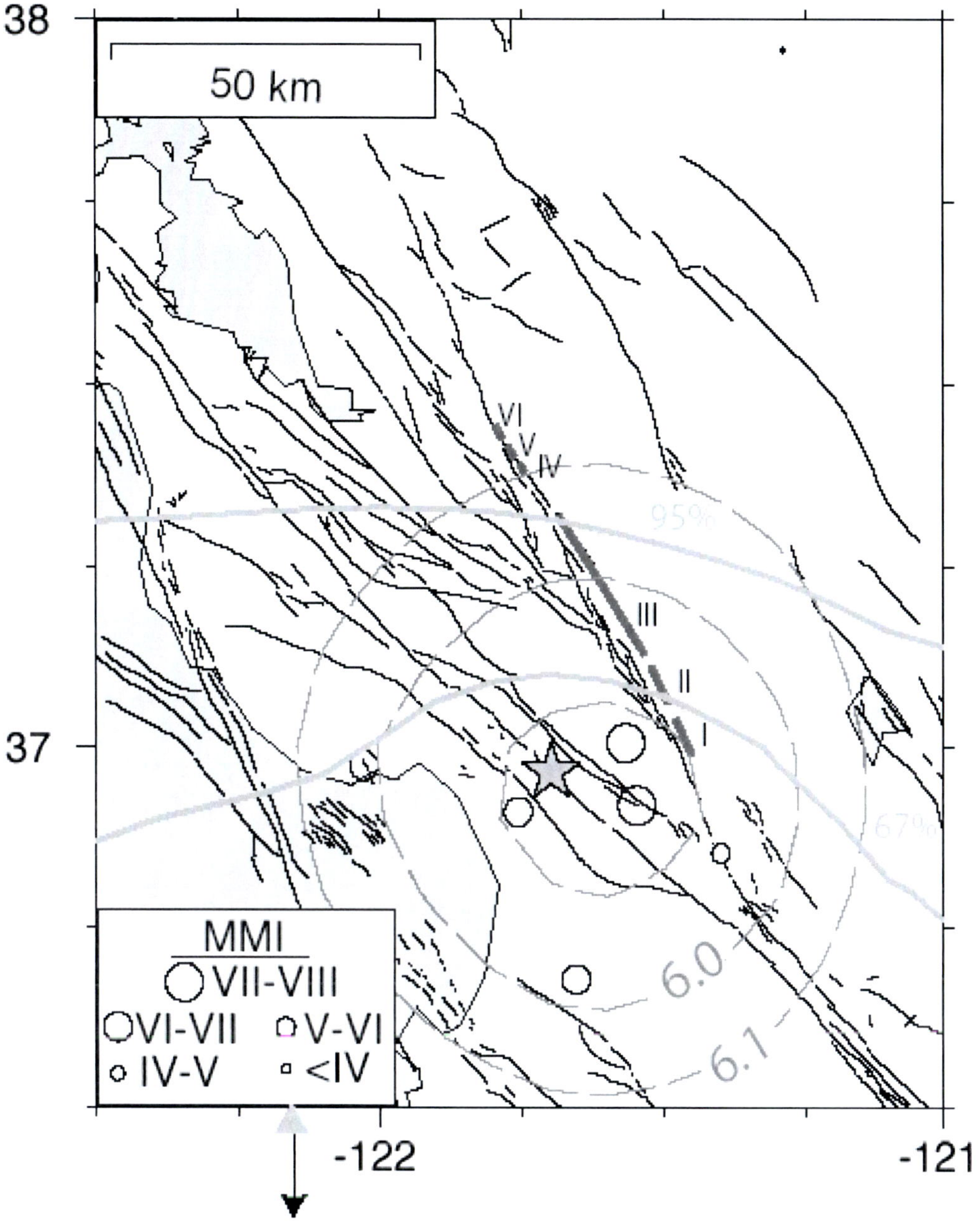

Figure 13. 30. March 1883 earthquake.
(MMI assignment-Site Correction) at 9 sites (some sites may be off the map) are shown as *circles* relative to Jenning's (1994) active faults (*black lines)*. The contours of M_I (*dashed red lines*) are the best estimates of **M** from the MMI data for assumed epicenters on that contour. The rms [M_I] contours corresponding to the 67% (innermost contour) and 95% confidence contour (outermost contour) for location (Bakun and Wentworth, 1999) are shown as *solid green lines*. The intensity center (Bakun, 1999) is a *green triangle*. The preferred location is the green star. Locations of Oppenheimer *et al.*'s (1990) zones I-VI on the central and southern Calaveras fault are shown as *violet lines*.

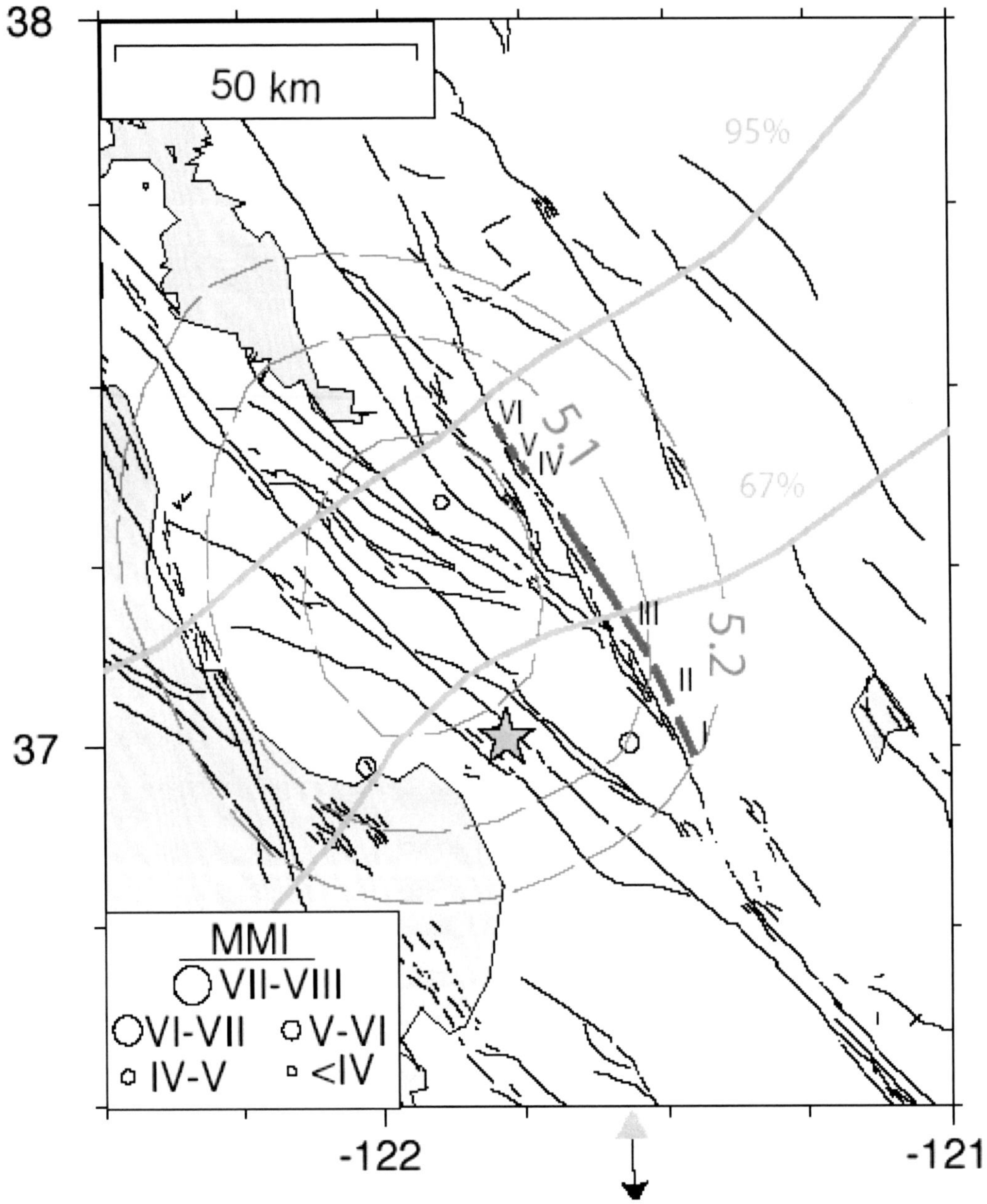

Figure 14. 17. September 1888 earthquake.
(MMI assignment-Site Correction) at 4 sites (some sites may be off the map) are shown as *circles* relative to Jenning's (1994) active faults (*black lines)*. The contours of M_I (*dashed red lines*) are the best estimates of **M** from the MMI data for assumed epicenters on that contour. The rms [M_I] contours corresponding to the 67% (innermost contour) and 95% confidence contour (outermost contour) for location (Bakun and Wentworth, 1999) are shown as *solid green lines*. The intensity center (Bakun, 1999) is a *green triangle*. The preferred location is the green star. Locations of Oppenheimer *et al.*'s (1990) zones I-VI on the central and southern Calaveras fault are shown as *violet lines*.

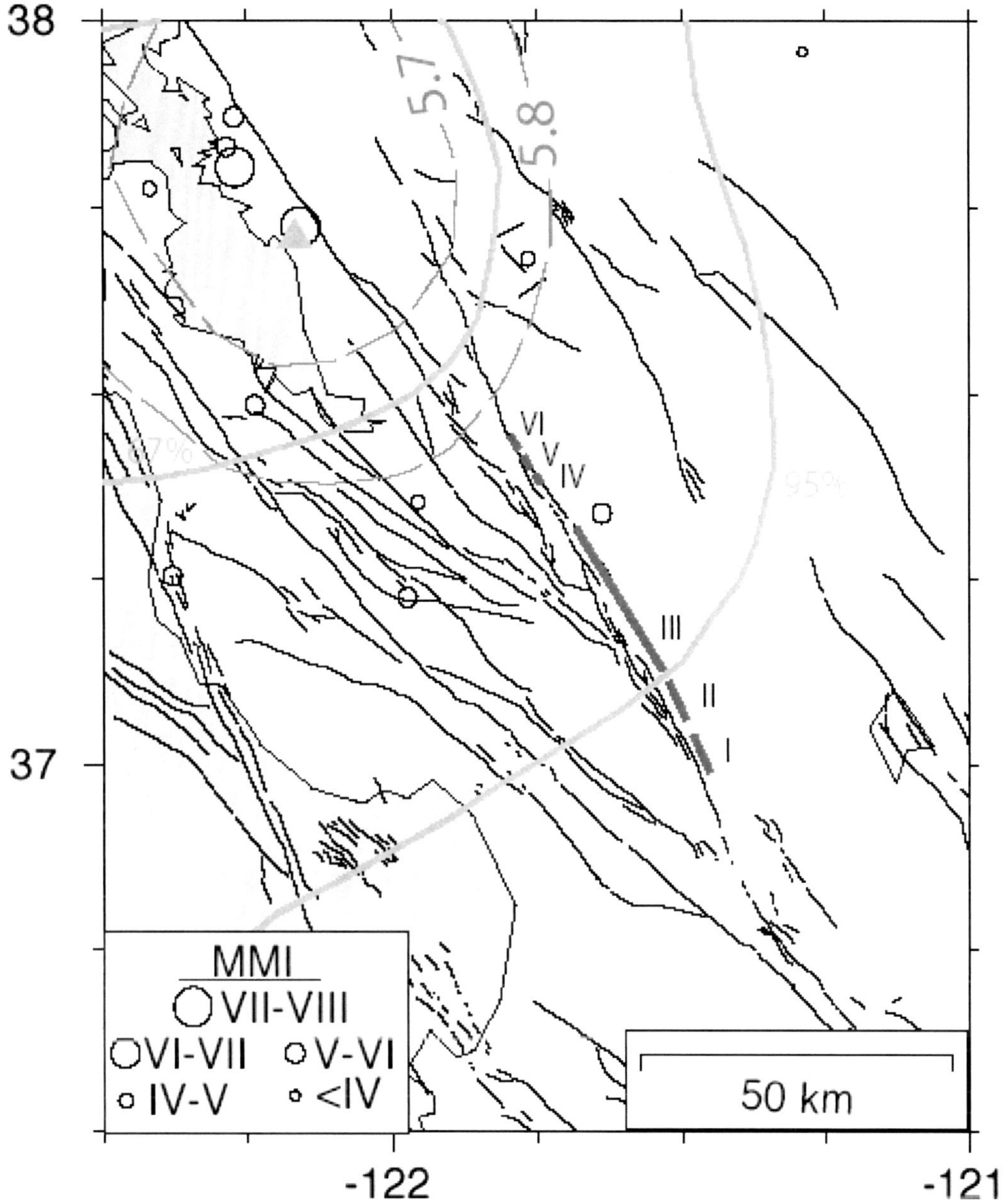

Figure 15. 31. July 1889 earthquake.
(MMI assignment-Site Correction) at 17 sites (some sites may be off the map) are shown as *circles* relative to Jenning's (1994) active faults (*black lines*). The contours of M_I (*dashed red lines*) are the best estimates of **M** from the MMI data for assumed epicenters on that contour. The rms [M_I] contours corresponding to the 67% (innermost contour) and 95% confidence contour (outermost contour) for location (Bakun and Wentworth, 1999) are shown as *solid green lines*. The intensity center (Bakun, 1999) is a *green triangle*. The preferred location is the intensity center. Locations of Oppenheimer *et al.*'s (1990) zones I-VI on the central and southern Calaveras fault are shown as *violet lines*.

24 April 1890

There are 29 MMI available for this event. Cracks were observed on the San Andreas Fault trace near the Pajaro River. The maximum intensity was VIII at Corralitos, Green Valley, Pajaro, San Juan Bautista, and Sargents (Toppozada *et al.*, 1981). The preferred location is on the San Andreas Fault near the intensity center (Figure 16). M_I is 6.3.

2 January 1891

MMI can be assigned at sixteen of the 31 localities for which felt reports (Toppozada *et al.*, 1981) are available. The maximum MMI VI was assigned at Alameda, Mt. Hamilton, San Jose, Boulder Creek, and Santa Cruz. Zones II-VI on the Calaveras fault are within the 67% confidence contour for location (Figure 17). M_I is 5.7. The preferred location is in zone V at the epicenter of the 31 Oct 2007 event.

13 November 1892

MMI can be assigned at eleven of the 28 localities for which felt reports (Toppozada *et al.*, 1981) are available. The maximum MMI VII was assigned at Hollister and Sargents where chimneys were thrown down. The preferred location is on the San Andreas Fault (Figure 18). M_I is 5.8.

20 June 1897

MMI can be assigned at 39 localities for which felt reports (Toppozada *et al.*, 1981) are available. The maximum MMI VIII was assigned at Mission San Carlos, Gilroy, and San Felipe. Fissures were reported on the Pacheco Pass road near San Felipe for the **M**6.3 1897 event (Toppozada *et al.*, 1981). The trace of the Calaveras fault crosses the Pacheco Pass Road near San Felipe, which is located near the north end of zone I. Surface slip during the 1979 Coyote Lake earthquake was observed near San Felipe (Reasenberg and Ellsworth, 1982). McClellan and Hay (1990) reported triggered right-lateral slip on the Calaveras fault at Highway 152 near San Felipe Lake at the time of the 1989 Loma Prieta earthquake. Fresh northwest-striking cracks, observed at Highway 152 after the 1984 Morgan Hill earthquake, were interpreted as triggered slip (Galehouse and Brown, 1987; Harms *et al.*, 1987). The fissures noted in 1897 in the same area may also have been triggered slip. Bakun's (1999) analysis is assumed. Zones I-III on the Calaveras fault are within the 67% confidence contour for location (Figure 19). M_I is 6.3. The preferred location is in zone II.

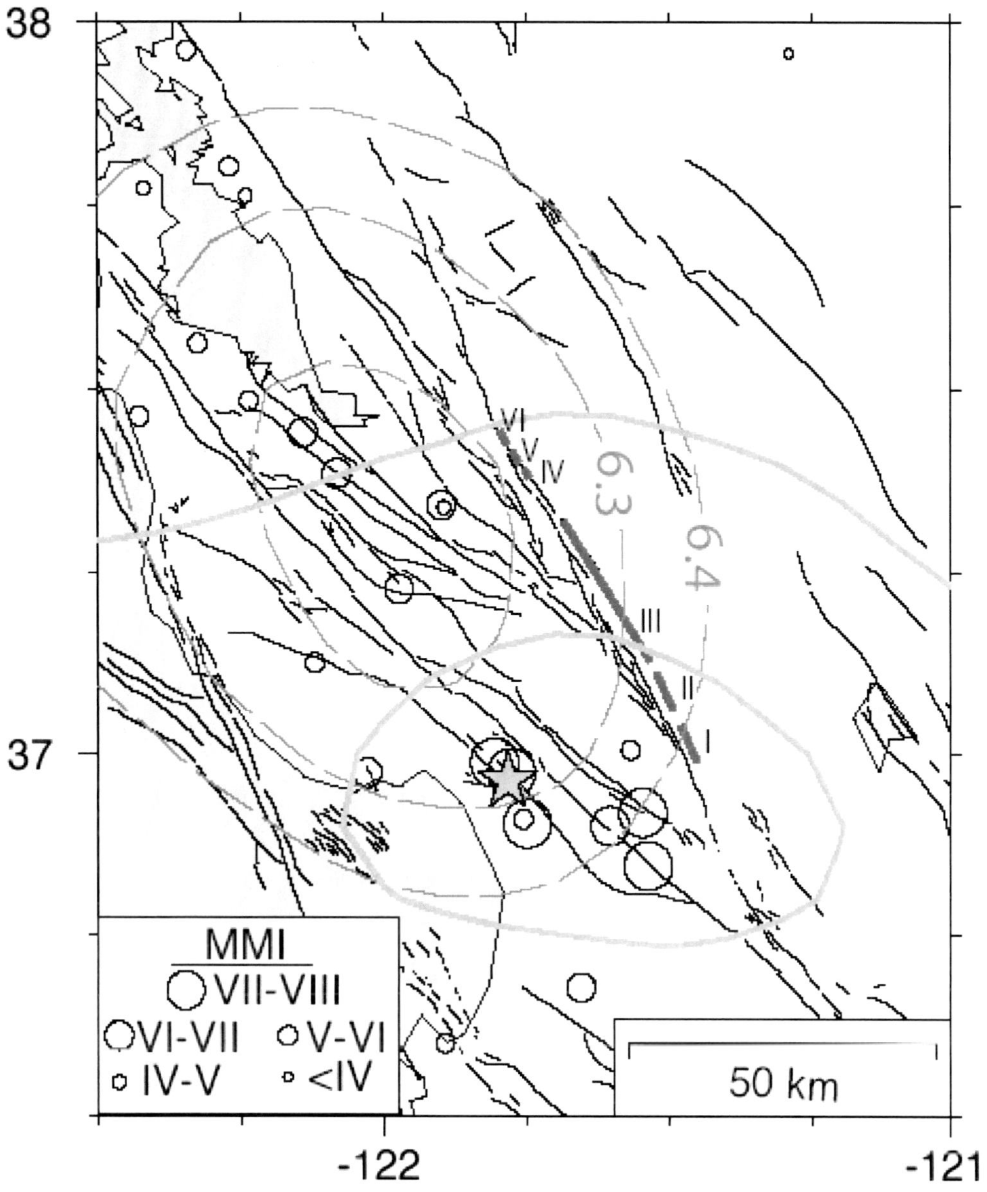

Figure 16. 24. April 1890 earthquake.
(MMI assignment-Site Correction) at 29 sites (some sites may be off the map) are shown as *circles* relative to Jenning's (1994) active faults (*black lines*). The contours of M_I (*dashed red lines*) are the best estimates of **M** from the MMI data for assumed epicenters on that contour. The rms [M_I] contours corresponding to the 67% (innermost contour) and 95% confidence contour (outermost contour) for location (Bakun and Wentworth, 1999) are shown as *solid green lines*. The intensity center (Bakun, 1999) is a *green triangle*. The preferred location is the green star. Locations of Oppenheimer *et al.*'s (1990) zones I-VI on the central and southern Calaveras fault are shown as *violet lines*.

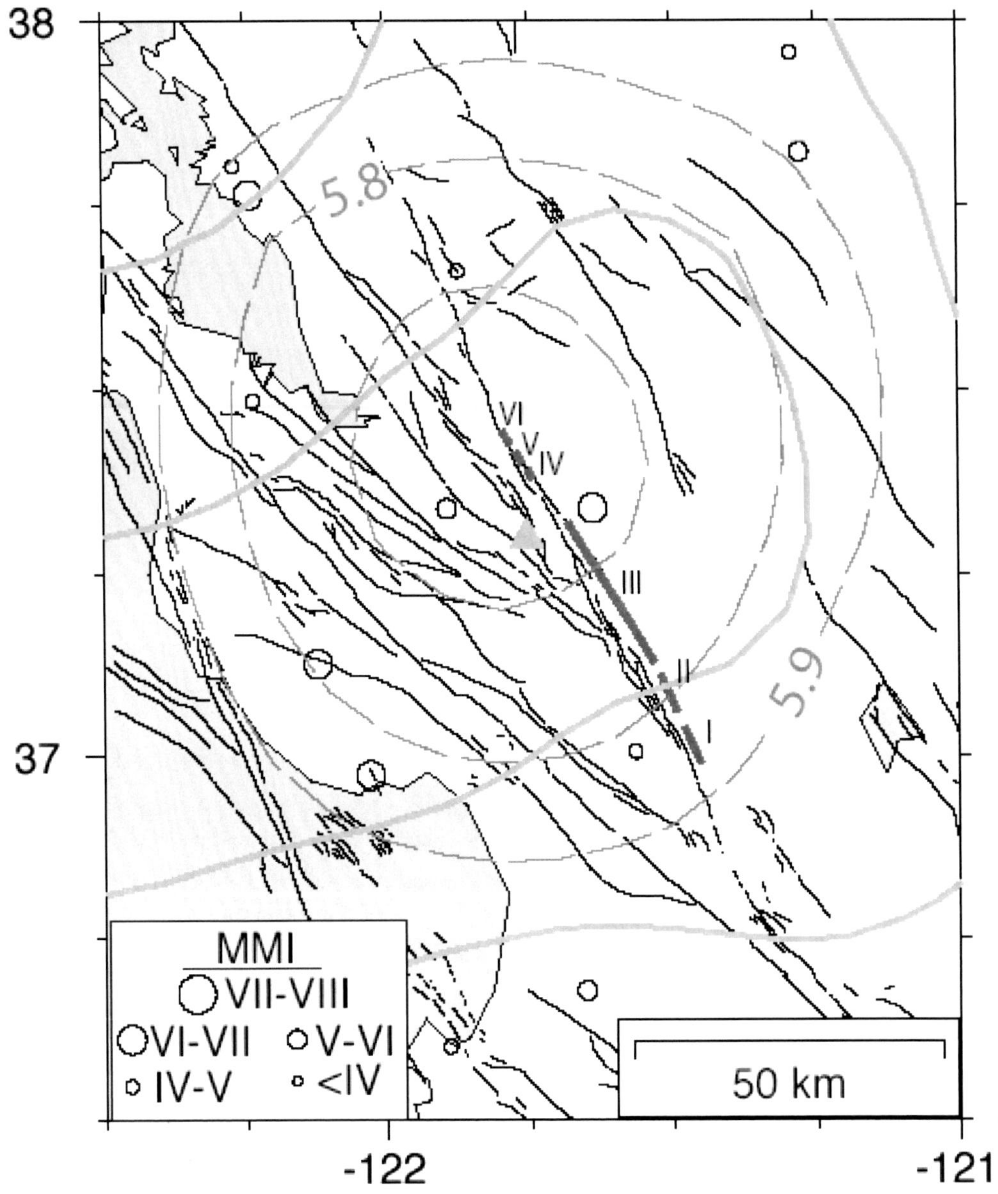

Figure 17. 2. January 1891 earthquake.
(MMI assignment-Site Correction) at 16 sites (some sites may be off the map) are shown as *circles* relative to Jenning's (1994) active faults (*black lines).* The contours of M_I (*dashed red lines*) are the best estimates of **M** from the MMI data for assumed epicenters on that contour. The rms [M_I] contours corresponding to the 67% (innermost contour) and 95% confidence contour (outermost contour) for location (Bakun and Wentworth, 1999) are shown as *solid green lines.* The intensity center (Bakun, 1999) is a *green triangle.* Locations of Oppenheimer *et al.*'s (1990) zones I-VI on the central and southern Calaveras fault are shown as *violet lines.*

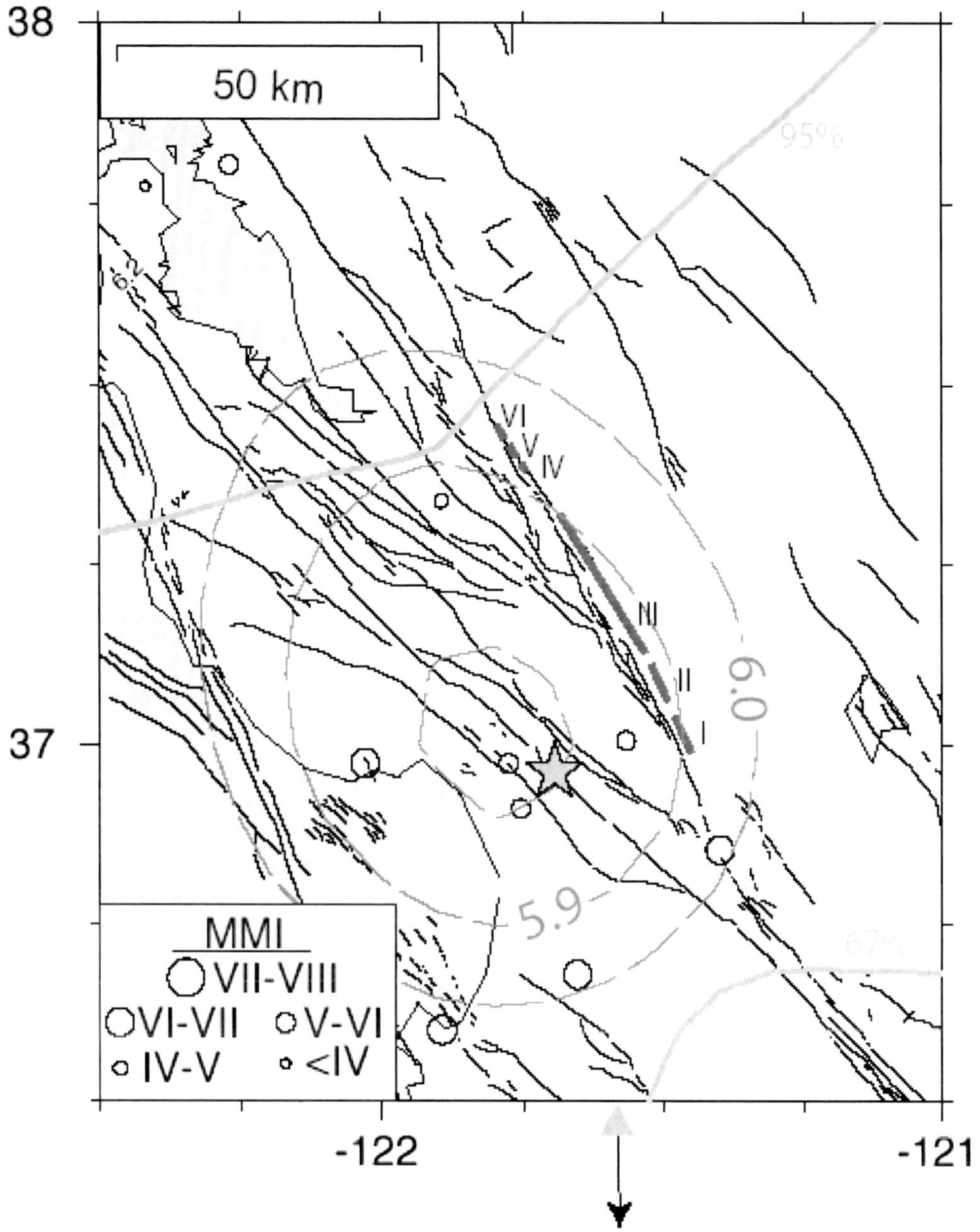

Figure 18. 13. November 1892 earthquake.
(MMI assignment-Site Correction) at 11 sites (some sites may be off the map) are shown as *circles* relative to Jenning's (1994) active faults (*black lines*). The contours of M_I (*dashed red lines*) are the best estimates of **M** from the MMI data for assumed epicenters on that contour. The rms [M_I] contours corresponding to the 67% (innermost contour) and 95% confidence contour (outermost contour) for location (Bakun and Wentworth, 1999) are shown as *solid green lines*. The intensity center (Bakun, 1999) is a *green triangle*. The preferred location is the green star. Locations of Oppenheimer *et al.*'s (1990) zones I-VI on the central and southern Calaveras fault are shown as *violet lines*.

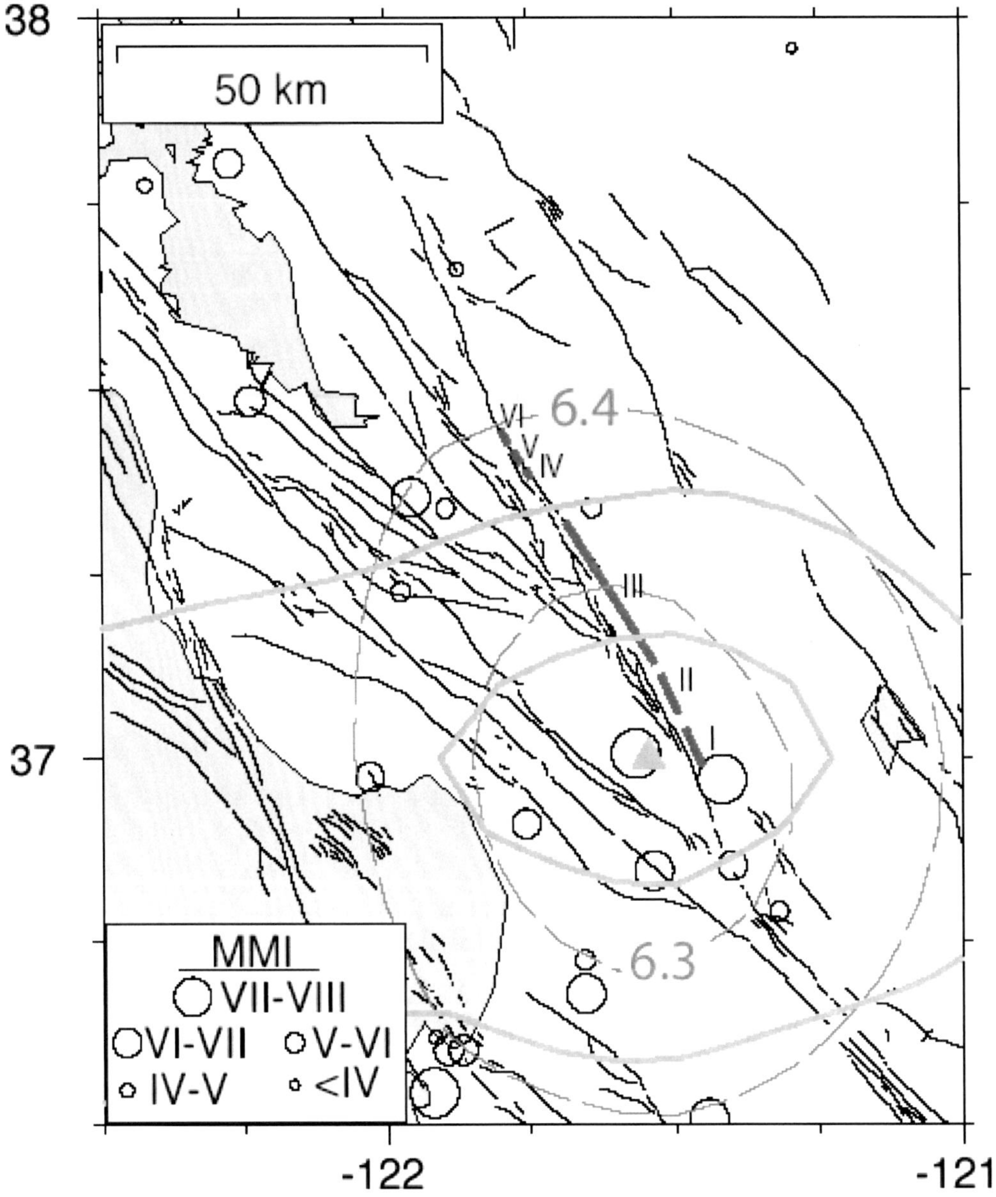

Figure 19. 20. June 1897 earthquake.
(MMI assignment-Site Correction) at 39 sites (some sites may be off the map) are shown as *circles* relative to Jenning's (1994) active faults (*black lines*). The contours of M_I (*dashed red lines*) are the best estimates of **M** from the MMI data for assumed epicenters on that contour. The rms [M_I] contours corresponding to the 67% (innermost contour) and 95% confidence contour (outermost contour) for location (Bakun and Wentworth, 1999) are shown as *solid green lines*. The intensity center (Bakun, 1999) is a *green triangle*. Locations of Oppenheimer *et al.*'s (1990) zones I-VI on the central and southern Calaveras fault are shown as *violet lines*.

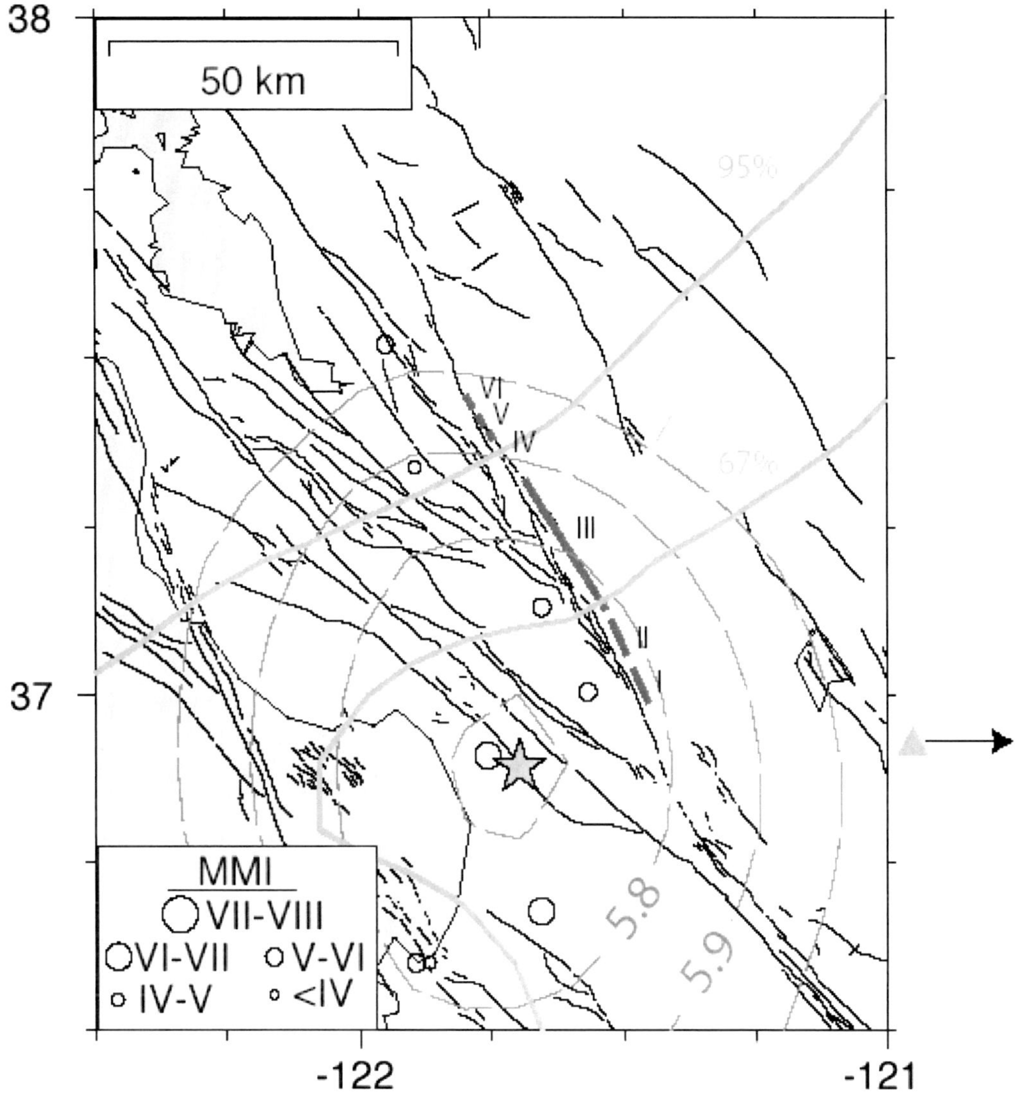

Figure 20. 6. July 1899 earthquake.
(MMI assignment-Site Correction) at 11 sites (some sites may be off the map) are shown as *circles* relative to Jenning's (1994) active faults (*black lines*). The contours of M_I (*dashed red lines*) are the best estimates of **M** from the MMI data for assumed epicenters on that contour. The rms [M_I] contours corresponding to the 67% (innermost contour) and 95% confidence contour (outermost contour) for location (Bakun and Wentworth, 1999) are shown as *solid green lines*. The intensity center (Bakun, 1999) is a *green triangle and id located off the map*. The preferred location is the green star. Locations of Oppenheimer *et al.*'s (1990) zones I-VI on the central and southern Calaveras fault are shown as *violet lines*.

6 July 1899

MMI can be assigned at twelve localities for which felt reports (Toppozada *et al.*, 1981) are available. The pattern of reports is anomalous in that high intensities were reported at Pleasanton, Alameda County, and near Watsonville but high intensities were not reported at sites between Pleasanton and Watsonville (Toppozada *et al.*, 1981). Townley and Allen (1939) suggested that two earthquakes occurred –a larger event near Watsonville and a smaller earthquake near Pleasanton. I discount here the account at Pleasanton. Zones I and II on the Calaveras fault are within the 67% confidence contour for location, but a location closer to the San Andreas Fault is preferred (Figure 20). M_I is 5.7.

30 April 1899

MMI can be assigned at 15 localities for which felt reports (Toppozada *et al.*, 1981) are available. The maximum MMI VII was assigned at Green Valley and Watsonville where chimneys were thrown down. The preferred location is on the San Andreas Fault (Figure 21). M_I is 5.8.

11 June 1903

MMI can be assigned at 26 localities for which felt reports (Toppozada *et al.*, 1981) are available. Zones II-VI on the Calaveras fault are within the 67% confidence contour for location (Figure 22). M_I is 6.1. The preferred location is at the intensity center.

3 August 1903

MMI can be assigned at 22 localities for which felt reports (Toppozada *et al.*, 1981) are available. Zones III-VI on the Calaveras fault are within the 67% confidence contour for location (Figure 23). M_I is 6.1. The preferred location is at the intensity center.

1 July 1911

MMI can be assigned at 31 localities for which felt reports (Toppozada *et al.*, 1981) are available. Bakun's (1999) analysis is assumed. Zones III-VI on the Calaveras fault are within the 67% confidence contour for location (Figure 24). Oppenheimer *et al.* (1990) argue that the 1911 event occurred on Zone III. M_I is 6.2. The preferred location is in zone III at the epicenter of the 1984 Morgan Hill earthquake.

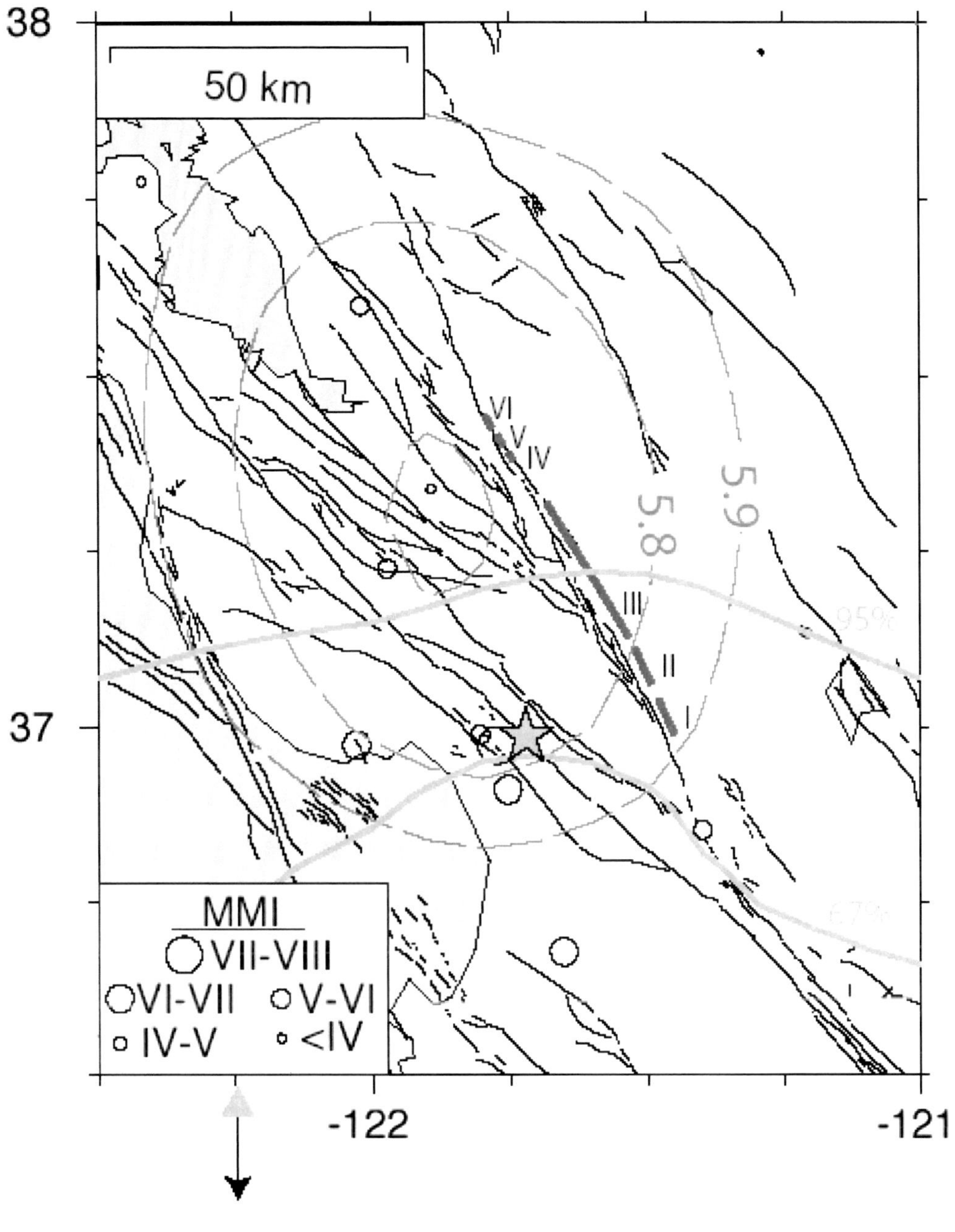

Figure 21. 30. April 1899 earthquake.
(MMI assignment-Site Correction) at 15 sites (some sites may be off the map) are shown as *circles* relative to Jenning's (1994) active faults (*black lines*). The contours of M_I (*dashed red lines*) are the best estimates of **M** from the MMI data for assumed epicenters on that contour. The rms [M_I] contours corresponding to the 67% (innermost contour) and 95% confidence contour (outermost contour) for location (Bakun and Wentworth, 1999) are shown as *solid green lines*. The intensity center (Bakun, 1999) is a *green triangle*. The preferred location is the green star. Locations of Oppenheimer *et al.*'s (1990) zones I-VI on the central and southern Calaveras fault are shown as *violet lines*.

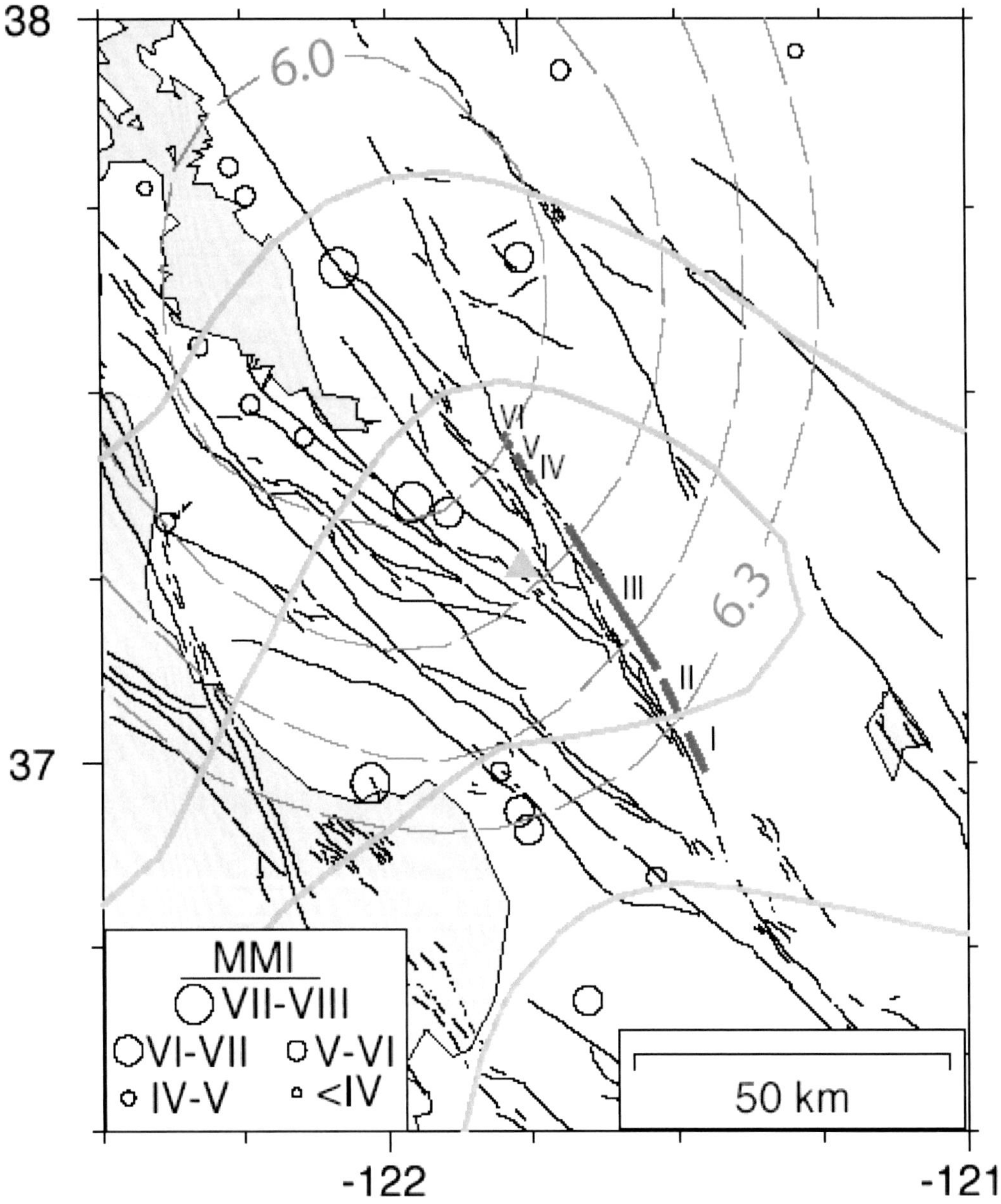

Figure 22. 11. June 1903 earthquake.
(MMI assignment-Site Correction) at 26 sites (some sites may be off the map) are shown as *circles* relative to Jenning's (1994) active faults (*black lines).* The contours of M_I (*dashed red lines*) are the best estimates of **M** from the MMI data for assumed epicenters on that contour. The rms [M_I] contours corresponding to the 67% (innermost contour) and 95% confidence contour (outermost contour) for location (Bakun and Wentworth, 1999) are shown as *solid green lines.* The intensity center (Bakun, 1999) is a *green triangle.* Locations of Oppenheimer *et al.*'s (1990) zones I-VI on the central and southern Calaveras fault are shown as *violet lines.*

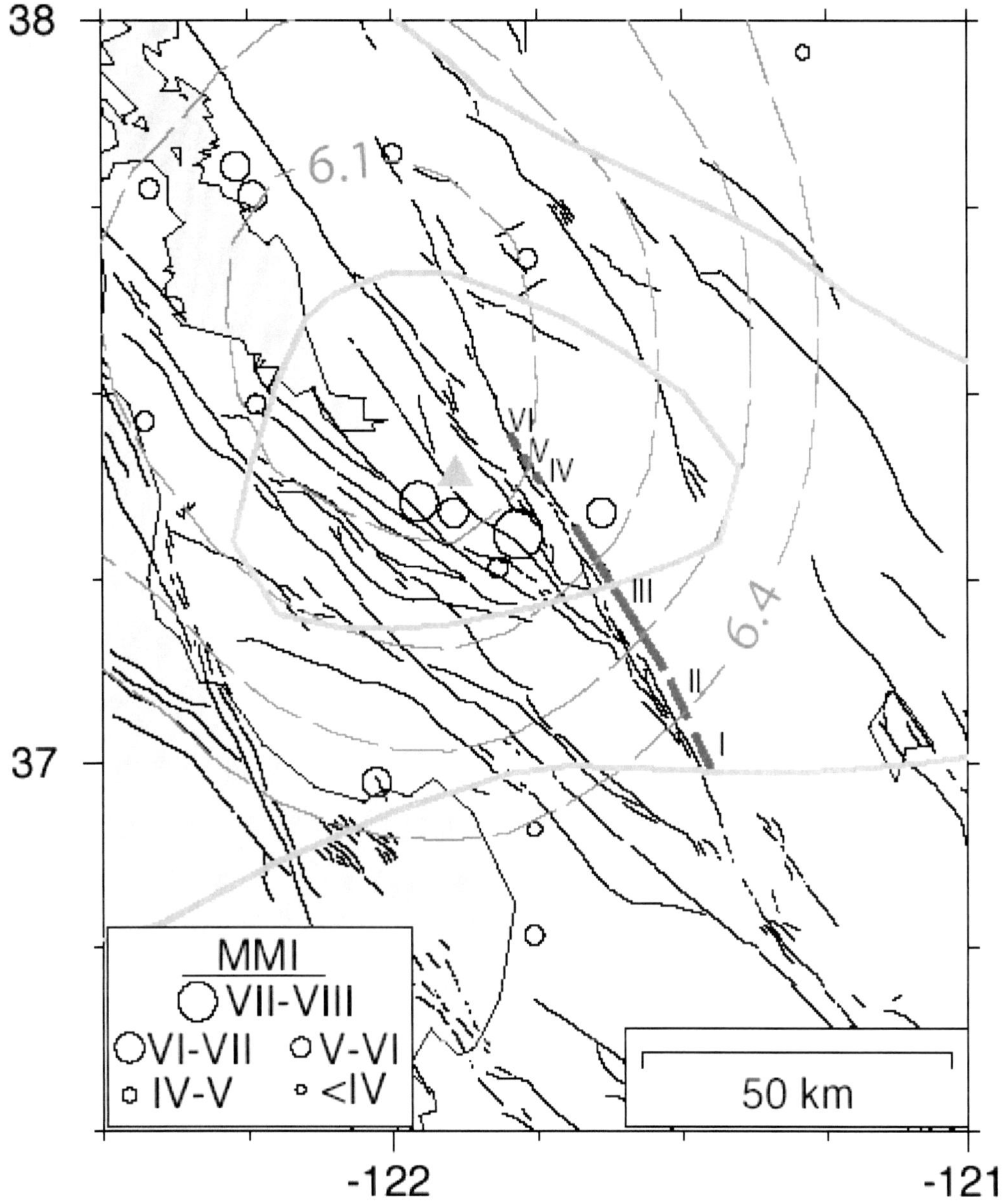

Figure 23. 3. August 1903 earthquake.
(MMI assignment-Site Correction) at 22 sites (some sites may be off the map) are shown as *circles* relative to Jenning's (1994) active faults (*black lines)*. The contours of M_I (*dashed red lines*) are the best estimates of **M** from the MMI data for assumed epicenters on that contour. The rms [M_I] contours corresponding to the 67% (innermost contour) and 95% confidence contour (outermost contour) for location (Bakun and Wentworth, 1999) are shown as *solid green lines*. The intensity center (Bakun, 1999) is a *green triangle*. Locations of Oppenheimer *et al.*'s (1990) zones I-VI on the central and southern Calaveras fault are shown as *violet lines*.

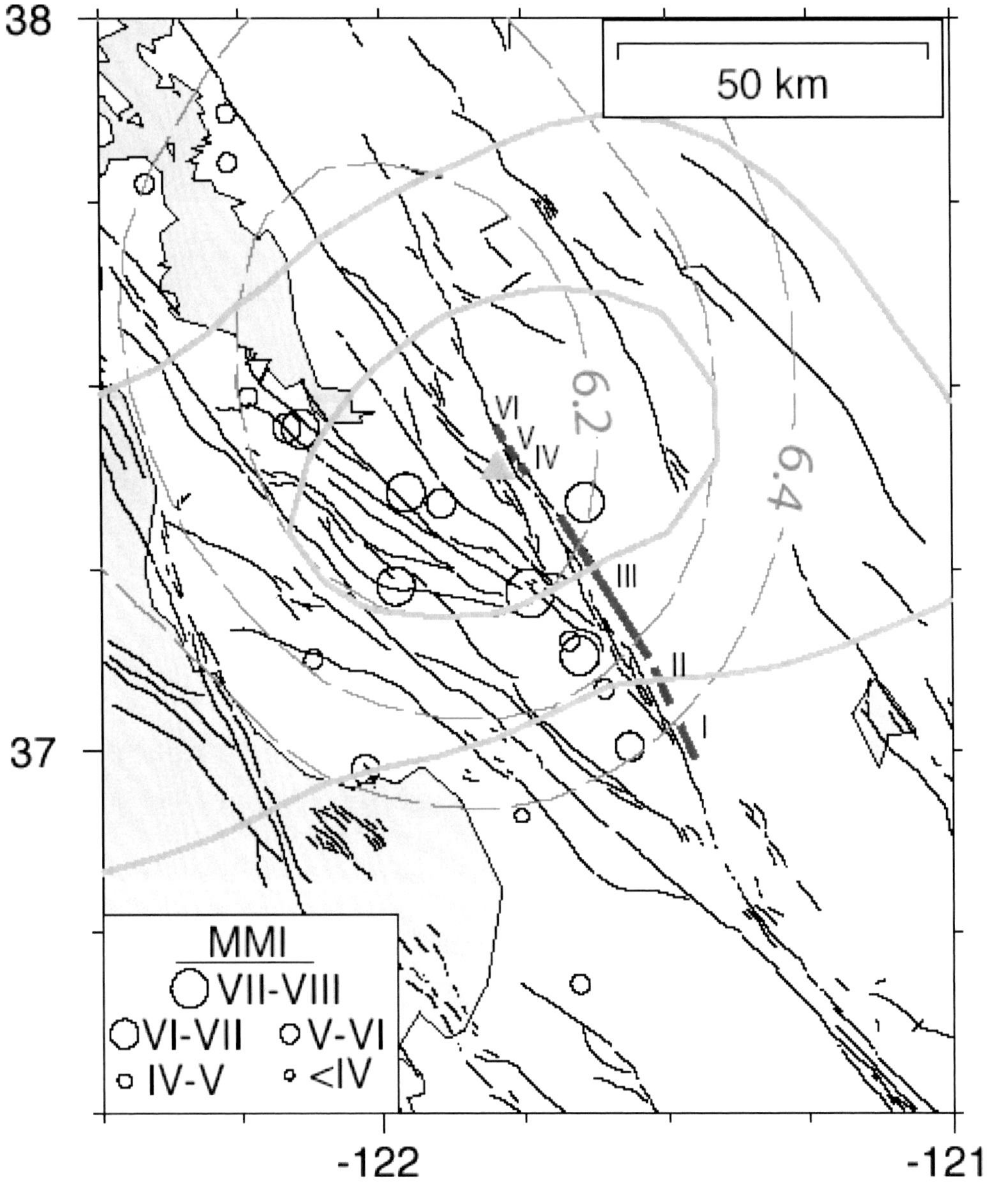

Figure 24. 1. July 1911 earthquake.
(MMI assignment-Site Correction) at 31 sites (some sites may be off the map) are shown as *circles* relative to Jenning's (1994) active faults (*black lines*). The contours of M_I (*dashed red lines*) are the best estimates of **M** from the MMI data for assumed epicenters on that contour. The rms [M_I] contours corresponding to the 67% (innermost contour) and 95% confidence contour (outermost contour) for location (Bakun and Wentworth, 1999) are shown as *solid green lines*. The intensity center (Bakun, 1999) is a *green triangle*. Locations of Oppenheimer *et al.*'s (1990) zones I-VI on the central and southern Calaveras fault are shown as *violet lines*.

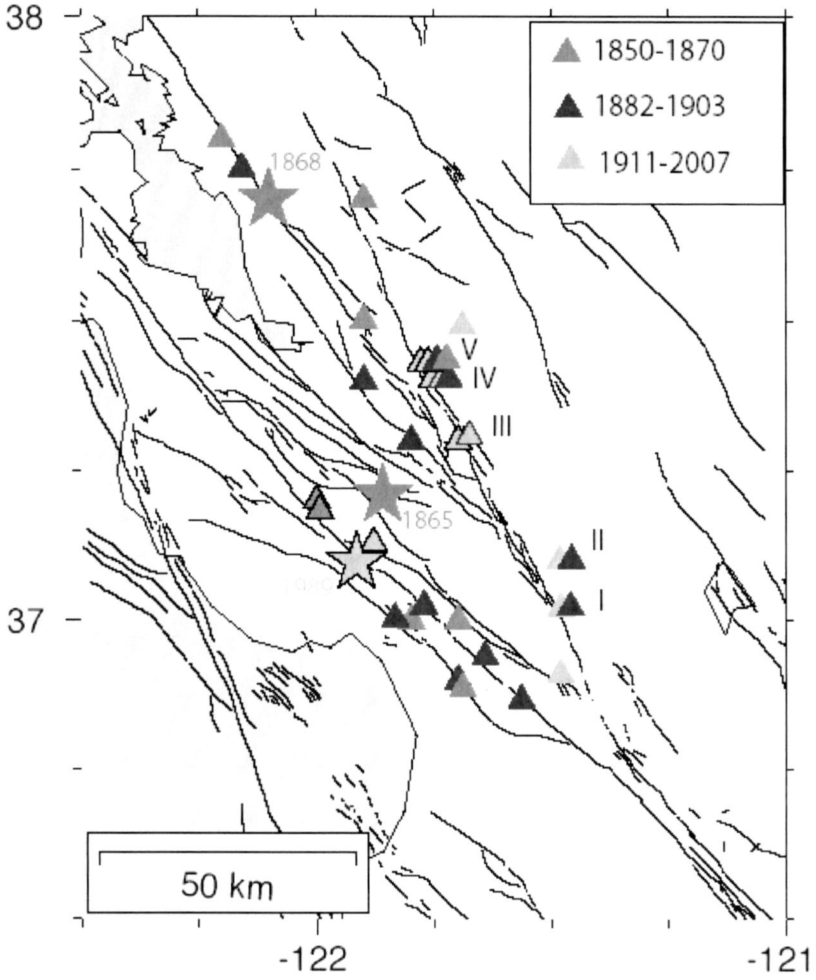

Figure 25. Preferred earthquake epicenters.
(Preferred locations in Table 1 are shown as *triangles* relative to Jenning's (1994) active faults (*black lines)*. The **M**6.5 1865, **M**6.8 1868, and **M**6.9 1989 earthquakes are plotted as stars. I-VI are Oppenheimer et al.'s (1990) zones of recurring earthquakes on the Calaveras fault.

CONCLUSIONS

The preferred locations and **M** for the historical earthquakes are summarized in Table 1 and Figure 25. Five of the 11 earthquakes in the 1850-1870 period apparently were on east Bay faults; the remaining six events were on west Bay faults. Seven of the 12 earthquakes in the 25 years before the 1906 earthquake and five of the 6 earthquakes in the 25 years before the 1989 Loma Prieta earthquake occurred on east Bay faults. That is, significant seismicity in the west Bay apparently preceded the large 1868 east Bay event and significant seismicity in the east Bay apparently preceded the large 1906 and 1989 west Bay events.

REFERENCES

Bakun, W. H. (1998). *Modified Mercalli intensities for some recent California earthquakes and historic San Francisco Bay region earthquakes*, U. S. Geol. Surv Open-file Report 98-584, 175 p.

Bakun, W. H. (1999). Seismic activity of the San Francisco Bay region, *Bull. Seism. Soc. Am*, *89*, 764-784.

Bakun, W. H. & Wentworth, C. M. (1997). Estimating location and magnitude from seismic intensity data, *Bull. Seism. Soc. Amer, 87*, 1502-1521.

Bakun, W. H. & Wentworth, C. M. (1999). Erratum to Estimating earthquake location and magnitude from seismic intensity data, *Bull. Seismol. Soc. Am, 89*, 557.

Bakun, W. H., Oppenheimer, D., Boatwright, J. & Simpson, R. (2008). Earthquakes on the Central and Southern Calaveras fault and Earthquake Forecasts: Part II, 1850- 1910 (abstract), *Seismol. Res. Lett.*, vol. *79*, p. 343.

Galehouse, J. S. & Brown, B. D. (1979). Surface displacement near Hollister, California, in Hoose, S, N., (Ed.), The Morgan Hill, California, earthquake of April 24, 1984, *U. S. Geol. Surv. Bull, 1639*, 69-72.

Harms, K. K., Clark, M. M. & others. (1987). The search for surface faulting, *in* Hoose, S, N., (Ed.), The Morgan Hill, California, earthquake of April 24, 1984, *U. S. Geol. Surv. Bull, 1639*, 61-68.

Jennings, C. W. (1994). Fault activity map of California and adjacent areas, California Geological Data Map Series, California Division of Mines and Geology, map no. 6.

McClellan, P. H. & Hay, E. A. (1989). Triggered slip on the Calaveras fault during the magnitude 7.1 Loma Prieta, California, earthquake, *Geoph. Res. Lett*, *17*, 2027- 2030.

Oppenheimer, D. H., Bakun, W. H. & Lindh, A. G. (1990). Slip partitioning of the Calaveras fault, California, and prospects for future earthquakes, *J. Geophys. Res*, *95*, 8483-8498.

Oppenheimer, D., Bakun, W., Uhrhammer, R., Boatwright, J. & Simpson, R. (2008a). Earthquakes on the Central and Southern Calaveras fault and Earthquake Forecasts: Part I, 1910-Present (abstract), Seismol. Res. *Lett.*, vol. *79*, p. 342.

Oppenheimer, D., Bakun, W., Uhrhammer, R., Boatwright, J. & Simpson, R. (2008b). The 2007 M5.4 Alum Rock Earthquake: Implications for Future Earthquakes on the Central and Southern Calaveras Fault, *J. Geophys. Res.* in preparation

Reasenberg, P. & Ellsworth, W. L. (1982). Aftershocks of the Coyote lake, California, earthquake of August 6, 1979; a detailed study, *J. Geophys. Res*, *87,* 10,637- 10,665.

Toppozada, T. R., Real, C. R. & Parke, D. L. (1981). Preparation of isoseismal maps and summaries of reported effects for Pre-1900 California earthquakes, *Calif. Div. Mines and Geol. Open-File Rept. 81-11 SAC,* 182 pp.

In: The Big Shake Implications of a Major Earthquake in... ISBN: 978-1-61728-512-7
Editor: Christopher D. Monahan

Chapter 6

The Hayward Fault: Is it Due for a Repeat of the Powerful 1868 Earthquake?

U.S. Geological Survey

On October 21, 1868, a magnitude 6.8 earthquake struck the San Francisco Bay region. Although the region was then sparsely populated, this quake on the Hayward Fault was one of the most destructive in California's history. Recent studies show that such powerful Hayward Fault quakes have repeatedly jolted the region in the past. U.S. Geological Survey (USGS) scientists describe this fault as a tectonic time bomb, due anytime for another magnitude 6.8 to 7.0 earthquake. Because such a quake could cause hundreds of deaths, leave thousands homeless, and devastate the region's economy, the USGS and other organizations are working together with new urgency to help prepare Bay Area communities for this certain future quake.

In the early morning of October 21, 1868, seismic waves from a powerful earthquake raced through the fog-shrouded San Francisco Bay region. Frightened people ran out of their homes, and cattle and even fire-engine horses panicked and bolted. Strong shaking lasted more than 40 seconds, devastating several East Bay towns. Brick buildings, walls, and chimneys were also shaken down in Oakland, San Francisco, Santa Rosa, and San Jose, and there was serious damage in Napa and Hollister. Numerous witnesses reported seeing the ground move in waves. Shaking was felt as far away as Nevada, and aftershocks rattled the Bay Area for weeks. Even though the region was only sparsely populated at the time, the 1868 quake killed about 30 people and caused great property damage. It still ranks as one of the most destructive earthquakes in California's history—but this is not the end of the story. The Hayward Fault will rupture violently again, and perhaps very soon.

Strong shaking during the 1868 Hayward Fault earthquake caused the second story of the Alameda County Courthouse in San Leandro to collapse and severely damaged the building (photo courtesy of the Bancroft Library, University of California). The inset photo shows the building before the quake (photo courtesy of San Leandro Public Library). The 1868 quake devastated several East Bay towns and caused widespread damage in the San Francisco Bay region.

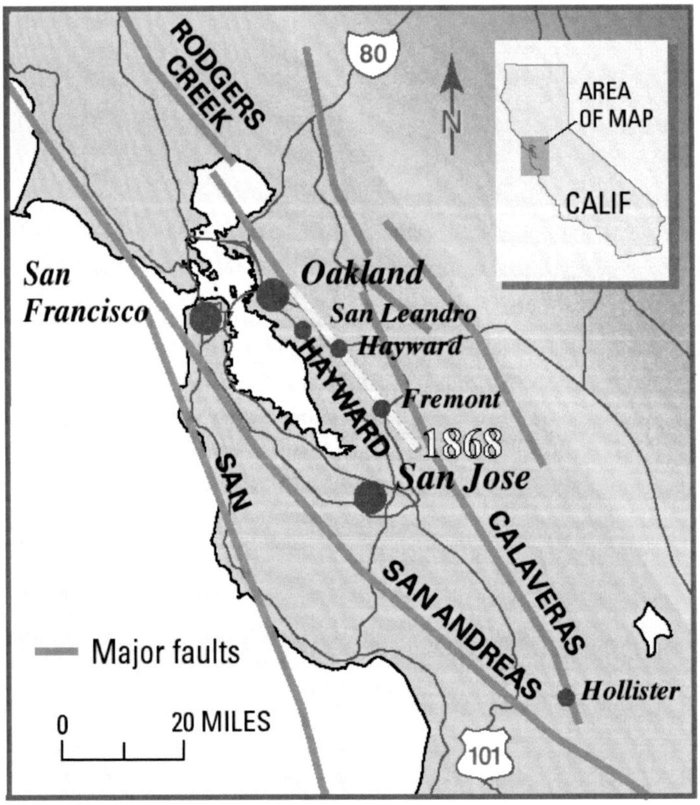

Major earthquake faults in the San Francisco Bay region. The section of the Hayward Fault that slipped and produced the 1868 Hayward earthquake is highlighted in yellow. Dot size indicates the present relative population sizes of cities.

THE EARTHQUAKE OF 1868

The 1868 earthquake on the Hayward Fault was the last of a decade-long sequence of seven increasingly strong quakes in the Bay Area. Surface rupture of the ground in 1868 was traced for 20 miles along the Hayward Fault, from Warm Springs in Fremont north to San Leandro. Historical land-survey data suggest that the fault broke as far north as Berkeley, with an average horizontal movement of about 6 feet (2 meters).

Shaking from the 1868 quake was the strongest that the new towns and growing cities of the Bay Area had ever experienced. Until overshadowed by the 1906 earthquake, the 1868 event was known as the "great San Francisco quake." The area of strongest shaking covered about 1,000 square miles (2,500 km2). San Francisco, then the largest city on the West Coast with a population of 150,000, suffered five deaths and experienced $350,000 in property loss (in 1868 dollars). Many brick walls, cornices, and other heavy architectural elements of buildings in the city fell, and the Custom House and several other structures built on landfill reclaimed from the former Yerba Buena Cove (today's Financial District) sustained severe damage. However, as in 1906, well-constructed buildings on firm ground sustained much less damage.

The most severe damage in 1868 was in East Bay towns. Almost every building in Hayward, then a town with about 500 residents, was wrecked or severely damaged—few places have paid so dearly to have an earthquake fault named after them. At San Leandro, with a population of about 400, the second floor of the Alameda County Courthouse collapsed, and many other buildings were destroyed. At Mission San Jose, in southern Fremont, the adobe church built in 1809 and other mission buildings were heavily damaged. Damage in Oakland, a town of about 12,000 with mainly wood-frame buildings, was much less than observed farther south at San Leandro and Hayward, but San Jose, a South Bay town of about 9,000, had many wrecked buildings and fallen chimneys.

THE HAYWARD FAULT IS CREEPING

The Hayward Fault is one of a number of "creeping" earthquake faults in the San Francisco Bay region. Ample evidence for the steady slip or "creep" of the fault is seen in roads, curbs, and buildings that are being progressively offset. U.S. Geological Survey and other scientists have shown that the two sides of the Hayward Fault are creeping past each other at the surface at a rate of about 1/5 inch (5 mm) per year. Faults like the Hayward Fault are near-vertical planes that extend deep beneath the surface. Creep generally appears to be limited to the topmost 3 miles (5 km) of a fault plane; below that depth, the fault is locked and building up stress. Creep and small earthquakes account for only about one-third of the long-term (thousands of years) movement on the Hayward Fault—the remaining two-thirds must happen in large earthquakes like the 1868 event.

Recently, scientists at the University of California at Berkeley began using a new technique called Interferometric Synthetic Aperture Radar (InSAR), which is based on satellite radar surveys of the Earth's surface, to monitor creep on the Hayward Fault. This technique allows scientists to determine which parts of a fault are creeping and at what rates, and which parts are locked all the way to the surface and thus may be capable of producing even stronger quakes.

Creep on the Hayward Fault has broken and offset this curb in the city of Hayward.

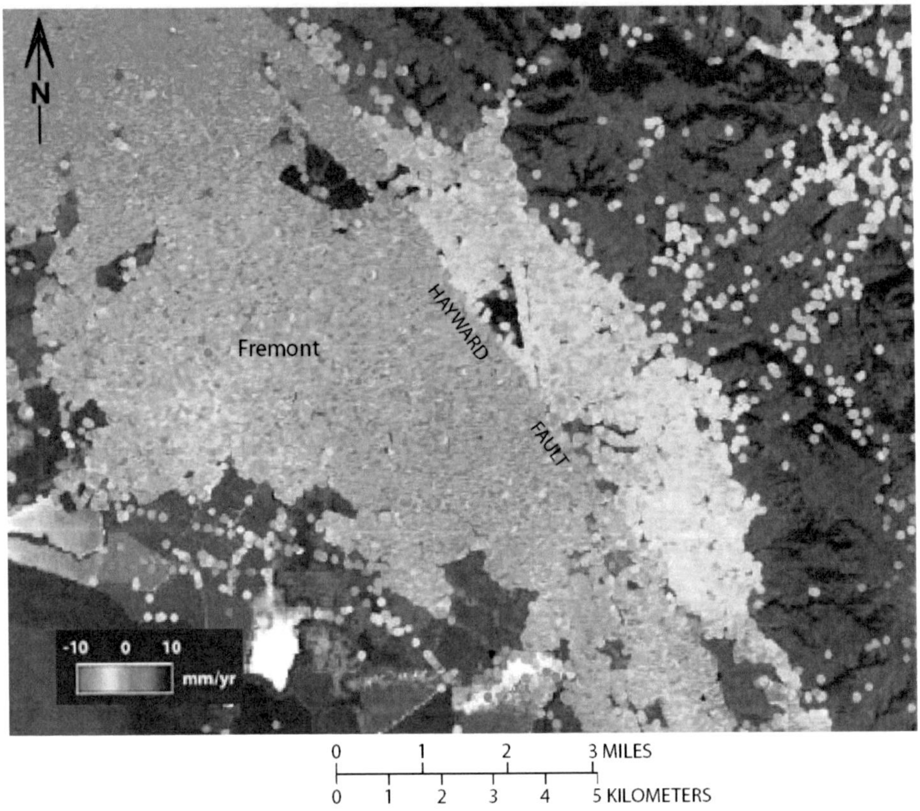

An InSAR image of part of the East Bay clearly shows creep on the Hayward Fault. Repeated passes by satellites to the south record the change in distance between the satel lite and recognizable points on the ground (dots). In general, warm colors mean the points are moving to the southeast (closer to the satellite), cool colors mean the points are moving to the northwest (farther away). The color scale indicates the rate of this movement in mil limeters per year (1 mm = 0.04 inch).

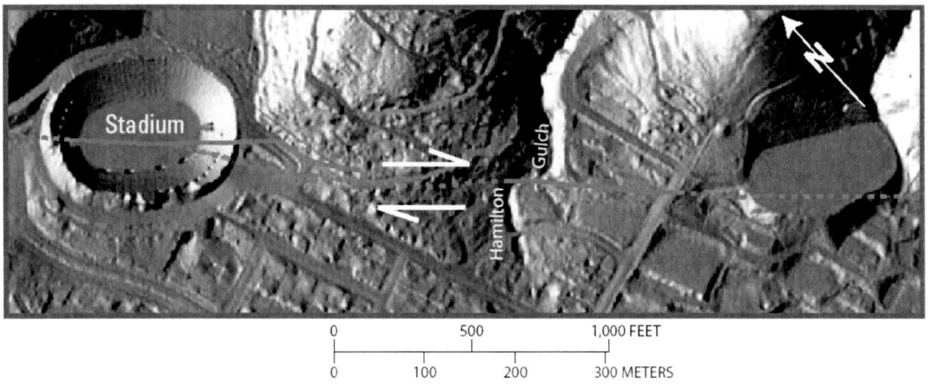

A filtered vertical laser image, taken using a technique called Light Detection and Ranging (LIDAR), of part of the Hayward Fault (red lines) in the City of Berkeley. The fault passes through the University of California Berkeley football stadium (left), and past earthquake movements have significantly offset Hamilton Gulch (center). Arrows show relative movement on the fault.

UNDERSTANDING THE 1868 QUAKE

Because seismographs did not exist then, there are no recordings of the 1868 earthquake. A committee that included scientists studied the quake at the time, but no report on the event was published. Much of what we know about the 1868 quake is documented in the famous 1908 Lawson Report on the 1906 San Francisco earthquake. As part of that report, the Hayward Fault was reexamined and some survivors of the 1868 quake were interviewed. The 1908 report also included photographs taken in 1868 of damage to buildings in San Francisco, San Leandro, and Hayward.

ANALOGY TO THE 1995 KOBE EARTHQUAKE

Studies of past earthquakes allow scientists to forecast the effects of future quakes. A similar quake to the anticipated next large Hayward Fault earthquake may be the magnitude 6.9 earthquake that struck Kobe, Japan, in 1995. The geography of Kobe, a port city built along the Nojima Fault on Osaka Bay, is strikingly similar to that of the East Bay, and the Nojima Fault and the Hayward Fault are both strike-slip faults of about the same length. The 1995 Kobe quake and subsequent fires caused more than 5,000 fatalities. Shaking and ground failures, including liquefaction (in which shaken soils behave like a liquid and cannot support structures), devas tated residences and infrastructure. Damage to the port facilities in Kobe resulted in a permanent loss of business, because some shipping companies relocated to other undamaged ports in Japan. More than a decade later, large segments of the lower income areas of Kobe have yet to be rebuilt.

Before 1995, construction codes and practices for residential buildings in Japan were not as stringent as those now in use, and Kobe was not as prepared for earthquakes as the Bay Area is now. Nonetheless, it is likely that a repeat of the 1868 Hayward earthquake will also produce significant loss of life and high levels of damage.

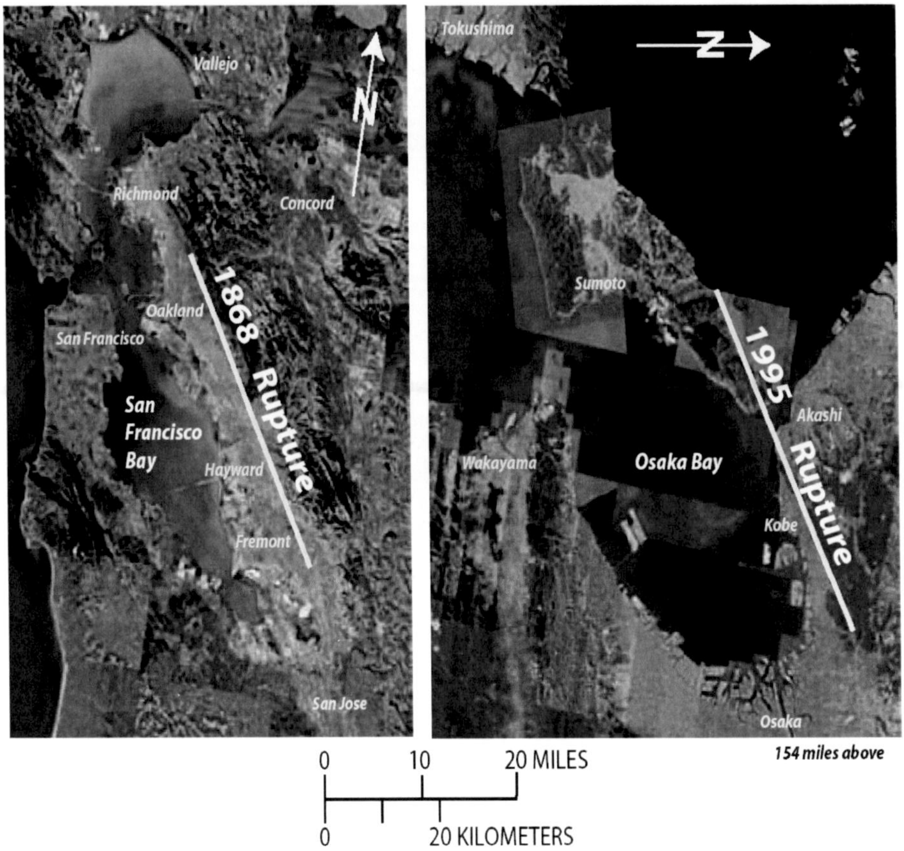

The urban geographic settings of the Nojima Fault of Kobe, Japan, and the Hayward Fault, California, are very similar, as were the lengths of rupture in their most recent strong earthquakes. The 1995 Kobe earthquake caused more than 5,000 deaths and widespread devastation. (satellite images from Google Earth.)

In recent years, U.S. Geological Survey (USGS) and other scientists have used the Lawson Report, as well as historical newspaper accounts and pioneer diaries and letters, to better understand the size and the effects of the 1868 earthquake. Evaluating the extent of damage throughout the Bay Area and the types of buildings that were damaged (brick, stone, and adobe) is critical to determining the local shaking intensity of the 1868 quake. Combined with modern knowledge of earthquake effects and geology, these historical data have been used to create a "ShakeMap" that depicts the inferred intensity of shaking throughout the region in 1868. The map shows that shaking was strongest in Hayward, Fremont, and San Leandro and weaker in Oakland, San Francisco, and San Jose. Scientists now estimate that the magnitude of the 1868 quake was 6.8. Comparing the ShakeMaps from the 1868 Hayward and 1989 (magnitude 6.9) Loma Prieta earthquakes shows that a repeat of the 1868 quake will produce much higher levels of shaking and damage in the East Bay than did the 1989 quake.

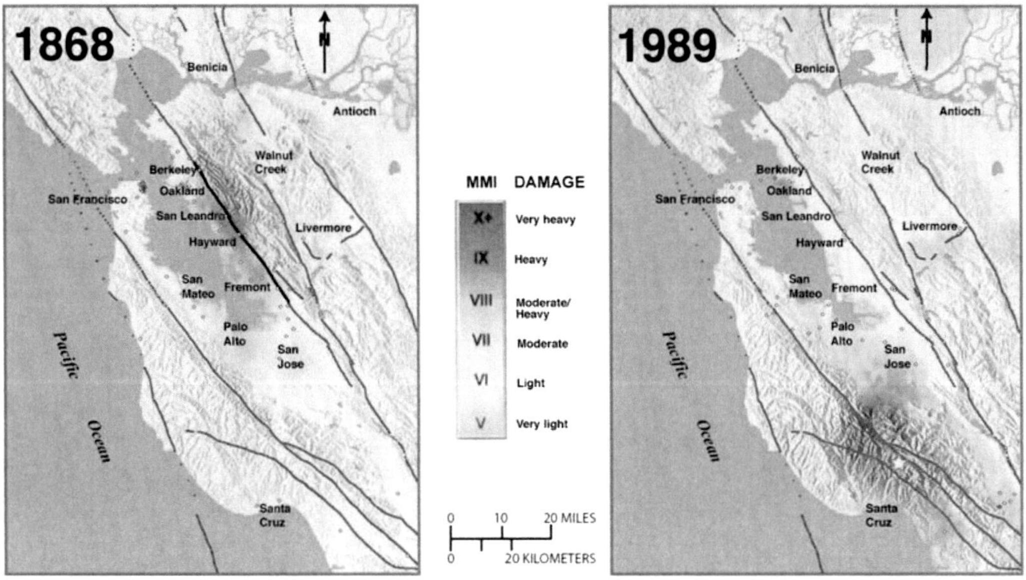

A ShakeMap showing the inferred intensity of ground shaking in the 1868 earthquake (MMI, or Modified Mercalli Intensity), compared to a ShakeMap for the 1989 magnitude 6.9 Loma Prieta earthquake. Red lines are earthquake faults; black line shows the portion of the Hayward Fault that ruptured in 1868; yellow star marks the epicenter of the 1989 quake. Diamonds show locations of damage reports (1868) and of seismic recordings (1989).

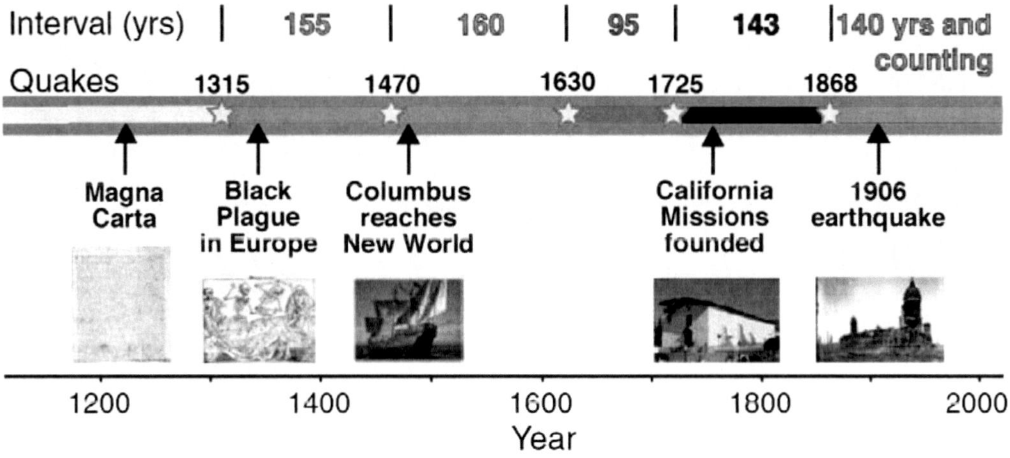

Study of excavations across the Hayward Fault has provided a record of strong earthquakes, all estimated at magnitude 6.3 or greater, going back hundreds of years—some historical events are shown for reference. The dates of earthquakes before 1868 are based on radiocarbon dating. Note that the interval between successive quakes has varied from 95 to 160 years (average 138 years), and it is now 140 years since the 1868 earthquake.

IMPACT OF THE 1868 HAYWARD QUAKE ON BUILDING PRACTICES

Engineers realized after the 1868 quake that structures in the Bay Area needed to be made more resistant to earthquakes. Buildings built in the 1870s for the fledgling University of California Berkeley campus along the Hayward Fault, as well as others in San Francisco, included special features designed to strengthen them against quakes. Many such buildings built in San Francisco between 1868 and 1906 fared well during the powerful shaking produced by the 1906 San Francisco earthquake.

Sadly, other important engineering lessons from the 1868 quake, though openly discussed at the time, were not heeded. The hazards of building on landfill ("made land") reclaimed from the San Francisco Bay and the admonition to "build no more cornices" were largely forgotten by the time of the 1906 quake.

IS THE HAYWARD FAULT OUR NATION'S MOST DANGEROUS?

Two factors combine to make the Hayward Fault very dangerous. The first is its location in the urban heart of the Bay Area. The Hayward Fault is the single most urbanized earthquake fault in the United States—in 1868 there were only 24,000 people living near the fault in Alameda County, now there are more than 2.4 million. Hundreds of homes and other structures are built directly on the fault's trace, and mass transit corridors, major freeways, and many roadways cross it at numerous locations. Also, critical regional gas and water pipelines and electrical transmission lines cross the fault. A September 2007 U.S. Bureau of Labor Statistics Report determined that more than 1.5 million people with combined annual wages exceeding $100 billion work at sites that would experience strong or very strong levels of shaking from the next powerful Hayward Fault quake.

A second factor making the Hayward Fault so dangerous now is that its most recent damaging earthquake was 140 years ago. USGS scientists have found evidence for 12 quakes on the southern Hayward Fault during the past 1,900 years. Ominously, the last five events (in 1315, 1470, 1630, 1725, and 1868) occurred at intervals of 95 to 160 years, with an average interval of 138 years.

THE NEXT MAJOR HAYWARD FAULT EARTHQUAKE

The 140th anniversary of the 1868 Hayward Fault earthquake will be observed this year (2008), and scientists are convinced that the Hayward Fault has reached the point where a powerful, damaging quake can be expected at any time. According to a 1996 Earthquake Engineering Research Institute report, the next major Hayward Fault quake is expected to cause significant loss of life and extensive damage to homes, businesses, and infrastructure, such as transportation and utilities. Several hundred thousand people are likely to be homeless after the quake.

As demonstrated by the aftermath of 2005's Hurricane Katrina, recovery from such catastrophic events can take years. According to another recent study, a repeat of the 1868 earthquake could cause economic losses (including damage to buildings and contents, business interruption, and living expenses) exceeding $120 billion, with more than 90% of both residential and commercial losses being uninsured. Also, damage to infrastructure and other long-term economic effects could substantially increase the total losses.

The 1868 Hayward Fault earthquake is a stark reminder of the awesome power that lurks beneath the Earth's surface in the San Francisco Bay region. Other large quakes, like the 1995 Kobe earthquake in Japan, offer sobering glimpses of the destruction that such events can inflict in a densely populated urban region. The population at greatest risk from a Hayward Fault earthquake is now 100 times greater than in 1868. The urban infrastructure in the Bay Area has been tested only by the relatively remote, 1989 magnitude 6.9 Loma Prieta earthquake. Modern earthquake science tells us that it is now urgent to prepare for the next magnitude 7 quake on the Hayward Fault. Preparing now can significantly reduce loss of life and property from that inevitable quake.

Prepare Now!!!

Information in several languages on earthquakes in the San Francisco Bay region and how to prepare for them can be found in USGS General Information Products 15, 41, and 42 . Order your copy from http://earthquake.usgs.

Chapter Sources

The following chapters have been previously published:

Chapter 1 – This is an edited, excerpted and augmented edition of a United States Congressional Research Service publication, Report Order Code RL33861, dated March 3, 2010.

Chapter 2 – This is an edited, excerpted and augmented edition of a United States Department of the Interior, U.S. Geological Survey Special Report, Circular 1324, dated 2008.

Chapter 3 – This is an edited, excerpted and augmented edition of a United States Department of the Interior, U.S. Geological Survey Fact Sheet, 2008-3027, dated 2008.

Chapter 4 – This is an edited, excerpted and augmented edition of a United States Department of the Interior, U.S. Geological Survey Open File Report, 2008-1150, CGS Preliminary Report 25, Version 1.0, dated 2008.

Chapter 5 – This is an edited, excerpted and augmented edition of a United States Department of the Interior, U.S. Geological Survey Open File Report, 2008-1151, v.1.1, dated 2008, revised 2009.

Chapter 6 – This is an edited, excerpted and augmented edition of a United States Department of the Interior, U.S. Geological Survey Fact Sheet, 2008-3019, dated 2008.

INDEX

A

accelerometers, 27, 30
accommodation, 206
accounting, 16, 283
acid, 161
actuality, 243
adjustment, 125, 212, 268, 283
advertising, 269
age, 87, 88, 120
agricultural sector, 226
agriculture, 226, 227, 273
Air Force, 163
airports, 171, 262, 264, 266, 268, 276
Alaska, 2, 9, 11, 21, 23, 75, 113, 178, 261, 295, 296, 298, 300
alcohol, 206
alcohol abuse, 206
algorithm, 250, 338
alternatives, 40, 70, 267, 268, 271, 275, 283
American Recovery and Reinvestment Act, viii, 2, 20
ammonia, 176
amplitude, 27, 34, 63, 64, 66, 69, 112, 120
ANS, 20
anxiety, 217, 218, 219
anxiety disorder, 218
architects, 3
arteries, 116, 316, 319
asbestos, 147, 148
assessment, 31, 32, 42, 43, 76, 125, 142, 153, 163, 170, 175, 176, 213, 232, 287, 293, 299
assets, 36, 113, 159, 289, 335
assignment, 342, 343, 344, 345, 347, 348, 349, 350, 351, 352, 354, 355, 356, 357, 358, 359, 361, 362, 363, 364, 365, 367, 368, 369, 370
assumptions, 124, 125, 256, 257, 283, 289
attacks, 213
authority, 6
authors, 126
automobiles, 39
availability, 40, 154, 266, 268, 275, 276, 291
averaging, 16
awareness, 175

B

background, 57, 60, 61, 77, 285
Bangladesh, 13
banks, 319
batteries, 138, 143, 161
beams, 160
behavior, 13, 22, 26, 28, 35, 37, 96, 108, 119, 129, 133, 199, 278
Beijing, 13
bending, 321
benzene, 157
bias, 69
bleeding, 243
blocks, 36, 38, 102, 156, 157, 169, 200, 209, 236, 243, 325, 329
bones, 329
bounds, 239, 240
breakdown, 236, 335
brittleness, 127
broadband, 21, 91
buffer, 78
building blocks, 221
building code, ix, 3, 4, 19, 37, 39, 44, 129, 133, 134, 135, 137, 144, 172, 292, 311, 316, 321
burn, 169, 198, 209, 210, 211, 212, 215, 239, 325
burning, 199, 200, 203, 207, 209, 210, 238

C

cabinets, 148
cables, 34, 83, 114, 116, 117, 120, 122, 123, 159, 160, 319
calibration, 125
campaigns, 44
Canada, 29, 266, 269, 275
canals, 116, 146
Caribbean, 14, 15, 21
cast, 37, 163
catastrophes, 42
cattle, x, 373
cell, 91, 160, 161, 162, 197, 267
cell phones, 197, 267
Census, 103, 104, 226, 247
ceramic, 37, 166
channels, 22, 35, 93, 95, 105, 146, 197, 298
chaos, 40, 271
children, 215, 323
Chile, vii, 1, 17, 18
chimneys, x, 341, 353, 354, 360, 366, 373, 375
China, 12, 13, 14, 24, 263, 268, 275
chlorine, 175, 176
Civil War, 221
classes, 36, 38, 88, 93, 95, 96, 97, 98, 102, 170, 176, 233, 237, 238, 239, 240, 286
classification, 88, 98, 176, 227
cleaning, 322
clients, 319
closure, 130, 268
clusters, 30, 133, 224
CNN, 198
Coast Guard, 276
coccidioidomycosis, 297
codes, ix, 4, 19, 64, 129, 137, 245, 279, 286, 307, 377
coherence, 60
collaboration, 65, 145, 172, 208, 288
collateral, 297
colleges, 331
combined effect, 238
commodity, 256, 271
communication, 30, 90, 116, 117, 120, 121, 122, 123, 146, 154, 155, 156, 165, 197, 210, 221, 260, 264, 265, 267, 284, 340
community, 31, 41, 42, 43, 135, 146, 159, 169, 175, 176, 197, 201, 202, 204, 206, 211, 236, 250, 270, 271, 288, 292, 310, 312, 316, 331, 334
compensation, 268, 287
competition, 144
competitive advantage, 224
compilation, 88, 90
compiler, 293
complement, 36, 113
complex interactions, 44
complexity, 60, 65, 77, 176, 242
compliance, 22
complications, 63, 147
components, 16, 21, 22, 44, 79, 91, 109, 129, 154, 161
composition, 256
comprehension, 200
Comprehensive Test Ban Treaty, 22
compression, 82, 119, 162, 163
computation, 23
computer simulations, 27, 32, 33, 291
computer systems, 32
computing, 93, 98, 102
concentration, 164, 176, 224, 226, 227, 261, 267, 271, 273, 304
conception, 21
concrete, 18, 36, 37, 114, 126, 133, 134, 138, 139, 147, 148, 158, 170, 171, 213, 214, 319, 321, 322, 324
conditioning, 37, 159, 160, 331
confidence, 341, 342, 343, 344, 345, 346, 347, 348, 349, 350, 351, 352, 353, 354, 355, 356, 357, 358, 359, 360, 361, 362, 363, 364, 365, 366, 367, 368, 369, 370
configuration, 127, 213
congestion, 157, 251, 264, 266, 269, 271
Congress, 1, 2, 3, 4, 6, 7, 8, 20, 28, 29
Congressional Budget Office, 14
consensus, 54, 108
conservation, 243, 252, 278
consultants, 122
consumer goods, 268
consumers, 163, 285
consumption, 249, 256, 262
contamination, 147, 330
contingency, 155, 276, 290
contour, 89, 341, 342, 343, 344, 345, 346, 347, 348, 349, 350, 351, 352, 353, 354, 355, 356, 357, 358, 359, 360, 361, 362, 363, 364, 365, 366, 367, 368, 369, 370
control, 39, 95, 128, 155, 157, 162, 163, 164, 176, 206, 209, 210, 216, 304
convergence, 13, 204, 210
conversion, 37

cooking, 322
cooling, 142, 149
coping strategies, 217
copper, 143
corporations, 208, 334
correction factors, 99
correlation, 95
costs, 2, 19, 25, 124, 128, 160, 170, 172, 174, 250, 251, 260, 263, 269, 274, 282, 284, 316, 331
covering, 20, 196, 197, 284
crack, 129, 147, 310, 322
creep, 375, 376
critical infrastructure, 3, 13, 155, 290
crops, 221
crude oil, 269
crystalline, 97
customers, 40, 43, 151, 163, 164, 242, 243, 244, 247, 248, 249, 264, 266, 275, 331
Customs Service, 208

D

danger, 139, 203, 205, 207, 210, 217
data collection, 310
data set, 54, 222, 253, 286
database, 35, 52, 66, 77, 82, 90, 97, 113, 122, 125, 271, 295, 297, 298, 299, 300, 301, 303, 304
dating, 379
death, ix, 145, 149, 210, 215, 216, 316
deaths, 12, 14, 17, 29, 39, 40, 109, 135, 210, 211, 212, 215, 216, 218, 292, 316, 322, 323, 373, 375, 378
decay, 112
decision makers, 203, 277
decision making, 290
decision-making process, 146
decisions, 19, 54, 206, 265, 290, 307, 333, 334
definition, 176, 244, 283, 286
deformation, 26, 74, 75, 76, 79, 82, 84, 93, 94, 95, 99, 117, 120, 123, 132, 150, 151, 157, 162, 163, 244, 284, 293, 296, 300, 302, 305
degradation, 129
delivery, 41, 145, 154, 178, 207, 248, 256, 258, 260, 276, 292
demand curve, 250, 251, 252
density, 9, 21, 35, 95, 168, 232, 236, 306
Department of Commerce, 22
Department of Homeland Security, 6, 208, 296
Department of the Interior, 383
deposits, 76, 86, 87, 88, 90, 95, 120, 304
depression, 217, 218, 219, 289, 335
designers, 133
destruction, 9, 29, 134, 214, 289, 381
detection, vii
detention, 146
diaphragm, 128
diesel fuel, 163
digital communication, 23
directors, 4
discharges, 163
discomfort, 207
disorder, 217, 218
displacement, 34, 62, 74, 75, 76, 77, 78, 79, 80, 82, 83, 84, 86, 99, 101, 104, 105, 107, 117, 119, 120, 121, 122, 123, 128, 151, 163, 372
disseminate, 4, 144
distortions, 169
distress, 217
District of Columbia, 9
donations, 204, 208
donors, 208
doors, 132, 144, 148, 269, 322
double counting, 238
draft, 63, 267
drainage, 79, 163
drinking water, 114, 147, 207
drought, 243
ductility, 127, 143
dumping, 147
duration, 34, 40, 63, 67, 69, 70, 71, 109, 112, 119, 120, 127, 282, 290, 323
Duration, 34, 69, 70
duties, 320

E

early warning, 199
earnings, 227, 228
earth, 13, 26, 27, 30, 74, 75, 95, 105, 113, 126, 170, 304, 324, 325
economic activity, 12, 16, 40, 221, 222, 233, 271, 273, 289, 335
economic growth, 283
economic systems, viii, 31, 43
economics, 298
editors, 296
Education, 173, 225, 281
elderly, 215
electricity, 36, 40, 113, 159, 196, 197, 205, 277, 278, 316, 328

email, 15, 23, 44
e-mail, 64
emergency management, 71, 175, 177, 208, 327
emergency planning, ix, 123, 130, 132, 133, 307
emergency preparedness, 221, 290
emergency response, viii, ix, 6, 21, 22, 23, 26, 31, 38, 40, 42, 43, 125, 130, 148, 165, 170, 175, 177, 198, 290, 292, 315, 316, 318, 324, 327
emotional state, 205
employees, 43, 148, 151, 165, 227, 233, 234, 245, 258, 270, 289, 331
employment, 222, 223, 224, 233, 257, 258, 270, 283
energy, 9, 13, 56, 63, 66, 69, 87, 112, 116, 129, 133, 319, 321
environment, 16, 38, 76, 83, 87, 88, 105, 124, 232, 236, 275, 278, 291, 297
environmental impact, 158, 275
equilibrium, 256
erosion, 90
estimating, 18, 23, 53, 57, 66, 77, 91, 95, 101, 175, 213, 286, 298
evacuation, 114, 132, 138, 139, 145, 146, 157, 200, 203, 204, 205, 206, 210, 276, 326
evolution, 27, 61
exaggeration, 200
exercise, 31, 41, 125, 145, 165, 222, 290
expenditures, 20
expertise, 36, 44, 113, 136
export ratio, 274
exports, 273
exposure, 16, 102, 103, 104, 105, 125, 217, 236, 284

F

fabric, 42
facies, 88, 304
faith, 331
family, 135, 167, 169, 201, 202, 205, 209, 215, 217, 232, 233, 239, 240, 241, 285, 286, 327, 328, 331
family members, 201, 205
fatigue, 203
fear, 199, 206, 207, 283, 329
fears, 201, 207
feedback, 31, 221, 262, 288
feet, 34, 84, 132, 146, 151, 158, 169, 170, 171, 212, 239, 265, 267, 276, 318, 321, 375
FEMA, 1, 2, 3, 4, 6, 7, 16, 18, 26, 29, 35, 49, 93, 99, 104, 113, 117, 123, 129, 130, 135, 147, 148, 175, 178, 201, 203, 205, 276, 296, 327, 336
fencing, 153
filtration, 143
finance, 270
financial institutions, 18
fire fighting, 209, 210, 325
fire suppression, 39, 177, 208, 209, 210, 215
7
firms, 277
first aid, 202
first responders, 114, 144, 156, 325
fish, 269, 272, 273
flexibility, 222
flood, 42, 88, 146, 323
flooding, 114, 145
flour, 322
fluid, 26
focusing, 66, 324
food, 39, 148, 176, 200, 204, 205, 206, 207, 265, 276, 277, 328, 329
food spoilage, 200
football, 377
fractures, 74, 76, 96, 129
fragility, 104
framing, 128
friction, 112, 319
fruits, 269
fuel, 123, 140, 148, 154, 156, 159, 162, 163, 166, 169, 201, 213, 236, 248, 264, 269, 321
full capacity, 40, 248
funding, 2, 3, 4, 7, 20, 22, 156, 290, 291
funds, 2, 20, 27, 28, 274, 284
furniture, 148, 320, 331

G

gangs, 270
gases, 166
gasoline, 155, 162, 163
General Motors, 269
generation, 67, 91, 125, 158, 242, 248, 298
geography, 116, 377
geology, 26, 30, 44, 102, 107, 108, 293, 305, 310, 378
goals, 137
goods and services, 39, 176, 221, 257, 277

government, 2, 21, 22, 24, 161, 171, 198, 205, 206, 227, 236, 270, 276, 284, 287, 291, 315, 329, 334, 335
grades, 79
grains, 272
grants, 27, 52
graph, 242, 243, 244, 251
gravitational force, 75
gravity, 8, 9, 147, 251
grids, 91
grounding, 143
groundwater, 124, 165, 243
groups, 4, 22, 23, 26, 29, 64, 66, 96, 125, 201, 202, 203, 307, 334
growth, 18, 226, 227, 275, 290, 306
growth rate, 226
guidance, 78, 276
guidelines, 86, 144, 148
Guinea, 303

H

Haiti, 1, 12, 14, 15, 17, 24, 25, 29
hands, 324
Hawaii, 2, 9, 17, 21, 23, 146, 261, 296, 304
hazardous materials, 175, 176, 203, 205, 215
hazardous substances, 176
head injuries, 212
health, 38, 44, 137, 138, 147, 176, 211, 218, 335
health care, 137, 138
health care system, 137
health effects, 176
heat, 26, 160, 209
heating, 138, 160, 248
height, 56, 128, 214
heroism, 204
heterogeneity, 62
highway system, 151, 323
highways, 13, 36, 38, 40, 80, 113, 114, 116, 123, 126, 150, 151, 152, 170, 210, 260, 263, 270, 271, 292
home value, 289
homeowners, 43, 284, 285, 287
Honda, 269, 271
hospitals, 39, 114, 137, 138, 139, 142, 143, 144, 148, 155, 174, 201, 202, 203, 206, 212, 323, 326, 329
hotels, 207
House, 4, 6, 7, 29, 302, 375
household income, 256
households, 26, 38, 207, 215, 245, 248, 252, 257
housing, 204, 207, 215, 247, 319
humidity, 167
Hurricane Katrina, 12, 42, 139, 147, 270, 289, 336, 380
hurricanes, 12, 19, 22, 28, 123

I

identification, 197, 205
identity, 197
image, 27, 376, 377
imagery, 86, 106, 119, 120
images, 8, 122, 201, 305, 318, 378
implementation, 4, 170, 289
imports, 136, 263, 267
incidence, 217
income, 16, 114, 206, 222, 223, 227, 230, 271, 377
indication, 69
indices, 253
individual action, 292
industrial sectors, 222, 252, 256, 295
industry, 161, 224, 225, 226, 227, 228, 233, 243, 256, 257, 258, 267, 270, 273, 286, 315
inelastic, 129
infancy, 134
information technology, 32, 33, 63, 66
initiation, 32, 61, 64, 70
initiation time, 61
injuries, 1, 14, 29, 39, 128, 133, 134, 148, 200, 202, 211, 212, 214, 215, 216, 217, 218, 316, 322, 329
injury, 202, 210, 211, 212, 213, 214, 215, 217
innovation, 205
insight, 277
insomnia, 217
inspections, 154, 156, 157, 158, 209
inspectors, 142, 153, 160, 203
instability, 117
institutions, 4, 21, 22, 30, 332
instruments, 21, 23, 27, 29, 203
insulators, 37, 166
insurance, 18, 124, 169, 201, 222, 270, 276, 283, 284, 285, 286, 287, 298, 307, 319, 330, 335
integration, 310
integrity, 135
intelligence, 202
intelligence gathering, 202
intensive care unit, 215
interaction, 232, 301
interactions, 43, 112, 315, 334

interagency committees, 6
interdependence, 221
interface, 163, 169, 210, 231, 250
interference, 112, 276
international trade, 270, 271
internet, 159, 264
interpersonal communication, 197
interval, 58, 379, 380
intervention, 276
interview, 197, 199
investment, 283, 289, 323
iron, 37, 163
isolation, 143

J

Japan, 17, 18, 29, 129, 131, 269, 275, 304, 377, 378, 381
jobs, 25, 227, 230, 252, 260, 270, 274, 322, 328, 330, 331
joints, 96, 138, 142, 143, 164
Jordan, 48, 53, 299
journalists, 204
judgment, 78, 88, 117, 124, 125, 126
jurisdiction, 135, 146
justification, 20

K

kidney, 329
kidney dialysis, 329
Korea, 275

L

LA Metro, 178
labor, 256, 262, 266, 268, 269, 270, 271, 273, 274, 275, 288, 305
labor force, 270
lakes, 112
land, 114, 207, 210, 267, 290, 375, 380
landfills, 147
landscape, 76
land-use, 290
law enforcement, 39, 176, 201, 329
laws, 35, 108, 135
leakage, 243, 247, 248
leaks, 142, 148, 165, 166, 172, 247
legislation, 28, 102, 127, 133, 270
lending, 66
lifespan, 30
lifetime, 15, 42
light rail, 154, 171
likelihood, 9, 10, 55, 75, 83, 86, 89, 95, 96, 103, 104, 137, 142, 205, 217, 218, 307, 310
limitation, 155, 165, 265
linear function, 67
linearity, 277
links, 161
liquid fuels, 126
loans, 287, 332
local authorities, 23
local government, 3, 204, 208
locus, 310
logistics, 261, 266, 270, 271, 275, 276
long distance, 78, 264
Louisiana, 289
love, 198
low risk, 9
lower prices, 269
lying, 16, 375

M

maintenance, 157, 276
management, 4, 21, 39, 152, 165, 175, 176, 204, 205, 208, 275, 290
manpower, 37
mantle, 27, 76
manufacturing, 233, 262, 274, 278
map unit, 87, 88, 93, 95, 96, 97
mapping, 35, 77, 86, 88, 96, 99, 117, 119, 121, 122, 232, 296, 298, 299, 300, 304
Margesson, Rhoda, 15
market, 269, 275, 278, 284
market share, 275
markets, 278
mass care, 205, 206, 207
matrix, 38, 177, 252
measurement, 90, 91
measures, 2, 3, 6, 28, 34, 36, 70, 71, 113, 126, 159, 203, 243
media, 14, 22, 24, 29, 196, 197, 198, 199, 200, 201, 203, 204, 208, 210, 247, 326, 329
medical care, 39, 140, 215
medical expertise, 202
medication, 206
memory, 19
mental health, 217, 218

metals, 272
Mexico, 11, 18, 35, 66, 129, 132, 213, 242, 267, 269, 270, 326
mobility, 205, 298
modeling, 29, 33, 40, 44, 54, 57, 58, 60, 64, 65, 66, 67, 70, 93, 133, 170, 175, 222, 277, 287, 288, 291, 293, 296
models, 10, 26, 33, 58, 64, 66, 67, 69, 91, 97, 98, 112, 125, 175, 178, 256, 277, 278, 288, 290, 291, 298
modernization, 20
money, 151, 204, 277, 328, 329, 334
morbidity, 288
morning, x, 134, 215, 318, 373
mortality, 39, 211, 215, 288
mothers, 215
motivation, 202
mountains, 38, 111, 129, 169, 292, 316, 322
MTS, 302
multiplier, 277, 278, 282, 291
multiplier effect, 282, 291
MWD, 82, 242, 243

N

narratives, 177
nation, 1, 3, 12, 20, 21, 28, 116, 198, 199, 207, 224, 226, 267, 275
National Bureau of Standards, 3
National Research Council, 93, 295, 299
natural disasters, 1, 8, 42, 44, 177, 217, 316, 334
natural gas, 34, 82, 114, 154, 157, 162, 163, 166, 205, 209, 247, 248, 264, 276, 319, 328
natural hazards, 6, 8, 22, 26, 28, 298
natural sciences, 3
negative consequences, 264
Netherlands, 303
network, 20, 21, 22, 29, 114, 137, 154, 155, 156, 161, 162, 250, 251, 252, 260, 268, 275, 296
news coverage, 198, 326
nodes, 159
North America, 14, 18, 27, 29, 222, 245, 260, 275, 310, 319
Northridge Earthquake, 128, 211, 300, 303
nucleation, 54
nurses, 143

O

Obama, President, 4
objectives, 3, 203, 221, 222, 288
observations, 6, 9, 76, 78, 93, 100, 129, 199, 340
Office of Management and Budget, 4
oil, 82, 114, 116, 119, 209, 269, 306, 322
older adults, 215
omission, 278
open spaces, 204, 205, 210
order, 53, 63, 66, 86, 88, 119, 125, 133, 145, 158, 166, 168, 170, 171, 172, 174, 196, 203, 204, 206, 211, 214, 215, 222, 232, 249, 277, 283, 336
ores, 272, 273
orientation, 63, 77, 96
oscillation, 84, 112, 296
overload, 154, 155, 197
overtime, 269

P

Pacific, 17, 18, 23, 27, 80, 156, 157, 158, 260, 264, 267, 310, 319
Pakistan, 12, 29
Panama, 267
parameter, 54, 125, 172
parameters, 23, 34, 54, 56, 58, 62, 67, 70, 71, 74, 86, 91, 96, 102, 103, 109, 123, 124
parents, 289
partnership, 27
payroll, 222, 223, 245, 278
penalties, 268
performers, 36
permit, 285
personal communication, 56, 122, 197, 226, 238, 264, 265, 266, 267, 268, 269, 284, 340, 341, 346, 352
Peru, 139, 264, 300, 336
pharmaceuticals, 268
Philippines, 17
photographs, 377
physical health, 217
physics, 31, 32, 33, 35, 53, 63, 64, 66, 67, 69, 70, 108, 113, 291
planning, 3, 4, 6, 26, 42, 43, 44, 109, 114, 145, 148, 155, 165, 177, 221, 227, 243, 276, 288, 289, 290, 302, 313, 316, 332, 334
plants, 114, 147, 166, 176, 209, 210, 269, 331
plausibility, 56

police, 151, 154, 155, 156, 276, 324
polymer, 166
polymers, 37
pools, 285
poor, 36, 66, 127, 133, 134, 203
poor performance, 133
population, 9, 10, 11, 17, 19, 24, 54, 124, 125, 159, 176, 177, 203, 206, 211, 215, 217, 218, 226, 247, 249, 252, 288, 374, 375, 381
population density, 159
population growth, 177, 226
population size, 374
port of entry, 262
ports, 114, 136, 137, 151, 170, 223, 227, 260, 262, 263, 264, 265, 266, 267, 268, 269, 270, 271, 273, 274, 275, 278, 288, 290, 317, 330, 377
prediction, 3, 22, 28, 43, 78, 297, 316
predictors, 217
premiums, 201, 285
preparedness, 3, 31, 41, 42, 43, 44, 211, 290, 292, 313, 316, 318, 334
pressure, 26, 76, 122, 163, 165, 167, 209, 210, 310, 325
prices, 278
private schools, 331
private sector, 44
probability, vii, 10, 15, 16, 19, 22, 28, 76, 83, 86, 93, 95, 99, 100, 101, 102, 103, 104, 105, 106, 109, 112, 133, 137, 298, 310, 311
probability density function, 109
process control, 77
producers, 283
production, 221, 256, 277, 278, 331
productivity, 289
profit, 30
program, viii, 2, 3, 4, 5, 6, 7, 8, 20, 21, 26, 27, 28, 32, 35, 38, 63, 90, 113, 127, 155, 161
progress reports, 8
propagation, 61, 63, 67
propane, 155
public awareness, 22
public domain, 125
public health, 297
public safety, 312
public schools, 135, 328
public service, 36, 113, 208
Puerto Rico, 11
pulse, 320
pumps, 161, 165
purification, 200

R

race, 326
radar, 375
radiation, 33, 63, 67
radio, 143, 148, 155, 156, 196, 197, 199, 200, 247, 327
rain, 307, 322
range, 15, 44, 66, 74, 79, 80, 81, 82, 83, 110, 117, 120, 133, 134, 150, 151, 175, 176, 200, 247, 265, 278
ratings, 139
raw materials, 274
reading, 44
reality, 35, 95, 109, 157, 328
reason, 13, 162, 247
recession, 283, 289
recognition, 44, 206
reconstruction, 15, 25, 123, 158, 200, 270, 283, 332
recovery, 7, 26, 39, 111, 150, 159, 165, 176, 201, 204, 208, 230, 250, 270, 276, 278, 287, 291, 332, 380
recreation, 204, 206, 328
recurrence, 32, 54, 63, 110, 302
redundancy, 165
refining, 288
regeneration, 335
regional economic impacts, 222
regulation, 247
rehabilitation, 135
reinforcement, 37
reinsurance, 284
relationship, 18, 23, 28, 55, 71, 77, 95, 172, 247
relative size, 9, 306
relatives, 205, 206, 207
relief, 24, 204
remote sensing, 77
rent, 133
repackaging, 175
repair interval, 167
research funding, 290
residuals, 68, 69
resilience, 28, 221, 222, 244, 262, 278, 283, 289, 290, 312
resistance, 13, 40, 129, 134, 156, 292, 324
resolution, 21, 60, 86, 97, 98
resources, 20, 40, 135, 149, 165, 168, 169, 201, 206, 207, 208, 209, 211, 276, 327, 330, 331
respirator, 138, 149
respiratory, 215

response time, 215
restaurants, 237, 278
restructuring, 4
retail, 232, 233, 243, 257
returns, 325
revenue, 223, 260, 261, 263, 269, 277
risk assessment, 125, 175, 290
risk management, 26
river basins, 35
robustness, 37, 292
routing, 147, 205
rural areas, 91, 215

S

safety, 3, 4, 28, 36, 39, 127, 135, 142, 144, 145, 151, 155, 156, 157, 196, 205, 206, 209, 316, 321
sales, 223, 277
satellite, 161, 375, 376, 378
saturation, 93, 120, 143, 159, 209
scaling, 63, 277
scatter, 64
school, ix, 25, 133, 148, 149, 204, 206, 238, 289, 315, 330, 331, 336
scientific knowledge, 26
scientific understanding, ix, 3, 22, 26, 308, 311, 315
scrublands, 145
search, 39, 149, 176, 199, 201, 202, 203, 204, 208, 296, 328, 372
searches, 202
searching, 196, 324
security, 39, 125, 159, 176, 205, 206, 243, 276, 329
sediment, 64, 84, 88, 95, 322
sedimentation, 305
sediments, 66, 69, 86, 321, 322, 333
seismic data, 21, 178
sensitivity, 234, 278
sensors, 30, 144
service provider, 161, 162
severity, 10, 217, 218
sewage, 114, 147
shape, 112, 317
shear, 14, 75, 82, 117, 138, 143
shear strength, 75
shelter, 26, 39, 148, 169, 176, 196, 204, 205, 206, 207, 210, 330
shock, 30, 111, 197, 199, 222, 231, 232, 277, 278, 288, 289, 322
short supply, 329
signals, 38, 153, 156, 205, 209
signs, 156
Silicon Valley, 273
simulation, 56, 64, 65, 67, 69, 70, 91, 109, 110, 111, 150, 170, 287, 299, 300
Singapore, 295
skills, 201, 202, 289
smoke, 198, 209, 215
software, 26, 91, 125, 172, 178, 208, 222
soil, 34, 63, 74, 77, 84, 95, 96, 97, 100, 101, 105, 119, 124, 165, 296, 297, 301, 304
solitude, 196
SPA, 51, 167
space, 135, 233, 268
spare capacity, 266
specialization, 268
spectrum, 129
speed, 58, 61, 62, 63, 156, 157, 158, 251, 264, 289, 291
stability, 128, 305
staffing, 39, 176
stakeholders, 42, 170, 288
standard deviation, 61, 67, 69
standards, 3, 4, 40, 41, 98, 137, 145, 292, 331
stars, 371
statistics, 79, 142, 248, 256, 306, 318
steel, 18, 37, 114, 124, 126, 129, 130, 131, 132, 133, 138, 139, 158, 163, 170, 171, 212, 213, 215, 319, 322
stimulus, 221
stock, 16, 29, 40, 101, 130, 222, 282, 289, 330
stockpiling, 165
storage, 143, 146, 154, 165, 176, 227, 242, 243, 247, 248, 265, 266, 268, 269, 276, 306
storms, 200
strain, ix, 14, 42, 54, 56, 75, 77, 95, 122, 315, 319
strategic planning, 42
strategies, 28, 210, 275
strength, 10, 75, 84, 95, 96, 128, 129, 143, 233, 310
stress, 26, 77, 203, 293, 323, 375
structural characteristics, 35, 124, 127
students, 320, 334
substitutes, 266
substitution, 256
summer, 90, 98
suppliers, 40, 269, 277, 283, 319, 331
supply chain, 262, 269, 270, 275
supply curve, 250, 251
suppression, 209, 222
surging, 159
surplus, 251

survival, 177
survivors, 42, 202, 204, 377
susceptibility, 75, 86, 88, 89, 90, 92, 93, 95, 96, 97, 98, 99, 102, 247, 300
suture, 211
switching, 114, 160
symptom, 217
symptoms, 218

T

Taiwan, 13
tanks, 138, 140, 142, 143, 161, 162, 163, 165, 306
targets, 116
tax base, 290
teachers, 331
team members, 202, 204
technology transfer, 8
telecommunications, 38, 143, 159, 160, 161, 162, 209, 232, 264, 288, 292, 319
telecommunications services, 288
telephone, 11, 23, 30, 44, 114, 160, 199, 316
television, 196, 197, 198, 199
tenants, 43, 283, 329, 332, 334
tensile strength, 127
tension, 82, 83, 163, 166
terminals, 260, 261, 268
terraces, 87, 120
territory, 322
terrorism, 217
Thailand, 13
thoughts, 196
threat, 14, 175, 176, 310
threats, 111, 276
threshold, 21, 176, 283, 286, 289
thresholds, 23
time constraints, 247
time periods, 137, 265
timing, 22, 222
Title I, 2, 6, 8
urban centers, 137

V

valley fever, 297
variability, 10, 56, 62, 66, 67, 69
vector, 62
vegetables, 269
vehicles, 151, 154, 157, 166, 202, 328
velocity, 33, 34, 64, 65, 69, 70, 71, 91, 124, 291
ventilation, 160
vessels, 269, 270
vibration, 160
victims, 39, 161, 177, 196, 197, 199, 200, 201, 202, 204, 206, 207, 323, 324, 326, 329, 336
Vietnam, 13
voice, 159
voluntary organizations, 208
Volunteers, 208
vulnerability, 1, 3, 24, 36, 113, 127, 135, 145, 175, 177, 264, 292, 296, 311

W

wage rate, 274, 275
wages, 227, 269, 380
waste water, 139
wastewater, 114, 139, 140, 142, 143, 147, 172, 176
water heater, 204
water quality, 165
water supplies, 207
watershed, 134
wave propagation, 66
wavelengths, 61
weakness, 29, 119
wealth, 4, 283
web, 130, 149, 248
welding, 129
wells, 89, 90, 139
wholesale, 232, 233
wildfire, 38, 42, 197
wildland, 169
wind, 6, 8, 169
windows, 37
winter, 19, 90, 116
witnesses, 24, 373
wood, 18, 135, 167, 168, 174, 319, 326, 375
workers, 35, 108, 199, 202, 204, 205, 225, 227, 228, 269, 270, 328
workplace, 325
World Trade Center, 198, 213
World War I, 283
worry, 145, 196
writing, 333